5th National Conference on Processing and Characterization of Materials

IOP Conference Series: Materials Science and Engineering
Volume 115

Rourkela, India
12 - 13 December 2015

Editors:

B.C. Ray
Animesh Mandal
A. Basu
D. Chaira
Alok Satapathy

ISBN: 978-1-5108-2231-3
ISSN: 1757-8981

Printed from e-media with permission by:

Curran Associates, Inc.
57 Morehouse Lane
Red Hook, NY 12571

CURRAN ASSOCIATES INC.
proceedings
.com

Some format issues inherent in the e-media version may also appear in this print version.

Printed by Curran Associates, Inc. (2016)

Institute of Physics
Dirac House, Temple Back
Bristol BS1 6BE UK

Phone: 44 1 17 929 7481
Fax: 44 1 17 920 0979

techtracking@iop.org

Additional copies of this publication are available from:

Curran Associates, Inc.
57 Morehouse Lane
Red Hook, NY 12571 USA
Phone: 845-758-0400
Fax: 845-758-2633
Email: curran@proceedings.com
Web: www.proceedings.com

Table of contents

Volume 115

5th National Conference on Processing and Characterization of Materials
12–13 December 2015, Rourkela, India

Accepted papers received: 5 February 2015
Published online: 7 March 2016

Preface

011001
OPEN ACCESS
5th National Conference on Processing and Characterization of Materials

011002
OPEN ACCESS
Conference Sponsors

011003
OPEN ACCESS
Peer review statement

Papers

012001
OPEN ACCESS
Cellular Automata Modeling of Decarburization of Metal Droplets in Basic Oxygen Steelmaking

Ankit and T. K. Kundu pg. 1

012002
OPEN ACCESS
Impact resistance and hardness modelling of Aluminium alloy welds using square-headed friction-stir welding tool

U. Sudhakar and Dr. J. Srinivas pg. 11

012003
OPEN ACCESS
Dielectric properties of epoxy resin fly ash composite

A Pattanaik, S K Bhuyan, S K Samal, A Behera and S C Mishra pg. 19

012004
OPEN ACCESS
Effect of Post-thermal shock on Prolonged Sea Water aged GFRP Composite

A. P. Chakraverty, U. K. Mohanty, S. C. Mishra and B. B. Biswal pg. 26

012005
OPEN ACCESS
Semi-Solid Processing by Electric Current During Sand Casting of Aluminium Alloys

Anjan Prodhan pg. 35

012006
OPEN ACCESS
Epoxy matrix composites filled with micro-sized LD sludge: wear characterization and analysis

Abhilash Purohit and Alok Satapathy pg. 49

012007
OPEN ACCESS
Studies on Effective Elastic Properties of CNT/Nano-Clay Reinforced Polymer Hybrid Composite

Arvind Kumar Thakur, Puneet Kumar and J. Srinivas pg. 57

012008
OPEN ACCESS
Effect of Nanosize Yittria and Tungsten Addition to Duplex Stainless Steel During High Energy Planetary Milling

A K Nayak, R Shashanka and D Chaira pg. 66

012009
OPEN ACCESS
Application of Hydrogen for the Reduction of Bauxite Mineral

B. R. Parhi, S. K. Sahoo, B. Bhoi, B. K. Satapathy and R. K. Paramguru pg. 75

012010
OPEN ACCESS
Effect of delamination on vibration behaviour of woven Glass/Epoxy composite plate-An experimental study

C K Hirwani, S S Sahoo and S K Panda pg. 82

012011
OPEN ACCESS
An experimental investigation on the effect of particle size on the thermal properties and void content of Solid Glass Microsphere filled epoxy Composites

Debasmita Mishra and Alok Satapathy pg. 96

012012
OPEN ACCESS
Physical, Mechanical and Water Absorption Behaviour of Coir Fiber Reinforced Epoxy Composites Filled With Al_2O_3 Particulates

Geetanjali Das and Sandhayarani Biswas pg. 103

012013
OPEN ACCESS
Analysis of Effects of Cutting Parameters of Wire Electrical Discharge Machining on Material Removal Rate and Surface Integrity

H. R. Tonday and A. M. Tigga pg. 113

012014
OPEN ACCESS
Free Vibration and Bending Behaviour of CNT Reinforced Composite Plate using Different Shear Deformation Theory

K Mehar and S K Panda pg. 120

012015
OPEN ACCESS
Effect of post-deposition annealing on the structural and electrical properties of RF sputtered hafnium oxide thin films

K C Das, S P Ghosh, N Tripathy, G Bose and J P Kar pg. 136

012016
OPEN ACCESS
Stress Ratio Effect on Ratcheting Behavior of AISI 4340 Steel

K Divya Bharathi and K Dutta pg. 140

012017
OPEN ACCESS
Effect of loading rate on tensile properties and failure behavior of glass fibre/epoxy composite

K K Mahato, M Biswal, D K Rathore, R K Prusty, K Dutta and B C Ray pg. 148

012018
OPEN ACCESS
Ratcheting fatigue behaviour of Al-7075 T6 alloy: Influence of stress parameters

Lala Amarnath, Antara Bhattacharjee and K. Dutta pg. 154

012019
OPEN ACCESS
Deformation of $Ni_{20}W_{20}Cu_{20}Fe_{20}Mo_{20}$ high entropy alloy for tensile followed by compressive and compressive followed by tensile loading: A molecular dynamics simulation based study

Md Meraj and S Pal pg. 161

012020
OPEN ACCESS
A Novel Technique for Making Cold Briquettes for Charging in Blast Furnace

M K Mohanty, S Mishra, B Mishra, S Sarkar and S K Samal pg. 168

012021
OPEN ACCESS
Ultrasonic spot welding of dissimilar materials: characterization of welded joints and parametric optimization

M P Satpathy and S K Sahoo pg. 179

012022
OPEN ACCESS
Fabrication of high-k dielectric Calcium Copper Titanate (CCTO) target by solid state route

N Tripathy, K C Das, S P Ghosh, G Bose and J P Kar pg. 188

012023
OPEN ACCESS
Development of Cu-E-Glass Fiber Composites by Powder Metallurgy Route

Pallabi Bhuyan, Harspreet Singh, Lailesh Kumar, Nidhi Sharma, Deepankar Panda, Deepanshu Verma and Syed NasmulAlam pg. 192

012024
OPEN ACCESS
High Temperature Mechanical Behavior of Aluminum- $Cu_{50}Zr_{50}$Metallic Glass Interface

P Gupta and N Yedla pg. 206

012025
OPEN ACCESS
Characterisation of Suspension Precipitated Nanocrystalline Hydroxyapatite Powders

P K Mallik, P.K. Swain and S.C Patnaik pg. 213

012026
OPEN ACCESS
Fabrication and Characterization of novel $W_{80}Ni_{10}Nb_{10}$ alloy produced by mechanical alloying

R. Saxena, A. Patra, S. K. Karak, A. Pattanaik and S. C. Mishra pg. 217

012027
OPEN ACCESS
Scheming of microwave shielding effectiveness for X band considering functionalized MWNTs/epoxy composites

S Bal and S Saha pg. 227

012028
OPEN ACCESS
Dry sliding wear behavior of epoxy composite reinforced with short palmyra fibers

Somen Biswal and Alok Satapathy pg. 236

012029
OPEN ACCESS
Water absorption behavior and residual strength assessment of glass/epoxy and glass-carbon/epoxy hybrid composite

S C Mohanty, B P Singh, K K Mahato, D K Rathore, R K Prusty and B C Ray pg. 243

012030
OPEN ACCESS
Production and Compressive Characterization of Aluminium MMC Foam Manufactured Using Dual Foaming Agent

S Haidar, S Ansary and A Rahman pg. 247

012031
OPEN ACCESS
Parametric Optimization of Simulated Extrusion of Square to Square Section Through Linear Converging Die

S. K. Mohapatra and K. P. Maity pg. 257

012032
OPEN ACCESS
Grain boundary character modification employing thermo-mechanical processing in type 304L stainless steel

S K Pradhan and S Mandal pg. 263

012033
OPEN ACCESS
Improvement of thermal properties of water and ethylene glycol by metallic and ceramic dispersion

Sripooja Mishra, Manaswini Chinara, Mohan Nuthalapati and A. Basu pg. 270

012034
OPEN ACCESS
Synthesis CNTs Particle Based Abrasive Media for Abrasive Flow Machining Process

Sonu Kumar, Q. Murtaza, R.S Walia, S. Dhull and P. K. Tyagi pg. 277

012035
OPEN ACCESS
Ultraviolet photodetection characteristics of Zinc oxide thin films and nanostructures

S P Ghosh, K C Das, N Tripathy, G Bose, D H Kim, T. I. Lee, J.M. Myoung and J.P. Kar pg. 283

012036
OPEN ACCESS
A Study On the effect of Surface treatment on the Physical and Mechanical properties of date-palm stem liber embedded epoxy composites

Satchidananda Tripathy, Janaki Dehury and Debasmita Mishra pg. 290

012037
OPEN ACCESS
Effect of pre-strain on ratcheting behavior of A668 Class D steel

S. Tripathy, A. K. Mondal and K. Dutta pg. 298

012038
OPEN ACCESS
Equilibrium state of anatase to rutile transformation for nano-structured Titanium Dioxide powder using polymer template method.

Tapabrata Dam, Sidhartha S. Jena and Dillip K. Pradhan pg. 303

012039
OPEN ACCESS
Study on blast furnace cooling stave for various refractory linings based on numerical modeling

T R Mohanty, S K Sahoo and M K Moharana   pg. 309

012040
OPEN ACCESS
Hygrothermal effects on the flexural strength of laminated composite cylindrical panels

Trupti R Mahapatra and Subrata K Panda   pg. 316

IOP Conf. Series: Materials Science and Engineering **115** (2016) 011001 doi:10.1088/1757-899X/115/1/011001

PREFACE

This volume contains selected full length technical papers from the oral presentations made during the 5th National Conference on Processing and Characterization of Materials (NCPCM) 2015 organized by Department of Metallurgical and Materials Engineering, NIT Rourkela, India on 12th-13th December, 2015. The first conference of the NCPCM series was held in December 2011. NCPCM 2015 has successfully carried the tradition of previous conferences. About 100 participants from various organizations across India have participated in the conference. The conference has attracted researchers, scientists and engineers from various R & D organizations, academic institutions and industries at a single forum. The interdisciplinary approach of the conference allowed the participants to look beyond their areas of activities. Besides oral presentations the conference also had poster and metallography sessions. The conference had four technical sessions. The themes of the sessions were: Materials Processing, Characterization, Materials Deformation and Materials Modelling and Simulation. In all, during the two day conference about fifty contributory talks along with four keynote lectures were presented. Out of these, forty peer-reviewed papers have been selected for publication in this volume of IOP Conference Series: Materials Science and Engineering.

We would like to thank all the contributors, members of the organizing committee, session chairs as well as colleagues and students who helped us with the preparation of the conference and particularly, with the preparation of this volume. We would also like to convey our heartiest gratitude to all the sponsors of NCPCM 2015.

February 2016 Editorial Board, NCPCM 2015

IOP Conf. Series: Materials Science and Engineering **115** (2016) 011001 doi:10.1088/1757-899X/115/1/011001

NCPCM-2015
IIM
NCPCM 2015
5th National Conference on
Processing & Characterization of Materials
December 12th - 13th, 2015
Our Sponsors

IOP Conf. Series: Materials Science and Engineering **115** (2016) 011002 doi:10.1088/1757-899X/115/1/011002

Our Sponsors

1. Board of Research in Nuclear Sciences (BRNS), India
2. Science and Engineering Research Board (SERB), Dept. of Science & Technology, Govt. of India)
3. OCL India Ltd., India

LIST OF COMMITTEES

Editorial Board

1. **Prof. B.C. Ray**
   Professor
   Department of Metallurgical and Materials Engineering
   NIT Rourkela, India, Pin 769008
2. **Prof. Animesh Mandal**
   Assistant Professor
   Minerals, Metallurgical & Materials Engineering
   Indian Institute of Technology Bhubaneswar, Bhubaneswar, Orissa, India, Pin-751013
3. **Prof. A. Basu**
   Associate Professor
   Department of Metallurgical and Materials Engineering
   NIT Rourkela, India, Pin - 769008
4. **Prof. D. Chaira**
   Assistant Professor
   Department of Metallurgical and Materials Engineering
   NIT Rourkela, India, Pin - 769008
5. **Prof. Alok Satapathy**
   Associate Professor
   Department of Mechanical Engineering,
   NIT Rourkela-769008, Orissa, India

Organizing Committee (NIT Rourkela, India)

Patron: Prof. S. K. Sarangi, Director, NIT Rourkela
Chairman: Prof. S. C. Mishra,
Head of The Department, Department of Metallurgical and Materials Engineering, NIT Rourkela
Convener: Prof. S. N. Alam
Co-Convener and Treasurer: Prof. Ajit Behera
Co-Convener: Prof. D. Nagarajan

Technical Committee (NIT Rourkela, India)

Prof. A. Basu
Prof. D. Chaira
Prof. K. Dutta
Prof. S. Pal
Prof. A. Mallick
Prof. S. K. Sahoo
Prof. N. Yedla
Prof. S. K. Karak
Prof. A. K. Mondal

Conference Advisory Committee

Prof. B. B. Verma
Prof. B. C.Roy
Prof. S. Sen
Prof. M. Kumar
Prof. S. Sarkar

Technical Advisory Committee

Prof. D. K. Tripathy (Pro-Vice Chancellor, KIIT University, Ex Director of NIT Warangal)
Mr. A. M. Pujari (DGM, CPTI, Rourkela Steel Plant)
Er. S.C. Padhi (Consulting Engineer)
Dr. B. Mishra (Principal Scientist, DISIR)
Prof. A. K. Mohanty (Ex-Principal, REC/NIT)
Dr. S. K. Mukherjee (Scientist, Institute Plasma Research, Gnadhinagar)
Dr. B. N. Ratha (Scientist-G, PIED, BARC Mumbai)
Dr. A. Pradhan (Principal Scientist, NML)
Prof. A. M. Tigga (Professor, NIT Jamshedpur)
Dr. A. K. Das (GM, Foundry & Forge Division, HEC, Ranchi)
Dr. B. Mallick (IOP, Bhubaneswar)

Peer review statement

All papers published in this volume of *IOP Conference Series: Materials Science and Engineering* have been peer reviewed through processes administered by the proceedings Editors. Reviews were conducted by expert referees to the professional and scientific standards expected of a proceedings journal published by IOP Publishing.

IOP Conf. Series: Materials Science and Engineering **115** (2016) 012001 doi:10.1088/1757-899X/115/1/012001

Cellular Automata Modeling of Decarburization of Metal Droplets in Basic Oxygen Steelmaking

Ankit[1], T. K. Kundu

Department of Metallurgical and Materials Engineering, Indian Institute of Technology Kharagpur, Kharagpur, West Bengal--721302, India

[1]E-mail: ankitv89@gmail.com

Abstract. In steelmaking, a supersonic jet is blown over the bath to refine the hot metal to produce steel. The refining process primarily consists of removal of impurities from the hot metal to a permissible level. The impact of oxygen jet on the surface of the hot metal bath results in ejection of droplets, which mix with slag and form emulsion. The formed emulsion plays an important role in refining reactions kinetics and understanding of this process is required todevelopimproved process control model for the steel industry. In this paper, cellular automata technique has been explored to simulate decarburization in emulsion caused by interfacial reactions between the metal droplets and slag. In the course of the work, a framework has also been developed to quantify the contribution of carbon monoxide, generated by decarburization, in bloating of metal droplets and formation of halo around the droplets. The model has incorporated diffusion and decarburization reaction based on probabilities to study the evolution of the system. Simulations with varying parameters have been performed and decarburization trends obtained are comparable with the experimentally determined data reported in literatures.

Keywords: Steelmaking, Decarburization, Cellular Automata, Kinetics

1. Introduction

The blast furnace - basic oxygen furnace (BF-BOF) is the most dominant routes for steel production due to high productivity and efficiency of the process [1]. In BOF steelmaking, the hot metal from the blast furnace is poured into a specially designed converter along with scrap and flux. The input hot metal is in the temperature range of 1300 - 1400°C and contains carbon, silicon, phosphorous and other impurities. The objective of the steel making process is to refine the hot metal by removing these impurities to a predefined target level by oxidising these impurities and then separating their oxides from the hot metal. The Carbon is removed in gaseous form whereas remaining impurity oxides are removed as slag. The oxidizing atmosphere is achieved by blowing a jet of supersonic oxygen over the metal bath surface as depicted schematically in Fig. 1. The amount of flux (CaO and MgO) is decided by the amount of impurities in the hot metal, mainly silicon and phosphorous. The scrap is added to utilize the excess heat generated by oxidation of impurity species. With time, the CaO dissolves and mixes with silica (formed by oxidation of silicon) and forms the slag. Slag mainly consists of SiO_2, CaO and FeO, and the important parameters for the refining process are basicity and oxidation potential of the slag. Also, upon impact of oxygen jet, metal droplets are ejected from the metal surface into the slag resulting in formation of emulsion. There have been varying estimates of the extent of reactions occurring in the emulsion, varying from insignificant to as high as 78% [2, 3]. Recent studies have highlighted the importance of reaction in emulsions and deeper understanding of the phenomenon occurring in the emulsion [4-8]. Mainly two phenomena have been observed when the metal droplets undergo reactions in emulsion; a) formation of halo around the droplet [9] and b) bloating or swelling of the droplet [10]. The exact conditions for occurrence of either or both of these phenomena are still a matter of further study. Mulholland et. al. [9] have carried out a detailed study of decarburization using X-ray fluoroscopy technique. They have observed formation of halo around the droplet, however, Sarkar [11] has expressed some uncertainties about whether the droplets nucleated inside the droplet or at the interface. Similarly, Molleseau and Fruehan [10] have studied the decarburization in the droplet and observed bloating of droplet. Brooks et. al. [12] have highlighted the role of bloating on the residence time of the droplet which in turn affects the decarburization kinetics of the bath. They have also developed a bloating model known as 'Bloated Droplet Theory' to estimate the degree of bloating of a droplet based on the decarburization rate [12]. The bloating of droplet increases the residence time of the droplet in the emulsion which affects the extent of reactions occurring in emulsion phase. However, conditions under which either of these two mechanisms will dominate is still not clear. To resolve the issue of selection of mechanism, Chen and Coley [13] have conducted a series of experiments. They have highlighted the role of oxygen potential and carbon concentration on the selection of mechanism. They also have highlighted the role of nucleation and growth of carbon monoxide bubbles in occurrence of bloating [14]. They have concluded that classical nucleation theory can explain the nucleation and growth of droplet but the value of interfacial surface tension values has to be around 40 times lower than that of its theoretical value. Kinetics studies have also been performed by Gaye and Riboud [15] using droplets and slag of different chemical composition and oxidizing power. In continuation, Wei Pan et. al, [16] have conducted similar experiments with different mass and have considered the effect of stirring also. Both of them have concluded that decarburization characteristics depends on the initial carbon content as well as oxidizing power of the slag. For modeling of decarburization, some empirical models as well as mathematical models have been proposed, however, only the model developed by Dogan et. al. [4-6] have distinguished between the decarburization occurring in emulsion and metal bath. Using bloated droplet theory they have estimated the residence time of the droplets in the slag and calculated the total decarburization in the emulsion was based on it. Sun [17] has derived and solved a set of partial differential equations to analyse the behaviour of species inside the droplet. The model is not limited to decarburization and have also analysed desiliconization and composition distribution of generated carbon monoxide.

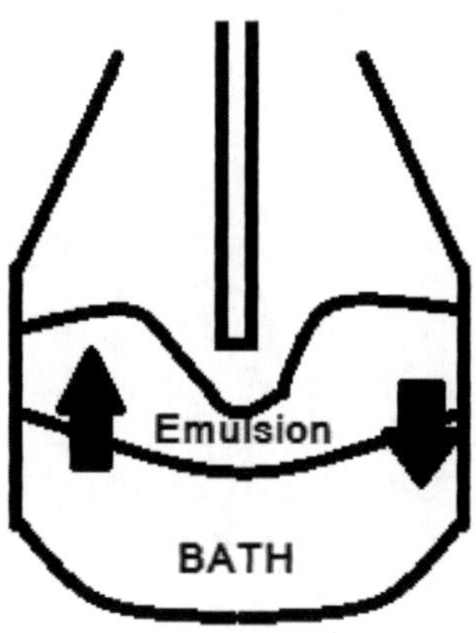

Figure 1. Schematic of a LD converter

These work clearly show that the behavior of metal droplets in emulsion is a very complex and it is difficult to model using traditional deterministic approach. So, use of cellular automata (CA) is proposed to develop a simulation method of decarburization reaction occurring in the emulsion. The ability of cellular automata to generate complex patterns using simple local rules makes it a very powerful technique for modeling physical and biological systems. The CA came in light by works of Von Neumann who was working on self-replicating biological systems [18]. Based on advice by Ulman, Von Neumann developed a two dimensional CA with 29 states capable of self-generating [19]. In subsequent decades, CA has been applied to urban planning [20], medical research [21, 22] and financial modelling [23, 24], thus, demonstrating its vast application base. In metallurgical context, CA has been extensively used to study recrystallization in steels but its application to reactions occurring during steelmaking operation is still unexplored as far as authors' knowledge. With regard to study of general chemical systems, Seybold and group [25] have demonstrated the use of stochastic approach with CA to simulate first order reactions including decay, opposing, sequential and irreversible reactions. Their work has highlighted the use of simple CA rules with a probability factor to simulate first order reactions. Further works have utilized the methodology to simulate acid dissociation [26]. Neuforth and co-workers [27] have used the CA to model kinetically and thermodynamically controlled processes. The work of Neuforth is then extended to model CSTR reactors [28]. On the front of diffusion reactions, Chopard and Droz [29], Weimar [30] and Adamatzky et. al. [31] have demonstrated the use of CA. A recent work by Ween et. al. [32] has attempted to describe the reaction between $CaCO_3$ and HCl using CA. They also have developed a PDE based model and validated it with experimental data. The authors also have noted that CA model can be related to physical time either exponentially or linearly.

In our present work, we have explored cellular automata approach to simulate decarburization in metal droplets under steelmaking conditions. The model is tested under different oxygen and carbon concentrations and an effort is also made to investigate situations that may cause halo and bloating of the droplet.

2. Theory of Cellular Automata

A cellular automata system is described by a i) lattice cells, ii) state of cells, iii) neighbourhood function and iv) transition function [33].The lattice can be visualized as a set of cells which form a grid. All transitions steps are operated upon these cells and the lattice provides platform for those transitions. Each of those cells has a state which defines its identity. The cells have a neighbourhood i.e. a methodology for selection of particles to form its neighbourhood. The neighbourhood commonly used are Newman's neighbourhood as shown in Fig. 2 or Moore's neighbourhood as shown in Fig. 3. The transition functions update the state of the cell. The transition function uses a rule table, which is based on the neighbourhood of the cell, to decide upon the appropriate transition. For example, one of the rules in the rule table can state that if a cell in the lattice with state 1 is surrounded by cells in Newmann's neighbourhood with state 0, then the state of the concerned cell should be changed to 0 from 1. The transition function, evaluates the cell for neighbourhood and if the condition is satisfied, the cell's state is updated from 1 to 0. The rules are generally simple and are designed based on the system under consideration.

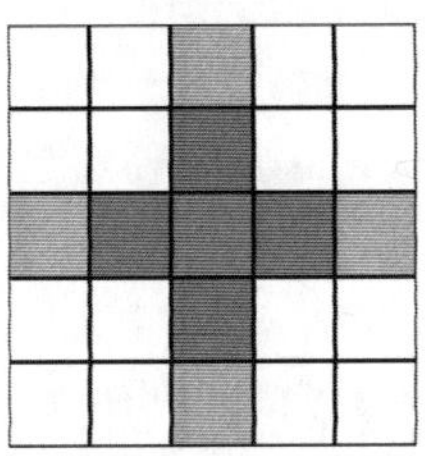

Figure 2. Newmann Neighbourhood

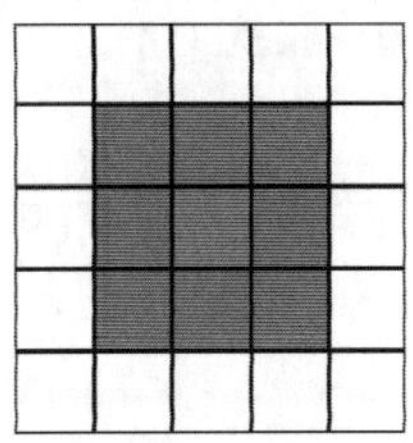

Figure 3. Moore's Neighbourhood

3. Model Development

In steelmaking, the overall reaction in slag phase can be divided into the following steps: a) migration of FeO from slag to the reaction interface, b) dissociation of FeO into Fe and O and c) reaction of O with carbon at the interface and lastly d) removal of the products. Similarly, reaction steps involved in metal phase are: a) migration of carbon from metal towards the reaction interface, b) oxygen diffusion through the interface into the metal droplet, c) reaction of carbon and oxygen and d) removal of products [16]. Sun [17] has simulated the concentration profile in the metal droplet and assumed that the rate controlling steps is the migration of FeO towards the slag interface and diffusion of species inside the metal droplet. The reaction at the interface can be assumed to be fast. Also, the formation of gas halo will add an additional barrier for migration of species to and from the reaction interface. However, in the present model, migration through gas halo is not considered. The objective here is to simulate the situations inside the droplets which may lead to either halo formation or bloating of droplets using cellular automata.

A lattice consisting of 100 x100 square cells is considered for the study. The species under consideration are carbon (C), oxygen (O), carbon monoxide (CO) and iron (Fe). A sample distribution with states is shown in Fig. 4.

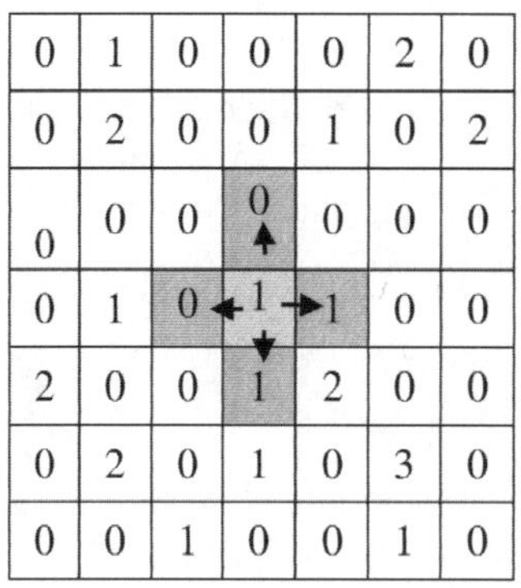

0	**Metal (Fe)**
1	**Carbon (C)**
2	**Oxygen (O)**
3	**Carbon monoxide (CO)**

Figure 4. Newman Neighborhood with state number for the species considered for decarburization. Possible transition positions are shaded and arrows are depicting a possible exchange during diffusion.

Out of 100x100 grid, carbon is distributed in 98x98 grid leaving 2 rows of cells from boundaries. The two cells from the boundary are filled with oxygen such that the mole fraction of O is maintained. In the system, iron is considered as solvent. The initial distribution of C in the lattice is given on the basis of it concentration as mole fraction. The mole fraction of O is calculated from the equilibrium relationship denoted by Equation 1, where (%FeO) is the FeO concentration in the slag and [%O] is the oxygen concentration in the metal [34].

$$\frac{[\%O]}{(\%FeO)} = (0.1T - 155.3) \times 10^{-4} \qquad (1)$$

Initially, CO presence is zero and the system is consisted with only of oxygen at the boundaries and carbon distributed throughout the 98x98 inner grid. The model can be divided into different subroutines, namely, diffusion, reaction and replenishment model. These subroutines are discussed in detail in the following sections.

3.1 Diffusion Model

Diffusion of species is calculated by random walk method. A species has the freedom to move in either of the cells in its Neumann neighbourhood. The possible events that can occur are i) the particle/species can move to any one cells among the four shaded cells (Newman neighborhoods) shown in Fig. 4 or ii) the particle may choose not to move at all. In the model under discussion, all these events are considered equi-probable and a random number is generated based on which one of the five events (movement in four directions and stationary state) are selected. The state of cells between which the movement occur are also get exchanged. The same process is repeated for oxygen but since the diffusivity of oxygen is two orders of magnitude higher than carbon, for one diffusion step of carbon, oxygen is allowed to diffuse 100 steps [35]. Also, in case of oxygen, the event for not moving is not considered, as under high concentration gradient and high diffusivity oxygen is highly probable to move. In this study, only carbon and oxygen are taken to be diffusing species however, exchange between CO and iron is not restricted. It is assumed that oxygen in the extreme boundary cells can 'jump' into the adjoining cells, if they are not blocked by CO.

3.2 Reaction Model

Once the diffusion subroutine is executed, the control is transferred to the reaction subroutine. The reaction rule is based on the state of neighbouring cells. The rule is defined as if the cell with state 1 (i.e. carbon) has a cell with state 2 (i.e. oxygen) in its Newman neighbourhood, then the reaction can occur and the state of cell is changed to 3 (i.e. carbon monoxide). Also, the cell with state 2, which participated in the reaction, is changed to 0 (denoting iron). If more than one cell with state 2 is present, one of them is randomly selected for the reaction. To summarize, one carbon cell reacts with one oxygen cell to form carbon monoxide. In the reaction subroutine, the lattice is divided into two zones A and B as shown in Fig.5. Zone A is around the boundary and is marked till 5 cells from the boundary while the remaining inner cells are in Zone B. It is assumed that carbon reacted in Zone A, and the formed CO here have a higher chance of contributing towards halo formation as it has a higher probability to escape into the boundary. On the contrary, carbon monoxide generated in Zone B has higher chance to contribute to bloating of the droplet as the bubbles formed here will have lower probability of escaping than Zone A bubbles. The demarcation is completely arbitrary due to absence of any validated study and difficulty in conducting the experiments to quantify the decarburization rate at different location in a droplet. The amount of carbon reacted in both the zones were recorded separately and stored for each time step.

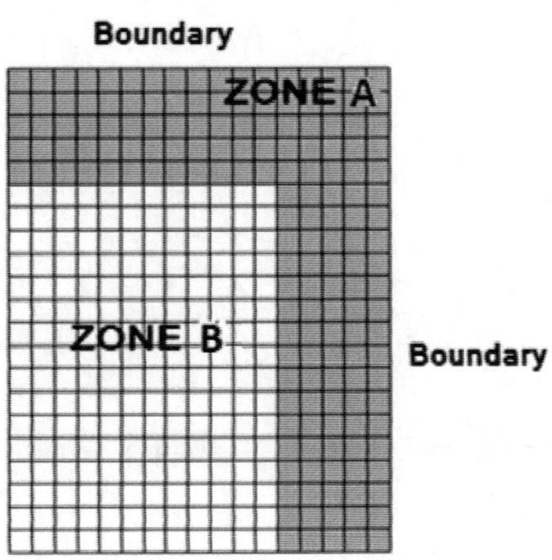

Figure 5. An upper right portion of grid under study depicting two zones with different rules

3.3 Replenishment Model

After the reaction step, as the oxygen is consumed, a count on oxygen consumption is maintained. As discussed earlier, the concentration of oxygen in metal is determined by the FeO in slag by equation 1. So, the used oxygen must be resupplied in order to keep the concentration constant. The consumed oxygen that have reacted, is resupplied by replacing the iron cells in second last boundary cells.

The model was run for 600 iterations with different oxygen potential and initial carbon content.

4. Results and Discussion

In the simulation, carbon and oxygen are allowed to diffuse and if the reaction condition as mentioned in section 3.2 is satisfied, the reaction is allowed to occur and carbon monoxide is generated. Here the initial carbon percentages (C_0) are taken as 3.8 wt. %, 1.7 wt. % and 0.8 wt. %. Similarly, different potentials of oxygen depending upon the % FeO in slag are taken. The FeO amount are taken as 10 wt. %, 20 wt. % and 30 wt. %. The plots of decarburization with varying initial carbon percentages are shown in Fig. 6 for 30 wt. % FeO. The effect of FeO content in the slag on decarburization curve is plotted in Fig. 7. The rate of decarburization decreased with decreasing FeO wt. % in slag. As discussed earlier, decarburization in Zone A and Zone B are calculated separately. Fig. 8 shows the percentage of decarburization occurred in Zone A (with respect to total carbon monoxide generated in both zones) for different initial carbon concentrations. Fig. 9 plots the same parameter with different FeO content in slag having 3.8 wt. % FeO content. In all the above simulations, we have defined Zone A till a distance of 5 boundary cells from the boundary. A simulation is then performed with this distance increased to 10 boundary cells in Zone A. The result for the same is shown in Fig. 10. A snapshot of lattice at 150th iteration of a simulation run is shown in Fig. 11. The background light green colour is of iron, the CO, carbon and oxygen is shown in yellow, light blue and dark blue respectively. The simulation results are further discussed in the following sections.

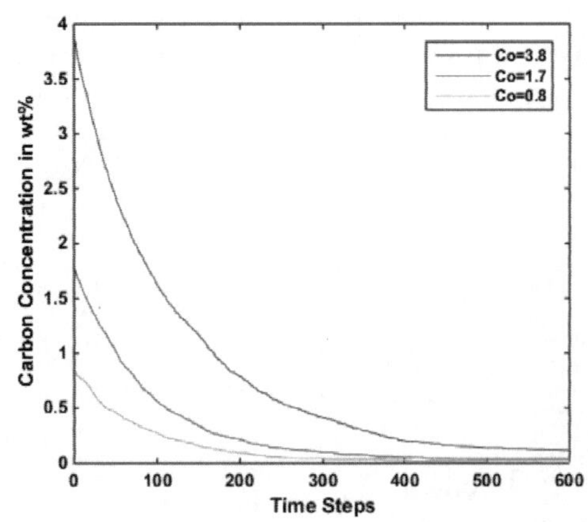

Figure 6. Decarburization with varying initial carbon concentration in droplet with 30 wt.% FeO in slag

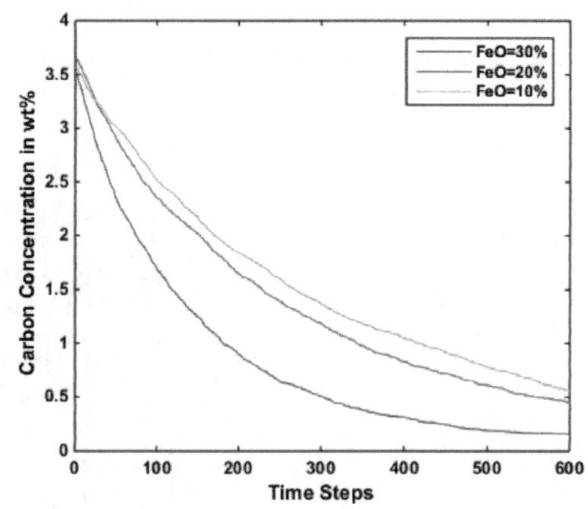

Figure 7. Decarburization of 3.8 wt.% carbon with varying FeO wt.% in slag

4.1 Effect of initial carbon concentration

In the study, higher decarburization rate is observed for systems with higher initial carbon contents, as shown in Fig. 6. In Fig. 8, the initial amount of carbon has a positive effect on the percentage of carbon monoxide formed in Zone A, when the oxygen potential is constant, though the increase of magnitude is not that large. This result is in agreement with the conclusion of Chen and Koley [13, 14]; according to their result, higher carbon content should favour halo formation i.e. higher percentage of carbon monoxide in Zone A. One of the reasons for the low increment can be attributed to the simplicity of the present model. As discussed above, the transport of species only inside the droplet is considered. The transport in slag and across the halo is not considered. Non-inclusion of this phenomenon may have increased the diffusion rate of oxygen resulting in faster diffusion of oxygen towards the centre in the model than in real phenomenon.

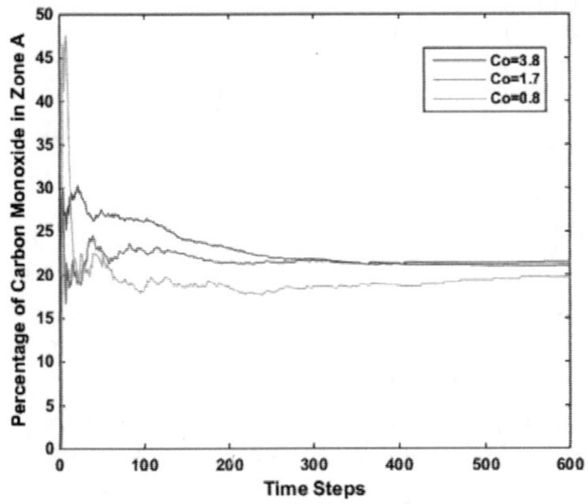

Figure 8. Percentage of CO in Zone A with varying initial carbon concentration

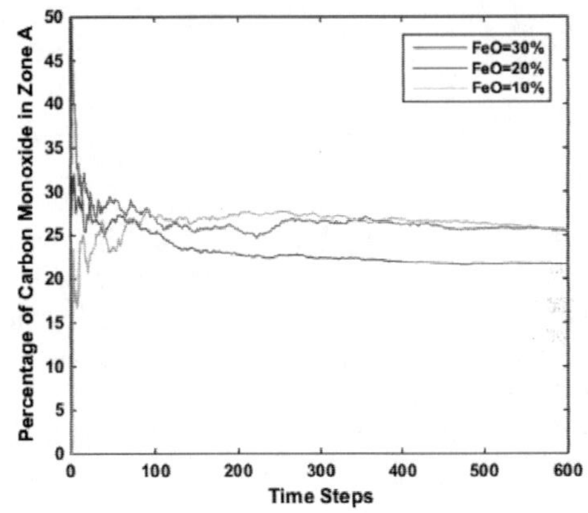

Figure 9. Percentage of CO in Zone A with varying slag FeO wt.% (C_0=3.8 wt.%)

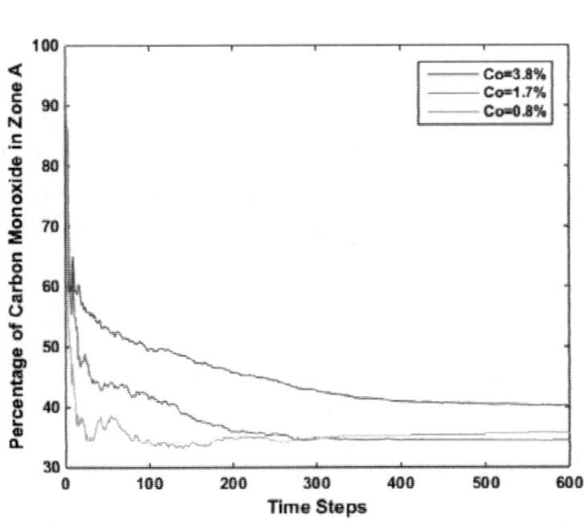

Figure 10. Percentage of CO in Zone A with varying initial carbon concentration and 10 cell zone A

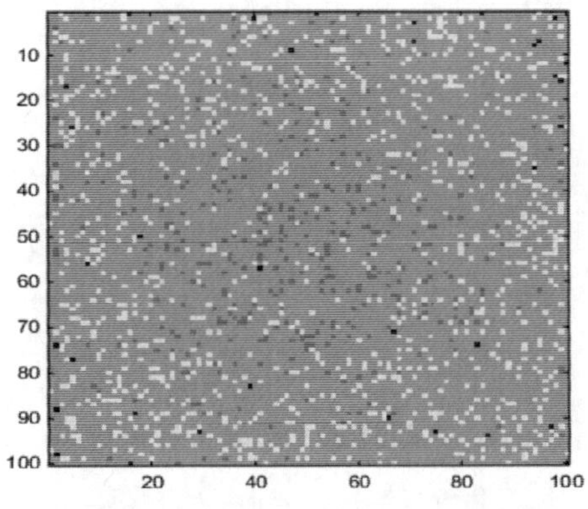

Figure 11. Intermidate state of a simulation, Co=3.8 wt.% FeO=30 wt.% at 150th time step

4.2 Effect of oxygen potential

The oxygen potential is defined as a function of FeO content in the slag using equation 1. In experimental studies by Gaye and Riboud [15], higher decarburization rate has been observed with increasing FeO content. Similar observation has been witnessed from the results of the present model. The Oxygen potential affected the decarburization rates and higher FeO wt.% in slag resulted in higher decarburization rates as shown in Fig.7. The effect on percentage of carbon monoxide in Zone A is shown in Fig. 9. The percentage of CO in zone A decreases with higher FeO in the slag; as with higher oxygen potential, higher amount of oxygen cells were present in the droplet. Due to high number of oxygen cells, more oxygen 'cells' were able to migrate towards the interior and react. In the case with varying carbon content, the difference in value is not considerable and the reason can be similarly attributed to the simplicity of the model and non-inclusion of transport in slag and across interface.

4.3 Effect of size of Zone A

Most of the above simulations (Fig. 6-9) are performed with size of Zone A limited to 5 boundary cells from the edge. In a simulation run, the Zone A boundary size is increased to 10 cells. As a result, the percentage of CO formed in Zone A increases as more cells come under it, as shown in Fig. 10.

Though, the values of percentage of CO in Zone A show some trends, but it is difficult to directly relate it to actual conditions as the selection of size of Zone A is completely arbitrary and it is also very difficult to validate. However, this can be one of the methods to demarcate the mechanisms of halo formation and bloating. Both of these mechanism involve formation of CO bubble but the behaviour of bubble is different. In gas halo mechanism, the bubble may easily escape forming a region between the slag and metal droplet whereas the bubble involved in bloating mechanism, may not escape and grow inside the droplet and may even cause bursting of droplet. However, the two parameters – the initial C concentration and the oxygen potential behave as reported in the earlier literature [13, 14]. The percentage of CO in Zone A can be studied in more detail when the model is augmented to include other phenomena like mass transfer kinetics across slag and gas halo, which are under development.

We have used cellular automata model successfully and highlighted how the complex behaviour of the decarburization process involving different phenomena can be simulated using simple local rules. In this model, the rules for diffusion and reaction are defined only locally to simulate the global complex behaviour. To understand the ease of implementation of rules, a case of limited oxygen mobility is simulated and the result is shown in Fig. 12. The result shows that if the diffusion or supply of oxygen inside the droplet is partially restricted, the reactions will happen mostly near the interface for a longer duration of time. The only change in this model is that the oxygen in the extreme boundary is not allowed to diffuse. This causes a drop in oxygen available for reaction in the central portion of the lattice, causing an increase in the percentage of carbon in Zone A.

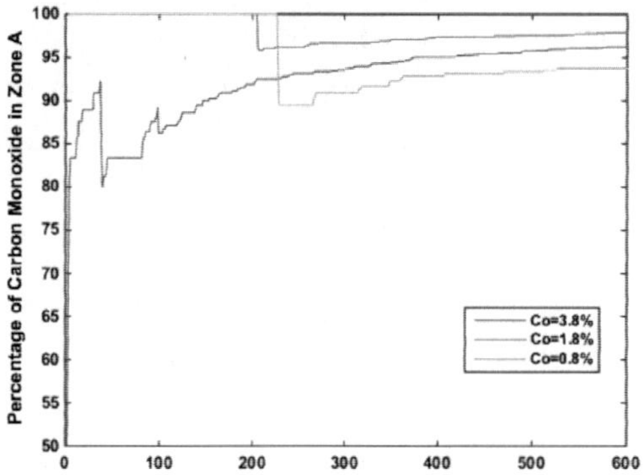

Figure 12. Percentage of CO in Zone A under restrictive supply of oxygen

The rules decided for the present model are unique in order to explain decarburization. The rules are open ended and are capable of incorporating further complexity to represent the actual phenomena occurring during steel making. Also, there can be more than one rules to simulate a certain phenomenon. For example, there are multiple ways to implement diffusion, like, using models by Chopard and Droz [29], or Weimar [30], as well as different rules for reactions as discussed by Kier et. al. [36]. The flexibility of making small changes at local level to reflect a complex global phenomena makes cellular automata a powerful tool to study complex processes like steelmaking as demonstrated in the present work. The authors expect that the current work may inspire other researchers to develop their own cellular automata rules for the decarburization and other complex phenomena in domain of process metallurgy.

5. Conclusions

In the present work, a cellular automata based model is developed to study decarburization in steelmaking process. The model is successful in capturing the effects of initial carbon concentration and FeO content of slag on decarburization. A study to evaluate the dominance of halo forming mechanism and bloating mechanism is also proposed based on the distance of formed CO from the boundary of the droplet. Further development of the model to include mass transfer in slag and across the gas halo is under progress.

References

[1] Ghosh A and Chatterjee A 2008 *Ironmaking and Steelmaking* (New Delhi: Prentice-Hall of India)
[2] Meyer H, Porter W, Smith G and Szekely J 1968 *Journal of Metals* **20** 35
[3] Kozakevitch P 1969 *Journal of Metals* **22** 57
[4] Dogan N, Brooks G and Rhamdhani M 2011 *ISIJ Int.* **51** 1086
[5] Dogan N, Brooks G and Rhamdhani M 2011 *ISIJ Int.* **51** 1093
[6] Dogan N, Brooks G and Rhamdhani M 2011 *ISIJ Int.* **51** 1102
[7] Urquhart R and Davenport W 1973 *Canadian Metallurgical Quarterly* **12** 507
[8] Millman M, Overbosch A, Kapilashrami A, Malmberg D and Bramming M 2013 *Transactions of the Indian Institute of Metals* **66** 525
[9] Mulholland E, Hazeldean G and Davis M 1973 *Journal of Iron and Steel Institute* **211** 632
[10] Molloseau C and Fruehan R 2002 *Metall and Materi Trans B,* **33** 335
[11] Sarkar R, Gupta P, Basu S, Ballal N B 2015 *Metall and Materi Trans B* **46** 961
[12] Brooks G, Pan Y, Subagyo and Coley K 2005 *Metall and Materi Trans B* **36** 525
[13] Chen E 2011 *Kinetic Study of Droplet Swelling In BOF Steelmaking*, PhD Thesis, McMaster University
[14] Chen E and Coley K 2010 *Ironmaking & Steelmaking* **37** 541
[15] Gaye H and Riboud P V 1977 *Metall and Materi Trans B* **8** 409
[16] Pan W, Sano M, Hirasawa M and Mori K 1991 *ISIJ Int.* **31** 358
[17] Sun H 2006 *ISIJ Int.* **46** 1560
[18] Von Neumann J and Burks A W 1966 *Theory of self-reproducing Automata (*Urbana: University of Illinois Press)
[19] Kari J 2005 *Theoretical Computer Science* **334** 3
[20] Batty M 2005 *Cities and Complexity (*Cambridge, Mass.: MIT Press)
[21] Kansal A, Torquato S, Harsh G, Chiocca E and Deisboeck T 2000 *Journal of Theoretical Biology* **203** 367
[22] Moreira J and Deutsch A 2002 *Advances in Complex Systems* **05** 247
[23] Wei Y, Ying S, Fan Y and Wang B 2003 *Physica A: Statistical Mechanics and its Applications* **325** 507
[24] Bartolozzi M and Thomas A 2004 *Physical Review E* **69** 046112
[25] Seybold P, Kier L and Cheng C 1997 *Journal of Chemical Information and Modeling* **37** 386

[26] Kier L, Cheng C, Tute M and Seybold P 1998 *Journal of Chemical Information and Modeling* **38** 271
[27] Neuforth A, Seybold P, Kier L and Cheng C 2000 *International Journal of Chemical Kinetics* **32** 529
[28] Parez-Terrazas J, Ibarra-Junquera V and Rosu H 2008 *Korean Journal of Chemical Engineering* **25** 461
[29] Chopard B and Droz M 1991 *Journal of Statistical Physics* **64** 859
[30] Weimar J 1997 *Parallel Computing* **23** 1699
[31] Adamatzky A, Martanez G and Mora J 2006 *International Journal of Bifurcation and Chaos* **16** 2985
[32] Van der Ween P, Baetens J and Baets B 2011 *J. Comput. Chem.* **32** 1952
[33] Toffoli T, Margolus N 1991 *Cellular Automata Machines* (Cambridge: MIT Press)
[34] Turkdogan E 2000 *ISIJ Int* **40** 964
[35] Kawakami M, Yokoyama S, Takagi K, Nishimura M and Kim J S 1997 *ISIJ Int.* **37** 425
[36] Kier L, Seybold P and Cheng C 2005 *Cellular Automata Modeling of Chemical Systems* (Netherlands: Springer).

Impact resistance and hardness modelling of Aluminium alloy welds using square-headed friction-stir welding tool

U.Sudhakar[1], Dr.J.Srinivas[2]

[1] Asst. professor, Department of Mechanical Engineering,GMR IT-Rajam,

[2] Associate Professor, Department of Mech.Engg., NIT Rourkela, Rourkela 769 008

Correponding author e-mail: mr.u.sudhakar@gmail.com

Abstract: This paper proposes modelling and optimization issues relating to friction-stir welding process of aluminium alloys. A specially prepared SS tool of square headed pin profile with cylindrical shoulder is used with a vertical milling machine. Effects of process variables including tool rotation and tool velocity on the weld performance are studied in terms of impact strength and hardness. Three different rotational motions and three welding speeds (feeds) of tool are considered at constant axial load (depth of cut) condition and altogether nine experiments are conducted on a vertical milling machine with specially prepared fixture. Each weld sample is then tested for its impact strength (IS) and hardness independently. A model is developed to correlate the relations between the hardness/impact strength with tool rotation and weld speed using neural networks. The optimized process conditions are predicted to improvise the impact strength and hardness of the weld. Further, the morphology of the weld is studied using SEM to know the material flow characteristics.

Keywords: Friction-stir welding, Joint toughness, impact strength, hardness, neural network model.

1 Introduction

Recently, several cost-effective and eco-friendly joining techniques have been evolved in material industries. Solid-state joining techniques like laser-welding, friction-welding, friction-cladding and friction-stir welding have replaced the conventional fusion welding processes for Aluminium alloys. Friction-stir welding (FSW) was invented in 1991 by welding institute in UK for welding dissimilar materials. It results in excellent weld strength, ductility and toughness. Here, a rotating tool with a specially designed shoulder and probe is pressed over the workpiece surface at the junction and traversed along the interface with a weld-speed. Due to friction between the rotating tool and work piece, the plastic deformation occurs and temperature rises. The generated heat softens the work piece and yield strength of base materials reduces. As tool is moved from one location to other, the momentary softening at each instant of time will form the weld. Main advantage of the process is that it is free from oxide formation and welding cracks as well as it results in low residual stresses due to relatively small input energy. During FSW, the weld quality is affected by various process and tool parameters. It results in changes of different mechanical properties of work-material.

There are several studies in the literature concerning mechanical characterization of friction stir welded joints. The effect of tool rotational and traverse speed on friction stir weldcharacteristics of AISI 430 ferritic stainless steels were studied by Bilgin and Mera [1]. Ahn et al. [2] examined the mechanical properties of friction stir welded 409L stainless steel. Heidarzadeh et al. [3] used a mathematical model for predicting the tensile properties of friction stir welded aluminum alloy joints. Rajakumar et al.[4] illustrated the response surface methodology to develop empirical relationship to predict grain size and hardness of weld nugget of AA6061 alloy weld as function of process parameters. Shojaeefard et al. [5] investigated the Artificial Neural Network (ANN) model for the mechanical properties offriction stir welded aluminum alloys butt joints. The ultimate tensile strength of joint of aluminum alloys werepredicted using Artificial Neural Network model. Elatharasan and Kumar [6] investigated the experimental analysis and optimization of process parameter on friction stir welding of AA 6061-T6 aluminum alloy. They modeled the ultimate tensile strength (UTS), yield-strength (YS), and percentage elongation in FSW process and analyzed through response surface methodology. Silva et al. [7] investigated the friction stir welded T joints optimization using Taguchi method. They suggested that the tool rotational speed was the most influencing factor in joint mechanicalproperties. Selvamani and Palanikumar [8] investigated optimizing the friction welding parameters to attain maximumtensile strength in AISI 1035 grade carbon steel rods. Birol and Kasman [9] studied effect of tool tilt angle and rotation on the ultimate strength and micro hardness of the friction stir welding of twin roll cast EN AW Al-Mn1Cu plates.Liu et al. [10] carried out FSW of aluminum alloy 6061-T6 alloy sheets with high strength steel and computed the effects of various process parameters on the ultimate strength of the weld. More recently Wang et al.[11] studied the microstructure and texture evaluations of weld formed by dissimilar aluminium alloys. Use of soft computing tools in modelling of friction-stir welding is another concept. Okuyucu et al.[12] shown the neural network application for friction-stir welding of Aluminium plates. Fratini et al. [13] presented the procedure of predicting the average grain size in friction-stir welding process using neural networks. BoldSaikhan et al.[14] used neural networks and Fourier transforms for real time evaluation of friction-stir welding. Bakara et al.[15] illustrated quality monitoring frame work for steel friction-stir welding using computational intelligence.

Modelling and optimization of mechanical properties from experimental studies is very much required to minimize the time-consuming experiments. Present paper deals with development of a process-model based on experimental data and further this model is employed for process optimization simultaneously to obtain better hardness and impact strength of the joint. Effects of the operating variables on impact strength and hardness of aluminium weld samples is initially studied. A neural network model is employed to obtain relationship between the input and output parameters, whose weight structure is adopted later in optimizing the process parameters. Results of the analysis are shown in the form graphs and tables.

2. Experimental work

This section describes the experimental work carried out on AA3003 alloy strips for welding along with description of the material characterization using impact testing machine and hardness tester.

2.1 Weld Preparation

The rolled plates of aluminium alloy AA3003 of 5 mm thick are cut in to required sizes (100 mm× 30 mm) by power hacksaw machine and square butt configuration is prepared. Initial joint configuration was obtained by securing plates in position using mechanical clamps. Non-consumable tool made of high speed steel was used in the weld joint preparation.To carry out the FSW process, a Column & Knee type Universal Milling Machine as shown in Fig.1 is employed. The cutting tools are initially fabricated on a lathe machine using stainless steel material. The work holding fixture was designed and tested before the experiments. The FSW tool consists of a 9mm diameter shoulder and square cross-section of side 4mm tool pin of thickness 3mm is considered as shown in Fig.2.

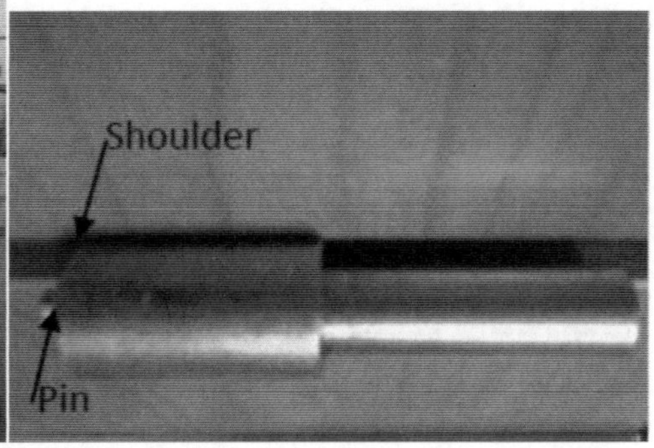

Figure 1. Universal milling machine **Figure 2**. Tool with square headed pin

It is made of stainless steel. All the experimental analysis is carried out at workshops of Mechanical Engineering with curtsey of GITAM University. The FSW tool was plunged into the pilot hole of size 3mm machined already in the work piece. The tool initially removes some material, and subsequently started to generate heat when the required plunge is provided. The tool was traversed through the length of the plate. The schematic diagram of the friction stir welding process is shown in Fig. 3.

Figure 3.Schematic of welding process **Figure4**. Samples for impact test

Welding was carried out with various tool rotations and weld speeds using full factorial experimental designs. Depth of cut was set at 3.5mm during welding. Standard test pieces were made for impact and hardness tests by cutting along the transverse direction. In this study, a total of nine experiments were carried out by varying two input variables in three levels. Table-1 shows the input variables considered.

Table-1 Parametric limits considered during experiment

Designation	Tool rotational Speed (TRS) RPM	Weld speed (mm/min)
A1	560	22.4
A2	560	31.5
A3	560	45.0
B1	900	22.4
B2	900	31.5
B3	900	45.0
C1	1400	22.4
C2	1400	31.5
C3	1400	45.0

2.2 Impact strength and hardness measurements

For impact and hardness tests, the samples were cut from weld specimens in cross direction. From each welded sample, two test specimens were prepared. Fig.4 shows the samples for impact test. The Charpy impact test was carried-out to find the impact strain energy and hence the strength. Fig.5 shows the variation of impact strength of the joints welded at different weld speeds and tool rotational speeds. In welding, the heat input plays an important role on the mechanical properties of the weldments. Vickers hardness was recorded by using a tester employing 100Kgf load. The measurements were made at weld nugget zone. Fig.6 shows the variation of hardness for the various tool rotations and weld speeds. It was observed that the hardness decreases with increase in heat input. For instance, atool rotation of 560 rpm at 45 mm/min weld-speed, the hardness value is low due to maximum heat input. For the same tool rotation, the weld hardness value is high at 22.4mm/min because of lower heat input.

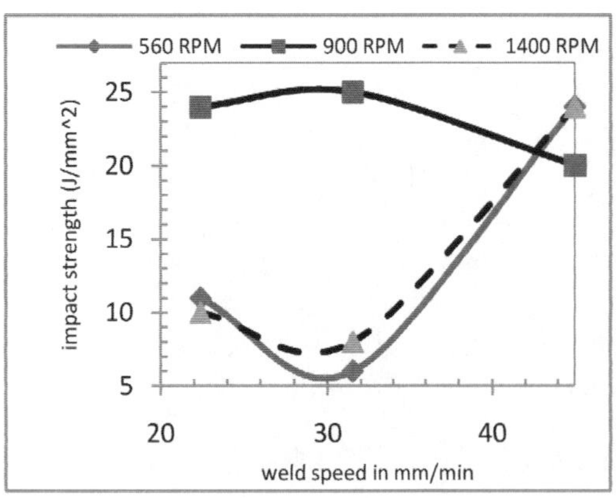

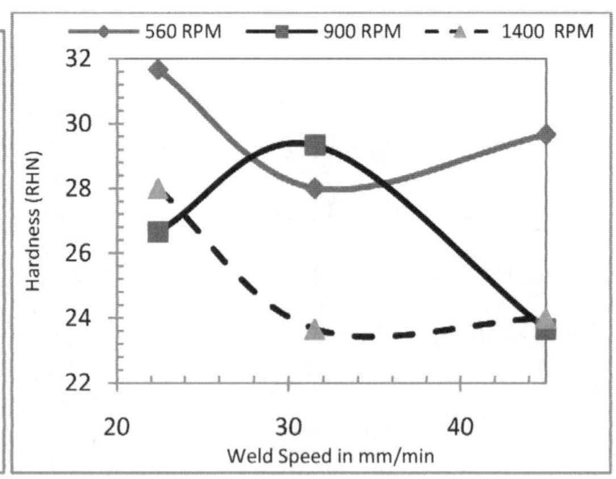

Figure 5.Impact Strength Measured**Figure 6**. Hardness Measured

3. Analysis and Modelling

The statistical analysis of the data was made in two phases. The first phase was concerned with the analysis of variance with the effect of process parameters and their interactions. The second phase was concerned with correlation of input parameters and properties of friction stir-weld. The simplest way to determine the most influencing parameters on impact strength and hardness is to obtain the idea from the

Main Effect plots as shown in Fig.7. The S/N response data are calculated using Minitab16 software(trail). The largest S/N response reflects the best response resulting in the lowest noise for impact strength and hardness. The lowest tool rotational speed (560 rpm) with lowest weld-speed (22.4 mm/min) is determined to be the best choices for obtaining highest impact strength and hardness. The plotsalso show that the tool rotation speed is most influential parameter on the impact strength and hardness.

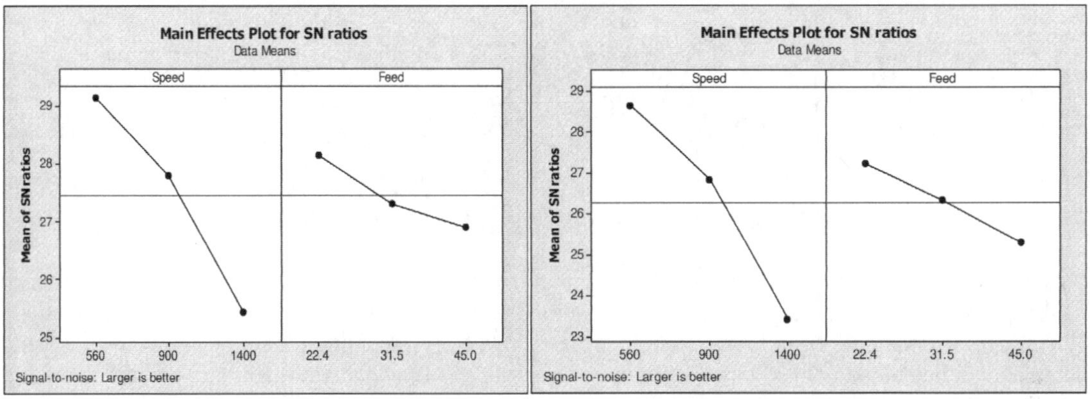

(a) (b)

Figure 7. Main effect plots of (a) Impact strength and (b) Hardness

The SEM micrographs of surface generated after the welding operation is shown in Fig.8. They indicate the weld uniformity at lower speed and feeds. The aluminium particles dispersion is found to be uniform in the alloy.More number of surface studies are required.

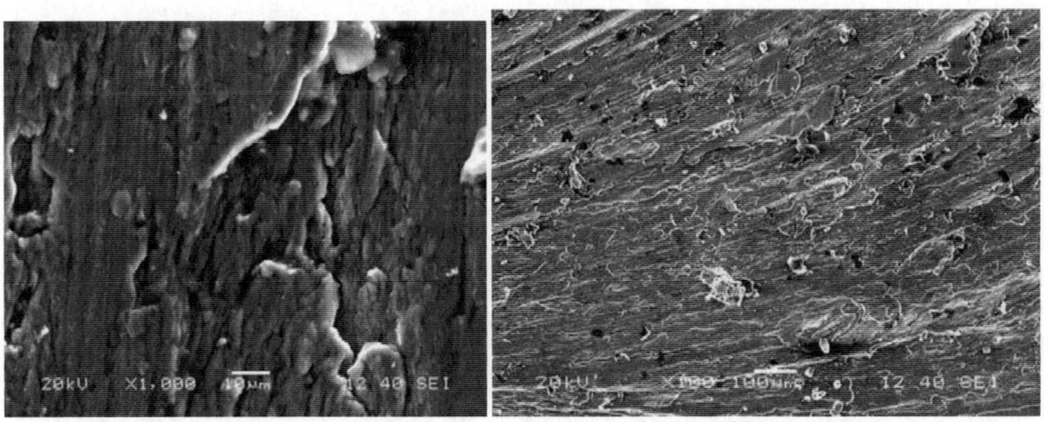

(a) At weld centre (b) At weld corners

Figure 8. SEM micrograph at the weld zone

3.1 Modeling data with network

Artificial neural networks are the computational models, which replicate the function of a biological network, composed of neurons and are used to approximate complex functions in various applications. The present system considered has three layers, which are input, hidden and output layers. The input layer consists of all the input factors including tool rotation and weld speed. Information from the input layer is then processed to hidden layer followed by the output, where the output vector iscomputed. A schematic description of the layers is given in Fig.9. In this work, supervised learning, in which the network is presented with inputs along with the target outputs together with adjusted weighting, are employed. One of the most popular learning-algorithms is the back-propagation algorithm. With a single hidden layer, Levenberg-Marquardt scheme is employed to minimize the square error between the data points obtained from experiments.

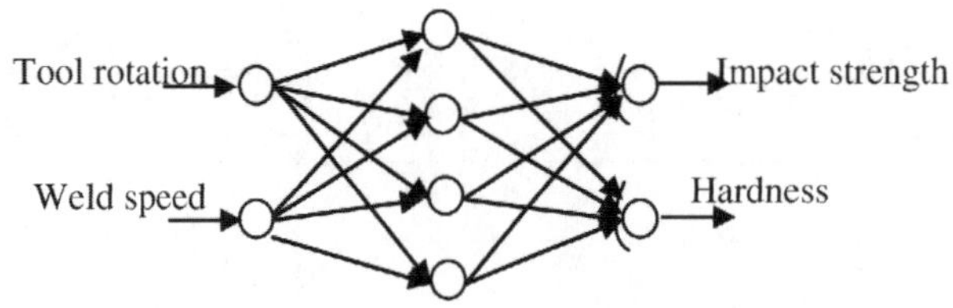

Figure9. Neural network employed in present task

Inputs and outputs have been normalized in the range of 0–1. Neurons in the input layer do not have transfer function, while in the other layers a logistic sigmoid (logsig) transfer function has been used. Both a built-in code and MATLAB toolbox are employed for modeling. After training the network successfully, it has been tested by using the known test data. Fig.10 shows the error convergence trend for various number of hidden-layer neurons.

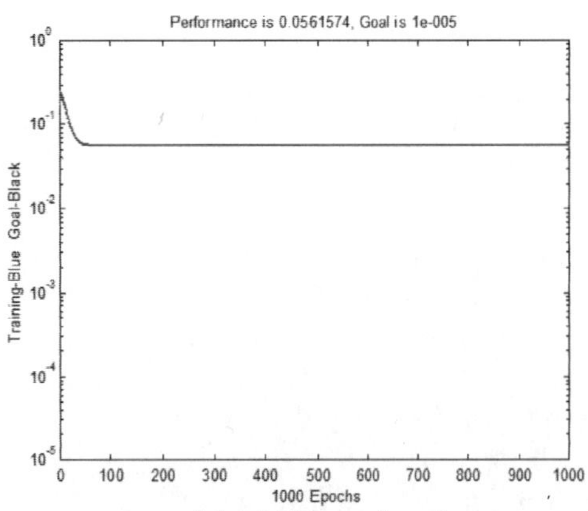

(a) hidden nodes=3

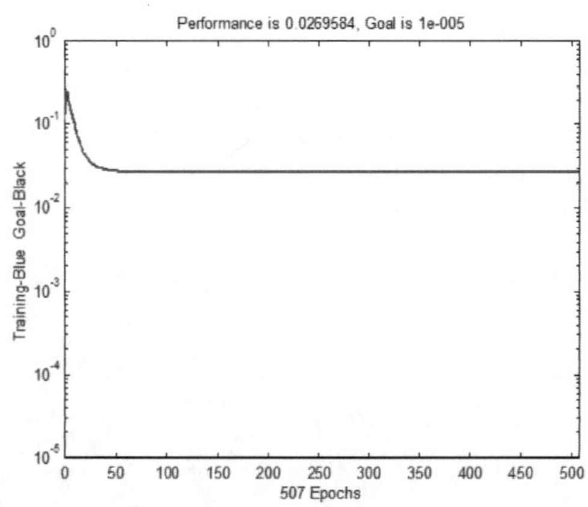

(b) hidden nodes=4

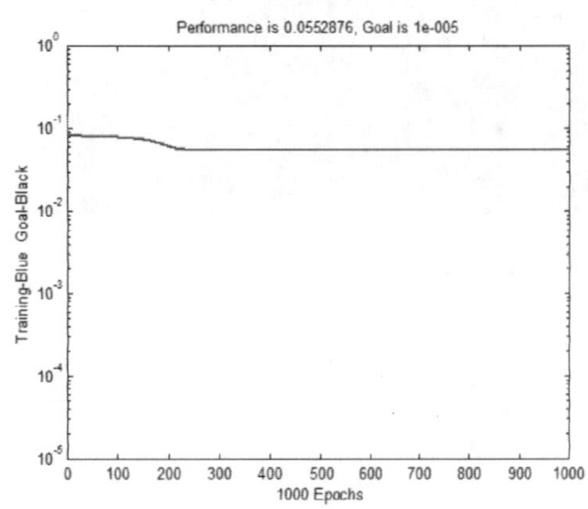

(c) Hidden Nodes=5

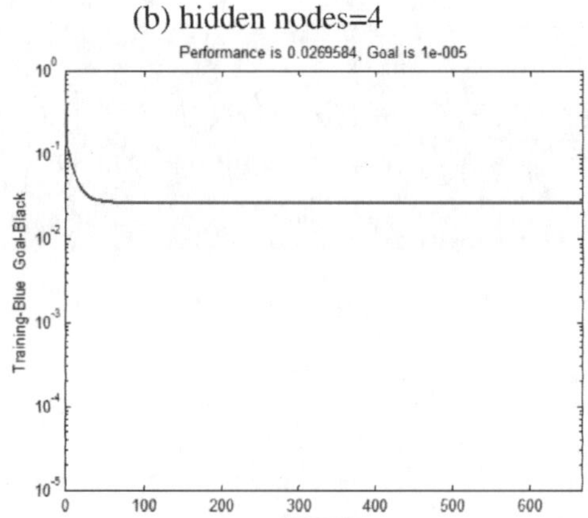

(d) Hidden Nodes=6

Figure 10. Mean-square error convergence

It is seen that 4 hidden layer architecture is best one and employed further. The mechanical properties of friction-stir welding of Al plates with tool rotation and weld speed as well as predicted new data obtained from the network after the training stage are compared and it is found that measured and predicted outputvalues were very close to each other. Fig.11 shows the process optimization module which uses the neural network weights for obtaining the impact strength and hardness.

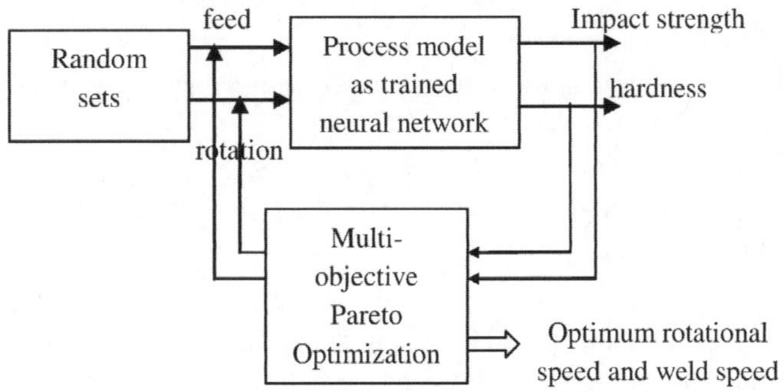

Figure 11. Process of estimation of optimum welding parameters

4. Conclusion

This work presented the outcomes of friction-stir welding of aluminum alloy plates using a tool with square headed pin profile. Nine welding conditions were considered by varying the tool rotation speed and weld speed using the vertical milling machine set-up. In all the cases, impact strength and hardness were predicted independently. The morphology was studied at the weld zone to understand the mixing ability of phases and weld defects including burrs. The optimum cutting states were predicted from analysis of variance test using Minitab software. Finally, the relation between the weld parameters and mechanical characteristics of weld was obtained using a three layer back propagation model. Use of this model towards prediction of optimal process parameters was briefed-out and can be treated as future scope of the work.

References

[1] B.M.Bilgin and C. Meran,' The effect of tool rotational and traverse speed on friction stir weldability of AISI 430 ferritic stainless steels'. Mater Des vol.33:pp.376–383, 2012.

[2] B.W.Ahn, D.H.Choi D.J.Kim and S.B.Jung, 'Microstructures and properties of friction stir Welded 409 L stainless steel using a Si3N4 tool', Mater Sci Eng A vol.532, pp.:476–479, 2012

[3] A.,Heidarzadeh, H.Khodaverdizadeh, A.Mahmoudi and E.Nazari, ' Tensile behavior of friction stir welded AA 6061-T4 aluminum alloy joints', Mater Des vol.37, pp.166–173, 2012.

[4] S.Rajakumar, C.Muralidharan and V.Balasubrahmanian, 'Statistical analysis to predict grain size and hardness of the weld nugget of friction-stir welded AA6061 Aluminium alloy joints', Int.J.Adv.Manuf.Technol. vol.57, pp.151-165, 2011

[5] M.H.Shojaeefard, R.A.Behnagh, M.Akbari, M.K.B.Givi and F.Farhani, 'Modelling and Pareto optimization of mechanical propertiesof friction stir welded AA7075/AA5083 butt joints using neural network and particle swarm algorithm', Mater Des vol.44:pp.190–198, 2013

[6] Elatharasan G, Senthil Kumar VS (2013) An experimental analysis and optimization of process parameteron friction stir welding of AA 6061-T6 aluminum alloy using RSM. Procedia Eng 64:1227–1234.

[7] Silva ACF, Braga DFO, Figueiredo MAV, Moreira PMGP (2014) Friction stir welded T-joints optimization. Mater Des 55:120–127.

[8] Selvamani ST, Palanikumar K (2014) Optimizing the friction welding parameters to attain maximum tensile strength in AISI 1035 grade carbon steel rods. Measurement 53:10–21.

[9] Y.Birol and S.Kasman, 'Effect of welding parameters on the microstructure and strength of friction stir weld joints in twin roll cast EN AW Al-Mn1Cu plates', J.Material Engg and Performance,vol.22, pp.3024-3033, 2013.

[10] X.Liu, S.Lan and J.Ni, 'Analysis of process parameters effects on friction stir welding of dissimilar aluminum alloy to advanced high strength steel', Materials and Design, vol.59, pp.50-62, 2014.

[11] B.Wang, B.Lei, J.Zhu, Q.Feng, L.Wang, D.Wu, 'EBSD study on microstructure and texture of friction-stir welded AA5052-O and AA6016-T6 dissimilar joint', Materials and Design, vol.87, pp.593-599, 2015.

[12] H.Ukuyucu, A.Kurt, E.Areakliuglu, ANN application to FSW of Al plates, materials & Design ,vol.28,pp78-84,2007.

[13] L.Fratini, G.Buffa, D.Palmesi, using ANN for prediciting the average grain size in FSW processer,comp & structure, Vol.87, pp 1166-1174,2009.

[14] E.boldsaikhan , E.M.Corwin , A.M.Logar, W.J.Arbegast Use of NM discrete FT for real time evaluation of FSWelding,Applied soft computing,Vol.11,pp.4839-4846,2011.

[15] A.Bakara, G.Panoutsos, S.Cater, A real time quality moniting frame work for steel FSW using computational intelligence, J.Man.processes,doi.10.1016/j.mapro.001,.2015

IOP Conf. Series: Materials Science and Engineering **115** (2016) 012003 doi:10.1088/1757-899X/115/1/012003

Dielectric properties of epoxy resin fly ash composite

A Pattanaik[1], S K Bhuyan[1], S K Samal[1], A Behera[1] & S C Mishra[1]

[1]Department of Metallurgical & Materials Engineering, National Institute of Technology, Rourkela-769008

E-mail: ashungr@gmail.com

Abstract: Epoxy resin is widely used as an insulating material in high voltage applications. Ceramic fillers are always added to the polymer matrix to enhance its mechanical properties. But at the same time, filler materials decreases the electrical properties. So while making the fly ash epoxy composite, it is obvious to detect the effect of fly ash reinforcement on the dielectric nature of the material. In the present research work, fly ash is added to four different weight percentages compositions and post-curing has been done in the atmospheric condition, normal oven and micro oven. Tests were carried out on the developed polymer composite to measure its dielectric permittivity and tan delta value in a frequency range of 1 Hz - 1 MHz. The space charge behaviours were also observed by using the pulse electroacoustic (PEA) technique. The dielectric strength and losses are compared for different conditions.

1. Introduction

The definition of dielectric relates to the permittivity of the material. The permittivity is the ability of a material to polarize with respect to an applied field. Physically, it means the greater the polarization developed from the material in an applied field of given strength, the greater the dielectric constant will be. In earlier days, inorganic materials like mica and silica are used to prepare these dielectric materials. But with the advance of polymeric materials, polymeric materials have taken the place of inorganic materials. This is due to easy process ability, durability and low cost as compared to metallic materials. Inorganic and ceramic materials, have higher thermal properties as compared to polymer composites. Polymers are good insulators but cannot withstand high temperature [1]. Inorganic and ceramic materials have higher dielectric constant than polymeric materials, but when they are affected by the atmospheric moisture, their dielectric property changes drastically. Composites having a high dielectric constant and high dielectric breakdown strength are used for applications of capacitors and electric energy storage devices [2-4]. Pb (Mg1/3Nb2/3) O3-$PbTiO_3$ (PMNPT), Pb (Zr, Ti) O3 (PZT) and $BaTiO_3$ (BT) have been used as fillers in polymers by various researchers to achieve maximum dielectric and piezoelectric properties [5-11].

Materials having both high and low dielectric are useful in electrical and electronics industries. Low dielectric constants are required as insulators where as high dielectric is required for polarizable media in capacitors. Introduction of small amounts of impurities makes the insulating and nonpolar materials store large amount of energy with a small applied electric field. In the present study effort has been made to utilize industrial waste like fly ash, fabricate a low cost polymer composite and utilize it in electrical and electronics industry. The fabricated composite is cured in different conditions to enhance the mechanical properties and simultaneously its dielectric properties have been evaluated at different frequencies and different treatment conditions.

2. Experimental details

2.1. Matrix and Filler material

Epoxy resin (LY 556), supplied by ATUL industries is used as the matrix material. It is a high viscous semisolid material whose chemical name is Bisphenol-A-Diglycidyl-ether. An amino group hardener (HY 951) is used for hardening the material. The thermosetting resin formed has high corrosion resistance, high thermal stability and good mechanical properties. Fly ash is collected from CPP-2 of Rourkela Steel Plant, Odisha, India. The SEM micrograph of fly ash show that these are spherical in nature which helps in proper bonding as the intermolecular gap is filled with a layer of epoxy material. A graph has been plotted between the particle size and the volume percentage of fly ash to determine the average particle size and it is found to be 27.26 μm as shown in figure 2. Being a ceramic particle, Si and Al are the main constituents present in the collected fly ash sample as revealed by the chemical compositional analysis as mentioned in Table 1.

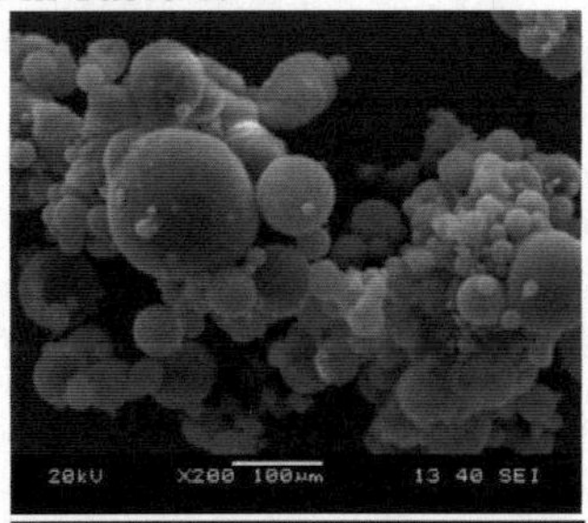

Figure 1. SEM micrographs of fly ash.

Table 1. Chemical composition of fly ash.

Constituents	Vol %
Fe_2O_3	8.1
MgO	1.14
Al_2O_3	24.98
SiO_2	55.85
P_2O_5	0.15
SO_3	1.16
K_2O	0.85
CaO	2.54
Na_2O	0.2
TiO_2	1.75
CO_2	1.56

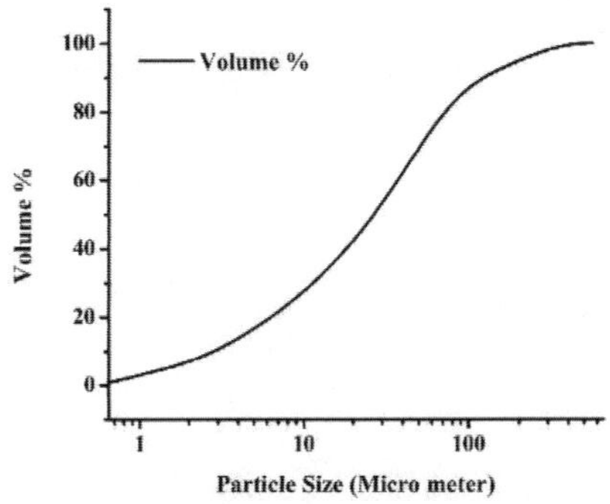

Figure 2. Particle size analysis of fly ash.

2.2. Composite fabrication
Epoxy resin and fly ash are mixed in four different weight percentages and put under an ultrasonic sonicator for 30 min with a pulse time of 5 Sec. Once it is complete, 10wt % of hardener is added to it and hand stirred gently to avoid trapping of bubbles and poured into plastic pipes of 15 mm diameter and 1inch height and allowed to cure at room temperature for 24 hrs. Once it is hardened, the plastic pipes were cut to remove the solidified cylindrical samples. Temperature, pressure, strain rate and even the environment may make an enormous difference in the mechanical response of polymers. So, controlled atmosphere is maintained while making of samples for uniformity. The cylindrical samples are post cured in oven and micro oven in predetermined conditions to enhance its glass transition temperature. The long cylindrical samples after curing are cut into thin slices with polished surfaces for dielectric test.

2.3. Thermal Analysis
Alternating low temperature differential scanning calorimetry is a thermo analytical technique in which the difference in the amount of heat required to increase the temperature of a given sample and that of a reference sample is measured as a function of temperature. It is normally used to find out the glass transition temperature of the samples. Mettler Toledo model no DSC 822 having a temperature range of -100^0c to $+400^0$c is used in this experiment. Tg (Glass Transition Temperature), shows the amount of curing due to the post curing treatments and changes in crystallinity of the material.

2.4. Dielectric Analysis
Solartron 1296 Dielectric Interface Instrument is used to measure the dielectric constant of the material. The measurement temperature range varies from 30^0 C to 150^0C and frequency from 100Hz to 1 MHz. To make the sample conductor both sides are coated with silver. The sample is placed between two parallel electrodes. Electric field produces polarization effect in the sample. As an alternating current is passed, it oscillates at the same frequency as that of the applied electric field, but has a phase angle shift δ.

Let C_p= Measured capacitance
C=capacitance= $€_0$ [A/d]
$€_0$=8. 8562×10$^{\wedge-12}$F/m= Dielectric constant of vacuum
€r=Relative permittivity
A=Area
D=Diameter=15mm

Dielectric constant=K=C_p/C

3. Result and Discussion:

3.1. Effect of frequency at different curing conditions on different curing conditions

0% FA

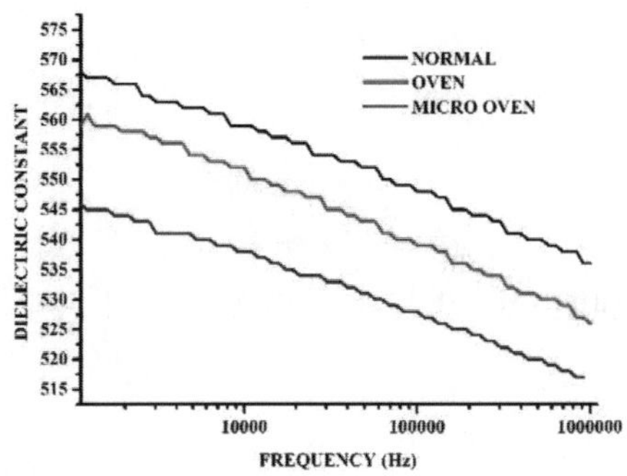

10% FA

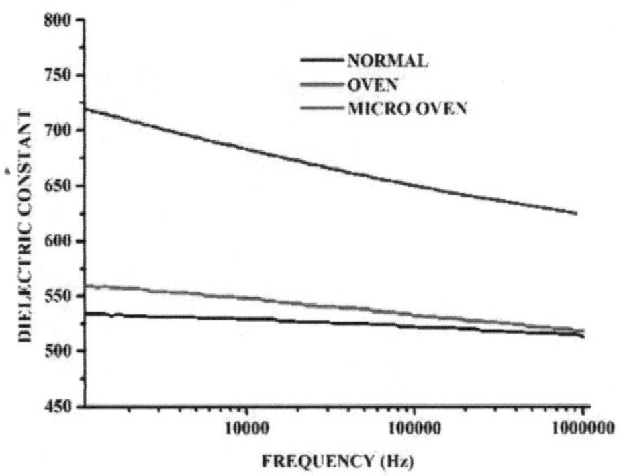

20% FA

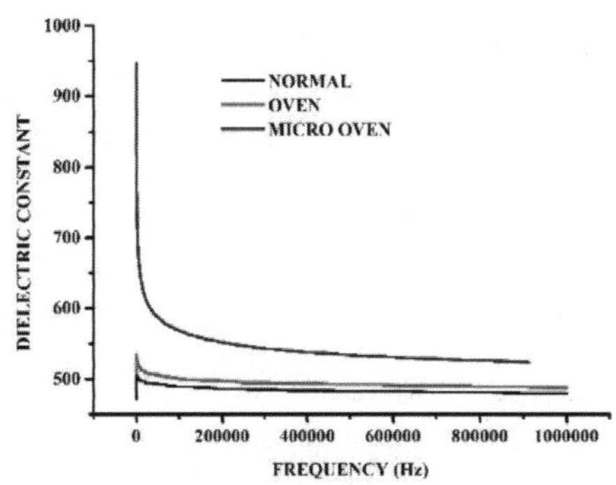

30% FA

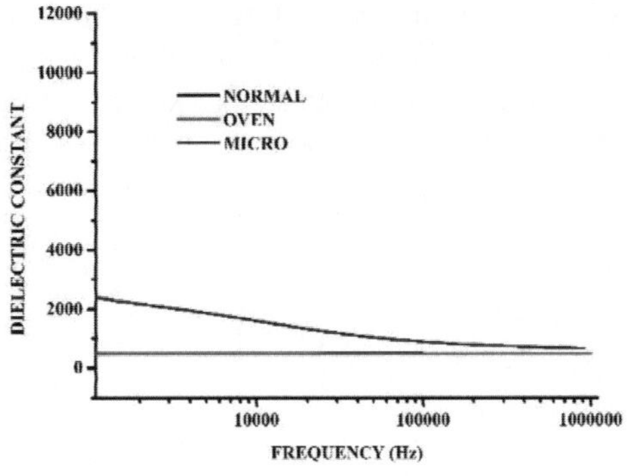

40% FA

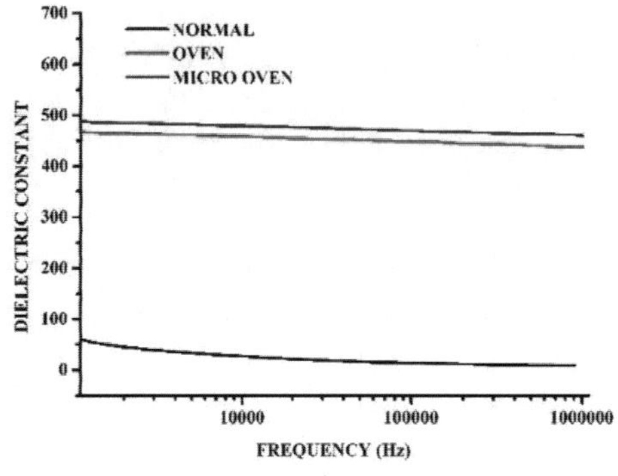

Figure 3. Variation of dielectric constant w.r.t frequency.

Dielectric constant with frequency have been plotted in figure 3 for different percentages of fly ash epoxy resin. In each figure into three different treatment conditions have been mentioned. From the figure, it is clear that the dielectric constant decreases with increasing frequency as it is the expected behaviour for most of the materials. As the frequency increases, charges become more random and start to oscillate out of phase with the applied voltage and will contribute to the alternating current causing decrease in K value. It is also seen that the dielectric constant is in decreasing order for normal, oven treatment and micro oven treatment only for 0% fly ash i.e. only epoxy resin condition. Once the fly ash has been added there is reverse of the order.

It may also be due to the dielectric relaxation which is the cause of anomalous dispersion. At higher frequencies the orientation of polar molecules along the direction of applied field is disturbed. The dielectric properties of a polymer are controlled by the charge conveyance. The polarization of a dielectric is contributed by ionic, electronic and dipole polarization. The electronic polarization happens amid a short interim of time of request of 10-20 sec, however more than for electronic polarization.

Free volume is additionally imperative for deciding the dielectric quality of the material. Free volume is resolved as the space that is not possessed by the polymeric material. The free volume connected with one mole of rehashed units of the polymer may be evaluated by subtracting the involved molar volume of rehashed unit [12].

3.2. *Effect of reinforcement of fly ash on dielectric constant*

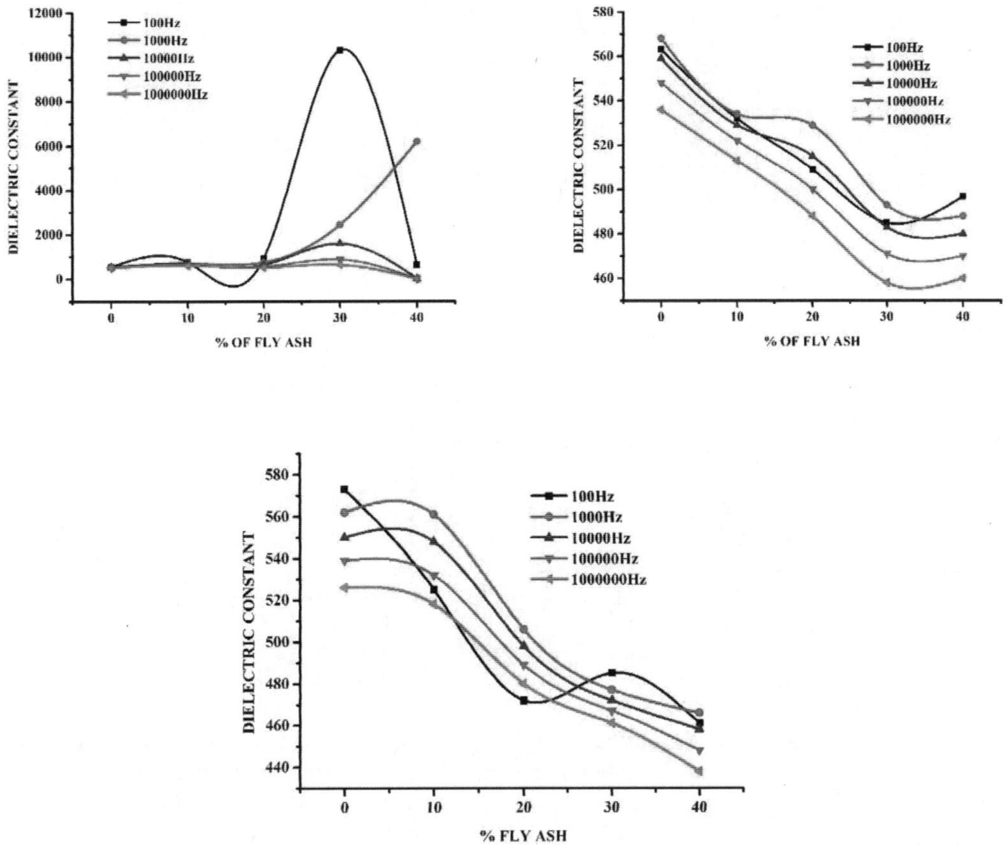

Figure 4. Variation of dielectric constant w.r.t. % of fly ash.

From the Figure 4, it is clear that dielectric constant is less with increase in frequency for a particular percentage of fly ash. The line drawn for 100 Hz frequency only deviate from following a trend. It may be due to the following reason.

Polymers are known to be a blend of shapeless and crystalline districts. In crystalline districts, the chains are methodical organized. Charge transporters are caught at the crystalline-formless interfaces make a huge commitment to the dielectric parameters at lower frequencies.

3.3. Effect of post curing on Tg

The glass transition temperature is the temperature where the material is converted into a glassy state material and its properties are changed. Graphs have been plotted between temperature and amount of heat absorbed. The DSC results are dependent on the heating rate and mass of the sample [13-15].

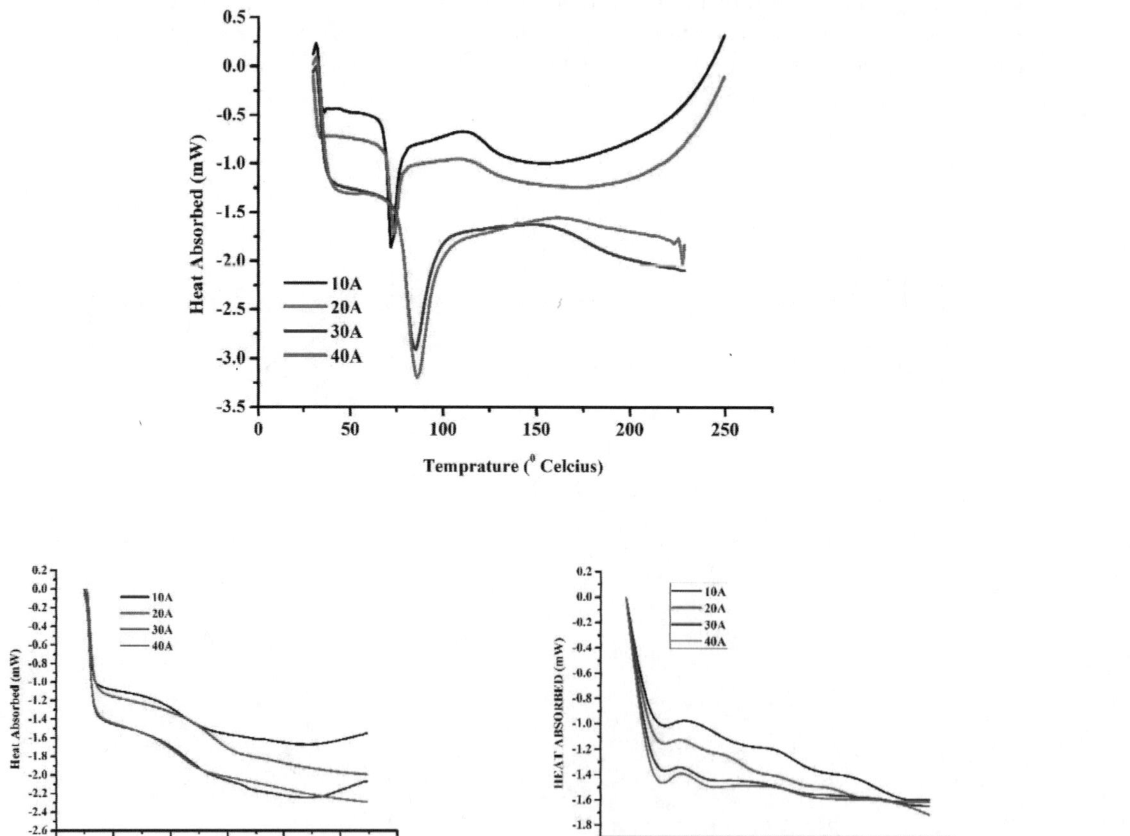

Figure 5. Variation of Tg for different curing condition.

Figure 5 show the T_g and the enthalpy of the material. The curves show the amount of energy required for the particular material to be completely cured. From the above graphs it is clearly evident that the T_g increases after treatment in oven and micro oven. As the T_g increases, the amount of crystallinity increases for the material which is proportional to the dielectric of the material. Figure 5 (a, b, c) shows the DSC curves for different percentages of fly ash for different curing conditions. In all the cases, it follows the trend of decreasing with increase in fly ash content. This may be due to the uniform coating of epoxy material over each particle of fly ash. As the matrix becomes continuous, the crystallinity, glass transition and the strength increases.

4. Conclusion

Polymers have been widely used in electrical and electronic industries due to its high value of flexural modulus in place of ceramic materials. Ceramic filler like fly ash increases the mechanical properties having good dielectric strength may be good choice of using it in electrical industries. Some of the results obtained from the experiment has been discussed as follows-

1. Post curing increases the crystallinity of the composite which increases the mechanical strength of the material.
2. Dielectric constant of fly ash epoxy composite decreases with increase in frequency.
3. Dielectric strength increases with post curing conditions.
4. Dielectric constant decreases with increase in % reinforcement of fly ash.
5. Fly ash epoxy polymer composite can be used in electrical and electronics industries as a good dielectric material.

References:

[1] Ren Y and Lam D C C 2008 Properties and microstructures of low-temperature-processable ultralow-dielectric porous polyimide films *J. Electron. Mater.* **37** 955–61

[2] Ranjan V, Yu L, Nardelli M B and Bernholc J 2007 Phase equilibria in high energy density PVDF-based polymers *Phys. Rev. Lett.* **99** 1–4

[3] Dias C J and Das-Gupta D K 1996 Ferroelectric ceramic polymer composite films for pyroelectric sensors *IEEE Trans. Dielectr. Electr. Insul.* **3** 393–6

[4] Kuo D H, Chang C C, Su T Y, Wang W K and Lin B Y 2001 Dielectric behaviours of multi-doped BaTiO3/epoxy composites *J. Eur. Ceram. Soc.* **21** 1171–7

[5] Das-Gupta D and Doughty K 1988 Polymer-ceramic composite materials with high dielectric constants *Thin Solid Films* **158** 93–105

[6] Kulek J and Hilczer B 1998 Dielectric , Piezoelectric and Pyroelectric Response of *Response* **32** 1079–82

[7] Bai Y, Cheng Z, Bharti V, Xu H S, Zhang Q M, Bai Y, Cheng Z, Bharti V, Xu H S and Zhang Q M 2016 High-dielectric-constant ceramic-powder polymer composites High-dielectric-constant ceramic-powder polymer composites **3804**

[8] Adikary S U, Chan H L W, Choy C L, Sundaravel B and Wilson I H 2002 Ceramic – Polymer Composites **62** 2161–7

[9] Dang Z M, Yu Y F, Xu H P and Bai J 2008 Study on microstructure and dielectric property of the BaTiO3/epoxy resin composites *Compos. Sci. Technol.* **68** 171–7

[10] Sekar R, Tripathi A K and Pillai R K C 1989 X-ray Diffraction and Dielectric Studies of a BaTiO 3 : PVDF Composite **5** 33–6

[11] Sánchez-jiménez P E, Pérez-maqueda L A, Diánez M J, Perejón A and Criado J M 2010 Mechanochemical preparation of BaTiO 3 – Ni nanocomposites with high dielectric constant *Compos. Struct.* **92** 2236–40

[12] Blythe T and Bloor D 2005 Electrical properties of polymers 1–475

[13] Dumas J P, Gibout S, Zalewski L, Johannes K, Franquet E, Lassue S, Bedecarrats J P, Tittelein P and Kuznik F 2014 Interpretation of calorimetry experiments to characterise phase change materials *Int. J. Therm. Sci.* **78** 48–55

[14] Jin X, Hu H, Shi X and Zhang X 2015 Energy asymmetry in melting and solidifying processes of PCM *Energy Convers. Manag.* **106** 608–14

[15] Yinping Z, Yi J and Yi J 1999 A simple method, the -history method, of determining the heat of fusion, specific heat and thermal conductivity of phase-change materials *Meas. Sci. Technol.* **10** 201–5

Effect of Post-thermal shock on Prolonged Sea Water aged GFRP Composite

A. P. Chakraverty*[, 1, 2], U. K. Mohanty[1,3], S. C. Mishra[1] and B. B. Biswal[4]

[1]Department of Metallurgical and Materials Engineering
National Institute of Technology, Rourkela-769008, Odisha, India
[2]Department of Physics
C. V. Raman College of Engineering, Janla, Bhubaneswar-752054, Odisha
[3]Department of Mechanical Engineering,
Institute of Technical and Educational Research
Sikhya O Anusandhana University, Bhubaneswar-751030, Odisha
[4]Department of Industrial Design
National Institute of Technology, Rourkela-769008, Odisha, India
Corresponding Author, email: anantac8@gmail.com

Abstract: The present investigation is an attempt of evaluating the suitability of glass fibre reinforced polymer composites to thermal shock treatments of various lengths of time subject to pre-immersion in sea water for 1 year. Mechanical properties like inter laminar shear strength (ILSS), stress at rupture, strain at rupture and modulus values are recorded by adopting 3-point bend test method. Mechanical properties show a general decreasing trend at higher durations of up and down-thermal shock exposure irrespective of showing initial non-equilibrium zig-zag trend. Glass Transition temperature (T_g) with respect to optimum durations of thermal shock treatment show considerable variation for the sample with minimum sea water immersion period. SEM fractographs of the thermally shocked specimens revealed the mode of failures like fibre pull out, fibre/matrix debonding, cusp formation indicating polymer crazing, matrix cracking, fibre breaking etc.

Key Words: GFRP composite, Thermal Shock, Inter laminar shear strength, Glass Transition Temperature (T_g), fibre pull-out

1. Introduction

Glass fibre reinforced polymer (GFRP) composites have gained tremendous attention in various marine applications like offshore pipes, ship structures, water storage vessels etc. [1]. The cost of installation and maintenance of these composite structures are reportedly lower than that of some conventional materials like steel and aluminium [2]. Out of these specific uses, the marine ship structures composed of GFRP composite laminates are subjected to immersion in sea water for years together necessitating accumulation of experimental data in this respect. Chakraverty *et al.* [3] have reported the response of E-glass fibre/epoxy layered composites to one year immersion in sea water at room temperature. Inter laminar shear strength (ILSS) of such immersed composites decreased up to 23% of that for as-cured one and glass transition temperature after initial decrease increased up to some extent. Polymeric composites during immersion (fully or partly) in sea water may lead to mechanical property deterioration by plasticization and swelling of polymer [4]. Also, various salt components of the sea water may annihilate the moisture absorption rate compromising some properties of the material subjected to sea water immersion [5]. On the basis of the above, the long term retention of structure related properties and stability under the sea water immersion condition is of significant concern from a prolonged service life point of view of these materials.

These components are supposed to meet fluctuating thermal conditions (thermal shock) during their use; specially, if there is a sharp change in temperature due to high speed collision or during weather change, change of temperature from one particular sea location to other during immersion period or simply due to a lightening in a rainy day. In such cases, thermal stresses due to temperature fluctuations concentrated around the defect tips across fibre/matrix interface would be catastrophic [6]. Added to it, temperature fluctuations from elevated to sub-ambient temperatures or vice-versa may influence the moisture desorption kinetics as well as the relaxation process in the matrix body [7]. Sometimes, huge thermal stresses due to higher thermal gradient may create irreversible mechanical degradation in the composite body. It is therefore, only pertinent that some durability data of GFRP composites with prolonged sea water immersion and concurrent thermal shocks be made available. The present study aims at evaluating the performance of the GFRP composite immersed in sea water for different lengths of periods when exposed to up and down thermal shocks for different durations.

2. Materials and methods

Conventional hand lay-up method [6] is adopted for the fabrication of GFRP composite comprising woven E-glass fibre and Epoxy resin (Lapox L-12). Diamine (Lapox K-6) is used as hardener/curing agent. The composite, after 48 hours of room temperature curing, was allowed for cutting in a diamond cutter for preparation of short beam shear (SBS) specimens (50 mm length X 12 mm breadth) as per the ASTM D 2344-84 (2013) standard [8]. The specimens are then heated in an electric oven up to 55-60°C of temperature and weighed; this process being repeated to ensure the complete removal of moisture during fabrication. Typical properties of Epoxy resin as provided by ATUL polymer division are cited from our previous article [3].

All the immersed composite specimens are allowed for thermal shock treatments. An electric oven and a cryogenic chamber (deep freezer) are maintained at +50°C and -40°C, respectively. The sea water immersed samples are allowed for up-thermal shock treatment (-40°C for 1, 2, 3, 4 and 5 hours in the cryogenic chamber, then suddenly put in an electric oven set at +50°C for 2 hours of fixed duration) and Down-thermal shock treatment (+50°C for 1, 2, 3, 4 and 5 hours in the electric oven, then suddenly put in cryogenic chamber set at -40°C for 2 hours of fixed duration).

The post-thermally shocked samples after 1 year of sea water immersion are allowed for 3-point bend test in INSTRON-1195 to determine inter laminar shear strength (ILSS). Stress at rupture, strain at rupture and modulus values of all treated composite samples are also recorded from the same test. ILSS is determined by using the following formula [9]

$$\text{ILSS} = 0.75 \times 1000 \frac{p_b}{bt} \text{ N/mm2 or (MPa)} \quad (1)$$

where, p_b = breaking load (load at rupture in kN)
b = width of the specimen
t = thickness of the specimen

Glass transition temperature (Tg) of the treated samples are recorded by adopting low temperature DSC test. The Tg measurement is performed on Mettler-Toledo 821 with intra-cooler using STAR software in DSC module. The scanning range is maintained as 30-150°C with the rate of 10°C/min. The fractographs of the samples treated for optimum durations of sea water immersion and thermal shock treatments are investigated by scanning electron microscope (SEM, JEOL, JSM-6480 LV) and the failure modes are divulged for corresponding treated samples.

IOP Conf. Series: Materials Science and Engineering **115** (2016) 012004 doi:10.1088/1757-899X/115/1/012004

3. Result and Discussion

3.1: Mechanical Properties

Fig. 1 represents the variation of ILSS with duration of up and down-thermal shocks to sea water immersed samples. ILSS values of all sea water immersed samples with up and down-thermal shocks show minor variations with an initial non-equilibrium zig-zag trend which may be due to non-equilibrium moisture desorption kinetics during the initial periods of thermal shock exposure.

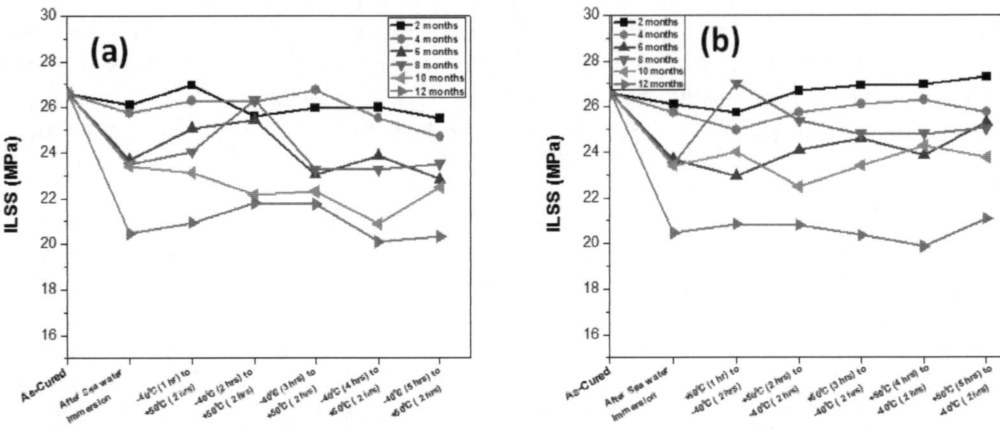

Fig. 1: Variation of ILSS for sea water immersed samples with time of exposure of (a) up-thermal shock (b) Down-thermal shock

Sea water immersion at room temperature for 1 year is insufficient to bring about an average response by the composite body, as a consequence of which locally confined swelling in the matrix or matrix/fibre interface causes complex deformation due to local in-homogeneity leading to non-equilibrium zig-zag variations in ILSS. The response to sea water immersion, therefore, may be due to a non uniform localised swelling, which can disturb the mechanical behaviour under load. In addition to this, different locations of the composite body response to the ILSS values differently. This may cause a zig-zag effect in the response.

On immersion of GFRP composites for a prolonged period of 1 year in sea water, its mechanical properties deteriorated as evidenced by the drop of ILSS values up to 23% [3]. Both up and down thermal shocks are found not to have caused any further deterioration in the mechanical behaviour (Figure 1). In other words, thermal shocks (up and down) have not created any remarkable excitements (matrix hardening due to -40°C and fine whisker forming by moisture desorption process due to +50°C [7]) as evidenced from no change in the ILSS values of the immersed samples. This signifies that the laminar strength degradation due to prolonged sea water immersion is irreversible and is not altered even after thermal shock treatments.

Fig. 2 and Fig. 3 illustrate the variations of stress at rupture and strain at rupture with varying time period of both types of thermal shock treatments. Both stress at rupture and strain at rupture values, despite of showing initial non-equilibrium zig-zag trend, decrease for the samples with maximum period of sea water immersion (1 year) at higher durations of up-thermal/down-thermal shock treatments. Up-thermal and down-thermal shock of higher durations cause decrease in stress at rupture values in general. Larger thermal gradient during higher durations of both up and down-thermal shocks are responsible for not allowing the accommodation of residual stresses generated due to thermal expansion mismatch between the matrix and the fibre [10]. This phenomenon becomes more prominent for the samples immersed in sea water for prolonged periods (10 and 12 months). This signifies the less requirement of stress at rupture for the GFRP composite samples with higher duration of sea water immersion followed by up/down-thermal shock for higher during +50°C and cryogenic matrix hardening during -40°C exposure. This competing phenomenon is expected to be more prominent for the samples with higher sea water immersion period. For this case, samples with higher moisture absorption could have responded the moisture desorption and cryogenic matrix hardening in a very prominent manner to show the strain misfit between fibre and epoxy polymer. This may give rise to a permanent mechanical deterioration by which stress requirement at rupture was found to be decreased compared to the value as observed during initial duration of both up and down-thermal shock treatments with a non-equilibrium zig-zag trend.

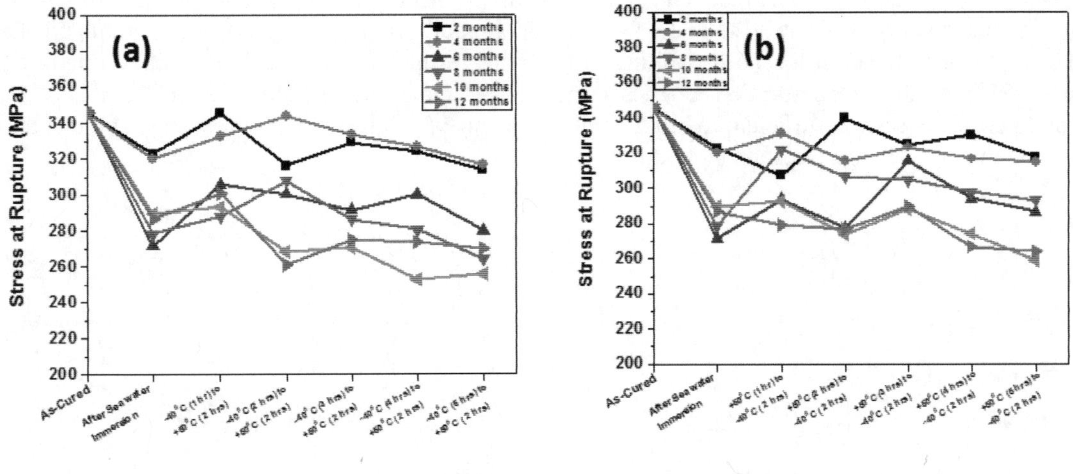

Fig. 2: Variation of stress at rupture for sea water immersed samples with time of exposure of (a) up-thermal shock (b) Down-thermal shock

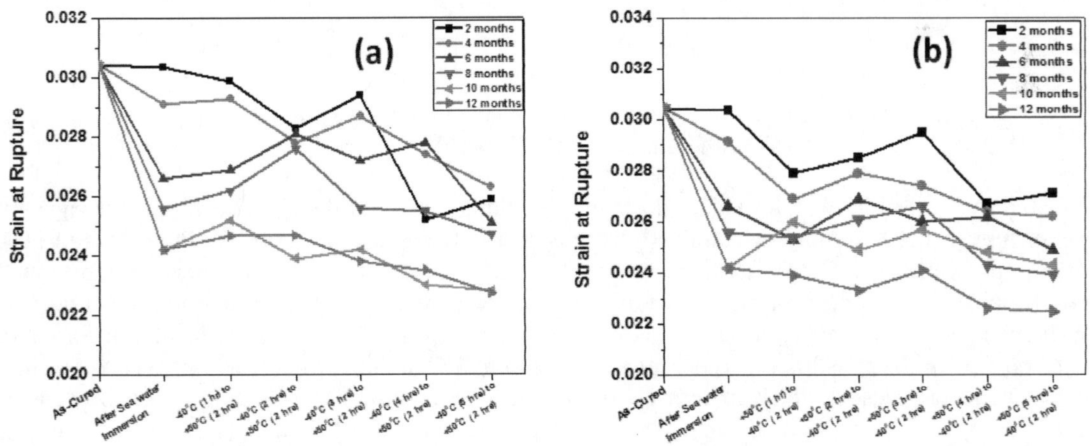

Fig. 3: Variation of strain at rupture for sea water immersed samples with time of exposure of (a) up-thermal shock (b) Down-thermal shock

Strain at rupture values show a general decreasing trend with higher periods of thermal shock exposure (Fig. 3). As explained earlier, higher durations of thermal shock treatments with larger thermal gradient are likely to result in thermal expansion mismatch creating strain misfit between fibre and matrix. This could have reduced the required strain at rupture values of the composite specimens subjected to mechanical loading. The GFRP composite samples after one year of sea water immersion show minor decreasing trend in strain at rupture with increasing durations of up/down-thermal shock treatments. This is because of the occurrence of irreversible mechanical damages in the composite samples due to prolonged sea water immersion. The permanent damage due to prolonged sea water immersion is not likely to allow any recovery of the strain property of the treated samples during the thermal shock exposures. The rate of decrease of strain at rupture is not very prominent for the samples with higher sea water immersion period. For this case, the mechanical deterioration occurred after sea water immersion is not responsible for any scope to compensate such decrease of strain at rupture during any of the thermal shock conditioning. This can indicate the less impact of thermal shock treatments to the strain property of GFRP composite sample with higher sea water immersion.

The variations of modulus values of the sea water immersed samples with varying periods of up and down-thermal shocks are presented in Fig. 4. Modulus values of all the samples increase at higher durations of up-thermal shock treatment. However, minor changes in the modulus values are observed with higher durations of down-thermal shock. The non-equilibrium zig-zag variations in modulus are likely to be the effects of localised swelling in the matrix (in some restricted areas only) due to non-homogeneous moisture desorption kinetics during initial durations of thermal shock treatments, the initial periods not being sufficient to bring about an average response of the composites to the external excitements by showing variation in modulus values.

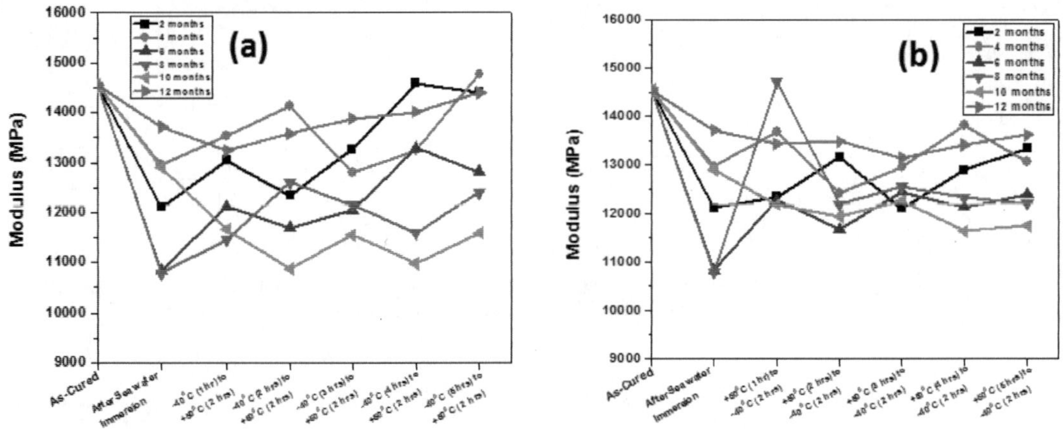

Fig. 4: Variation of modulus for sea water immersed samples with time of exposure of (a) up-thermal shock (b) Down-thermal shock

The increasing trend in modulus values during higher durations of up-thermal shock puts the signature of moisture induced swelling, which finally results in a lowering of the strain at failure. This result is well in agreement with the variation of strain at rupture values of the immersed samples with up-thermal shock treatment. In the present case, the increased durations of exposure to -40°C under the up-thermal shock treatment, enhance the moisture to ice transformation. This process of transformation to ice dominates the moisture desorption process due to +50°C of duration 2 hours only. Thus, swelling of the matrix is caused, which in turn results in increase of modulus values [5]. For the samples immersed in sea water for maximum period (1 year), moisture distribution is quite homogeneous leading to the resultant overall swelling in the matrix. This may cause continuous increase in modulus with increasing durations of up-thermal shock treatment.

Modulus values are observed to vary within minor ranges during higher durations of down-thermal shock treatment. In the present case, increasing durations of +50°C cause the formation of fine whiskers/ inter-connected pores. These processes are likely to neutralize the effect of freezing of moisture due to exposure to sub-ambient temperature of -40°C for 2 hours only. This may lead to minor resultant swelling in the matrix. Thus, minor variations in modulus values are observed for the immersed samples with increasing durations of down-thermal shock treatment.

3.2: Glass Transition Temperature (T_g)

The T_g values as obtained from the DSC thermograms are plotted for different thermal shock treatments and illustrated in Fig. 5. Fig. 6 represents the DSC thermograms of the immersed samples after the exposures to up-thermal and down-thermal shock treatments for optimum durations. The T_g variation plot indicates the minimal ranges of variation for all immersed samples (except for 2 months of immersed one) subject to post thermal shock treatments including both up and down-thermal shocks.

The samples subjected to 2 months of sea water immersion show remarkable variations in T_g values with respect to up and down-thermal shock treatments. The thermal stress due to minimum/maximum thermal gradient cause a combined effect pertaining to fine whisker formation due to desorption of moisture during +50°C and cryogenic hardening during -40°C [7]. These two competitive phenomena influence the extent of free space constituting voids or free volumes in the polymer matrix [6]. In the present case, the particular sample, with less absorbed moisture, respond to the above competitive phenomena in such a way that the cryogenic hardening dominates over the fine whisker formation. Thus, the free volume decreases with increase of mechanical locking [11]. Hence, the T_g for the sample show higher values after the thermal shock treatments for shorter duration of sea water immersion.

The immersed samples with immersion period greater than 2 months, however, show minor changes in T_g values with negligible fluctuations. Under such conditions the higher extents of moisture gain would result in more/less balancing of the effect due to the two competitive phenomena (whisker formation and matrix hardening) which is responsible for minimal variations of T_g values.

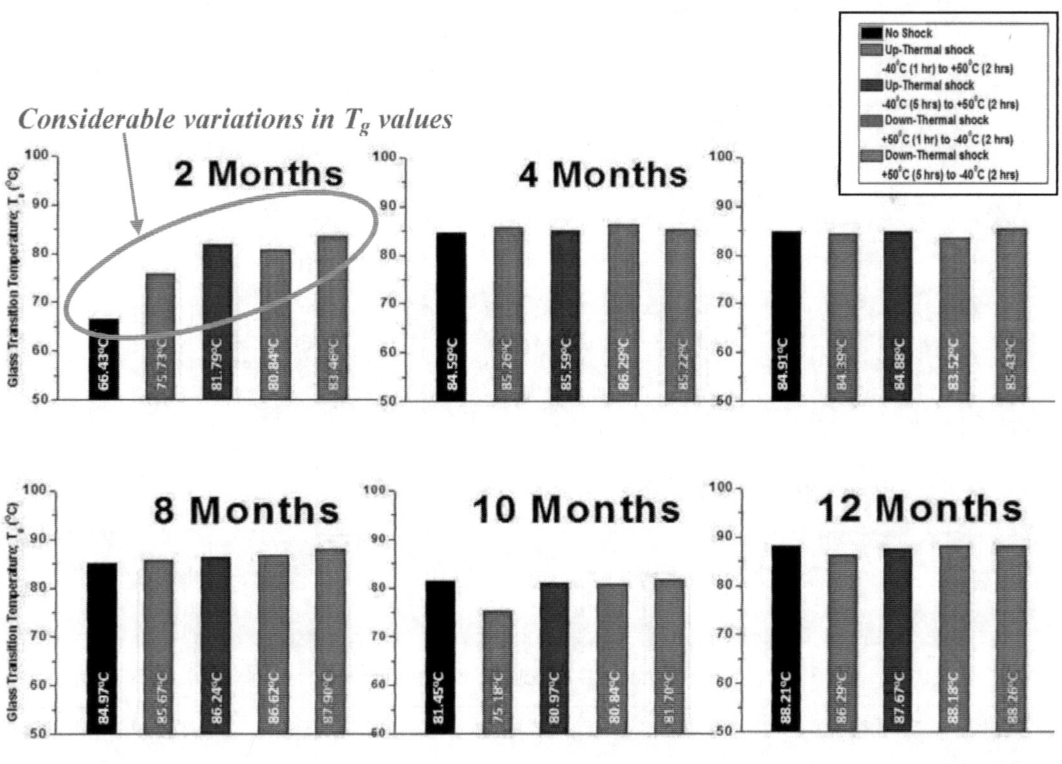

Fig. 5: Variation in T_g values with optimum durations of up and down-thermal shock treatments for sea water immersed samples

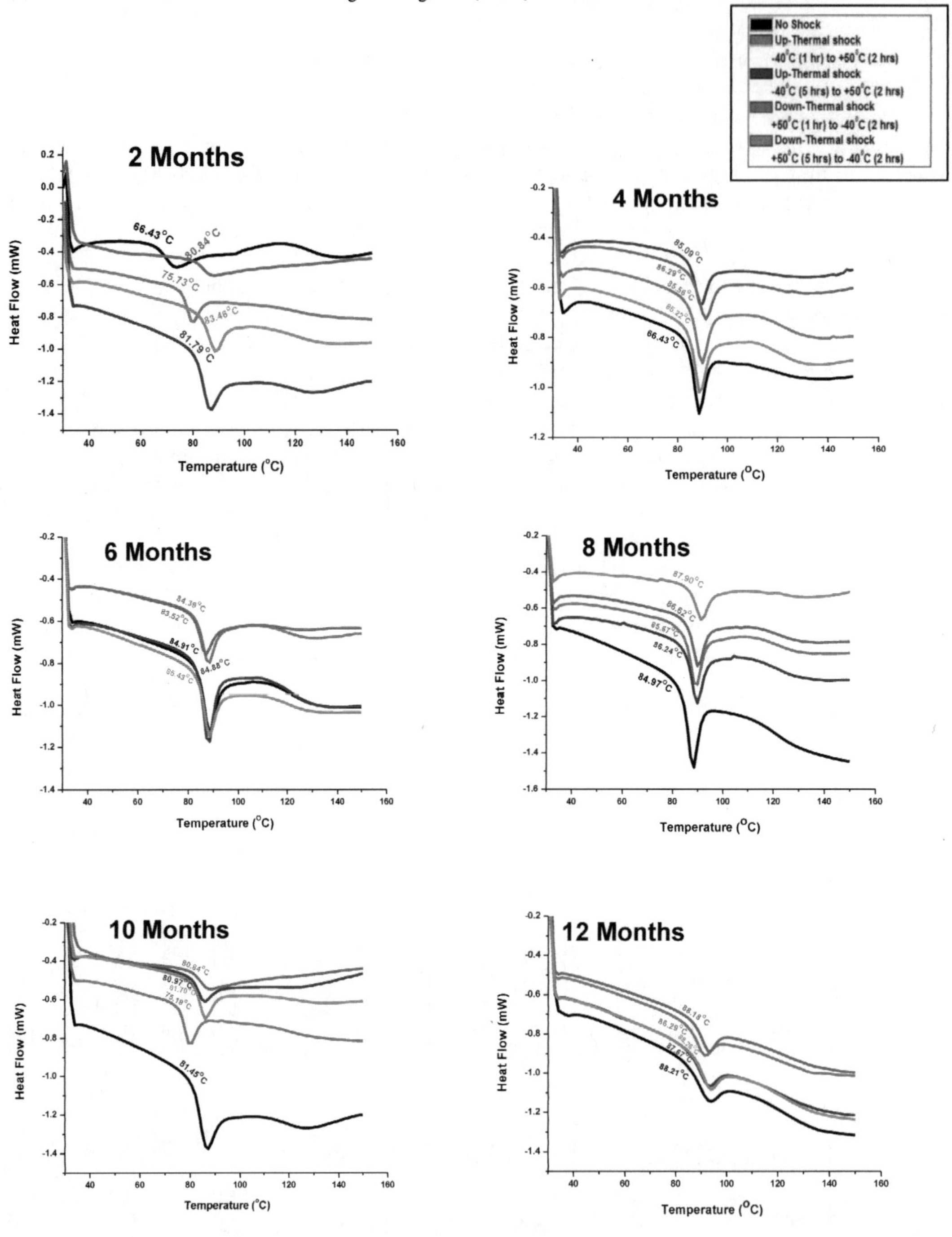

Fig. 6: DSC thermograms of sea water immersed composite samples subject to thermal shock treatments

3.3: SEM Fractographs

The fractographs as revealed in Fig. 7 and Fig. 8 represent the failure modes associated with the fractured composite specimens subject to sea water immersion of extreme periods (2 and 12 months) and optimum (minimum and maximum) durations of up-thermal and down-thermal shock treatments. The fractographic images revealed some chief mode of failures like fibre pull out, fibre/matrix debonding, cusp formation indicating polymer crazing, matrix cracking, fibre breaking etc.

Fractographs, in some cases clearly show fibre imprints as a result of debonding of fibre from the matrix. The debonded matrix body sometimes show some crazing mark (cusp) signifying the stress concentrations for prior failure under load. Fibre breaking and matrix cracking are mostly visible in case of the immersed samples subjected to maximum durations of up-thermal and down-thermal shocks. During fibre pull-out, fibre breaking or matrix cracking can be observed depending on the load sustaining capacity of both the components.

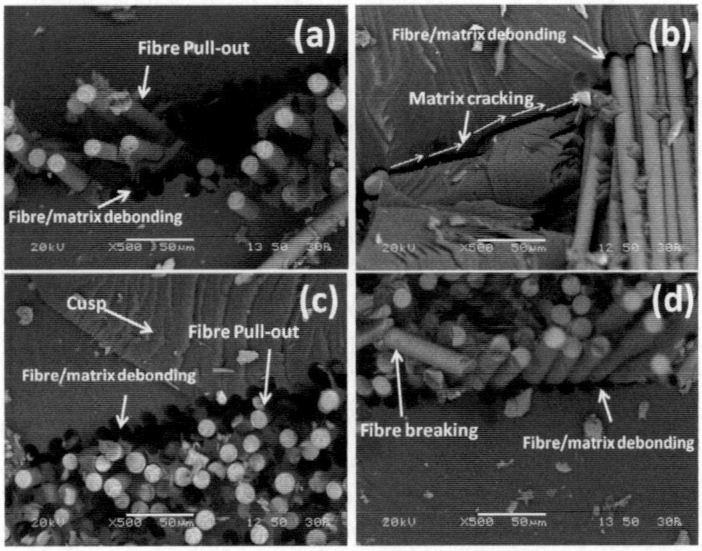

Fig. 7: SEM fractographs for 2 months sea water immersed samples
(a) up-thermal shock [-40°C (1 hour) to +50°C (2 hours)] (b) up-thermal shock [-40°C (5 hours) to +50°C (2 hours)] (c) down-thermal shock [+50°C (1 hour) to -40°C (2 hours)]
(d) down-thermal shock [+50°C (5 hours) to -40°C (2 hours)]

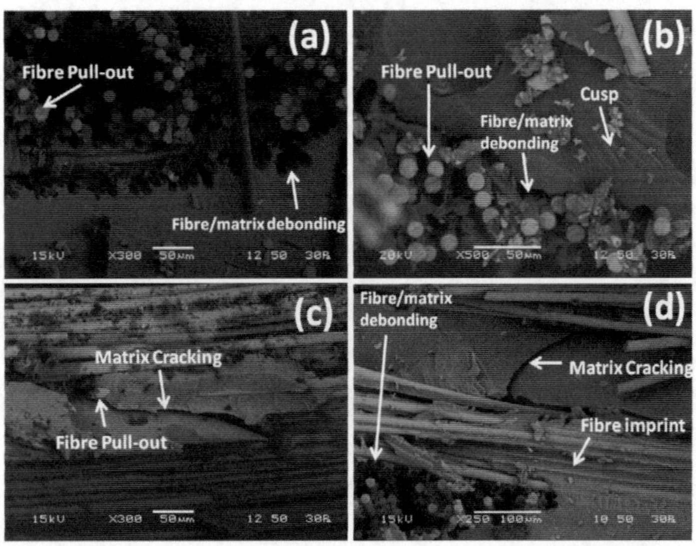

Fig. 8: SEM fractographs for 1 year sea water immersed samples
(a) up-thermal shock [-40°C (1 hour) to +50°C (2 hours)] (b) up-thermal shock [-40°C (5 hours) to +50°C (2 hours)] (c) down-thermal shock [+50°C (1 hour) to -40°C (2 hours)]
(d) down-thermal shock [+50°C (5 hours) to -40°C (2 hours)]

4. Conclusions

Some following conclusions as drawn from the present findings are mentioned below.

(i) Thermal shock treatments to the sea water immersed samples cause initial zig-zag trend in mechanical property variation with a general lowering trend for the higher durations.

(ii) T_g values for the thermally shocked samples show small variations. However, for the samples immersed for 2 months, Tg values are increased for both up and down-thermal shocks.

(iii) SEM fractographs of the thermally shocked GFRP specimens revealed the mode of failures like fibre pull out, fibre/matrix debonding, cusp formation indicating polymer crazing, matrix cracking, fibre breaking etc.

References

[1] Boisseau A, Davies P and Thiebaud F 2012 *Appl. Compos. Mater.* 19 1259
[2] Gellert EP and Turley DM 1999 *Composite Part A* 30 1259
[3] Chakraverty AP, Mohanty UK, Mishra SC and Satapathy A 2015 *IOP Conf. Series: Mat Sc & Engg.* 75 012029
[4] Rutowska M, Krasowska K, Heimwoska A, Steinka E and Janik H 2002 *Polym. Degrad. & Stab.* 76 233
[5] Mourad AI, Abdel-Magid BM, EI-Maddawy T and Grami ME 2010 *Appl. Compos. Mat.* 17 557
[6] Chakraverty AP, Mohanty UK and Biswal BB 2012 *Emer. Mat. Res.* 1 263
[7] Mishra G, Mohapatra SR, Behera PR, Dash B, Mohanty, UK and Ray BC 2010 *Aircraft Engg & Aerospace Technol* 82 258
[8] ASTM D 2344/D 2344 M-13 2013 Standard Test Method for Short beam strength of Polymer matrix composite Materials and Their Laminates ASTM International, West Conshohocken, PA
[9] Wei B and Cao H, Song S, 2011 *Corrosion Science* 53 426
[10] Chakraverty AP, Mohanty UK, Mishra SC and Biswal BB 2014 *Adv. Sci. Lett.* 20 671
[11] Siriruk A, Penumadu D and Weitsman Y 2009 *Compos Sc & Technol* 69 821

Semi-Solid Processing by Electric Current DuringSand Casting of Aluminium Alloys

Anjan Prodhan

MEF Division, National Metallurgical Laboratory, CSIR, Jamshedpur – 831 007, Jharkhand, India

E-mail: ap@nmlindia.org, Phone: +919430186450

Abstract. This study reports the effect of DC and 50Hz AC treatment (ECT) on aluminium or aluminium alloys during solidification in sand moulds i.e., at their semisolid state. Castings, with different geometry, were made in open or closed sand moulds. It is observed that ECT (a) reduces dissolved gas, (b) reduces internal shrinkage and (c) metal mould reactions in castings. It is also observed that the AC treatment is more effective compared to DC treatment. ECT changes the movement of solidification front. The optical microstructures of ECT samples are quite similar to the samples treated in other semisolid processing methods.

Keywords: Semi Solid Sand Casting, Electric Current Treatment, Aluminium Alloys, Sand Casting, GrainRefinemnt

1. Introduction

Normally, molten metal is treated with inoculants to refine or modify cast microstructures to improve properties of cast components. Due to multiple recycling of scraps, the inoculants form debris and discarded permanently. Alternatively, molten metal is treated in its semisolid state (i.e., in between the liquidus and solidus temperatures of the alloy) to refine/modify cast microstructures. Campbell [1] reviewed the effect of vibration during solidification. Flemings [2] reported that the increased convection and slow cooling leads to formation of spherical grains that possess rheological properties in the semisolid state. They behave thixotropically and viscosity can be varied over a wide range. The metal structure and its rheological properties are retained after solidification and partial remelting. The semisolid alloys can be formed by semisolid metal forming process to produce metal components. Several authors [3-6] reported the effect of different types of electromagnetic stirring (EMS) during solidification of aluminium alloys. During EMS, the strong fluid flow (i) detaches the dendrites from the solid liquid interface, (ii) fragments and (iii) carries them into the melt to form slurry. In rotating EMS (REMS)[3], thestirring increases cooling rate, therefore, dendritic arm spacing is reduced due to undercooling. Additionally, the grain size is also reduced because the fragmented dendrite debris acts as nucleation centres. The linear EMS(LEMS)[4], operates in the same way as induction furnace, the melt turns to slurry consisted of fragmented dendrites globular solid phase surrounded by lower melting point (eutectic) phase. The shape of the fragmented dendrites changes with variation of the frequency of EMS [5-6]. Low frequency AC favours large equiaxeddendrites. Dendrites are fragmented and refined by intermediate frequency AC (up to 1.5 kHz) but highfrequency AC (10kHz) promotes columnar dendrite formation. These materials, when partially remelted, offers less resistance to flow even at high solid fraction. This property is used to process the material in pressure die casting or forging (known as SSM). Electro-hydro-dynamics (EHD) or electro-fluid-dynamics (EFD) [7 - 9] is the study of the dynamics of

Electrically charged fluids. The external electric field, when applied on a fluid, induces fluid motion as the fluid bears a net electrostatic charge. It induces both normal and tangential forces on a fluid interface. The flow rate is linear in the electric field. The force acting on the fluid is given by the equation:

$$F = I\,d\,/\,k \quad [1]$$

Where,
F is the resulting force (Newtons)
I is the current flow (ampeares)
d is the distance between electrodes (meters)
k is the ion mobility co-efficient of the dielectric fluid (m^2/Volt sec).

In general, the phenomena relates to the direct conversion of electrical energy into kinetic energy. The author [10], reported the effect of ECT during solidification of 99.5% Al. The melts (in semisolid state) are (a) treated in linear electromagnetic field (LEMS), (b) treated with 50 Hz AC electric current (ECT) and then compared with the sample without any treatment. It is observed that both LEMS and ECT (a) reduces the freezing temperature, (b) increases the heat transfer rate but (c) the LEMS process consumes more energy compared to ECT. Therefore, further studies [11-14] were conducted with ECT (AC / DC) on different compositions of Al-alloys and patented [15]. These studies revealed that the ECT on aluminium alloys:
i) reduces shrinkage pipe formation (LM-0),
ii) modifies columnar structures to equiaxed (LM-0),
iii) refines eutectic silicon (LM-6 & LM-25),
iv) increases volume fraction of primary α-phase and reduces volume fraction of eutectic phase (LM-6, LM-25, LM-30)
v) reduces pinhole porosity (to minimum) at an intermediate range of current but it increases with increased current.
vi) improves machinability of LM-6 and LM-30 alloys due to refinement of eutectic silicon and primary silicon cuboids.

1.1. Objectives

Normally, SSM is done in pressure die-casting route. This process is quite complex and requireshuge infrastructures that involve cost. Moreover, the flexibility of making shaped casting is limited in SSM. The objective of this study is to develop a cheaper SSM method for sand-casting foundries. This study explores:
i) The effect of ECT (DC / AC) on removal of internal shrinkage or gas porosity in castings.
ii) Microstructural modifications and reduction of metal-mould reactions.
iii) Energy consumption for this process

2. Materials and Methods

2.1 Castings made in open sand moulds

These moulds were made using bentonite and moisture as binders. During mould preparation (for ECT) thepattern was placed inside of a steel frame (i.e. mould box). A graphite rod (~ 0.01 m dia) was placed at one corner of the pattern and the frame was filled with sand mixture and rammed. The mould was properly vented. The extended graphite rod was cut after keeping a gap of 0.025 m from the sand surface for electrical connections. The mould box (along with pattern and sand) turned upside down and the wooden pattern was removed. The inside part of graphite rod becomes visible from the top. The bottom part of the graphite rod was connected to power supply unit. One mould was made without any graphite connector for normal casting.

Commercial purity (99.5%Al; LM - 0) aluminium (~ 1 Kg.) was melted in a medium frequency air induction furnace with graphite lining. The melt was superheated (~ 750°C) and poured into a preheated

(~ 300°C) clay graphite crucible without any degassing treatment. A graphite rod was used to clean oxide layer formed on the melt-surface and poured in sand mould and solidified.

For ECT, the moulds with graphite connectors were used. The graphite rod was connected to one pole of either DC or AC power supply unit by insulated copper wire. The other pole was connected to another graphite rod with copper wire. After pouring liquid metal into mould, the second graphite rod was connected to molten metal. Then the power supply unit was switched on and its voltage was adjusted (manually) to flow desired amount of current through the liquid metal. The power supply was switched off (manually) at the end of solidification. This process is repeated with different current intensity.

2.2. Study on castings made in closed (Cope and Drag) sand moulds
The moulding process, the charge material and the melting process were same as mentioned above. In two-part moulds, the graphite rods were embedded in both moulds. The electrodes were placed along the bodydiagonal of the pattern. Fig. 1 shows electrical connection to the electrode embedded at cope and pouring ofmolten aluminium in a mould. The electricity was switched on only after complete filling of the mould and the current was adjusted accordingly. However, it was disconnected automatically as the casting contracts after solidification and detached from the electrode. Except sample 1, other samples were treated with increased AC. All samples were cut along the marked line to see the effect of current treatment on porosity and internal shrinkage.

Fig.1. Melt is poured in a closed sand mould for electric current treatment

2.3. Study on hollow cylindrical castings
Trials were also conducted on hollow cylindrical castings with length 0.175 m X 0.075 m (i.d.) X 0.025 m (wall thickness) as shown in Fig. 2. LM-4 (Al-4Cu-5Si-0.8Fe-0.2Mg) alloy ingot pieces were melted in a gas fired melting unit and superheated to ~ 800°C. The melts were not degassed. Riser free castings were made in sodium-silicate-CO_2 bonded sand moulds with bottom pouring arrangements. Except sample 0, all other castings were treated with varying AC current. The details of electrical connection and application of electric current were mentioned in 2.1. After removal of casting (Fig. 2a) from sand moulds, ~ 0.05 m from the top of each casting was cut (Fig. 2b & c) for visual inspection of both inside and outside surfaces. Then ~ 0.025 m thick sections were cut from each ring for estimation of (i) external shrinkage, (ii) gas porosity and (iii) microstructural studies. Samples for hydrogen gas analysis were cut from each casting (0.003 X 0.003 X 0.01 m^3) and analysed in a LECO hydrogen determinater (RH - 402). Three samples were etched in a solution consisted of 10 ml HCl, 30 ml HNO_3, 5 ml HF and 55 ml ethyl alcohol (dehydrated). Specific gravity of these samples was measured by water-immersion method.

3. Results

3.1. Study on castings made in single (Open) sand mould

Fig 3 shows visual appearance of the castings. Sample 1 was cast without any treatment. Samples 2–5 were treated with 2–5A AC respectively. Similarly, samples 6–9 were treated with 2–5 A DC respectively. All castings were sectioned through its middle along their length (as marked on sample 2) to see the effect of electric current on internal shrinkage.

Fig. 2: (a) Hollow cylindrical casting with 0.075 m (I. D.), 0.025 m wall thickness and 0.175 mlong. Sections from the top part of castings (b) without current treatment, (c) with currenttreatment

IOP Conf. Series: Materials Science and Engineering **115** (2016) 012005 doi:10.1088/1757-899X/115/1/012005

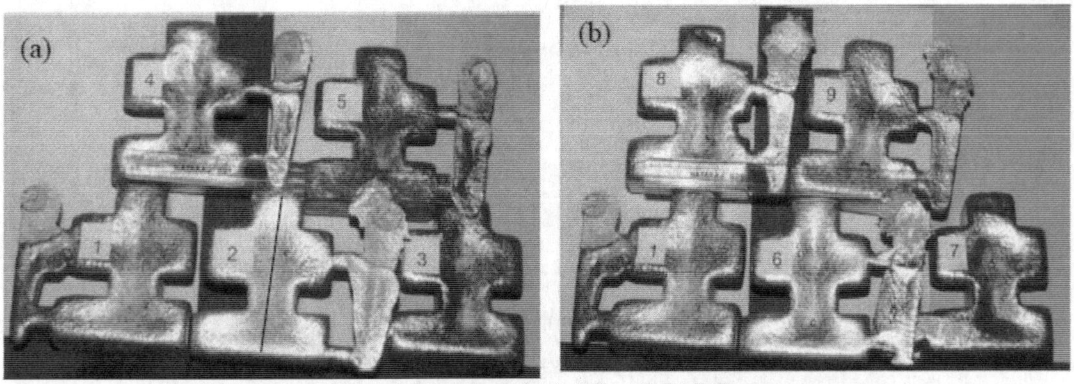

Fig. 3: Open castings with increased (a) AC and (b) DC current treatment.
Castings were sectioned along the black line and shown in the next figure

Fig. 4 shows the cross sections of castings without and with ECT (DC / AC). The untreated sample (Fig. 4a) shows that the internal shrinkage extends along horizontal and vertical directions. Figs. 4b - e show the internal shrinkage of AC treated samples. It is observed from the figures that

i) Low intensity AC eliminates shrinkage in a casting (Fig. 4b)
ii) It increases with increased current (Fig. 4c-e).
iii) It extends along the current flow direction.
iv) The inside surface of the shrinkage is smooth.

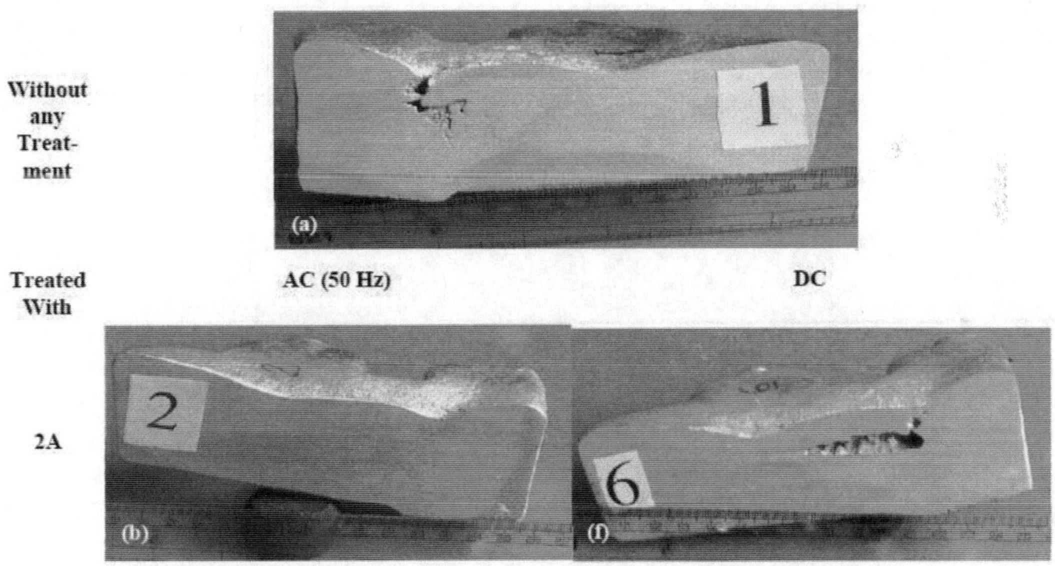

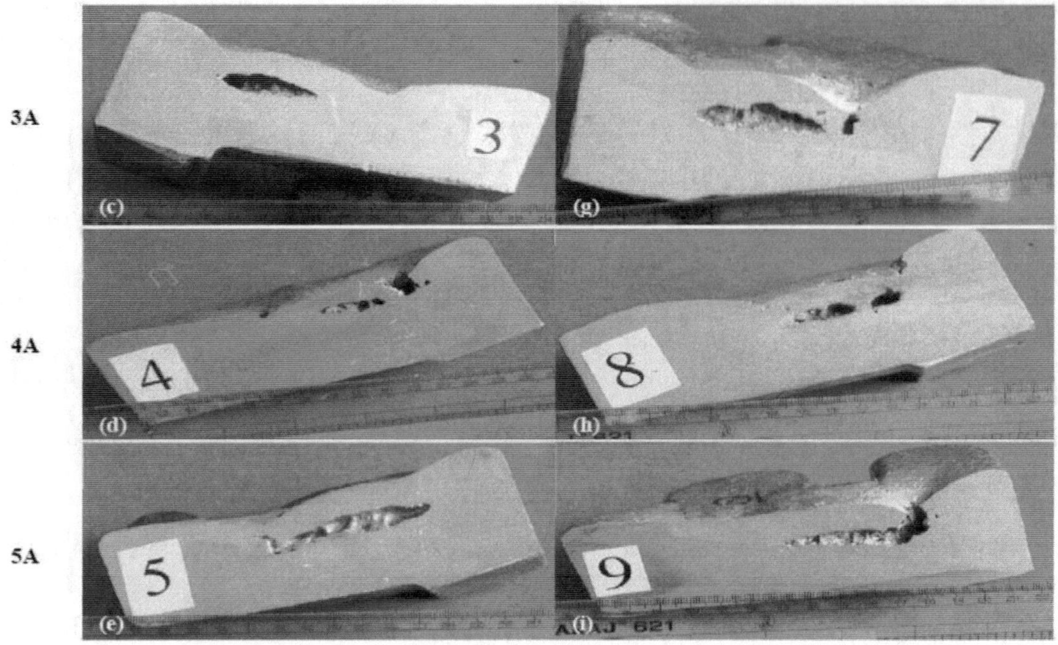

Fig. 4: Cross section of the castings without and with variation of AC and DC current

However, the effect of DC treatment on internal shrinkage is shown in Figs. 4f – i. The figures show that:
i) The shrinkage cannot be eliminated by DC treatment.
ii) It also extendeds along the current flow direction
iii) The inside surface of shrinkage is rough.

3.2. *Study on castings made in closed (Cope and Drag) sand moulds*

Fig. 5 shows a set of castings (~ 0.025 m dia) made in closed box sand moulds. All samples were sectioned along the marked line as shown (on sample 5).

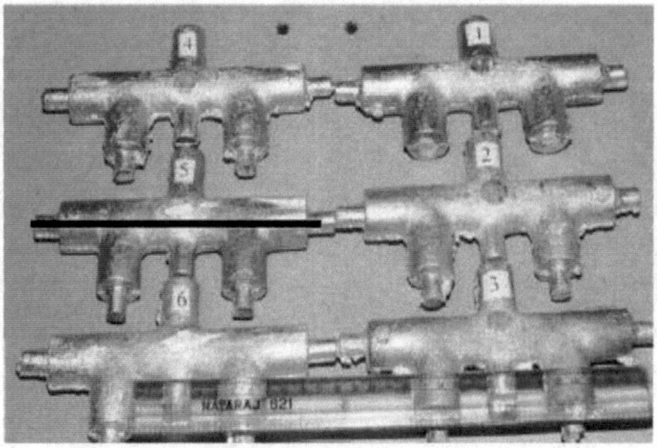

Fig. 5: Sample castings without and with 50 Hz AC
Castings were sectioned along the black line and shown in the next figure

Fig. 6 shows the cross sections of AC treated castings. The untreated sample (No. 1) shows gas porosity and shrinkage marked as a circled zone 'A' (Fig. 6a). However, low intensity current treatment (sample 4, Fig. 6b) reduces both shrinkage and gas porosity to minimum. But at higher current (sample 6, Fig. 6c) the gas porosity (marked as 'A') reappears at casting sections. The figure also shows that the gas porosities are aligned parallels on casting section i.e., the direction of current flow.

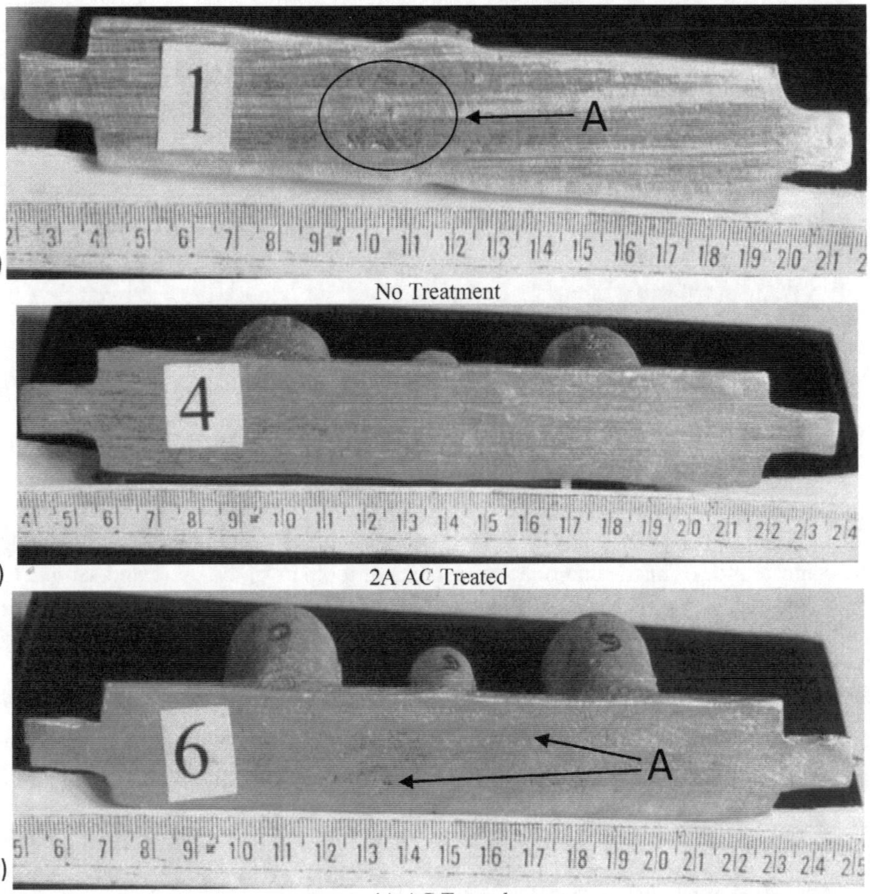

(a) No Treatment

(b) 2A AC Treated

(c) 4A AC Treated

Fig. 6: The cross section of castings without (1) and with ECT (4 & 6); "A" indicates the shrinkage and gas porosity.

3.3. *Study on hollow cylindrical castings*

Fig. 7 shows typical shrinkage profile of untreated and AC treated samples. The untreated sample (No. 0) shows that more shrinkage at inner (core) side of the casting compared to the outer (wall) side of the casting. However, in current treated samples (No. 2, 3, 5 & 6), the extent of shrinkage is same irrespective of the sides.

IOP Conf. Series: Materials Science and Engineering **115** (2016) 012005 doi:10.1088/1757-899X/115/1/012005

Fig. 7: Shrinkage at the top of castings sections. 0: No treatment; 2: 2A AC, 3: 3A AC, 5: 5A AC and 6: 6A AC treatment. The left of each sample is the inner surface andthe right of each sample is the outer surface of castings.

Fig. 8 shows the appearance of inside and outside surfaces of top part of the untreated sample (No. 0). The inside surface of the casting (Fig. 8a) appears dull due to sand sticking but the outside surface (Fig. 8b) is freefrom such defect.

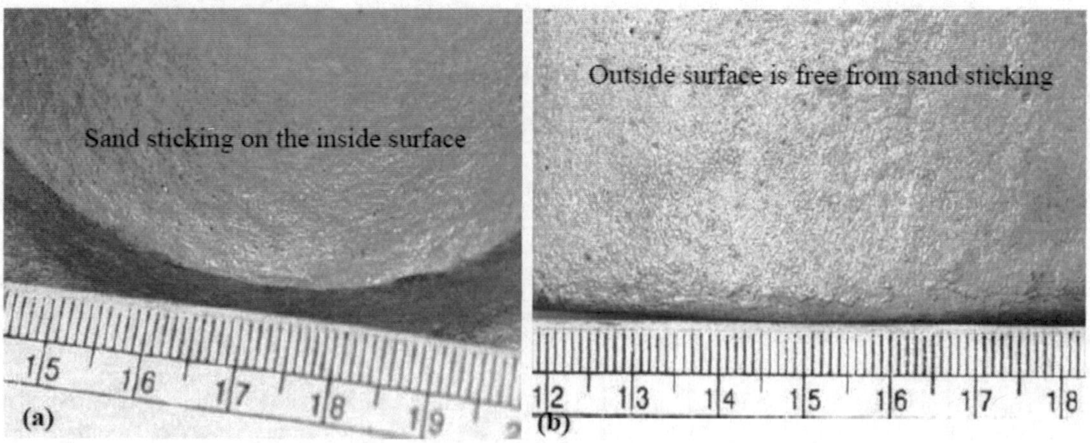

Fig. 8: Casting surfaces of untreated samples (No. 0): (a) the inside surface and (b) the outside surface.

Fig. 9 shows that the inside surface of sample 2 (with 2A ECT) is free from sand sticking. Fig 10 shows typical parallel lines on casting surface on sample 5 (with 5A ECT).

IOP Conf. Series: Materials Science and Engineering **115** (2016) 012005 doi:10.1088/1757-899X/115/1/012005

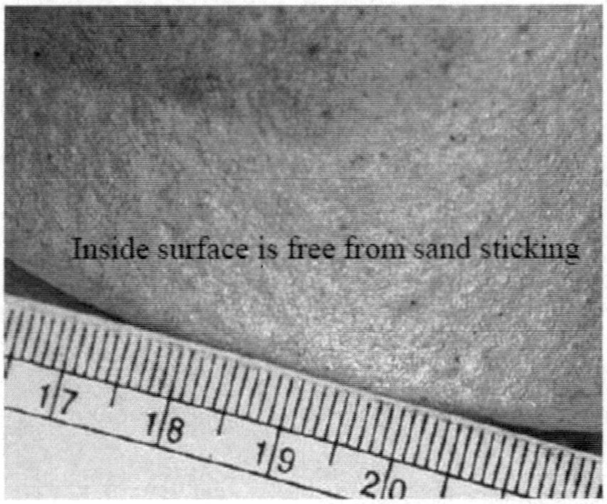

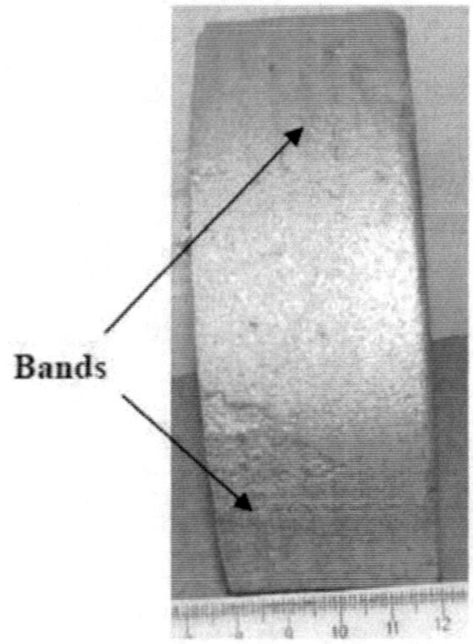

Fig. 9: Inside casting surface of sample 2. Fig. 10: Bands on casting surfaceSample 5

Fig. 11(a-e) show the effect of increased current on (a) gas porosity and (b) hydrogen content in castingsmade without and with ECT. The untreated sample (No. 0; Fig. 11a) shows maximum porosity. However, low intensity ECT (Fig. 11b) significantly reduces gas porosity (sample 2). But it reappears with increased treatment current (Fig. 11c-e).

(a) Hydrogen: 3.1 ppm

(b) Hydrogen: 1.0 ppm(c) Hydrogen: 1.2 ppm

(d) Hydrogen: 1.5 ppm **(e) Hydrogen: 1.9 ppm**

Fig. 11: Samples show variation in hydrogen gas porosity in castings without and with AC currenttreatment: (a) Sample 0 (no current), (b) Sample 2 (2A), (c) Sample 3 (3A), (d) Sample 5 (5A) and (e)Sample 6 (6A).

Fig. 12 shows optical microstructures of samples without and with current treatment. The untreated sample (No 0, Fig. 12a) shows coarse dendrites. But low intensity ECT (Fig. 12b, sample 2) fragments primary dendrites. Some fraction of dendrites becomes globular. The fragmented dendrites become coarser by increased current intensity (sample 3, Fig. 12c).

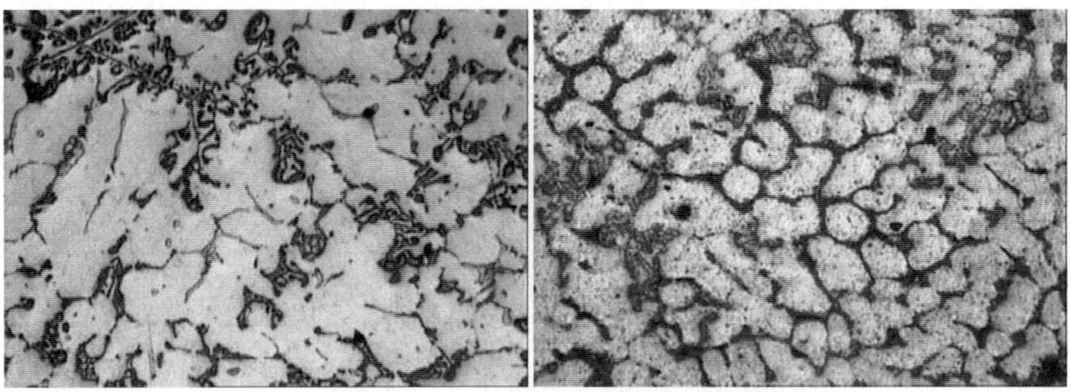

(a) Specific gravity: 2.55g/cc; 100μm l-------l(b) Specific gravity: 2.71g/cc; 100μm l-------l

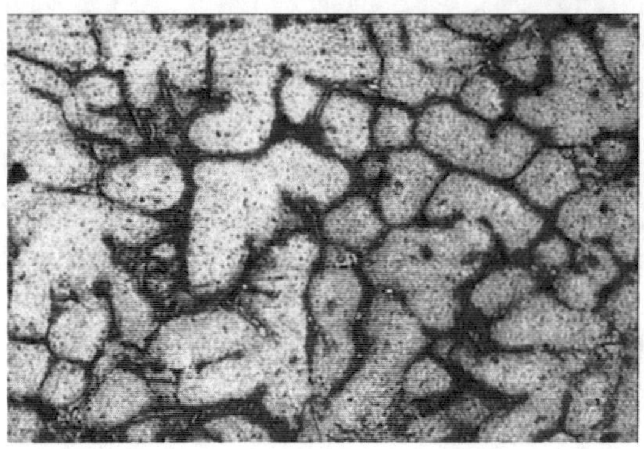

(c) Specific Gravity 2.67g/cc; 100μm l-------l

Fig. 12: Optical microstructures of the samples (without degassing) treated without and with AC current.

(a) Sample 0 (no current), (b) Sample 2 (2A) and (c) Sample 3 (3A).

Table I shows a comparative data of the energy consumption for different semisolid casting processes withaluminium alloys.

Table I: A comparison of energy consumption in different semisolid processes

Reference	Method	Weight of Metal, Kg	Wattage, Watts	Time, Sec	Energy Consumption	
					Total, Wh	Specific Wh/Kg
Maity[22]	Mechanical Stirring	1	750	900	187.5	~ 185
Mondal[23]	Do	3	5000	600	375	~125
Dutta etal [4]	LEMS	10	750	1800	833	~ 83
Thisstudy	ECT	3.7	40	500	5.55	~ 1

4. Discussion

This study compares the effect of EHD [7-9] or ECT [10-17,19,21] during solidification (i.e., at the semisolid state) of 99.5% Al and LM-4 Al-alloy in sand moulds. It also reports the effect of ECT on castings with different geometry, energy consumption, modification of cast microstructures etc.

4.1. Study on castings made in open sand moulds

It is evident from Fig. 4 (a-i) that ECT (AC/ DC) plays a significant role in the formation of internal shrinkagein castings. Normal casting (No. 1, Fig. 4a) undergoes shrinkage along vertical and horizontal directions. This is due to the movement of solidification fronts from the bottom as well as from the mould walls. However, 2A AC ECT produces shrinkage-free castings (Fig. 4b). The ECT increases the EHD [7-9] force that increases flow of molten metal. It feeds internal shrinkage that forms during the progress of solidification in a casting. In EHD, the current flow enhances the fluid flow along the peripheral region of the conducting fluid. Therefore, heat transfer rate is increased [10] that suppresses the freezing point and favours early solidification of melt [11-13]. However, Fig. 4(c–i) show that increased ECT produces elongated shrinkage along the current flow direction. This is in addition to the external shrinkage of samples as shown in Fig. 3. This is the effect of increased fluid flow along the peripheral region of the casting. Here the top surface of the casting (in contact with air) solidified earlier rather than feeding the melt into the shrinkage cavity. Dutta et al [4], reported similar phenomena when castings made with linear electro-magnetic stirring (LEMS). They reported early solidification of the upper part of the casting compared to untreated casting, therefore, the entrapped liquid below the upper solidified layer forms internal shrinkage. The nature of inside surface of shrinkage changes with type of power supply (AC or DC). The AC treated samples (Fig. 4c–e) shows smooth surface, which indicates entrapment of dissolved gas. It may be concluded that the dissolved gas diffuses to the shrinkage zone of the casting. It is observed that increased ECT increases heat input due to Joules heating (I^2R) [21] and increases solidification time and gas entrapment [11-13]. So, the volume of shrinkage cavity increases at increased current.

4.2. Study on castings made in closed (Cope and Drag) sand moulds

Based on the previous observations, this study is limited to melt treatment with AC. Fig. 1 shows that SSM by ECT is also possible in closed sand moulds. It is observed that the ECT is effective as long as the metal remains at its semi-solid state [10-13]. Castings made in closed sand mould are the typical examples of optimum use of electric energy. The circuit completes when the liquid metal touches both top and bottom electrodes connected to power supply. It is automatically disconnected due to contraction of solid casting. Therefore, there is no need to switch on and off electric current in individual mould. The untreated sample (No 1, Fig. 6a) contains both gas porosity and internal shrinkage located at the central zone "A" of the casting section. This is a common phenomenon in normal castings; however, these defects are minimized at 2A AC treatment (sample 4, Fig. 6b). This is due to the optimized fluid flow (by ECT/ EHD) [7-9] that fills up the interdendritic areas during solidification. Secondly, this process removes the dissolved gas also. Further increase of current increases the intensity of fluid flow [7-9], as well as, heat input (I^2 R) to the molten metal [19]. This heat input increases thesolidification time [11-13] and dissolved gas. Therefore, the dissolved gas (Sample 6, Fig. 6c) cannot be removed at higher intensity ECT. Fig. 6c shows that the gas porosity is aligned parallel along the current flow direction [10]indicating that the entrapped hydrogen gas is a part of fluid flow during solidification. The hydrogen is not separated from the melt to form shrinkage cavity in casting. It is obvious that (i) 2A AC is the optimum current for these casting and (ii) maximum utilization of electric energy is possible in closed sand mould.

4.3. Study on hollow cylindrical castings

ECT (AC) is also conducted on hollow cylindrical castings (Fig. 2). The shrinkage profile at the top of the untreated casting (No 0; Fig. 7) shows that the extent of shrinkage at the outer side of the casting is less compared to the inner (or core) side of the casting. This may be explained as during solidification, the solidified layer forms along the surfaces adjacent to the mould/ core walls. In untreated casting (Fig. 2b), due to contraction, the outer surface of the casting gets detached from mould wall. However, the inner surface of the casting adheres to the core. Therefore, the heat transfer rate from the outer surface of the casting is reduced compared to the inner surface of the casting. So, the inner wall side of sample shows more shrinkage compared to its outer side.

The shrinkage profiles of the top part of ECT castings are shown in Fig. 7 (No. 2 – 6). It is similar on both sides of each casting. This indicates that during solidification, the solidified layers are detached from both mould and core walls. A comparison between Fig. 8 and 9 reveals this phenomenon. In untreated casting, the outer surface of the casting (Fig. 8b) is free form sand sticking compared to the inner surface of the casting (Fig. 8a). However, the inner side of current treated sample (Fig. 9) is free from sand sticking. This may be explained with the schematic diagram (Fig. 13). In conventional solidification (Fig. 13b), the solidification starts from both the bottom and along the sidewall of the mould. The electric current treatment changes the location of solidification front (Fig. 13c). These phenomena may be explained as the current flow induces both normal and tangential forces at the mould-melt interface [7-9]. The solidified metal layer along mould walls is washed away and comes in contact with liquid pool [3-6]. The nuclei are partially remelt in contact with hot metal before they deposit on the solidification front i.e. at the bottom due to increased density of solidified metal (Fig. 13c). Liao et al [19] reported similar phenomena. Thus the solidification front moves along the vertical direction.

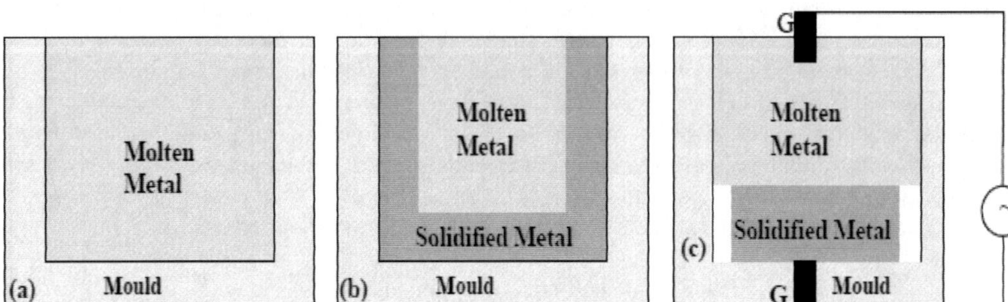

Fig. 13: Schematic diagram of movement of solidification front without and with electric current treatment: (a) metal in a sand mould, (b) conventional solidification and (c) solidification with electric current. G: Graphite connector

Sample 5 shows typical bands on casting surface (Fig. 10) but these are not observed in untreated and low intensity AC treated samples. This may be explained as the effect of deposition of coarse grain at the solid liquid interface. It is observed [11–13] that current treatment beyond a critical limit increases solidification time and the grain size. The electric energy increases heat input as $I^2 \times R$ [21]; [I = current and R = resistance of molten bath]. This additional heat takes longer time to dissipate and causes grain coarsening. Those coarse grains settle at the bottom i.e., at the solidification front. This layer contracts after freezing. Meanwhile, another layer is started forming just above the solidified layer that causes band formation (Fig. 13c). To summarise, (i) the current treatment changes the movement of solidification front along vertical direction only (ii) the metal mould reaction is reduced due to detachment of solidified metal from mould walls, (iii) due to formation of coarse grains at higher treatment current, typical bands appears on the casting surface.

4.4. Gas porosity

Another effect of ECT is the removal of dissolved Hydrogen from molten Al-alloy. Fig. 11 (a-e) shows the variation of gas porosity in castings treated with increased current intensity. Naturally, porosity is the maximum for untreated sample (Fig. 11a) but it is significantly reduced by low intensity AC (Fig. 11b). Ahmed et al [16-17] mentioned that flow of current through molten metal creates a condition similar to "cold plasma" where both electrons and positive ions take part in transfer of electrical charge [18]. In this process, H in melt forms H+ ions and transported from the zone of higher metallostatic pressure to lower metallostatic pressure and escapes to theatmosphere [12-13]. On the contrary, gas content in casting gradually increases with increased current beyond a critical current (Figs. 11c - e). Similar phenomenon is also reported in the other studies [11-13]. The reason is that increased current flows through melt increases fluid flow [7-9], as well as, heat input to the melt and thereforeincreases solidification time[12,13,21] that causes gas entrapment.

4.5. Optical microstructures

Fig. 12(a-c) shows optical microstructures of samples without and with ECT. The untreated sample (Fig. 12a) shows coarse dendrites of primary α-phase. Low intensity current (2A AC) treatment changes the microstructures to a mixture of fragmented dendrites/ globular primary phase. However, increased current (sample 3, Fig. 12c) increases volume fraction and size of the globular α-phase compared to sample 2. It isobserved that microstructures developed by EMS [4] and Rheo-processing [20] routes are very much similar to microstructures of ECT samples (Fig. 12b & c).

4.6. *Energy consumption*

It is observed from the Table I that the specific energy consumption is minimum for electric current treatment compared to electromagnetic stirring [4] or mechanical stirring [22-23] methods. This indicates the energy conversion efficiency is significantly high in ECT compared to LEMS or mechanical stirring. In ECT [10-15] /Plasma [16-19] / EHD [7-9] the electric energy is directly converted to kinetic energy to move charged ions or electrons in liquid metal. However, in LEMS /REMS [3-6] the electric energy is first converted to magnetic energy which then converted to kinetic energy to stir liquid metal. Therefore, ECT is an energy saving process compared to LEMS/ RMS or mechanical stirring.

5. Conclusions

1. At an optimum intensity, (50 Hz AC) electric current treatment (ECT) eliminates internal shrinkage incastings, however, it increases with increased current.
2. AC treatment is more effective compared to DC treatment.
3. Treatment with increased AC current increases gas porosity.
4. DC treatment causes interdendritc shrinkage formation but AC treatment favours gas entrapment at thecasting sections.
5. Optimum utilisation of energy is possible during ECT in closed box sand casting.
6. During casting by ECT, the solidification front moves along vertical direction
7. ECT reduces metal-mould reactions.
8. There is a good similarity between optical microstructures of ECT samples and conventional

SSM samples.

9. At increased ECT, typical casting bands are observed on casting surface due to formation of coarse grains.
10. ECT is more energy efficient compared to compared to LEMS / REMS.

References

[1] J. Campbell, Effects of vibration solidification, *International Metals Review*, 26(2) (1981) 71–108.
[2] M. C. Flemings,Behaviour of metal alloys in the semisolid state, *Metallurgical Transactions B*, 22B (1991)269 – 293.
[3] S. Ordonez S, O. Burtos, R. Colas, *International Journal of Metal Casting*, 3(3)(2009)37-41.
[4] P. Dutta, K. S. S. Murthy, *Indian Foundry Journal*, 55(7) (2009) 33 – 39.
[5] S. C. Lim, E. P. Yoon, J. S. Kim,*Journal of Materials Science Letters*, 16 (1997) 104-109.
[6] Y. Mizutani, T. Tamura, K. Miwa,*Materials Transactions (Japan)*, 48(3) (2007) 538-43.
[7] Electrohydrodynamics, http://en.wikipedia.org/wiki/Electrohydrodynamics
[8] Electrohydrodynamics, www.andrew.cmu.edu/user/dcprieve/Notes/Electrohydrodynamics.doc,
[9] Electrohydrodynamic Multiphase Flow, http://www.me.wpi.edu/research/MFG/Research/
[10] A. Prodhan, D. Sanyal, Cooling curves analysis of aluminium solidified without and with magnetic or electric field, *Journal of Materials Science Letters*, 16 (1997) 958-961.
[11] A. Prodhan, C.S. Sivaramakrishnan, A. K. Chakrabarti, Solidification of Aluminum in Electric Field, *Metallurgical and MaterialsTransactions B*, 32B (2001) 372-378.
[12] A. Prodhan, A method for degassing and microstructural modification of aluminium alloys by electric current treatment, *AFS Transactions,*116 (2008) 233 – 246.
[13] A. Prodhan, Effect of electric current (plasma) treatment on LM-6 Al-Si alloy, *AFS Transactions,*117 (2009) 63 – 77.
[14] K. Chrianjeevi, A. Prodhan: *Indian Foundry Journal,*53(6) (2007) 45-52.
[15] A. Prodhan, Indian Patent No: 226266 dt. 16.12.2008, http://ipindia.nic.in/ipr/patent/journal_archieve/Journal_2009/ pat_search_012009/ official_journal_02012009_part_ii.pdf; Page: 292/342.
[16] S. Ahmed, R. Bond, E. C. Mckannan,*Advanced Materials & Processes*, 140 (4) (1991)30-38.
[17] S. Ahmed, E.C. McKannan, *Materials Science and Technology*, 10 (1994) 941-946.
[18] Plasma-The fundamental state of matter, www.electric-cosmos.org/electricplasma.htm
[19] X. Liao, Q. Zhai, J. Luo, W. Chen, Y. Gong, *ActaMaterialia*, 55(7) (2007) 3103–3109.
[20] J. S. Sunitha, N. Barekar, K. Biswas, B.K. Dhindaw: *Trans. 57th Indian Foundry Congress*, Kolkata, India,February13-15, (2009) 252-256.
[21] R. Ambarder, D.G.R. Sharma, O. Pravakar, *Transactions IIM*, 47 (1994) 131-137.
[22] P. C. Maity, DAV Institute of Engineering and Technology, Kabirnagar, Jalandhar, Punjab, India, PersonalCommunication
[23] D. Mondal, National Institute of Technology, Durgapur, India, Personal communication.

Epoxy matrix composites filled with micro-sized LD sludge: wear characterization and analysis

Abhilash Purohit and **Alok Satapathy**

Department of Mechanical Engineering, National Institute of Technology, Rourkela, 769008, India
mr.abhilash.purohit@gmail.com, alok@nitrkl.ac.in

Abstract. Owing to the very high cost of conventional filler materials in polymer composites, exploring the possibility of using low cost minerals and industrial wastes for this purpose has become the need of the hour. In view of this, the present work includes the development and the wear performance evaluation of a new class of composites consisting of epoxy and micro-sized LD sludge. LD sludge or the Linz-Donawitz Sludge (LDS) are the fine solid particles recovered after wet cleaning of the gas emerging from LD convertors during steel making. Epoxy composites filled with different proportions (0, 5, 10, 15 and 20 wt %) of LDS are fabricated by conventional hand lay-up technique. Dry sliding wear trials are performed on the composite specimens under different test conditions as per ASTM G 99 following a design of experiment approach based on Taguchi's orthogonal arrays. The Taguchi approach leads to the recognition of most powerful variables that predominantly control the wear rate. This parametric analysis reveals that LDS content and sliding velocity affects the specific wear rate more significantly than normal load and sliding distance. Furthermore with increase in LDS content specific wear rate of the composite decreases for a constant sliding velocity. The sliding wear behavior of these composites under an extended range of test conditions is predicted by a model based on the artificial neural network (ANN).

Keywords: Epoxy composites; LD Sludge; Characterization; Sliding Wear; Artificial Neural Network (ANN)

1. Introduction

Polymers have found wide range of applications in various engineering fields due to their high strength and low density as compared to monolithic and metal alloys. For weight sensitive uses, polymers are the most suitable materials because of their lightweight, but their high cost sometimes becomes the bounding factor for many applications. Therefore easily available and low cost fillers are often used to reduce the cost of the element. Examination of the effect of such filler addition is necessary to ensure that the mechanical properties of the composites are not affected adversely by such addition. Till date large numbers of materials are used as fillers in epoxy composites [1]. The main objective of addition of fillers in to polymer can therefore be branched into two basic categories; first, to reduce the price of the element and second, to enhance the thermal, mechanical or tribological properties. There have been various reports on the use of materials such as alumina and silica mixed into polypropylene [2-3] and polyethylene [4-5]. But very few attempts have indeed been made to utilize low cost minerals and industrial wastes in preparing particulate-reinforced polymer composites. However, this is emerging as a subject of extensive research in recent years.

An important aspect of particulate reinforced polymer composites is that the properties of the materials can be varied by controlling the filler percent and polymer combinations. A proper combination of polymer matrix and reinforcing filler material can lead to the formation of a composite with enhanced properties comparable to or even better than those of conventional metallic materials [6]. These composites materials are desired because of their high corrosion resistance, low density, low cost and ease of fabrication [7-8]. The low-cost filler like red mud has been used in some earlier studies with different polymer matrices such as polyester [9], polypropylene [10] and nylon [11] to study the wear and mechanical properties. Patnaik *et al.* [13-14] studied erosion wear response of the glass-fiber reinforced fly ash filled polyester composites using Taguchi's design-of-experiment method. Many attempts have also been made in the past on the utilization of blast furnace and steel slags [12]. Padhi *et al.* [15] studied the tribo-performance analysis of blast furnace slag filled with polymer composites while Pati *et al.* [16] studied the utilization of LD slag in erosion resistant coatings.

It has been found that about 2 to 4 tonnes of wastes are being generated per tonne of steel produced. The various solid wastes in the form of sludges and slags that are produced from steel industry are, blast furnace flue dust and sludge, blast furnace slag (BFS), LD converter slag, LD sludge etc. LD sludge particles are the fine solid particles recovered after wet cleaning of the gas emerging from LD convertors during steel making. The LDS in solid lump form collected from Rourkela Steel Plant contains average particle size of about 70-100 μm with a true density of 1.5 g/cm^3. Detailed characterization of LDS is reported by Das *et al.* [12]. They found that LDS comprised of FeO, Fe_2O_3, CaO, MgO, SiO_2, Al_2O_3 and small quantity of P and MnO.

Although few attempts have been made on the utilization of LD sludge [12] in the past, no work has so far been reported on tribo-mechanical performance of LDS produced from steel plant. In view of this, the present work includes the investigation of dry sliding wear behaviour using a pin-on-disc set up for epoxy filled with LDS with an objective to find ways to use these locally available inexpensive industrial wastes as a substitute for highly expensive fillers for wear resistance applications. Taguchi technique and Artificial Neural Network (ANN) have been used in an integrated manner in this work for parametric analysis of the wear process and to predict the sliding wear response of epoxy-LDS composites within and beyond the experimental limits.

2. Experimental details

2.1. Composite Fabrication

Linz-Donawitz sludge (LDS) collected from Rourkela steel plant, India is sieved and particles with a size range of 70–100 μm are obtained. Composites are prepared by mixing these particles with low temperature curing epoxy resin (LY556) which is used as the matrix phase. Prior to LDS addition hardener (HY951) is properly mixed with epoxy resin in 10:1 ratio by weight as recommended. Epoxy resin has density of 1.1g/cm^3 and modulus of 3.42GPa. To prepare 10mm diameter and 250mm length cylindrical samples, glass tubes are used, which are uniformly coated with wax at the inner surface. The dough (epoxy resin mixed with LDS) is then poured in to the glass tubes. Five samples are made with different compositions (0, 5, 10, 15 and 20 wt %) of LDS. The tubes are kept at room temperature for 24 hours and then the glass tubes are broken and samples are collected.

2.2. Sliding Wear Test

Wear tests are carried out under dry sliding conditions to evaluate the performance of these composites in a pin-on-disc type friction and sliding wear test rig (supplied by DUCOM) as per ASTM G 99. The composites slide against a disc made up of hardened steel (EN-32, hardness 72 HRC, surface roughness 0.6 μm Ra). Samples are fitted to the holder and sliding rpm is set to the desire value before applying the load. All other parameters like sliding velocity, track radius and normal load are set before starting the test. In a pin-on-disc test the sample is held stationary and the steel disc is rotated at a required speed while a lever mechanism is used to apply the normal load. A series of tests using Taguchi's L_{25} orthogonal array are conducted with five sliding velocities of 63, 125,190, 250, and 315 cm/s under five different normal loadings of 5, 10, 15, 20, and 25 N. After each trial, the specimen is cleaned with acetone and a precision electronic balance with accuracy ±0.1mg is used to measure the weight loss from the composite and the specific wear rate (mm^3/N-m) is then expressed on 'volume loss' basis as:

$$W_s = \frac{\Delta \mathrm{m}}{\rho\, \mathrm{t}} * V_s F_n \quad \textbf{(1)}$$

where Δm is the mass loss in the test duration (g), ρ is the density of the composite (g/mm^3), t is the test duration (s), V_s is the sliding velocity (m/s), and F_n is the average normal load (N).

2.3. Experimental Design

Design of experiment (DOE) is a powerful tool used for analyzing and modelling the influence of control factors on performance output. The selection of the control factors is the most vital stage in the design of experiment. In the initial stage a number of factors are considered and subsequently the less important factors are eliminated leaving behind only the more important factors. Table 1 gives the various operating conditions under which wear tests are carried out. Four parameters, viz., sliding velocity, normal load, sliding distance and filler content each at five levels, are considered in this study in accordance with L_{25} orthogonal array design. The experiments are conducted at room temperature. Use of Taguchi L_{25} reduced the number of experiments from 4^5= 1024 conventional runs to just 25 runs thereby saving a lot of experimental time and cost. The signal to noise (S/N) ratio for minimum wear rate can be expressed as 'smaller is the better' characteristic and can be calculated as the logarithmic transformation of the loss function as shown below.

$$\frac{S}{N} = -10\log\frac{1}{n}\sum y^2 \quad \textbf{(2)}$$

where, y is the observed data and n is the number of observations. For minimization of wear rate, with the above S/N ratio, 'Lower is better' (LB) characteristic is suitable.

Table 1. Levels of the variables (control factors) used in the experiment

Control factors	Levels					
	I	II	III	IV	V	Units
Sliding velocity (A)	63	125	190	250	315	cm/sec
Normal load (B)	5	10	15	20	25	N
Sliding distance (C)	500	1000	1500	2000	2500	m
LDS content (D)	0	5	10	15	20	wt %

2.4. Prediction Using Artificial Neural Network

Artificial Neural Network (ANN) is an approach motivated by the biological neural system and has been used to solve a wide variety of issues in distinct fields. It finds various applications in the field of medical science, classification and control of dynamic system, image and speech recognition etc. To develop a functional model, the multiple-layered networks with differential transfer functions are trained by using the back-propagation ANN. In the present study ANN serves as a very helpful tool for predicting the specific wear rate as it is very difficult to get a mathematical formulation for the wear behaviour. This proposed approach not only yields a sufficient understanding of the effects of process parameters but also produces an optimal parameter setting to ensure that the composites exhibit the best wear performance characteristics [17].

Table 2. Taguchi Experimental design (L_{25} orthogonal array)

Test Run	Sliding Velocity A (cm/sec)	Normal Load B (N)	Sliding Distance C (m)	LDS Content D (wt %)	Sp. Wear Rate (10^{-9} mm^3/N-m)	S/N Ratio (dB)
1	63	5	500	0	1.612	-4.14730
2	63	10	1000	5	1.325	-2.44432
3	63	15	1500	10	0.873	1.17972
4	63	20	2000	15	0.786	2.09155
5	63	25	2500	20	0.688	3.24823
6	125	5	1000	10	1.053	-0.44857
7	125	10	1500	15	0.982	0.15777
8	125	15	2000	20	0.788	2.06948
9	125	20	2500	0	1.736	-4.79099
10	125	25	500	5	1.442	-3.17931
11	190	5	1500	20	0.864	1.26973
12	190	10	2000	0	1.823	-5.21573
13	190	15	2500	5	1.518	-3.62544
14	190	20	500	10	1.192	-1.52553
15	190	25	1000	15	1.118	-0.96884
16	250	5	2000	5	1.615	-4.16345
17	250	10	2500	10	1.316	-2.38512
18	250	15	500	15	1.214	-1.68437
19	250	20	1000	20	0.986	0.12246
20	250	25	1500	0	2.012	-6.07256
21	315	5	2500	15	1.396	-2.89771
22	315	10	500	20	1.117	-0.96106
23	315	15	1000	0	2.226	-6.95050
24	315	20	1500	5	1.834	-5.26799
25	315	25	2500	10	1.528	-3.68247

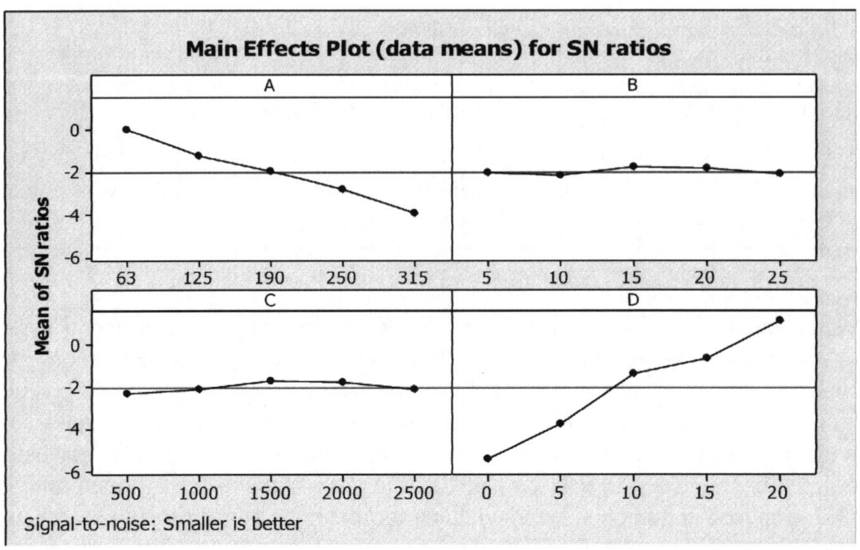

Figure 1. Effect of control factors on sliding wear rate of Epoxy-LDS composites

IOP Conf. Series: Materials Science and Engineering **115** (2016) 012006 doi:10.1088/1757-899X/115/1/012006

Table 3. Response table for S/N Ratios (Epoxy-LDS composites)

Level	A	B	C	D
1	-0.01442	-2.07746	-2.29951	-5.43542
2	-1.23832	-2.16969	-2.13795	-3.73610
3	-2.01316	-1.80222	-1.74667	-1.37239
4	-2.83661	-1.87410	-1.78013	-0.66032
5	-3.95195	-2.13099	-2.09020	1.14977
Delta	3.93752	0.36747	0.55285	6.58518
Rank	2	4	3	1

Table 4. Comparison of Experimental Results with ANN Predicted Values

	Specific wear rate (10^{-9} mm^3/N-m)		
Test run	Experimental	ANN Predicted	Error %
1	1.612	1.610	0.074
2	1.325	1.210	9.481
3	0.873	0.922	5.375
4	0.786	0.798	1.583
5	0.688	0.775	11.312
6	1.053	1.060	0.728
7	0.982	0.863	13.783
8	0.788	0.785	0.328
9	1.736	1.548	12.087
10	1.442	1.406	2.539
11	0.864	0.830	4.065
12	1.823	1.816	0.363
13	1.518	1.426	6.450
14	1.192	1.232	3.273
15	1.118	1.036	7.892
16	1.615	1.638	1.410
17	1.316	1.321	0.426
18	1.214	1.194	1.602
19	0.986	1.061	7.070
20	2.012	1.951	3.077
21	1.396	1.309	6.621
22	1.117	1.129	1.114
23	2.226	2.145	3.754
24	1.834	1.892	3.068
25	1.528	1.724	11.412

3. Results and discussion

3.1. Dry Sliding Wear Analysis

The specific wear rates calculated on the basis of the test results using equation (1) and their corresponding S/N ratios obtained by Taguchi L_{25} orthogonal array for all the 25 experiments are presented in Table 2. The overall mean of the S/N ratio is found to be -1.593 dB for epoxy based composites reinforced with LD sludge. The analyses are completed using commercial software specifically used for design of experiment applications known as MINITAB 14. From the S/N ratio output, it is observed that among all the four factors considered for the analysis, the specific wear rate is mostly affected by the LDS content in the epoxy composite followed by the sliding velocity, sliding distance and normal load. The analysis shows that at sliding velocity (A) of 63 cm/s, normal load (B) of 15 N, sliding distance (C) of 1500 m and LDS content (D) of 20 wt%, the specific wear rate is minimum as evident from Figure 1. The relative effects of the control factors on specific wear rate are evident from the response table (Table 3).

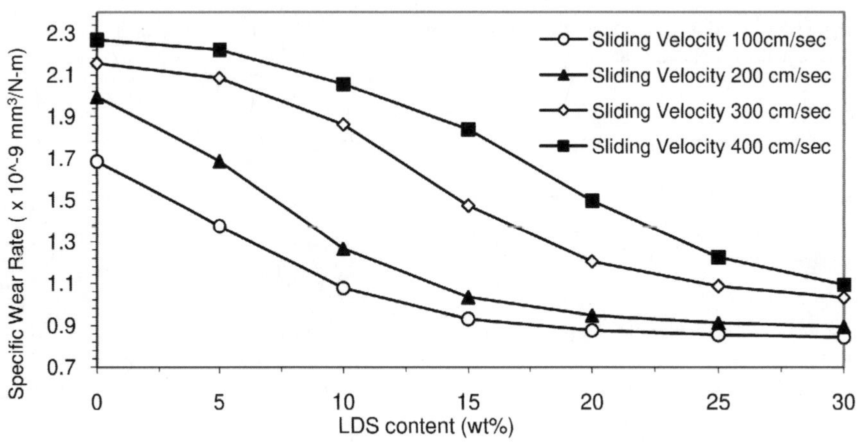

Figure 2. ANN prediction of variation in specific wear rate with LDS content for epoxy composite

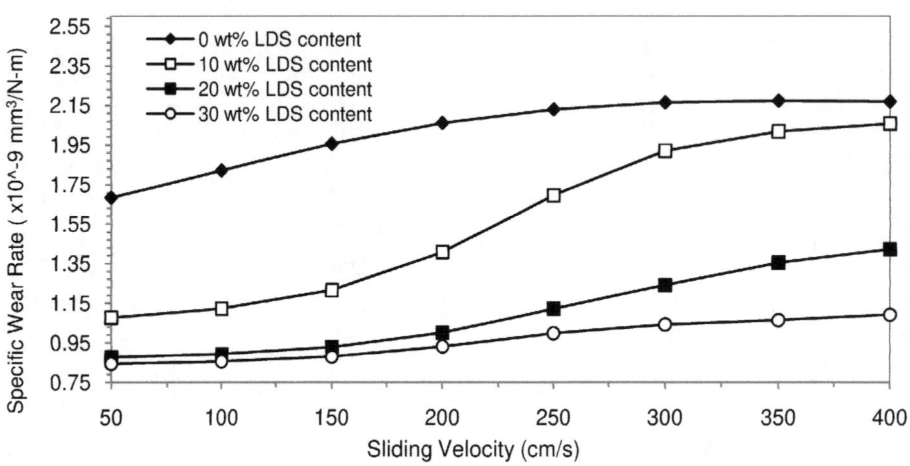

Figure 3. ANN prediction of variation in specific wear rate with sliding velocity for Epoxy composites

3.2. Artificial Neural Network for Prediction

Artificial Neural Network (ANN) is a very convenient tool to predict the output and input pattern of nonlinear problems such as sliding wear. In the present analysis LDS content, sliding velocity, sliding distance and normal load are taken as the four input parameters. The values of input parameters are normalized so that they lie in between 0 and 1. The output layer which represents the specific wear rate has one neuron. A number of networks are explored by varying the values of input training parameters and the one with least error criterion is finally selected for training. A comparison between the experimental values and the predicted values is performed and the results are shown in Table 4 with their corresponding error percentages.

The effects of two most influencing factors on the specific wear rate i.e. LDS content and sliding velocity are shown in Figure 2 and Figure 3 respectively, in which the simulated specific wear rates for various tests using ANN are presented. With increase in sliding duration, formation of wear debris of different sizes and shapes take place between the specimen and the counter surface. It is justifiable that with increase in sliding velocity with or without applied load, the epoxy resin softens due to generation of frictional heat. As a result the brittle LDS particle having sharp edges can easily tear the matrix phase of the composite and eventually get aligned along the sliding direction. The wear behaviour of the composites again depends on shape, size, hardness and brittleness of the LDS particles. Furthermore wear resistance of composites is enhanced due to presence of higher filler volume. Benefits of filler content presence are increased hardness, strength, and decrease in polymerization shrinkage, thermal expansion and contraction, water sorption, softening, staining and finally improved wear resistance. Increasing the filler content offers characteristics like bulk curing with less polymerization shrinkage, decreased wear and peackability to the composite resin. LDS filled epoxy composites show a decrease in the wear rate with the increase in the LDS content, which acts as obstacle to shear deformation during the sliding conditions. With a rise in the concentration of LD sludge the declining trend in the wear rate is significant but up to a limit.

4. Conclusions

This work suggests the use of LD sludge, an industrial waste as a potential filler material in the making of polymer composites. The compatibility of LDS particles in epoxy resin is found to be fairly good. Incorporation of LDS has resulted in substantial improvement in the wear resistance of neat epoxy. This work shows that dry sliding wear characteristics of these composites can be successfully analyzed using Taguchi experimental design. The analysis revels that LDS content and the sliding velocity are the most predominant control factors affecting the wear rates of the composites. This work further shows that artificial neural networks have been gainfully employed for the prediction of wear response of these composites in a parameter space wider than the experimental domain.

References

[1] Katz H S and Mileski J V 1987 Handbook of Fillers for Plastics.
[2] Mareri P Bastide S Binda N and Crespy A 1998 *Composites Science and Technology* 58 747–752.
[3] Ja rvela P A. and Ja rvela P K 1996 *Journal of Materials Science* 31 3853–3859.
[4] Rusu M Sofian N and Rusu D 2001 *Polymer Testing* 20 409–417.
[5] Barta S Bielek J and Dieska P 1997 *Journal of Applied Polymer Science* 64 1525–1530.
[6] Jang B Z 1994 *Advanced Polymer Composites*: Principles and Applications. ASM International.
[7] Zhu K and Schmauder S 2003 *Computational Material Science* 28 743–748.
[8] Tavman I H 1997 *Powder Technol*ogy 91 (1997) 63–67.
[9] Patnaik A and Satapathy A 2008 *Journal of Reinforced Plastics and Composites* 27 1039–1058.
[10] Wong K and Truss R W 1994 *Composites Science and Technology* 52 361–368.

[11] Srivastava V K and Shembekar P S 1991 *Journal of Materials Science* 26 6693–6698.
[12] Das B Prakash S Reddy P S R and V N Misra 2007 *Resources, Conservation and Recycling* 50 40–57.
[13] Patnaik A Satapathy A Mahapatra S S and Dash R R 2009 *Journal of Reinforced Plastics and Composites* 28 513–536.
[14] Patnaik A Satapathy A Mahapatra S.S Dash R.R 2008 *International Polymer Processing* XXIII 192–199.
[15] Padhi P K and Satapathy A 2013 Society of *Iridologists and Lubrication Engineers* 56 789-796.
[16] Pati P K and Satapathy A 2015 *Polymers for Advanced Technologies* 26 121–127.
[17] Kadi H E 2006 *Composite Structures* 73 1–23.

Studies on Effective Elastic Properties of CNT/Nano-Clay Reinforced Polymer Hybrid Composite

Arvind Kumar Thakur[1], Puneet Kumar[2],J. Srinivas[3]
[1] Master student, [2] Research scholar, [3]Associate Professor
Department of Mechanical Engineering, National Institute of Technology Rourkela, Rourkela-769008, Odisha, India

*E-mail: srin07@yahoo.co.in, Phone: +91661-2462503

Abstract. This paper presents a computational approach to predict elastic propertiesof hybrid nano-composite material prepared by adding nano-clayplatelets to conventional CNT-reinforced epoxy system. In comparison to polymers alone/single-fiber reinforced polymers, if an additional fiber is added to the composite structure, it was found a drastic improvement in resultant properties. In this regard, effective elastic moduli of a hybrid nano composite are determined by using finite element (FE) model with square representative volume element (RVE). Continuum mechanics based homogenization of the nano-filler reinforced composite is considered for evaluating the volumetric average of the stresses and the strains under different periodic boundary conditions.A three phase Halpin-Tsai approach is selected to obtain the analytical result based on micromechanical modeling. The effect of the volume fractions of CNTs and nano-clay platelets on the mechanical behavior is studied. Two different RVEs of nano-clay platelets were used to investigate the influence of nano-filler geometry on composite properties. The combination of high aspect ratio of CNTs and larger surface area of clay platelets contribute to the stiffening effect of the hybrid samples. Results of analysis are validated with Halpin-Tsai empirical formulae.

Keywords: Nano-clay, CNT, Hybrid composite, FEM, Halpin-Tsai, RVE.

1. Introduction

Polymers with suitable and proper nano filler have emerged as an advanced polymeric composites system which shows prospective applications in the field of automotive, aerospace, packaging and construction industries [1]. Compared to conventional and micro fillers, carbon nanotubes (CNTs) reinforced polymers have gained increased interest in both industrial and academic fields, specifically utilizing their high aspect ratio and tremendous mechanical strength of CNTs [2-4]. But practical point of view CNT reinforced polymer composite materials have limited applications because of poor load transfer through interfacial region and agglomerations of CNTs [5-6]. Surface treatment and chemical functionalization are few effective techniques to enhance the interfacial adhesion and dispersion ability of CNTs [7-8]. Similarly, polymer materials reinforced with Nano fillers such as layered silicates (clay) [9] are prominent substitute for conventional composite materials. Intuitively, polymer/clay Nano composite provide best performance with high degree of clay exfoliation. With the time, rapid advancement have taken place in material science and technology and developed new generation multi-scale, multifunctional nanostructured hybrid polymer composite reinforced by CNTs and other nano reinforcements. Therefore, it becomes necessary to investigate the reinforcing mechanism of hybrid polymer composite which provides basic foundation for modeling and design of these high performances composite. Clay-CNT/polymer matrix hybrid composite [10-14] is one of the most widely used materials due to their light weight, long durability, high strength, chemical resistant, and so forth. Great deals of analytical and numerical works have been carried out to investigate the mechanical properties of CNTs reinforced polymer [15-16] and clay reinforced polymer [17] composite separately. Among the analytical techniques employed to obtain effective properties of composite, most of the researchers used the rule of mixture and Halpin-Tsai equations to modeling nanocomposite. However, few modifications have been employed to traditional theories to account the irregular geometry and distribution of nanostructured reinforcement at nano-scale for hybrid composite. Till to date no work found in literature related to modeling of clay-CNTs/polymer hybrid composite, however few articles are investigated numerical analysis of silica nanoparticle-MWCNT hybrid composite [18]. Jia eta al. [19] prepared a novel nanostructure hybrid (SiO2-MWCNTs) polymer composite, in which nano silica particle are grown over CNTs. Rahmanian et al.[20] demonstrate the FE modeling of carbon nanotube silica reinforced epoxy composite considering that CNTs are grown over micro silica particle. In most of work related to hybrid composite finite element method is an efficient and preferable numerical approach for investigating the reinforcing effect of different nano fillers on polymer composite. This method demonstrates the continuum representation of nano filler structure and polymer matrix.

Since CNTs/nano-clay hybrid showed prominent reinforcement on polymer matrix than pure CNTs as reported in previous work, there is a great potential in preparing new generation high performance composite. In present work, finite element method is used to predict the elastic behavior of epoxy composite reinforce by CNTs and nano-clay platelets. Specifically, to accurately describe geometry and reinforcing mechanism of CNTs and nano-clay platelets, the traditional theories of the rule of mixtures and Halpin–Tsai equations were properly corrected and a finite element approach based on nano-scale representative volume element (RVE) model was built up. To better understand the stiffening effect of CNTs and nano-clay nanoparticles in hybrid polymer composite, the tensile modulus for CNT/nano-clay/epoxy composites predicted by three phase Halpin–Tsai equations, were compared with FEM outcomes.

2. Modeling of hybrid composite

Modelling of CNT and nano clay reinforced hybrid composite is presented in two steps. In first part illustrate the micromechanical modelling and second one numerical modelling (FEM).

2.1.Micromechanical modeling

To predict the mechanical properties of three phase polymer composite, Halpin-Tsai model and theory of micromechanics were applied in hierarchy. The rule of mixture are used to obtained the effective elastic modulus of effective clay particle and effective CNT fiber (as shown in figure 1) which include clay, CNT and interphase as follow,

$$E_{CP} = E_{clay}\alpha + E_{\text{int}}(1-\alpha) \quad (1)$$

$$E_{CF} = E_{cnt}\beta + E_{\text{int}}(1-\beta) \quad (2)$$

Where E_{CP} and E_{CF} are the effective elastic modulus of clay particle and CNT fiber. E_{clay}, E_{cnt}, α, β are the elastic modulus and volume fractions of clay and CNT respectively.

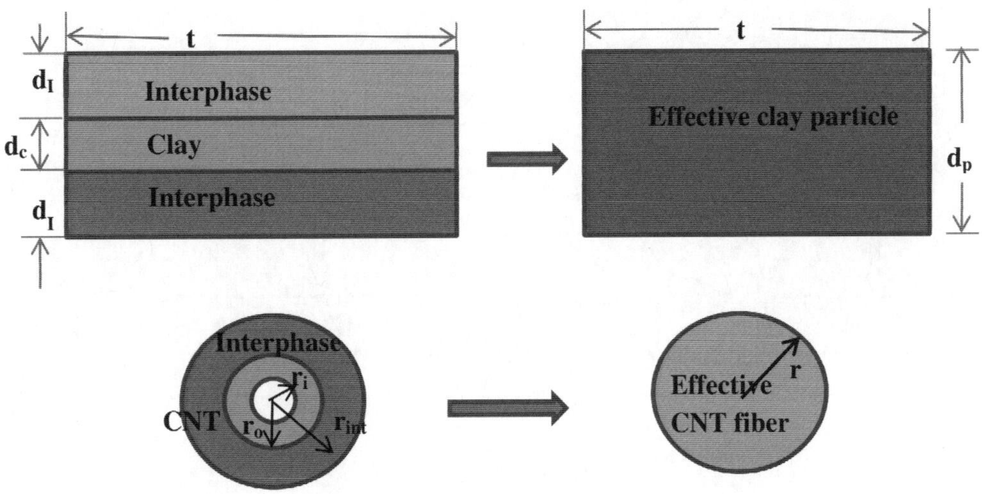

Fig. 1:- Representation of effective Clay particle and effective CNT fiber

To evaluate the elastic modulus of hybrid composite by combining effective particle and polymer (epoxy) by using modified Halpin-Tsai model as follow.

$$\frac{P}{P_m} = \frac{1+(\varepsilon_1 \eta_{CP} V_{CP} + \varepsilon_2 \eta_{CF} V_{CF})}{1-(\eta_{CP} V_{CP} + \eta_{CF} V_{CF})} \quad (3)$$

Where, P is the elastic property of hybrid composite and P_m is elastic properties of polymer matrix.

$$\eta_{CP} = \frac{(P_{CP}/P_m)-1}{(P_{CP}/P_m)+\varepsilon_1} \quad (4)$$

$$\eta_{CF} = \frac{(P_{CF}/P_m)-1}{(P_{CF}/P_m)+\varepsilon_2} \quad (5)$$

$$V_m + V_{CP} + V_{CF} = 1 \quad (6)$$

Where V_m, V_{CP}, and V_{CF} are the volume fraction of polymer matrix, clay particle and CNT fiber respectively. ε_1, ε_2 are the reinforcement efficiency parameters for clay particle and CNT fiber. For, longitudinal elastic modulus $\varepsilon_1 = (2l/d_p)$, $\varepsilon_2 = (2l/d)$ and for transverse elastic modulus $\varepsilon_1 = \sqrt{3}\log(l/d_p)$, $\varepsilon_2 = 1$. Where $d_p = d_c + 2*d_I$, $d = 2r$ are total thickness of effective clay particle and diameter of effective CNT fiber respectively. Figure 2 demonstrate the overall modeling procedure of CNT/nano-clay reinforced hybrid composite using micromechanical modeling.

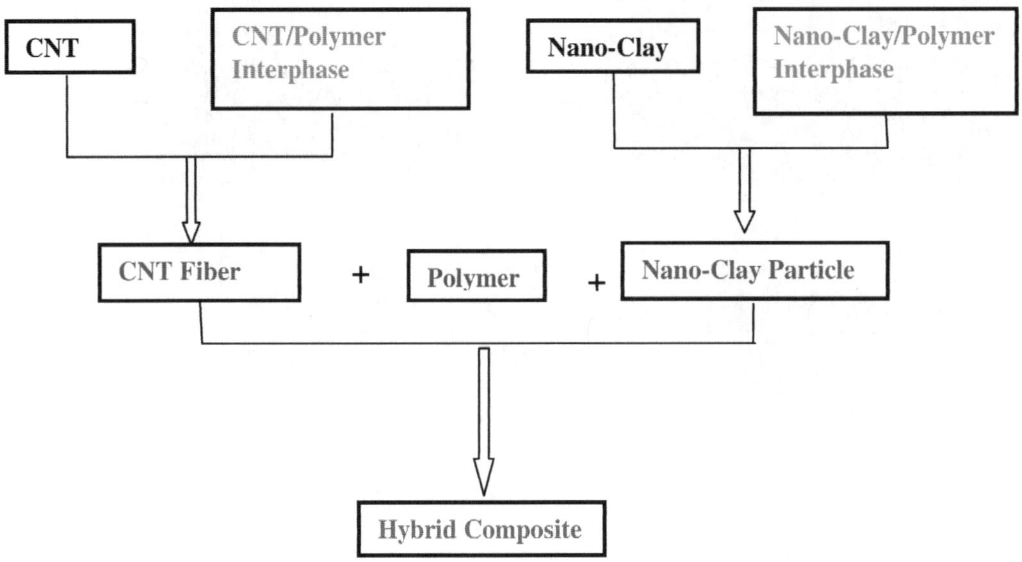

Fig. 2:- Modeling procedure of CNT/nano-clay reinforced hybrid composite

2.2. Constitutive relations and RVE model

For a transversely isotropic composite, the material behavior is based on only five independent constants. This concept is particularly ensured for regular fiber arrangement. In this work arbitrary fiber distributions are considered which results in transversely isotropic properties. By considering effective stiffness coefficient and average stress-strain value, the constitutive equations for the homogenized composite can be expressed as

$$\begin{Bmatrix} \overline{\sigma_1} \\ \overline{\sigma_2} \\ \overline{\sigma_3} \\ \overline{\sigma_4} \\ \overline{\sigma_5} \\ \overline{\sigma_6} \end{Bmatrix} = \begin{bmatrix} C_{11} & C_{12} & C_{12} & 0 & 0 & 0 \\ C_{12} & C_{22} & C_{23} & 0 & 0 & 0 \\ C_{12} & C_{23} & C_{22} & 0 & 0 & 0 \\ 0 & 0 & 0 & \frac{1}{2}(C_{22} - C_{23}) & 0 & 0 \\ 0 & 0 & 0 & 0 & C_{66} & 0 \\ 0 & 0 & 0 & 0 & 0 & C_{66} \end{bmatrix} \begin{Bmatrix} \overline{\varepsilon_1} \\ \overline{\varepsilon_2} \\ \overline{\varepsilon_3} \\ \overline{\gamma_4} \\ \overline{\gamma_5} \\ \overline{\gamma_6} \end{Bmatrix} \quad (7)$$

A representative volume element model can be used for the calculation of effective elastic coefficients by applying the appropriate periodic boundary conditions under the assumption of periodicity of fiber arrangement. Figure 3 shows the RVE model for CNT/nano-clay reinforced polymer composite.

Fig. 3:- RVE of CNT/nano-clay reinforced polymer

The main advantage of the method is to replace the original composite with globally homogenized equivalent medium with same strain energy stored. To find out effective elastic coefficients such load cases with different boundary conditions must be applied that for a particular load case only one value in the strain field vector is non-zero and all other becomes zero. Then from corresponding column, the effective coefficients can be determined using calculated average stress value corresponding to unit strain.

2.3.Finite element modeling

A three dimensional multi-field elements is used for finite element calculations using FE package ANSYS. ANSYS Parametric Design Language (APDL) coding is used for modeling and applying the constraint equations. An algorithm was written in APDL for automated generation of RVE with aligns CNT and clay inside the polymer matrix. First 2-D model is generated in X_2-X_3 plane with circle as CNT and rectangle as clay and meshed with PLANE82 element. Further FE mesh is extruded in X_1 direction with SOLID185, 3-D element for meshing of RVE. Fibers surfaces are connected by Boolean operation with the aim of predicting the modulus of resin matrix reinforced with periodic nano-reinforcement of clay-CNTs. An interphase region is also considered here between CNT/polymer and clay/polymer to demonstrate the imperfect load transfer phenomenon at interfaces. The interphase region is meshed with random material properties lies between fiber and polymer matric properties. The generation of RVE can be controlled by some input parameters like size of RVE, CNT and nano-clay diameters for desired volumefraction. A certain minimum distance must be ensuring to generate suitable meshing of each part of model. From the numerical analysis, the effective elastic parameters of compositional material were estimated by relating boundary conditions.The calculation of effective coefficients in order to evaluate the overall stiffness matrix [C] of hybrid composite, RVE is subjected to an average strain. The six components of strain are applied by imposing the following boundary conditions on the displacement components.

$$u_i(a_1,x_2,x_3)-u_i(-a_1,x_2,x_3)=2a_1\varepsilon_{i1} \quad \begin{array}{l}-a_2\le x_2\le a_2\\ -a_3\le x_3\le a_3\end{array}$$

$$u_i(x_1,a_2,x_3)-u_i(x_1,-a_2,x_3)=2a_2\varepsilon_{i2} \quad \begin{array}{l}-a_1\le x_1\le a_1\\ -a_3\le x_3\le a_3\end{array}$$

$$u_i(x_1, x_2, a_3) - u_i(x_1, x_2, -a_3) = 2a_3\varepsilon_{i3} \quad \begin{matrix} -a_1 \le x_1 \le a_1 \\ -a_2 \le x_2 \le a_2 \end{matrix} \tag{8}$$

Here $2a_j\varepsilon_{ij}$ represents the applied displacement necessary to enforce a strain ε_{ij} over a distance $2a_j$. The strain applied on boundary results in complex state of strain inside the RVE. So, volume average strain in the RVE equals to the applied strain, i.e.

$$\overline{\varepsilon_{ij}} = \frac{1}{V}\int_V \varepsilon_{ij} dV = \varepsilon_{ij} \tag{9}$$

Considering hybrid composite as homogeneous material, the average stress-strain relationship can be written as

$$\overline{\sigma_\alpha} = C_{\alpha\beta}\overline{\varepsilon_\beta} \tag{10}$$

By choosing a unit applied strain in one direction and imposing periodic boundary conditions on other directions, stress field can be computed. Whose average value over the volume gives the required components of stiffness matrix, one column at a time;

$$C_{\alpha\beta} = \overline{\sigma_\alpha} = \frac{1}{V}\int_V \sigma_\alpha(x_1, x_2, x_3)dV \tag{11}$$

Where α, β are varying 1 to 3. Gauss-Legendre quadrature can be used for evaluating the volume integrals of a finite element. Commercial ANSYS have such type of capability to evaluate the average stress over the volume, element by element. Three loading cases are imposed to evaluate the all elastic coefficient of [C] for transversely isotropic composite. Figure 4 shows the meshing of RVEs used in finite element analysis.

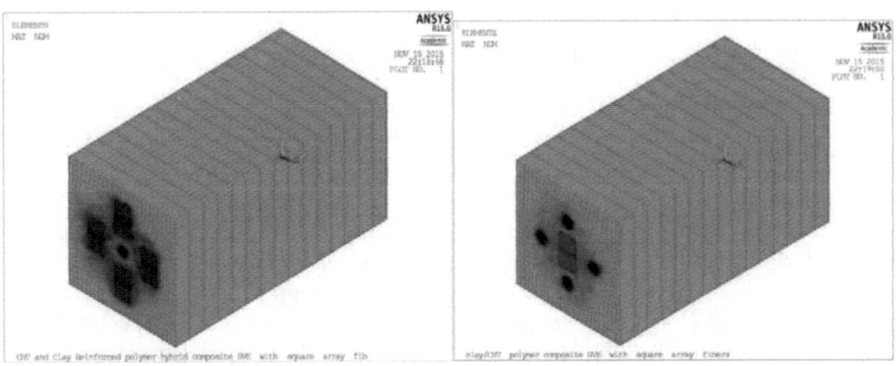

(a) (b)

Fig. 4:- Meshing of (a) 1 CNT and 4 clay based RVE (b) 1 clay and 4 CNT based RVE

3. Results and discussion

By considering interphase parameters, elastic behavior of nanocomposite reinforced by CNTs and nano-clay plateletsis first analyzed. The advantages of nano-clay in CNT- reinforced polymer composite can be explained with reduction of interfacial reinforcing area of CNTs and enhancing the interfacial adhesion and dispersion of CNTs in polymer. In Fig.5 stress and strain distributionswere shown to explain the elastic behavior of CNT/nano-clay reinforced hybrid composite.

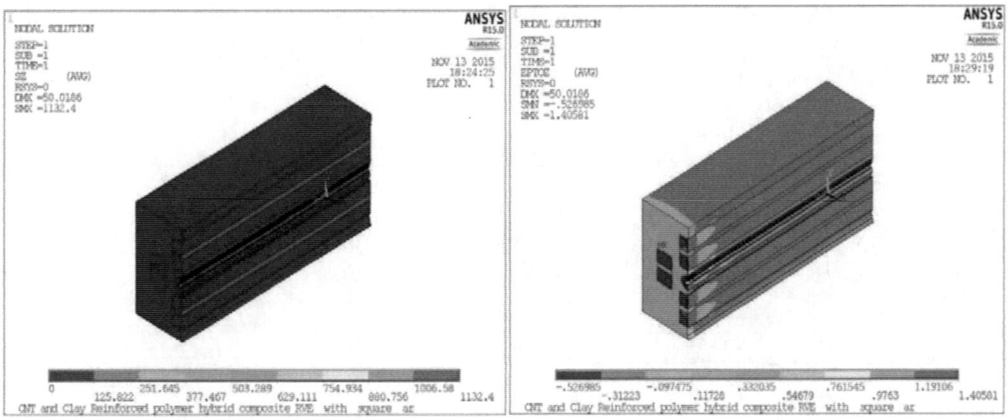

(a) (b)

Fig. 5:- (a) Stress and (b) strain distribution across cross-section of a RVE

The elastic and geometric property of each phase of RVE is listed in table 1. For simplification of calculation RVE length is considered as 50 nm.

Table 1 Material property of all phases in RVE

Material	Elastic Modulus (GPa)	Poisson's ratio (μ)	Geometry
Epoxy	2.026	0.4	25x25 nm
CNT	1054	0.25	r_i=0.315 nm, r_o=0.650
Clay (MMT)	178	0.28	t=4 nm, d_c=1 nm
CNT/polymer interphase	16.10	0.4	r_{int}=1.404 nm
Clay/polymer interphase	16.10	0.4	d_I=3 nm

A case study has been illustrated to reveal the importance of nano-clay reinforcement in CNT reinforced composite. Table 2 shows the elastic properties of CNT/clay hybrid composite considering four types RVE. The finite element modeling results are validated with micromechanical modeling results.

Table 2 Elastic properties of hybrid composite with variation of number of clay platelet

RVE with	E_L	E_T	G_T	υ_L	υ_T
1 CNT	3.667	2.2119	0.738	0.395	0.5054
1 CNT +1 Clay	5.3245	2.420	0.7868	0.3938	0.5382
1 CNT + 2 Clay	7.0019	2.626	0.8468	0.3938	0.5508
1 CNT + 3 Clay	8.6448	2.8338	0.9092	0.3928	0.5582
1 CNT +4 Clay	10.2846	3.0908	0.9929	0.3917	0.5564
Three phase Halpin-Tsai	9.8932	2.9931	0.9865	0.3899	0.5423

In Table 3, elastic moduli are investigated considering constant volume fraction of CNTs inside hybrid composite. It can be concluded from predicted results that four CNTs with one clay based RVE gives better results as compared to one CNT with four-clay RVE.

Table 3 Elastic properties of hybrid composite with variation of Number of CNTs

RVE with	E_L	E_T	G_T	υ_L	υ_T
1 Clay	3.6576	2.3079	0.7719	0.3952	0.4949
1 Clay +1 CNT	5.3236	2.4137	0.7838	0.3948	0.5397
1 Clay + 2 CNT	6.9921	2.4899	0.7963	0.3944	0.5634
1 Clay + 3 CNT	8.6552	2.5700	0.8164	0.3940	0.5753
1 Clay +4 CNT	10.3201	2.6521	0.8381	0.3935	0.5820

4. Conclusions

A nano structured hybrid composite composed of clay nano particle and carbon nanotubes were studied using finite element modeling. For numerical investigations through finite element approach, a square RVE was selected to demonstrate the hybrid composite structure. Predicted values from FE modeling were compared with three-phase Halpin-Tsai model. It was found that the elastic modulus from numerical modelling is in close agreement with micromechanics based results. This work provides basic overview of modeling of hybrid composite; furthermore studies have to be done by considering the variation of interphase properties and random distribution of nano-clay and CNTs. It is also planned to prepare CNT/nano clay polymer composite specimen and perform tensile and compressive test to validate the predicted numerical data.

References

[1]. K. Lau, C. Gu, D. Hui, A critical review on nanotube and nanotube/nanoclay related polymer composite materials, *Composites: Part B* 37 (2006) 425–436.

[2]. T. Hayashi, M.Endo, Carbon nanotubes as structural material and their application in composites, *Composites: Part B* 42(2011) 2151–2157.

[3]. M.T. Kim, K.Y. Rhee, J.H. Lee, D. Hui, A. K.T. Lau, Property enhancement of a carbon fiber/epoxy composite by using carbon nanotubes, *Composites: Part B*42 (2011) 1257–1261.

[4]. J. Srinivas P. Kumar,Analysis of elastic properties of transversely isotropic CNT-reinforced polymers, *Nano Science and Nano Technology*8 (2014) 291-297.

[5]. M.S.P. Shaffer, X. Fan, A.H. Windle, Dispersion and packing of carbon nanotubes, *Carbon* 36 (1998) 1603–1612.

[6]. S. Wang, R.Liang, B.Wang, C. Zhang, Load-transfer in functionalized carbon nanotubes/polymer composites, *Chemical Physics Letters* 457 (2008) 371–375.

[7]. S.W. Kim, T. Kim, Y.S. Kim, H.S.Choi, H.J.Lim, S.J.Yang, C.R.Park, Surface modifications for the effective dispersion of carbon nanotubes in solvents and polymers,*Carbon* 50 (2012) 3–33.

[8]. P.C. Ma, J.K.Kim , B.Z. Tang, Effects of silane functionalization on the properties of carbon nanotube/epoxy nanocomposites, *Composite Science and Technology*67 (2007) 2965–2972.

[9]. S. P. Pereira, G. Scocchi, R. Toth, P. Posocco, D. R nieto, S. Pricl, M. Fermeglia, Multiscale modeling of polymer/clay nanocomposites, *Journal of Multiscale Modelling*3(2011) 151-176.

[10]. S. Wang, R.Liang, B.Wang, C. Zhang, Reinforcing polymer composites with epoxide-grafted carbon nanotubes, *Nanotechnology*19 (2008) 085710.

[11]. V. Levchenko, Y. Mamunya, G. Boiteux, M. Lebovka, P. Alcouffe, G. Seytre, E. Lebedev, Influence of organo-clay on electrical and mechanical properties of PP/MWCNT/OC nanocomposites, *European Polymer Journal* 47 (2011) 1351–1360.

[12]. L. Liu and J. C. Grunlan, Clay Assisted Dispersion of Carbon Nanotubes in Conductive Epoxy Nanocomposites, *Advanced Functional Materials* 17 (2007)2343–2348.

[13]. B. L. Silva, F. C. Nack, C. M. Lepienski, L. A. F. Coelho, D. Becker, Influence of intercalation methods in properties of clay and carbon nanotube and high density polyethylene nanocomposites, *Materials Research* 17 (2014) 1628-1636.

[14]. Z. Wang, C. Xu, Y. Zhao, D. Zhao, Z. Wang, H. Li, K. Lau, Fabrication and mechanical properties of exfoliated clay–CNTs/epoxy nanocomposites, *Materials Science and Engineering A* 490 (2008) 481–487.

[15]. Y. J. Liu and X. L. Chen, Evaluations of the effective material properties of carbon nanotube-based composites using a nanoscale representative volume element, *Mechanics of Materials* 35 (2003) 69–81.

[16]. P. Kumar and J. Srinivas, Numerical evaluation of effective elastic properties of CNT-reinforced polymers for interphase effects, *Computational Material Science*88(2014) 139–144.

[17]. N. Sheng, M.C. Boyce, D.M. Parks, G.C. Rutledge, J.I. Abes, R.E. Cohen, Multiscale micromechanical modeling of polymer/clay nanocomposites and the effective clay particle,*Polymer*45 (2004) 487–506.

[18]. L. M. Jr., G. Dai,Hybrid and hierarchical nanoreinforced polymer composites: Computational modelling of structure–properties relationships, Composite *Structures* 117 (2014) 156–168.

[19]. X. Jia, B. Liu, L. Huang, D. Hui, X. Yang, Numerical analysis of synergistic reinforcing effect of silica nanoparticle–MWCNT hybrid on epoxy-based composites, *Composites Part B: Engineering* 54 (2013) 133–137.

[20]. S. Rahmanian, A.R. Suraya, B. Roshanravan, R.N. Othman, A.H. Nasser, R. Zahari, E.S. Zainudin, The influence of multiscale fillers on the rheological and mechanical properties of carbon-nanotube–silica-reinforced epoxy composite, *Materials and Design*88 (2015) 227–235.

Effect of Nanosize Yittria and Tungsten Addition to Duplex Stainless Steel During High Energy Planetary Milling

A K Nayak, R Shashanka, D Chaira*

Department of Metallurgical and Materials Engineering, National Institute of Technology Rourkela, Rourkela-769008, Odisha, India

*E-mail: chaira.debasis@gmail.com, Phone: +91(661)2462561

Abstract. In this present investigation, elemental powders of duplex stainless steel composition (Fe–18Cr–13Ni) with 1 wt. % nano yittria and tungsten were milled separately in dual drive planetary mill (DDPM) for 10 h to fabricate yittria dispersed and tungsten dispersed duplex stainless steel powders. The milled powder samples were characterized by X-Ray diffraction and scanning electron microscopy (SEM) to study the size, morphology and phase evolution during milling. The gradual transformation from ferrite to austenite is evident from XRD spectra during milling. The crystallite size and lattice strain of yittria dispersed duplex stainless steel after 10 h milling were found to be 7 nm and 1.1% respectively. The crystallite size of tungsten dispersed duplex stainless steel was 5 nm. It has been observed from SEM analysis that particles size has been reduced from 40 to 5 μm in both cases. Annealing of 10 h milled powder was performed at 750°C for 1 h under argon atmosphere to study phase transformation in both yittria and tungsten dispersed duplex stainless steel. The XRD analysis of annealed stainless steel depicts the phase transformation from α-Fe to γ-Fe with the formation of oxides of Y,Fe and Cr. The differential scanning calorimetry analysis was conducted by heating the milled powder from room temperature to 1200°C under argon atmosphere to investigate the thermal analysis of both the stainless steel powders.

Keywords: Mechanical alloying; DDPM; Annealing; Phase transformation; DSC

1. Introduction

Demands of stainless steel has been increased in various fields because of its properties like high corrosion resistance, good toughness, low thermal expansion, high energy absorption, weld-ability and high strength [1]. Based upon their crystalline structure and properties, stainless steels can be divided as austenitic stainless steel, ferritic stainless steel, martensitic stainless steel and duplex stainless steel.
Duplex stainless steel contains almost equal proportions of ferrite and austenite phases in its structure which has made it to have very good toughness, high corrosion resistance, low thermal expansion, high energy absorption, weld-ability and high strength compared to single phase austenitic and ferritic stainless steel, hence used in chemical, oil, petrochemical, marine, nuclear power, paper and pulp industries [1-4].

The structural components in super thermal plants and nuclear reactors are subjected to high operating temperatures. Oxide dispersion strengthened (ODS) steels are considered most suitable materials because of their ability to retain mechanical strength and resist oxidation/corrosion under extreme conditions of temperature and pressure [5]. Rare earth oxides, particularly yittria is known to increase the creep resistance and oxidation resistance. The properties of the ODS steel is dependent upon the size and distribution of the oxide phase in a given proportion.

Phaniraj et al. prepared mechanically alloyed austenitic steels (Fe–20Ni–14Cr–2.5Mo–2Mn–2.5Al-wt. %) dispersed with 0, 0.5 and 5 wt. % yittria and tested their cyclic oxidation behavior at 800°C in air. They found that addition of yittria improves the resistance to scale spallation but higher yittria content does not affect the oxidation rate of the alloy significantly [6]. Xu et al. synthesized ODS 304 austenitic steel (0.35 wt. % Y_2O_3) by mechanical milling and consolidated by hot isostatic pressing for 3 h at 1423 K under a pressure of 200 MPa. They concluded that the tensile and yield strength of ODS alloy were greatly improved compared with 304 austenitic steel, but the ductility reduced due to coarse oxide particles both at room temperature and high temperatures [7]. Torralba et al. also investigated the mechanical response of ODS ferritic steels (pre-alloyed grade of Fe–20Cr–5Al and Fe–14Cr–5Al–3W with the addition of Ti and Y_2O_3 as reinforcements) by mechanical alloying followed by spark plasma sintering at 1200°C and 64 MPa. They reported enhanced tensile properties of ODS alloys as compared to pre-alloyed alloys [8]. Olaniran et al. fabricated oxide dispersed duplex stainless steel (2205) dispersed with partially stabilized Zirconia (3 mol. % yittria) by hot press sintering machine for 30min at 30MPa at the temperature 1100°C in argon gas environment and demonstrated that addition of chromium and nickel can enhance densification and corrosion property of duplex stainless steel composites [15].

On the other hand, tungsten is used for preparation of super duplex stainless steel which exhibits better pitting corrosion resistance and mechanical properties. Tungsten is an effective element for solid solution strengthening for providing the required mechanical properties and for enhancing the stability of the ferrite phase. However, the addition of tungsten can result in a loss of ductility [8-10]. Xu et al. added tungsten in the ODS steel to stabilize the tensile property at high temperatures [7].

In the present work, an effort has been made to synthesize yittria dispersed and tungsten dispersed (both 1 wt. %) duplex stainless steel (Fe-18Cr-13Ni) through mechanical alloying of elemental powder particles in dual drive planetary mill (DDPM).

The dual drive planetary mill (DDPM) is a specially designed planetary mill meant for the preparation of nano-structured powders in bulk amount. Shashanka et al. showed that for synthesis of nano-structured stainless steel powders a specially designed DDPM required only 10 h as compared to 40 h milling in Fritsch pulverisette planetary mill because of its high acceleration field [12]. The DDPM consists of a gyratory shaft and cylindrical jars. Both shaft and jars rotate simultaneously with different speeds and in opposite directions. The details of mill fabrication and mechanism can be found elsewhere [12].

2. Materials and Methods

Powder compositions of Fe–18Cr–13Ni (wt. %) was selected from Schaeffler diagram to prepare duplex stainless steel by high energy planetary milling. Elemental powders of Fe (99.5% pure), Cr (99.8% pure), Ni (99.5% pure) was milled with 1 wt. % Y_2O_3 (99.5% pure) and 1 wt. % W (99.7% pure) separately in dual drive planetary mill (DDPM) for 10 h to obtain yittria dispersed and tungsten dispersed duplex stainless steel powders. The milling media of the DDPM consists of 1 kg stainless steel balls of 8 mm diameter. The milling was conducted at room temperature with 6:1 ball-to-powder weight ratio in both jars under toluene atmosphere (wet milling) to prevent oxidation. In the DDPM, the angular velocity of the vials and the supporting main shaft were 620 and 275 rpm respectively. Milled powder samples were then characterized by X-ray diffraction (XRD) in a Philips PANalytical diffractometer using filtered CuKα-radiation (λ = 0.1542 nm) followed by crystallite size and lattice strain calculation using

Williamson–Hall method. Powder morphology was investigated by scanning electron microscopy (SEM) using JEOL JSM-6480LV. Annealing of yittria and tungsten dispersed duplex stainless steel powders was carried out at 750 °C for 1 h under argon atmosphere to study the phase transformation. Thermal behavior of final powder samples were studied using differential scanning calorimetry (DSC) Netzsch, Germany. During DSC study, the powder samples were heated from room temperature to 1200°C at a heating rate of 10 °C/min in argon atmosphere.

3. Results and discussion

3.1. XRD analysis of milled powders

Fig. 1 (a) and (b) presents XRD patterns of tungsten dispersed and yittria dispersed duplex stainless steel respectively milled for various times. The XRD patterns of starting powder show sharp crystalline diffraction peaks of elemental Fe, Cr and Ni, while peaks corresponding to W or Y_2O_3 are not detectable owing to their low concentration. It also can be noticed that the peaks corresponding to the starting materials tend to broaden or disappear by increasing milling time. Such behaviour is attributed to considerable increase in structural defects, increased lattice strain, refined grain size and migration of Cr and Ni into lattice of Fe to form solid solution during milling [11]. The collisions and friction during milling resulted in the increase of the local temperature which facilitated diffusion of alloying atom into the matrix [7]. Increase in milling time increases the phase transformation from α-Fe to γ-Fe and after 5h both austenite and ferrite peaks resolve in to (111) and (110) planes of Fe as shown in Fig. 1 (a) and (b). Both 10h milled samples show highly broadened diffraction peaks. It has been widely accepted that during mechanical alloying, the enthalpy of the components increases due to defects introduction and increased internal energy. This could act as the driving force for solid-state amorphization and phase transformation during MA [11].

Fig. 2 (a) and (b) shows the XRD spectra of 10 h milled samples of tungsten dispersed and yittria dispersed duplex stainless steel scanned slowly (1°/min) in the range of 40 to 55°. In this figure, tungsten dispersed duplex stainless steel showed distinct peaks of austenitic and ferrtic phase while intensity of austenitic one is more, which confirms transformation from ferritic to austenitic phase. But in case of Y_2O_3 dispersed duplex stainless steel, there is no distinct austenitic and ferritic peaks. Hence it may be concluded that tungsten helps in rapid transformation from ferrite to austenite than yittria. It is also found that limited transformation from ferrite to austenite takes place during milling as milling energy is not sufficient for transformation. However, annealing of milled powder accelerates the transformation [2,12, 13].

3.2. Crystallite size and lattice strain calculation

The broadening of X-Ray diffraction peaks occurs mainly due to instrumental errors, decrease in particle size and increase in lattice strain during milling. The crystallite size and lattice strain can be calculated using Williamson-Hall method by de-convoluting the size and strain by calculating XRD peak width as a function of 2θ [2,12]. The Williamson-Hall equation is given as follows:

$$\beta \cos\theta = \frac{0.94\lambda}{D} + 4\eta \sin\theta \quad \textbf{(1)}$$

Where, β is the full width half maxima (FWHM), D is the crystallite size and η is the lattice strain.

The crystallite size and strain of tungsten and yittria dispersed duplex stainless steels were calculated from Williamson–Hall method and represented graphically in Fig. 3 (a) and (b) respectively. From the graph it is confirmed that as milling time increases, the crystallite size decreases and it attains a saturation level after 10 h, where further refinement of crystallite size is quite difficult. But lattice strain increases continuously with milling due to continuous contact between powder–ball-jar surfaces. The crystallite size of tungsten dispersed and yittria dispersed duplex stainless steel decreases from 101 to 5 nm and from 99 to 7 nm respectively. Similarly, the lattice strain increases from 0.36 to 0.7 % for 5 h tungsten dispersed and from 0.5 to 1.1 % for 10 h yittria dispersed duplex stainless steel milled samples respectively.

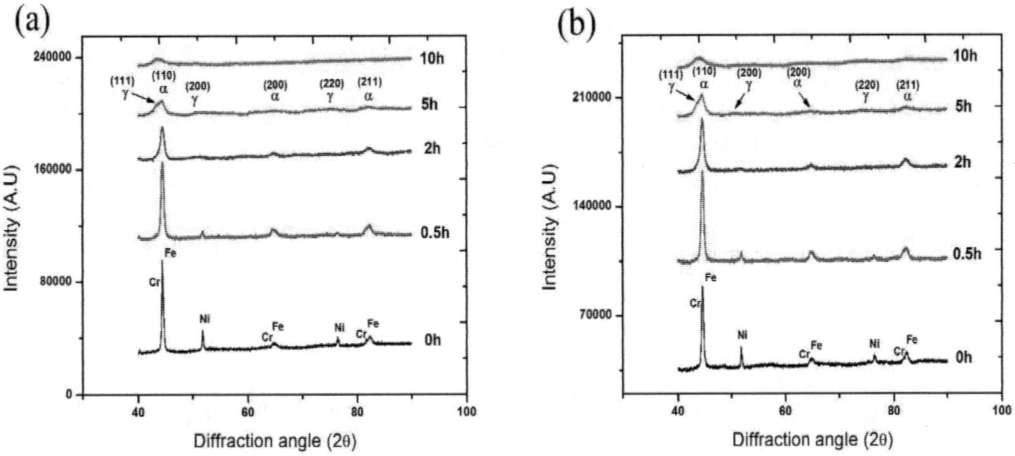

Fig. 1:- XRD spectra of (a) tungsten dispersed and (b) yittria dispersed duplex stainless steel from 0 to 10 h

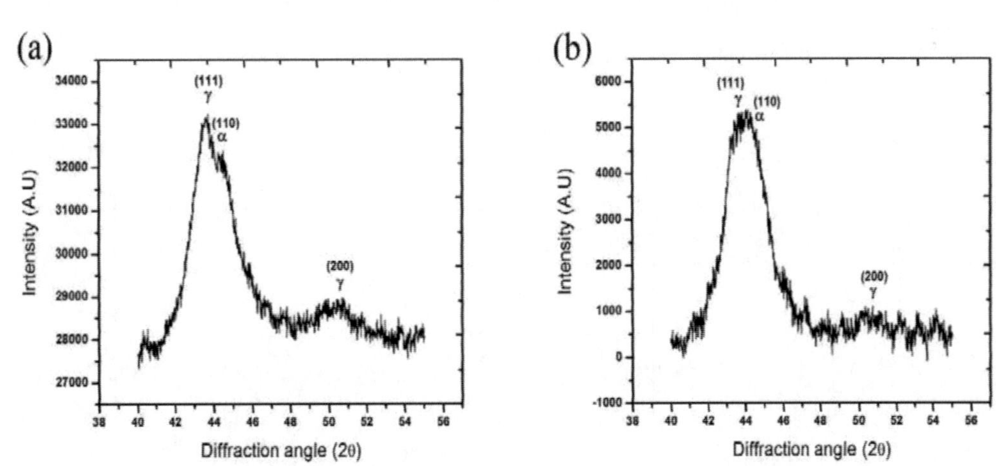

Fig. 2:- Slow scan range XRD spectra of (a) tungsten dispersed and (b) yittria dispersed duplex stainless steel after 10 h of milling

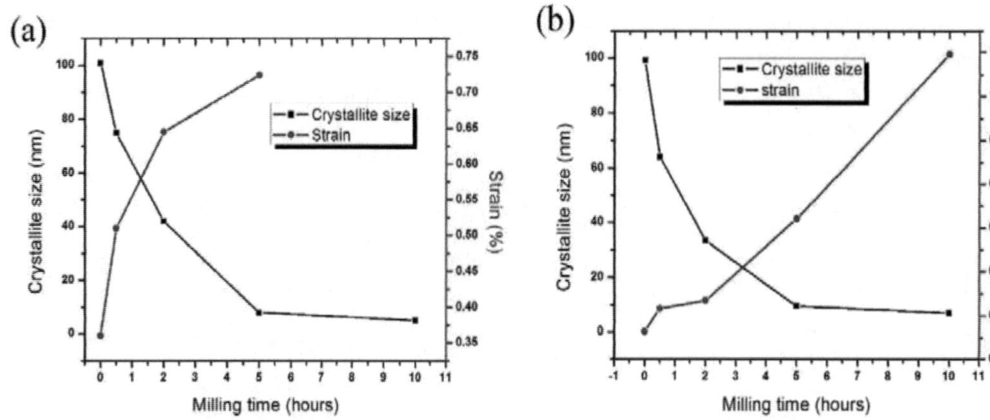

Fig. 3:- Variation of lattice strain and crystallite size with milling times of (a) tungsten dispersed and (b) yittria dispersed duplex stainless steel

3.3. XRD analysis of annealed sample in argon atmosphere

Fig. 4(a) and (b) shows the XRD traces of 10 h DDPM milled 1wt. % tungsten dispersed and 1wt. % yittria dispersed Fe–18Cr–13Ni powders after subsequent heat treatment at 750°C for 1 h in argon atmosphere followed by furnace cooling to room temperature. The XRD spectra after annealing in argon atmosphere exhibit the strong diffraction peaks of austenite (γ) and peaks of ferrite (α) and weak oxide peaks in both cases.However, annealed yittria dispersed sample showed more oxide peaks than tungsten dispersed due to steel containing yittria oxide.The presence of Y_2O_3produces higher amount of oxides containing Y, Fe and Cr as also referred by Phaniraj et al. [6].The intensity of austenitic peakis higher as compared to 10h milled powder as phase transformation is favoured during annealing due to supply of heat energy to the powders[12].

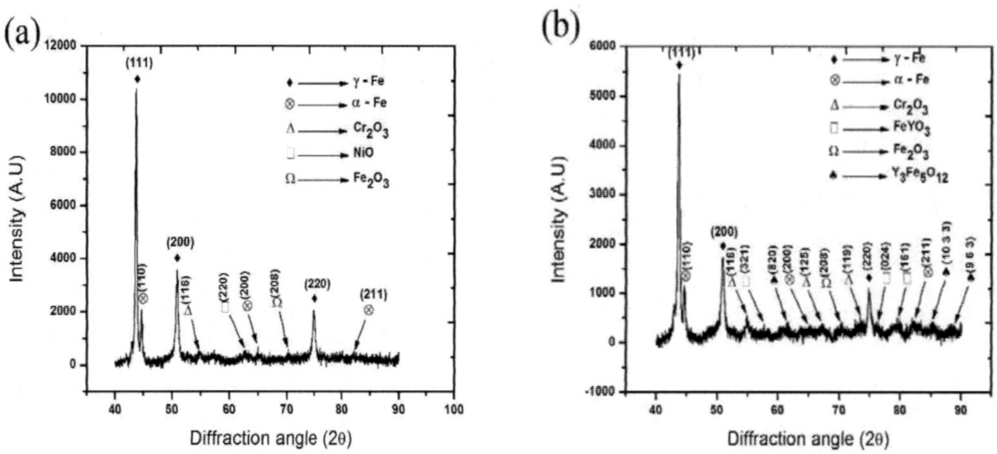

Fig. 4:- XRD traces of annealed sample of 10h milled (a) tungsten dispersed and (b) yittria dispersed duplex stainless steel heat treated in argon atmosphere at 750°C for 1h

3.4. Scanning electron microscopy

Fig. 5 (a) and (b) shows the SEM micrographs of 0 h and 10 h milled tungsten dispersed duplex stainless steel powders and Fig. 4 (c), (d) represents SEM micrographs of yittria dispersed duplex stainless steel milled for 0 h and 10 h respectively. From the figures, it is clear that before milling elemental powder particles were large and irregular in shape of size 40-50 μm, but after 10 h of milling particles become fine and spherical of size 5-10 μm. At the start of milling, the elemental powder particles become flat flakes due to ductile nature of Fe. As the milling proceeds, two or more flakes get cold welded together and this leads to increase in particle size. The composite particles at this stage have a characteristic layered structure consisting of various combinations of the starting constituents [14]. With further increase in milling time the particles get work hardened and fragmentation results into fine and spherical stainless steel powders.

Fig. 5:- SEM images of 1wt. % tungsten dispersed duplex stainless steel milled for (a) 0h (b) 10h and 1 wt. % yittria dispersed duplex stainless steel for (c) 0h and (d) 10h respectively

3.5. Thermal analysis

Fig. 6 (a) and (b) shows the DSC graphs of 10 h milled tungsten dispersed and yittria dispersed duplex stainless steels heated with the heating rate of 10°C/min respectively. It shows exothermic peaks with peak temperature of 755 and 486°C for tungsten dispersed and yittria dispersed duplex stainless steel respectively. The exothermic peaks represent crystal growth, lattice strain release and amorphous to crystalline phase transition during annealing [12]. In the DSC graphs, Tp stands for the peak temperature of the broad exothermic peak which is the crystallization temperature. Sherif et al. found a single exothermic peak corresponding to an amorphous-crystalline phase transformation with crystallization temperature 970°C with an enthalpy change of crystallization 16.88 kJ/mol during DSC analysis of mechanically alloyed Fe-18Cr-8Ni powders. They confirmed it with bright field imaging (BFI) which showed the formation of large polycrystalline grains with sharp grain boundaries [16]. The change in enthalpy is found to be 192.79 KJ/mol and 157.92 KJ/mol from DSC analysis of tungsten dispersed and yittria dispersed duplex stainless steel respectively. Shashanka et al. reported exothermic peak at 841°C with an enthalpy change of 62.35 KJ/mol by performing DSC analysis (heating rate 10°C/min) of partial amorphous powder of duplex stainless steel (Fe-18Cr-13Ni) prepared by milling in DDPM [2]. It also can be seen from DSC graph of yittria dispersed stainless steel that weak exothermic peaks are present at around 750°C. The appearance of these peaks may be due to the formation of oxides of Y, Fe and Cr which is evidenced by the XRD plots of annealed samples of 10 h milled yittria dispersed duplx stainless steel powders. Kimura et al. observed exothermic peaks at 927°C during DSC analysis of mechanically milled Fe-24Cr-15Y_2O_3 powder mixture due to the formation of $YCrO_3$ and Y_2O_3 [17]. Lei Dai et al. performed DSC analysis of 100 h mechanically milled Fe-9Cr-15Y_2O_3 powder mixture and showed exothermic peak formed at 620°C due to growth of Y_2O_3 nano-crystals [18].

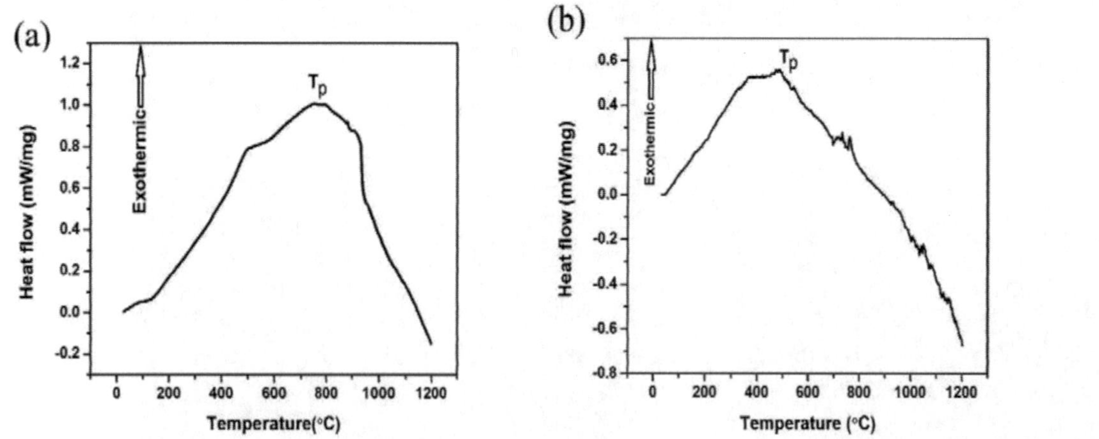

Fig. 6:- DSC graphs of (a) tungsten dispersed and (b) yittria dispersed 10 h milled duplex stainless steel after heating to 1200°C under argon gas

IOP Conf. Series: Materials Science and Engineering **115** (2016) 012008 doi:10.1088/1757-899X/115/1/012008

4. Conclusions

From the present study the following conclusions can be drawn:

1. The irregular and large elemental powders reduced to fine and spherical particles of size 5-10 μm after 10h of milling in both tungsten and yittria dispersed duplex stainless steels.
2. The XRD spectra of 10h milled samples showed highly broadened peaks due to high lattice strain and reduced particle size during milling. The lattice strains were 0.7 % upto 5h milling for tungsten dispersed and 1.1 % upto 10h milling for yittria dispersed duplex stainless steels. The crystallite sizes of 10h milled samples were 5 nm and 7nm for tungsten dispersed and yittria dispersed duplex stainless steels respectively.
3. Annealing in argon atmosphere enhanced ferrite to austenite transformation almost equally in both cases.
4. DSC analysis showed broadened exothermic peaks caused by reduction of lattice strain, crystal growth and crystallization of amorphous milled powder during heating.

References

[1] R. Shashanka, D. Chaira, Optimization of milling parameters for the synthesis of nano-structured duplex and ferritic stainless steel powders by high energy planetary milling, *Powder Technology* 278 (2015) 35–45.

[2] R. Shashanka, D. Chaira, Development of nano-structured duplex and ferritic stainless steels by pulverisette planetary milling followed by pressureless sintering, *Materials Characterization* 99 (2015) 220–229.

[3] L.A. Dobrza´nski, Z. Brytan, M.A. Grande, M. Rosso, E.J. Pallavicini, Properties of vacuum sintered duplex stainless steels, *Journal of Materials Processing Technology* 162–163 (2005) 286–292.

[4] R. Shashanka, D. Chaira, B.E. Kumara Swamy, Electrocatalytic Response of Duplex and Yittria DispersedDuplex Stainless Steel Modified Carbon Paste Electrode in Detecting Folic Acid Using Cyclic Voltammetry, *Int. J. Electrochem. Sci.,* 10 (2015) 5586 – 5598.

[5] S.K. Karak, J.DuttaMajumdar, Z.Witczak, W.Lojkowski, I.Manna, Microstructure and mechanical properties of nano-Y_2O_3 dispersed ferritic alloys synthesized by mechanical alloying and consolidated by hydrostatic extrusion, *Materials Science & Engineering* A580 (2013) 231–241.

[6] M.P. Phaniraj, Dong-Ik Kim, Jae-Hyeok Shim, Young Whan Cho, Cyclic oxidation of yttria dispersed austenitic stainless steels, *Corrosion Science* 52 (2010) 3573–3576.

[7] Yingli Xu, Zhangjian Zhou, Ming Li, Pei He, Fabrication and characterization of ODS austenitic steels, *Journal of Nuclear Materials* 417 (2011) 283–285.

[8] José M. Torralba, Luz Fuentes-Pacheco, Nerea García-Rodríguez, Mónica Campos, Development of high performance powder metallurgy steels by high-energy milling, *Advanced Powder Technology* 24 (2013) 813–817.

[9] T. Narita, S. Uka, S. Ohtsuka, M. Inoue, Effect of tungsten addition on microstructure and high temperature strength of 9Cr ODS ferritic steel, *Journal of Nuclear Materials* 417 (2011) 158–161.

[10] F. Abe, H. Araki, T. Noda, M. Okada, Microstructure and toughness of Cr-W and Cr-V ferritic steels, *Journal of Nuclear Materials* 155–157 (1988) 656–661.

[11] T. Haghir, M.H. Abbasi, M.A. Golozar, M. Panjepour, Investigation of α to γ transformation in the production of a nanostructured high-nitrogen austenitic stainless steel powder via mechanical alloying, *Materials Science and Engineering A* 507 (2009) 144–148.

[12] R. Shashanka, D. Chaira, Phase transformation and microstructure study of nanostructured austenitic and ferritic stainless steel powder prepared by planetary milling, *Powder Technology* 259 (2014) 125–136.

[13] M.H. Enayati, M.R. Bafandeh, Phase transitions in nano-structured Fe–Cr–Ni alloys prepared by mechanical alloying, *Journal of Alloy and Compounds* 454 (2008) 228–232.

[14] C. Suryanarayana, Mechanical alloying and milling, *Prog. Mater Sci.* 46 (2001) 1 – 184.

[15] O.Olaniran, B. O.Adewuyi, J. A. Omotoyinbo, A. E. Afolabi, D. Folorunso, A. Adegbola, E. Igbafen, Development of oxide dispersion strengthened 2205 duplex stainless steelcomposite, *Leonardo Electronic Journal of Practices and Technologies* ISSN 1583-1078 (2015) 129-140.

[16] M. Sherif El-Eskandarany, H.A. Ahmed, Morphological and structural studies of amorphous $Fe_{74}Cr_{18}Ni_8$ alloy prepared by the rod-milling technique, *Journal of Alloys and Compounds* 216 (1994) 213-220.

[17] Yuuji Kimura, Setsuo Takaki, Shinichi Suejima, Ryuji Uemori, Hiroshi Tamehiro, Ultra Grain Refining and Decomposition of Oxide during Super-heavy Detormation in Oxide Dispersion Ferritic Stainless Steel Powder, *ISIJ International* Vol. 39 (1 999) 176-182.

[18] Lei Dai, Yongchang Liu, Zhizhong Dong, Size and structure evolution of yttria in ODS ferritic alloy powder during mechanical milling and subsequent annealing, *Powder Technology* 217 (2012) 281–287.

IOP Conf. Series: Materials Science and Engineering **115** (2016) 012009 doi:10.1088/1757-899X/115/1/012009

Application of Hydrogen for the Reduction of Bauxite Mineral

B. R. Parhi[1], S. K. Sahoo[1], B. Bhoi[2], B. K. Satapathy[3] and R. K. Paramguru[4]

[1]Department of Metallurgical & Materials Engg, NIT Roukela, Odisha, India, 769008
[2]Institute of Minerals & Materials Technology (IMMT), Bhubaneswar, Odisha, India, 751013
[3]R & D Centre, National Aluminium Company Ltd (NALCO), Bhubaneswar, Odisha, India, 751013
[4]Department of Mechanical Engineering, Kalinga Institute of Industrial Technology (KIIT), Bhubaneswar, Odisha, India, 751024

*E-mail:brporg@gmail.com

Abstract. Reduction of oxides present in bauxite through hydrogen was investigated in the present study. The bauxite samples were subjected to reduction through molecular hydrogen and hydrogen plasma at 650°C and 800°C with different flow rates of hydrogen for different time periods respectively. The samples, after the reduction processes, were characterized by X-ray diffraction technique and chemical analysis. It was observed that the oxides of iron present in bauxite were only undergone through reduction while other oxides remain unreduced. An attempt was then made to separate pure Al_2O_3 present in bauxite samples through acid leaching process.
Keywords: Bauxite, Reduction, Hydrogen, Plasma Processing, Leaching.

1. Introduction

Extraction of metals from their ores (mainly oxide ores) is primarily made through the process of carbothermic reduction [1, 2]. Such reduction process produces a great lot of CO_2 (carbon dioxide) which is responsible for greenhouse effect. Also carbon used in the reduction process has a limitation of future supply. Now-a-days research and development programs are dedicated to find out innovative solutions to minimize the emissions of CO_2 from different industries. It has been reported that hydrogen in place of carbon for the reduction process has several advantages [3,4]: (a) hydrogen is readily available, being the most abundant element in the universe; (b) the reaction of metal oxides with hydrogen is far more environmentally friendly, as the by-product being water rather than carbon dioxide (CO_2) or carbon monoxide (CO); (c) Theoretically very less amount of hydrogen compared to carbon is required for the reduction of metal oxides – only 6 tons of hydrogen in place of 18-36 tons of carbon is required for 112 tons of iron production; (d) Hydrogen in the form of plasma (consisting of H and H^+) is a very powerful reductant that can reduce most metal oxides at temperatures below the melting point of metals; (e) A substantial decrease in the reduction temperature can be expected by using atomic/ionic hydrogen for extraction of metals; this can especially be significant for reduction of stable oxides, such as Al_2O_3, MgO, SiO_2 etc. Some studies have been reported for the reduction of metal oxides using hydrogen and hydrogen plasma respectively in the literature [4-10].

Extraction of aluminium from bauxite is initially purified by Bayer's process which is then processed by Hall-Heroult electrolytic process [1,2]. Use of carbon during extraction of aluminium from bauxite is not an exception. In general, significant amount of energy is consumed in Hall-Heroult process by using carbon [11]. High quality carbon anodes are used in Hall-Heroult process and these are consumed during reduction of Al_2O_3 in the process. This consumption of carbon anodes need frequent replacement which makes the process expensive. Also smooth escape of carbon monoxide and carbon dioxide requires sufficient distance between the anode and cathode in the electrolytic cell which involves large energy consumption. The collection and the processing of effluent gases for minimizing environmental pollution are difficult and expensive. Keeping the above in mind, an attempt was made to investigate the reduction of bauxite using both hydrogen and hydrogen plasma in the present study. The reduced bauxite was then characterized by X-ray diffraction (XRD) technique and chemical analysis. The present study summarizes the possible scopes for the reduction of bauxite mineral through hydrogen.

2. Materials and Methods

2.1 Materials and Sample Preparation

Bauxite samples of chemical composition (in wt. %) shown in Table 1 were obtained from National Aluminium Company Ltd., India. The bauxite samples were subjected to ball milling in a planetary ball mill for 1 hour. Pellets of 20 mm diameter and 4 mm thick were prepared from the bauxite powders by uniaxial cold compaction machine for hydrogen plasma processing. The powdered samples and the pellet samples were then calcined at 700°C for 4 hours to remove the moisture and volatile matters present in the samples. It may be noted that a series of calcination experiments were performed at different temperatures and time periods to remove entire moisture and volatile matters present in the samples. It was observed that the samples calcined at 700°C for 4 hours was sufficient to remove almost entire moisture and volatile matters present in the samples.

Table 1. Chemical composition (in wt. %) of bauxite used in the present study

Al_2O_3	Fe_2O_3	TiO_2	SiO_2	LOI
43.0	23.0	3.0	4.0	22.5

2.2. Reduction by Hydrogen

The calcined samples were subjected to reduction through molecular hydrogen and hydrogen plasma at SLM Metal (P) Ltd, Rourkela and IMMT, Bhubaneswar respectively. Hydrogen gas was propelled at 650°C in a furnace (Fig.1 (a)) with a flow rate of $1.5m^3/hr$ for 1hr 15min and $2m^3/hr$ for 1hr 30min respectively.

Hydrogen plasma processing was carried out in a microwave assisted hydrogen plasma reactor (Fig.1 (b)). The bauxite pellets were treated with hydrogen plasma at a temperature of 800°C and pressure of 100 torr with hydrogen flow rates of 400sccm, 450sccm and 500sccm for 1hr 30min, 1hr 45min and 2hrs respectively.

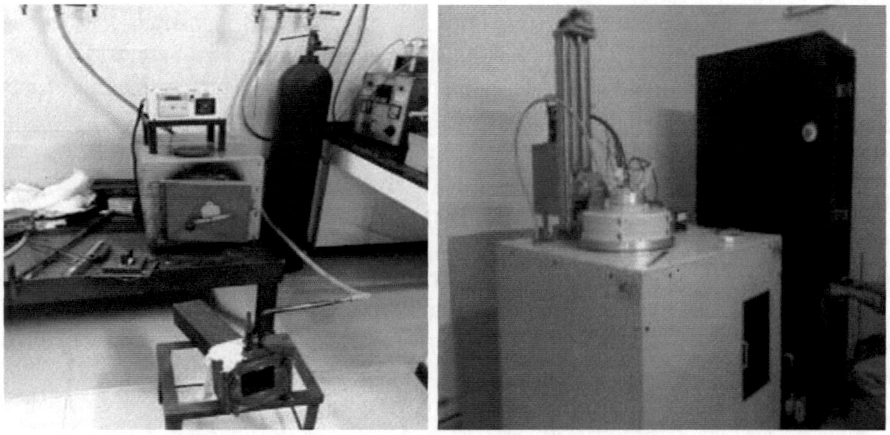

(a) (b)

Fig.1 (a) Muffle furnace for molecular hydrogen processing of bauxite samples and (b) Microwave hydrogen plasma reactor

2.3. X-Ray Diffraction (XRD)

A Panalytical PW 3040X'pert MPDsystem was used for XRD analysis. Bauxite samples before and after reduction processes were characterized by the XRD.

3. Result and Discussion

Fig.2 shows the XRD plots of calcined bauxite powders before and after reduction processing by molecular hydrogen. It can be seen that, the samples after reduction processing had dominant iron (Fe) and Al_2O_3 peaks (Fig. 2(b) and (c)). However, the percentage reduction was higher in samples subjected to reduction processing of higher hydrogen flow rate as evident from the peak intensities of Fe. Although TiO_2 and SiO_2 were not observed in XRD plots of the bauxite samples, presence of TiO_2 and SiO_2 were observed in bauxite samples before and after reduction processing through chemical analysis.

XRD plots of reduced bauxite pellets through hydrogen plasma processing are shown in Fig.3. The figures also showed non-reduction of Al_2O_3 present in the bauxite samples. Both Fe and Al_2O_3 peaks were observed in the samples. It may also be observed that the reduction percentage was different for different flow rates of hydrogen. However, increasing flow rates of hydrogen and time of reduction processing didn't always increase the reduction percentages (Fig.3). The chemical analysis of reduced bauxite samples processed by hydrogen plasma had also shown the presence of TiO_2 and SiO_2 in the samples.

From Ellingham diagram it can be found that the standard free energy change of formation (ΔG^o) of $Fe_2O_3$is -95Kcal/mol which is very less as compared to that of other oxides like to Al_2O_3, SiO_2 and TiO_2present in bauxite (ΔG^o for SiO_2, TiO_2 and Al_2O_3 are -203 Kcal/mol, -205 Kcal/mol and -252 Kcal/mol respectively) [1,2]. Hence, the reducibility of Fe_2O_3through hydrogen is higher compared to that of SiO_2, $TiO_2$and Al_2O_3. It was also observed that flow rate of hydrogen in plasma processing didn't affect reducibility (Fig. 3). It was, therefore, attempted to separate Al_2O_3 from reduced bauxite samples rather than increasing the process parameters like hydrogen flow rate, temperature and time. Further, reduction of Al_2O_3 after separation through hydrogen plasma is in progress by the authors of the present study.

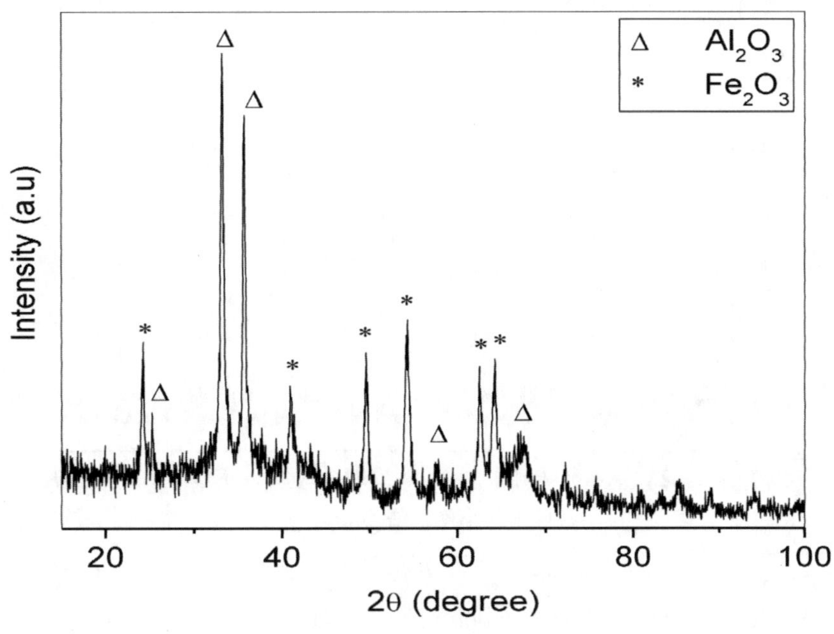

(a)

IOP Conf. Series: Materials Science and Engineering **115** (2016) 012009 doi:10.1088/1757-899X/115/1/012009

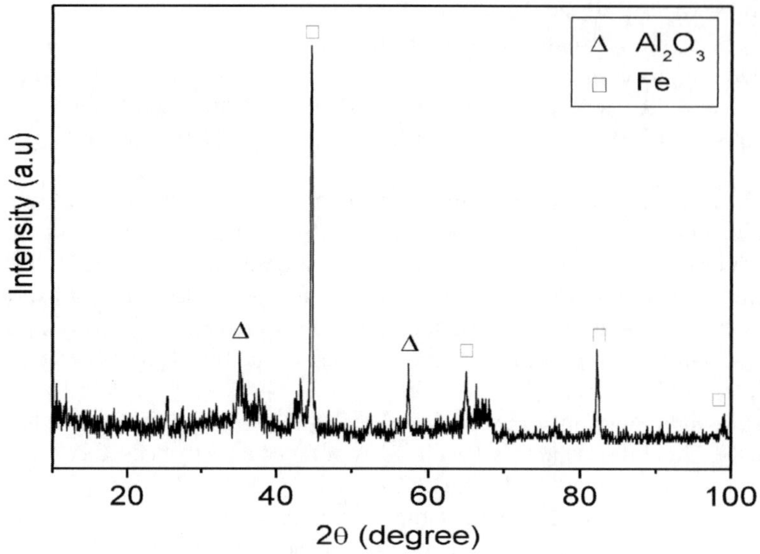

(b)

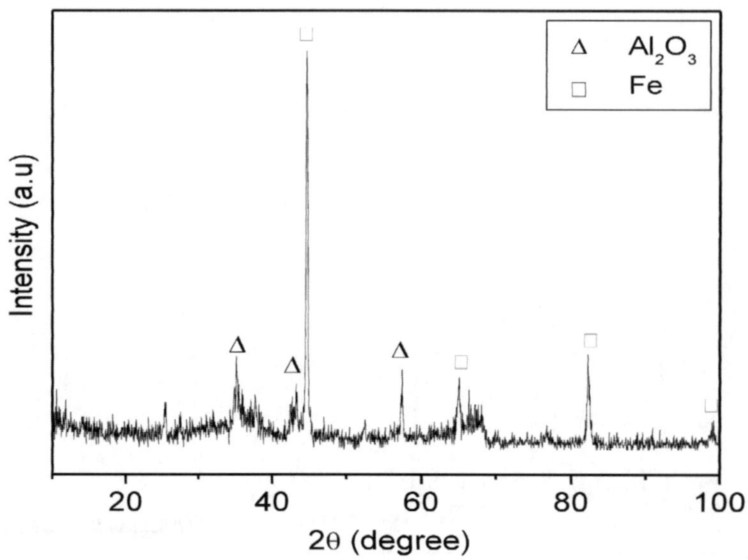

(c)

Fig.2 XRD plots of calcined bauxite before and after reduction processing by molecular hydrogen; (a) Before reduction processing, (b) After reduction processing at 650°C with hydrogen flow rate of 1.5 m^3/hr for 1hr 15min and (c) After reduction processing at 650°C with hydrogen flow rate of 2 m^3/hr for 1hr 30 min.

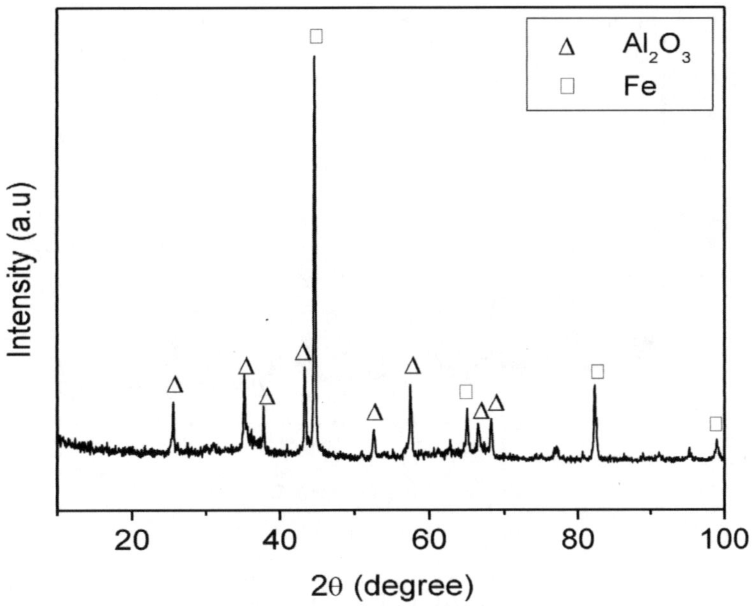
Al2O3
Fe
Intensity (a.u)
20
40
60
80
100
2θ (degree)

(a)

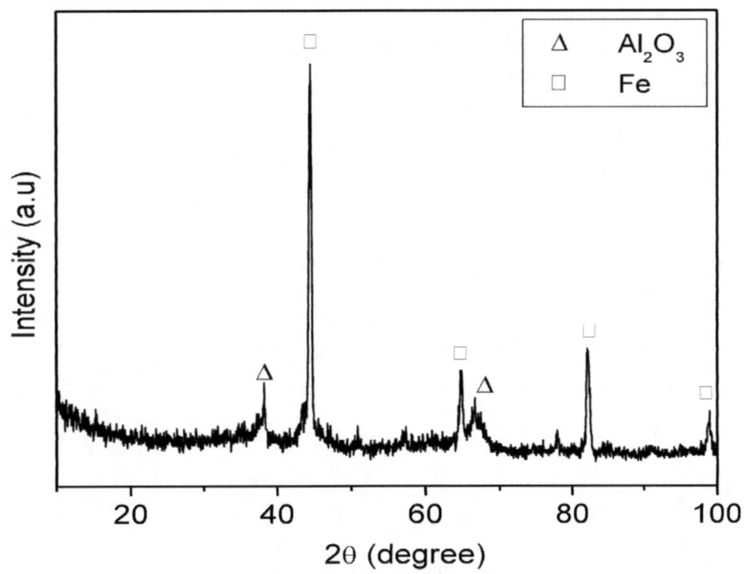
Al2O3
Fe
Intensity (a.u)
20
40
60
80
100
2θ (degree)

(b)

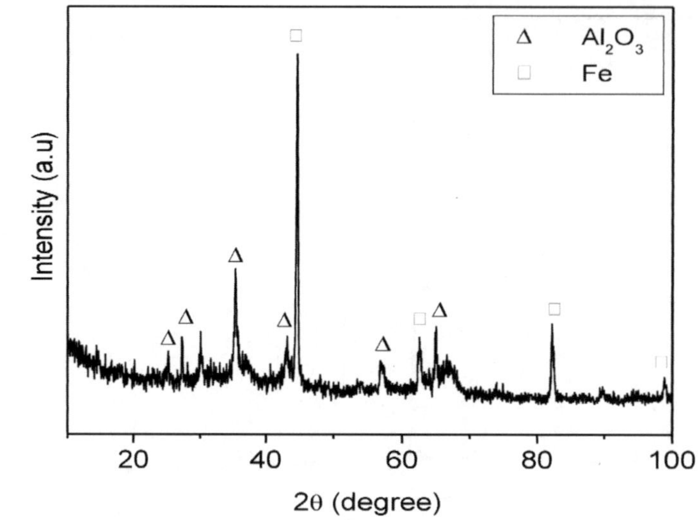

(c)

Fig. 3 XRD plots of calcined bauxite after reduction processing by hydrogen plasma at 800°C and 100 torr; (a) Hydrogen flow rate of 400sccm for 1 hr 30 min, (b) Hydrogen flow rate of 450sccm for 1 hr 45 min and (c) Hydrogen flow rate of 500sccm for 2hrs.

Separation of Al_2O_3 from reduced bauxite was made through acid leaching by using concentrated hydrochloric acid. Acid leaching of reduced bauxite produced a filtrate containing chlorides of iron and a residue containing Al_2O_3 along with other minor oxides. The filtrate was subsequently be treated with sodium hydroxide for the precipitation of iron hydroxide ($Fe(OH)_3$) which was converted to Fe_2O_3on further heating. The residue after acid leaching was analysed by XRD and the analysis is shown in Fig.4. A dominant Al_2O_3 peaks can be observed from the figure. The chemical analysis of the separated Al_2O_3 after leaching is shown in Table 2. A significant amount of $Al_2O_3$was recovered by the process of leaching. Further reduction of Al_2O_3 obtained after acid leaching is an on-going research of the present authors.

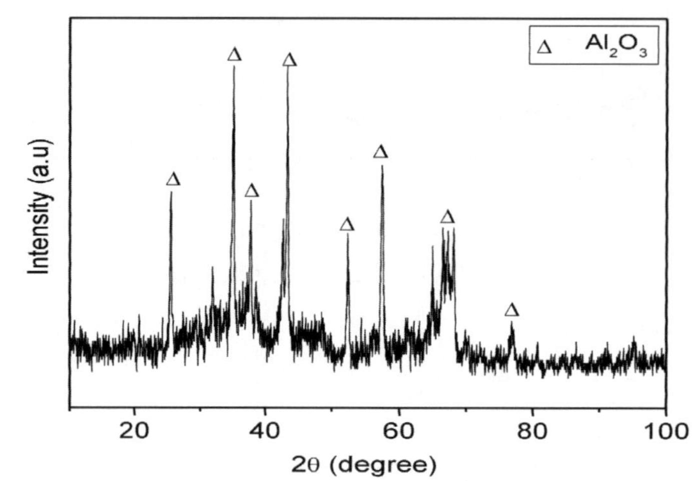

Fig.4 XRD plot of the residue after filtration of the reduced bauxite leaching.

Table 2. Chemical composition (in wt. %) of the residue after filtration of the reduced bauxite leaching.

Al_2O_3	Fe_2O_3	TiO_2	SiO_2	LOI
89.10	2.10	2.95	3.59	2.26

4. Summary

Reduction of bauxite through hydrogen was investigated in the present study. The bauxite samples were subjected to reduction through molecular hydrogen at 650°C at hydrogen flow rate of 1.5 m^3/hr for 1hr 15 min and 2 m^3/hr for 1 hr 30 min respectively. The samples were also reduced by hydrogen plasma (consisting of atomic and ionic forms of hydrogen) at temperature of 800°C and pressure of 100 torr with hydrogen flow rates of 400 sccm, 450 sccm and 500 sccm for 1 hr 30 min, 1 hr 45 min and 2 hrs respectively. In both the reduction processes, it was found that only iron oxide present in bauxite was reduced to iron while other oxides remain unreduced. It was also observed that the flow rate of hydrogen had significant effect on the reduction percentages. Acid leaching could separate Al_2O_3 from reduced bauxite in the present study.

Acknowledgements

The authors would like to thank NALCO for financial support. They also would like to thank Prof.B. K. Mishra of IMMT, Bhubaneswar and Mr. M. Agarwal of SLM Metal (P) Ltd, Rourkela for conducting the reduction processing of bauxite samples. Prof. S. C. Mishra, Dept. of Metallurgical & Materials Engg., NIT Rourkela, has initiated the work and the authors would like to thank him for his encouragement to carry out the experimental works.

References

[1] H.S. Ray, R.Sridhar, K.P.Abraham, Extraction of Nonferrous Metals*(Affiliated east-west press private limited)* 2008.

[2] C.Bodsworth, The Extraction and Refining of Metals *(CRC Press)* 1994.

[3] W. Zhang, D.R. Sadedin, M. A.Reuter, J. C. McCallum, The deoxidization of partially oxidized titanium by hydrogen plasma, *Mater. Forum,* **31** (2007) 76-83.

[4] A. A. Bergh, Atomic Hydrogen as a Reducing Agent, *Bell System Tech. Journal* (1965) 261-271.

[5] B.Bhoi, B.K.Mishra, R.K.Paramguru, S.K.Pradhan, P.S.Mukherjee, S.Sahoo, S. Priyadarshini, P.Rajput, S.K.Das, Green process for the preparation of direct reduced iron *(US Patent application publication: US2013/0047782A1)* 2013.

[6] R.A.Palmer, T.M. Doan, P.G. Lloyd, B.L. Jarvis, N.U.Ahmed, Reduction of $TiO_2$with Hydrogen Plasma, *Plasma Chemistry and Plasma Processing,* **22** (2002)335.

[7] D.Wagner, O.Devisme, F.Patisson, D.A. Ablitzer, *The Minerals, Metals & Materials Society (TMS)* **2**(2006)111.

[8] H.L.Gilles, C.W.Clump, Reduction of iron ore with hydrogen in a direct current plasma jet, *Industrial and Eng. Chemistry Process Design &Development* **9** (1970) 194.

[9] T.Watanabe, M.Soyama, A.Kanzawa, A.Takeuchi, M.Koike, Reduction and Separation of Silica-Alumina Mixture with Argon-Hydrogen Thermal Plasmas, *Thin solid Films* 345(1999)161-166.

[10] T.Maki, H.Kura, H.Ishida, T.Kaneko, R.Hatakeyama, M.Takahashi, T.Ogawa, Effect of hydrogen plasma treatment on reduction process of iron oxide nanoparticles,*2nd Int. Symp. On Advanced Magnetic Materials and Applications, Journal of Physics: Conference Series (IOP Science publishing limited)* 266 (2011) 012120.

[11] C.K.Gupta, Chemical Metallurgy: Principles and Practice, *Wiley-VCH, Weinheim* (2003) 811.

Effect of delamination on vibration behaviour of woven Glass/Epoxy composite plate-An experimental study

C K Hirwani[1], S S Sahoo[2] and S K Panda[3]

Department of Mechanical Engineering, National Institute of Technology Rourkela India.

Email[1] : chetanhirwani111@gmail.com

Abstract. We have analysed the free vibration responses of the laminated composite plate with delamination numerically and validated with subsequent experiment. In order to compute the numerical frequencies, the delaminated composite plate is modelled with two sub-laminate approaches in the commercial finite element package (ANSYS) using ANSYS parametric design language code in ANSYS environment. For the experimental analysis, the woven Glass/Epoxy composite plate is fabricated using hand layup method with the desired delamination. The natural frequencies of the delaminated plate are also computed experimentally with the help of the vibration analyser (NI-CDAQ) and validated by comparing with the simulation result. Further, the simulation model is extended for various design parameter and discussed in detail.

Keywords: Delaminated composites plate; Vibration; Experimental; ANSYS

1. Introduction

Composite materials have their own advantages of high strength and stiffness to weight ratio. These materials are not only satisfy the high strength criterion but also suitable for weight sensitive application. Delamination is one of the reason of failure of these material which occurs due to different loading, imperfect fabrication and edge effect. Delamination plays the significant role in the reduction in stiffness which seriously affect the vibration characteristics. So it is very important to predict the effect of delamination on vibration characteristics of the laminated composites. Few important work on the delaminated composites are discussed in the following line to make the article self-standing.

Borst and Remmers [1] proposed a method for the modelling of delaminated composite assuming the degraded properties using the damage law. A two sub-laminates model of delaminated composite plate is developed by Rajendran and Song [2] in ANSYS environment to perform the buckling analysis. Finite element model (two dimensional) of delaminated beam is presented by Çallio˘glu et al. [3] using finite element software package ANSYS to analyse the vibration characteristics. Tenek et al. [4] studied the vibration behaviour of the delaminated composites plate using finite element based on the three-dimensional theory of linear elasticity. Zheng and Sun [5] model the delamination by using triple plate model at two different thickness location to analyse the delamination interaction. Ju et al. [6] developed a finite element model to analyse the effect of delamination size, delamination location and number of delamination on the vibration characteristic of the delaminated composites plate using first ordered shear deformation theory (FSDT). Numerical investigation on vibration characteristic of the delaminated composites plates have been presented by Nanda and Sahu [7] using finite element method based on FSDT. Parhi et al. [8] presented a finite element model based on first-order shear deformation theory (FSDT) to analyse the vibration behaviour of a plate with arbitrary located multiple delaminations.

Experimental as well as finite element modelling study of the vibration of laminated composites plate with circular delamination is conducted by Hou and Jeronimidis [9]. Panda et al. [10] performed experimental and numerical study on vibration analysis of woven fiber Glass/Epoxy delaminated composite plates. For numerical study finite element formulation is prepared in the MATLAB environment. Mohanty et al. [11] presented the experimental, as well as numerical investigation on the buckling behaviour of multiple, delaminate composites plates.

From the literature review, it is clear that limited number of work on the finite element modelling and the subsequent experimental validation are reported. The aim of the present work is to compute the fundamental frequencies of the delaminated composite plate numerically using the commercial FE package (ANSYS) through the ANSYS parametric design language (APDL) code and compared with previously published literature. In addition to that, the laminated composite plate of the woven glass fiber is fabricated using the hand lay-up method with desired delamination and the experimental frequencies are computed through NI-CDAQ. Finally, the effect of different design parameters (thickness ratio and support conditions) on the frequency responses are computed and discussed in detailed.

2. Theoretical Formulations

2.1. Mathematical model for without delamination

In this analysis, the laminated composite plate composed of the finite number of orthotropic layers of uniform thickness is considered as in Fig. 1. A simulation model has been developed in ANSYS using APDL code. For the discretization purpose, Shell 281 element has been chosen from ANSYS element library. The shell element has eight nodes, more than other shell elements and hence it is suitable for analysing thin to moderately thick shell structure. The plate displacement field kinematics is modelled using the FSDT:

$$\begin{aligned} u(x,y,z,t) &= u_0(x,y) + z\theta_x(x,y) \\ v(x,y,z,t) &= v_0(x,y) + z\theta_y(x,y) \\ w(x,y,z,t) &= w_0(x,y) + z\theta_z(x,y) \end{aligned} \tag{1}$$

where, t is the time and u, v and w are the displacements of any point along the x, y and z coordinate axes respectively. u_0, v_0 and w_0 are corresponding displacements of a point on the mid-plane and θ_x and θ_y are the rotations of normal to the mid-surface, i.e., z=0 about the y and x-axes, respectively.

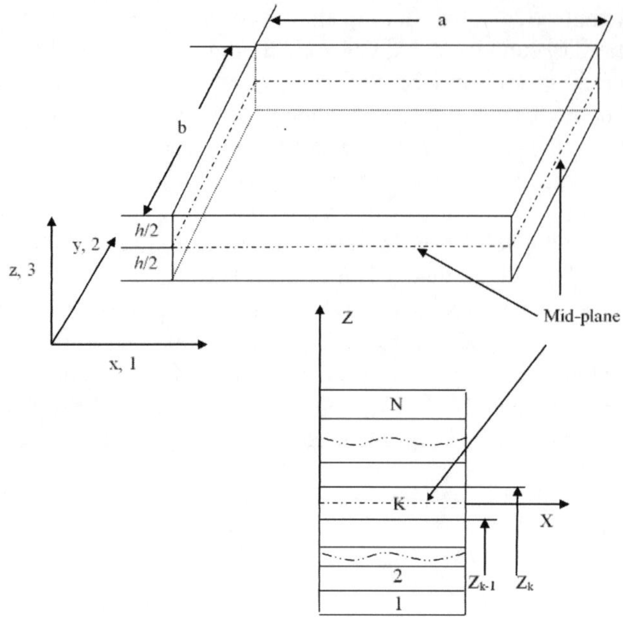

Fig. 1:- Geometry and stacking sequence of the laminated composite plate.

The stress-strain relations for any k^{th} lamina oriented at an arbitrary angle ϕ about any arbitrary axes are given by:

$$\{\sigma\} = \left[\overline{Q}_{ij}\right]\{\varepsilon\} \tag{2}$$

where, $\{\sigma\}$, $\left[\overline{Q}_{ij}\right]$ and $\{\varepsilon\}$ are the stress tensor, reduced stiffness matrix, and strain tensor, respectively. The stress tensor further modified in force form as:

$$\{F\} = [D]\,\{\varepsilon\} \tag{3}$$

where, [D] is the property matrix.

2.2. Finite Element Formulation

The displacement fields for the assumed kinematic models are expressed in terms of desired field variables, and the models are discretized using suitable FEM steps. The displacement vector '*d*' at any point within the mid-plane is given by:

$$d = \sum_{i=1}^{n} N_i(x, y) d_i \tag{4}$$

where, $\{d_i\} = \left\{u_{0_i}\ v_{0_i}\ w_{0_i}\ \theta_{x_i}\ \theta_{y_i}\ \theta_{zi}\right\}^T$ is the nodal displacement vector and N_i is the corresponding interpolating function for the 'i^{th}' node.

The strain vector can be expressed in the matrix form after introducing the FEM steps and conceded as:

$$\{\varepsilon\} = [T]\{\overline{\varepsilon}\} \tag{5}$$

where, $[T]$ is the thickness coordinate matrix and $\{\bar{\varepsilon}\}$ is the mid plane strain vector which can be further reduced as:

$$\{\bar{\varepsilon}\} = [B_L]\{d\} \tag{6}$$

where, $[B_L]$ is a general strain displacement relation matrix.

The total strain energy of the laminate can be expressed as:

$$U = \frac{1}{2}\iint \left[\int_{-h/2}^{+h/2} \{\varepsilon\}^T \{\sigma\} dz \right] dxdy \tag{7}$$

Equation (7) can be rewritten by substituting the strain and stress tensors as:

$$U = \frac{1}{2}\iint \left(\{\bar{\varepsilon}\}^T [D] \{\bar{\varepsilon}\} \right) dxdy \tag{8}$$

where, $[D] = \int_{-h/2}^{+h/2} [T]^T [\bar{Q}_{ij}][T] dz$

The kinetic energy of the laminate can be expressed as:

$$T = \frac{1}{2}\int_V \rho \{\dot{\delta}\}^T \{\dot{\delta}\} dV \tag{9}$$

where, ρ and $\{\dot{\delta}\}$ are the mass density and the global velocity vector, respectively.

The stiffness matrix ($[K]$) and the mass matrix ($[M]$) can be further expressed as:

$$\begin{aligned} [K] &= \int_A \left(\sum_{k=1}^{n} \int_{z_{k-1}}^{z_k} [B_L]^T [D][B_L] dz \right) dA \\ [M] &= \int_A \left(\sum_{k=1}^{n} \int_{z_{k-1}}^{z_k} [N]^T [N] \rho dz \right) dA \end{aligned} \tag{10}$$

2.3. Mathematical model for the delaminated panel

Now the earlier laminated model is extended further for the delaminated cases and for that a typical laminate with 'p' number of delamination as shown in Fig. 2. The local coordinate system for the element is x', y' and z' which is similar to that used segment without delamination. The displacement field of the delaminated element is assumed to be of the following form relative to its own local coordinate system.

$$\begin{aligned} u'(x', y', z', t) &= u_0'(x', y') + z'\theta_x'(x', y') \\ v'(x', y', z', t) &= v_0'(x', y') + z'\theta_y'(x', y') \\ w'(x', y', z', t) &= w_0'(x', y') + z'\theta_z'(x', y') \end{aligned} \tag{11}$$

where, t is the time and u', v' and w' are the displacements of any point along the x', y' and z' coordinate axes, respectively and it is similar to that of the without delamination. u_0', v_0' and w_0' are corresponding displacements of a point within the mid-plane and θ_x' and θ_y' are the rotations of normal to the mid-surface, i.e., $z'= 0$ about the x' and y' axes, respectively.

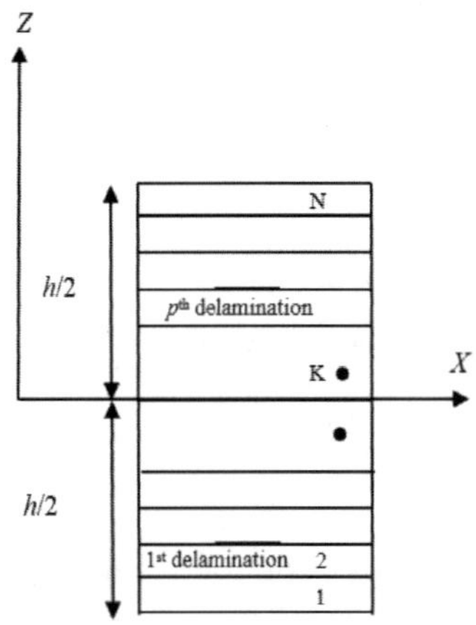

Fig. 2:- Laminated plate with delamination.

The corresponding element stiffness and mass matrices can be obtained as:

$$[K^{'}]=\int_A(\sum_{k=d+1}^{d+p}\int_{z_{k-1}^{'}}^{z_k^{'}}[B_L]^T[D][B_L]dz^{'})dA$$
$$[M^{'}]=\int_A(\sum_{k=d+1}^{d+p}\int_{z_{k-1}^{'}}^{z_k^{'}}[N]^T[N]\rho dz^{'})dA \quad (12)$$

where, p is the number of layers in the delaminated element, and two delaminations which form the bottom surface and the top surface of the element occur between layer d and layer $d + 1$, and layer $d + p$ and layer $d + p + 1$, respectively. The other symbols have the same meanings as before. The nodal displacement vector is $d_i^{'} = \{u_{0i}^{'}\ v_{0i}^{'}\ w_{0i}^{'}\ \theta_{xi}^{'}\ \theta_{yi}^{'}\ \theta_{zi}^{'}\}$.

2.3.1 Displacement continuity conditions

In the previous sections, the elements in the segment without delaminations and delaminated segment are derived separately, and all the variables in the displacement field are independent. However, at the boundary connecting these two segments, the continuity conditions for the displacement field must be satisfied.

Fig. 3. shows the elements for the case where the delaminated section is on the right. Consider one of the delaminated elements, element *m*, in the delaminated segment, where element 1 is an element in the laminated segment on the left of the element *m*. The common nodes 1, 4 and 8 are at the connecting boundary. Let the element stiffness and element mass matrices and element nodal displacement vectors be denoted by [*k*], [*m*] and $\{d\}$ for element 1, and [*k*'], [*m*'] and $\{d^{'}\}$ for element *m*.

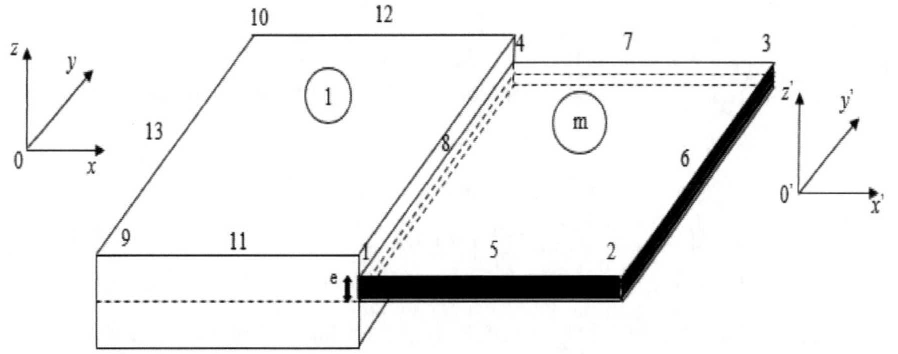

Fig. 3:- Elements at the connecting boundary.

For common nodes 1, 4 and 8, let

$$\{d_i\} = \{u_{0_i}\ v_{0_i}\ w_{0_i}\ \theta_{x_i}\ \theta_{y_i}\ \theta_{zi}\}^T \quad i=1, 4, 8 \text{ for element 1} \tag{13.a}$$

$$\{d_i^{'}\} = \{u_{0i}^{'}\ v_{0i}^{'}\ w_{0i}^{'}\ \theta_{xi}^{'}\ \theta_{yi}^{'}\ \theta_{zi}^{'}\}^T \quad i=1, 4, 8 \text{ for element m} \tag{13.b}$$

In order to satisfy the displacement continuity conditions at the connecting boundary, the following relations must hold:

$$\left.\begin{aligned} u_{0i}^{'} &= u_{0i} + e\theta_{xi} \\ v_{0i}^{'} &= v_{0i} + e\theta_{yi} \\ w_{0i}^{'} &= w_{0i} + e\theta_{zi} \\ \theta_{xi}^{'} &= \theta_{xi} \\ \theta_{yi}^{'} &= \theta_{yi} \\ \theta_{zi}^{'} &= \theta_{zi} \end{aligned}\right\} \quad \text{for i=1,4 and 8} \tag{14}$$

where e is the distance between the mid-plane of element 1 and the mid-plane of element m, and is positive in the sense indicated in Fig. 3.

Therefore,

$$\{d_i\} = [\lambda]\{d_i^{'}\} \qquad \text{for i=1, 4 and 8} \tag{15}$$

where, $[\lambda] = \begin{bmatrix} 1 & 0 & 0 & e & 0 & 0 \\ 0 & 1 & 0 & 0 & e & 0 \\ 0 & 0 & 1 & 0 & 0 & e \\ 0 & 0 & 0 & 1 & 0 & 0 \\ 0 & 0 & 0 & 0 & 1 & 0 \\ 0 & 0 & 0 & 0 & 0 & 1 \end{bmatrix}$

Using equation 11, the following transformation of element m can be written as

$$\{d^{'}\} = [T]\{\bar{d}^{'}\} \tag{16.a}$$

where,

$$\{\bar{d}^{'}\} = [d_1^{'}\ d_2^{'}\ d_3^{'}\ d_4^{'}\ d_5^{'}\ d_6^{'}\ d_7^{'}\ d_8^{'}]^T \tag{16.b}$$

$$[T] = Diag[\lambda \, I \, I \, \lambda I \, I \, I \, \lambda] \tag{16.c}$$

where, the matrix [I] is a 8 × 8 unit matrix.

Now, the transformations for the element stiffness and element mass matrices of element m can be expressed as

$$\begin{aligned}[\bar{k}'] &= [T]^T [k'][T] \\ [\bar{m}'] &= [T]^T [m'][T]\end{aligned} \tag{17}$$

The displacement continuity conditions have been incorporated into $[\bar{k}']$ and $[\bar{m}']$, and the nodal displacement $\{d_i'\}$ is replaced by $\{d_i\}$ for i = 1, 4, 8. The matrices $[\bar{k}']$ and $[\bar{m}']$, are subsequently used to assemble the global stiffness and global mass matrices. Obviously, a similar treatment can be carried out for cases when the connecting boundaries of the two segments are at the other edges.

2.4. Governing equations

The final form of governing equation of free vibrated composite plate is obtained by using Hamilton's principle and expressed as:

$$\delta \int_{t_1}^{t_2} (T - U) dt = 0 \tag{18}$$

Substituting Equations (4), (8) and (9) into equation (18), the final form of the equation will be conceded as:

$$[M]\{\ddot{d}_i\} + [K]\{d_i\} = 0 \tag{19}$$

where, $\ddot{d}_i$ is the acceleration, d_i is the displacement

Neglecting the required matrices, the eigenvalue form of the governing equation to obtain the natural frequency of the system is conceded as:

$$([K] - \omega^2 [M])\Delta = 0 \tag{20}$$

where, ω and Δ are the natural frequency and the corresponding eigenvector, respectively.

The Equation 20 is solved by using the following sets of support conditions to avoid any rigid body motion as well as reduce the number of unknowns.

Simply supported (SSSS):

$$v_0 = w_0 = \theta_y = \theta_z = 0 \text{ at x=0 and a;}$$

$$u_0 = w_0 = \theta_x = \theta_z = 0 \text{ at y=0 and b;} \tag{21}$$

Clamped (CCCC):

$$u_0 = v_0 = w_0 = \theta_x = \theta_y = \theta_z = 0 \text{ at x=0 and a; y=0 and b;} \qquad (22)$$

3. Result and discussion

The free vibration responses of the composite plate with delamination is computed using developed FE code in ANSYS environment. The validation of the proposed model is examined by comparing the responses with those available published results. Consequently, the woven Glass/Epoxy plate is fabricated with the help of hand layup method and the desired delamination also introduced by adding Teflon tape in between the laminates. The natural frequencies of the fabricated delaminated plate are obtained experimentally and validated with the simulation model. The effect of various parameters such as thickness ratios and different support conditions are computed using the present simulation model.

3.1. Validation and comparison study

The finite element model developed in the present study is validated by comparing the results with those available previously published literature. The free vibration responses of a twenty layered cantilever square composite plates are computed using M1 material and geometrical properties as in Table 1. same as to the reference considering three cases namely Case-I, Case-II and Case-III. The Case-I represents natural frequencies of the laminated plate and the Case-II and the Case-III are the square delamination of size *a*/2 and 3*a*/4, respectively. The delamination is located at the middle of the plate or at the mid-plane of the structure. The natural frequencies of the delaminated plate are computed for different lamination schemes and compared with Parhi et al. [12] and presented in Table 2. It can be seen that the results are showing very good agreement with that of the references for each case as presented in the table.

Table 1. Material properties of different composite delaminated plate.

Properties	MATERIAL-1 (M1)	MATERIAL-2 (M2)
Young's modulus x direction (E_l)	172.5GPa	7.366GPa
Young's modulus y direction (E_t)	6.9GPa	5.568GPa
Young's modulus z direction (E_z)	6.9GPa	5.568GPa
Shear modulus (G_{lt})	3.45GPa	2.79GPa
Shear modulus (G_{tz})	1.38GPa	1.395GPa
Shear modulus (G_{lz})	3.45GPa	2.79GPa
Poisson's ratio (υ_{lt})	0.25	0.17
Poisson's ratio (υ_{lz})	0.25	0.17
Poisson's ratio (υ_{tz})	0.25	0.17
Density (ρ)	1600 Kgm^{-3}	1100 Kgm^{-3}

Table 2. Natural frequency (Hz) of cantilever composite plate with centrally placed delamination.

Lamination scheme	Case no.	Natural Frequency (Hz)	
		Present	Parhi et al. [12]
$(60^0/-60^0)_{10}$	Case I	9.528	9.5
	Case II	8.3184	8.43
	Case III	6.4981	6.84
$(90^0/-90^0)_{10}$	Case I	6.7217	6.71
	Case II	6.3129	6.12
	Case III	5.02	5.0168

3.2. Preparation of the woven Glass/Epoxy delaminated composite plate

Glass/Epoxy composite plate specimens were fabricated using hand layup method. Very firstly, four layers of woven Glass fibers sheets were cut out in required size. On a rigid flat plywood platform, a plastic sheet was placed as a mold releasing sheet and a releasing agent i.e. a thin film of polyvinyl alcohol was applied to it. Two layers of Glass fibers sheets were glued by applying a gel coat (a mixture of Epoxy and Hardener) with the help of a brush, and a smooth steel roller was rolled over the laminate to minimize the void content in the laminates. Delamination was introduced by placing Teflon film of an area, 6.25% of the total laminate area in the centre of the prepared laminate. Complete care was taken in the casting of the plates with the presence of delamination. On top of it, other two layers of the Glass fiber sheets were glued with gel coat. The process of hand layup was continued for all layers of reinforcement before the gel coat fully hardened as shown in Fig. 4 (a). After completion of all the layers, again a plastic sheet sprayed with polyvinyl alcohol was covered at the top of the last ply. Now the hot-press (Fig. 4 (b).) is employed for pre-curing of the laminate where the temperature is maintain 60°C for half an hour and then for two days it is left in atmosphere condition for post curing. After proper curing of laminates, the release sheets were detached.

Fig. 4 (a):- Woven Glass/Epoxy delaminated plate after hand layup.

Fig. 4 (b):- Hot press

3.3. Material property evaluation

The mechanical properties say, Young's modulus, shear modulus and Poisson's ratio in principal material directions of the delaminated composite plate has been obtained for the computation of the desired natural frequency of the plate structure. For that the unidirectional tensile test was conducted via Universal Testing Machine (UTM) INSTRON 1195 at National Institute of Technology, Rourkela on the specimens in the longitudinal direction, transverse direction and at an angle of 45° inclined to the longitudinal direction of the plate. The ASTM standard: D 3039/D 3039M is used for the evaluation of Young's modulus. The tensile test was conducted with a recommended rate of loading, i.e., 1 mm/minute as shown in Fig. 5 (a). Tensile test specimen is shown in Fig. 5 (b). including grip length. The material properties, obtained and presented in Table 1. as M2 material properties. Poisson's ratio is taken as 0.17 as in Crawley et al. [13]. Similarly, the shear modulus was determined using the following formula as in Jones [14]:

$$G_{lt} = \frac{1}{\frac{4}{E_{45}} - \frac{1}{E_l} - \frac{1}{E_t} - \frac{2\upsilon_{12}}{E_l}}$$

Fig. 5 (a):- UTM INSTRON 1195.

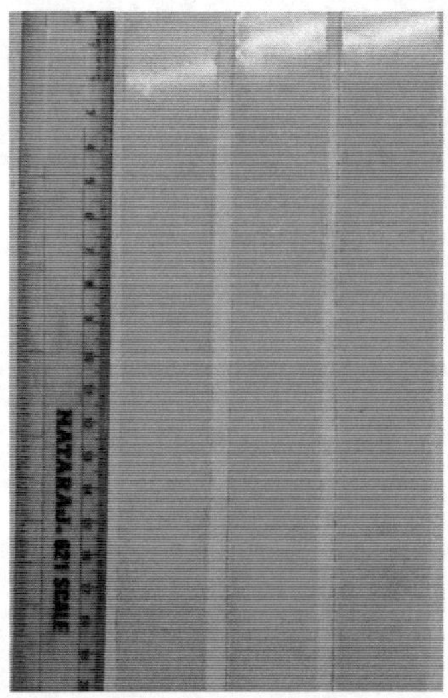

Fig. 5 (b):- Glass/Epoxy laminated composite specimen after tensile test.

3.4. Modal test

Now, the free vibration responses of the woven Glass/Epoxy delaminated composite plate is analyzed experimentally using the CDAQ-9178 (National Instruments) for the plate of dimension $a = b = 15$cm and $h = 0.12$cm with delamination of a size of $a/4$ at the midpoint of the laminate. The various components in the experimental set-up are shown in Fig. 6 (a) and (b). The vibration responses of the plate are recorded under CFFF support condition, with the help of a fixture via an accelerometer mounted on the plate. The plate is excited with the help of an impact hammer on any random points, and the desired data are received through the CDAQ-9178, which is a compact data acquisition system of USB chassis comprising of the eight channels. These signals are again processed with the help of LABVIEW software. In LABVIEW, a virtual instrument (VI) program circuit is developed which primarily comprises of three components: a back panel, a front panel and a connector panel. These three component helps in supply of the input and show the output on the computer screen Fig. 6(c). depicts

IOP Conf. Series: Materials Science and Engineering **115** (2016) 012010 doi:10.1088/1757-899X/115/1/012010

the block diagram of the LABVIEW software for the data acquisition of the desired signal and the subsequent analysis. The acceleration signals are passed through the power spectrum module of the LABVIEW for fast Fourier transformation to obtain the frequency domain responses. The peaks of the frequency response spectrum give the natural frequencies of vibration for different modes. Finally, the experimental results are compared with those the simulation model and presented in Table 3. The results of the simulation model are in decent agreement with the experimental results.

Fig. 6 (a):- NI CDAQ 9178.

Fig. 6 (b):- Experimental set-up.

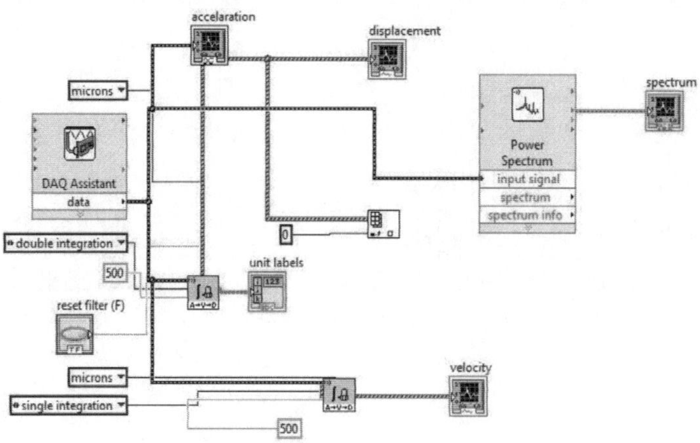

Fig. 6 (c):- Block diagram of the LABVIEW for experimental data recording.

Table 3. Natural frequency (Hz) of cantilever woven Glass/Epoxy composite plate with centrally placed delamination.

Mode no.	Natural Frequency (Hz)	
	Present	Experimental
1	21.053	19.5
2	55.079	44.5
3	133.13	126.5
4	167.96	163
5	199.5	196.5

3.5. Numerical illustrations

Based on the validation and the comparison study, it is evident that the presently developed simulation model is capable to compute the free vibration responses of the delaminated composite quite accurately. Now, to highlight the importance of the simulation model for the analysis of the delaminated structure responses are computed for different design parameters like the thickness ratios and the support conditions. Based on the responses, the quantitative understanding of the free vibration behaviour of the delaminated composite plates are discussed in the following lines. The responses are computed for twenty layered cross-ply composite plates with Case-II type of delamination placed centrally in the mid laminate having M1 material properties as in Table 1. unless otherwise stated.

3.5.1. Effect of support condition on vibration response of delaminated composited plate

The natural frequencies of the delaminated plate has been computed for different support conditions and the aspect ratios (a/b=1, 2, 3, 4 and 5) with b/h=100. In general, the support conditions constrained the degrees of freedom of the structure, and it is well-known fact that the increase in the number of constraints makes the structure stiffer. The effect of four types of support condition (CCCC, CFCF, SSSS and CFFF) on the delaminated plate is computed and presented in Fig. 7. It is interesting to note that the responses are increasing with increase in number of constrained DOF.

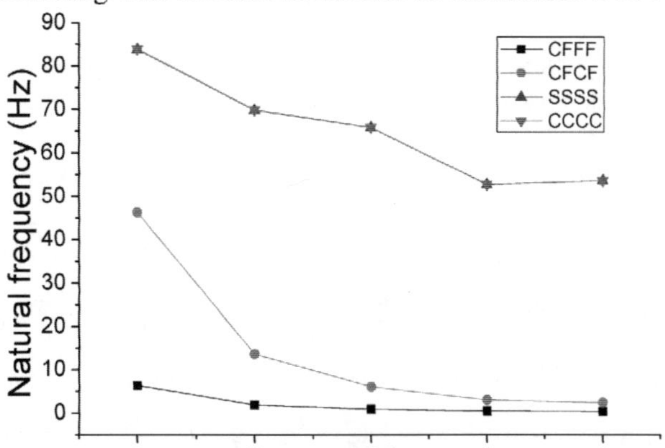

Fig. 7:- Natural frequency of twenty layered cross-ply composite plate with central delamination of case-II for varying support conditions.

3.5.2. Effect of thickness ratio on free vibration response of delaminated composited plate

It is well known that the thickness ratio of any structure or the structural component increases then the structure becomes thinner, and the structural stiffness of the laminated panel is inversely proportional to that. The free vibration response has been examined using simulation model for the clamped delaminated plate by varying its thickness ratios (b/h=100, 125, 150, 175 and 200). The first natural frequency of the plates is computed for different aspect ratio (*a*/*b*=1, 2, 3, 4 and 5) and presented in Fig. 8. It is clearly observed that the natural frequency decreases as the thickness ratio increases and the responses are within the expected line.

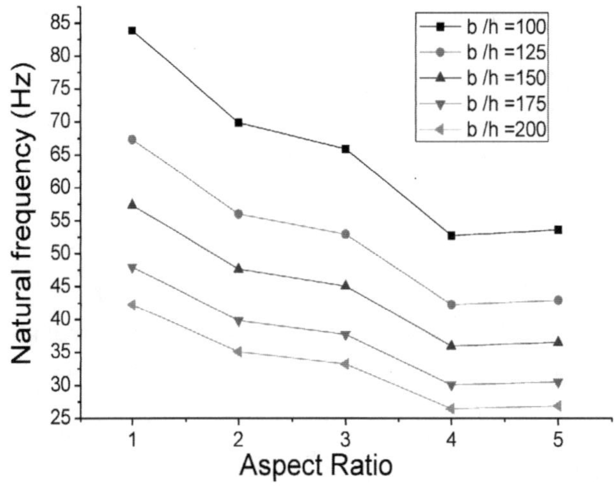

Fig. 8:- Natural frequency of twenty layered cross-ply composite plate with central delamination of case-II for varying thickness ratio.

4. Conclusion

In this present article, the free vibration responses of the delaminated composite plate have been numerically using the commercial FE package (ANSYS) and validated subsequently with that of the available numerical results. Further, the mechanical properties of the fabricated delaminated composite plate are computed experimentally, and the desired frequencies are also obtained experimentally with the help of in-house vibration analyser via CDAQ-9178. The frequencies obtained using the simulation, and the experimental results are compared for the validation purpose as well. Finally, few examples are solved numerically with the help of the present simulation model to highlight the effect of design parameter on the frequency responses. The parametric study indicated that the natural frequencies of the delaminated composite plate increases with the decrease in thickness ratio and increase in constrained.

5. References

[1] Borst R and Remmers J J C 2006 *Composites Science and Technology* **66** 713–722

[2] Rajendran S and Song D Q 1998 *Proceedings of 2nd Asian ANSYS User Conference*

[3] Çallio˘glu H, Atlihan G and Topçu M 2012 *Advanced Composite Materials* **21** 11-27

[4] Tenek L H, Henneke E G and Gunzburger M D 1993 *Composite Structures* **23** 253-262

[5] Zheng S and Sun C T 1998 *Engineering Fracture Mechanics* **59(2)** 225-240

[6] Ju F, Lee H P and Lee K H 1995 *Composites Engineering* **5(2)** 195-209

[7] Nanda N and Sahu S K 2012 *International Journal of Pressure Vessels and Piping* **98** 111-118

[8] Parhi P K, Bhattacharyya S K and Sinha P K 2000 *Journal of reinforced plastics and composites* **19(11)** 863-882

[9] Hou J P and Jeronimidis G 1999 *Composites: Part A* **30** 989–995

[10] Panda H S, Sahu S K and Parhi P K 2013 *Composite Structures* **96** 502–513

[11] Mohanty J, Sahu S K and Parhi P K 2013 *Journal of Structural Integrity* **4** 240-257

[12] Parhi P K, Bhattacharyya S K and Sinha P K 1999 *Aircraft Engineering and Aerospace Technology* **71** 451 – 461

[13] Crawley E F 1979 *Journal of Composite Materials* **13** 195-205

[14] Jones R M 1975 Mechanics of composite materials *Taylor and Francis*

An experimental investigation on the effect of particle size on the thermal properties and void content of Solid Glass Microsphere filled epoxy Composites

Debasmita Mishra[1], Alok Satapathy[2]

[1]Department of Mechanical Engineering, VSSUT, Burla, Odisha, 768018
[2]Department of Mechanical Engineering, NIT, Rourkela, Odisha, 769008

E-mail - debasmita.mishra01@gmail.com

Abstract This paper investigates about the thermal characterization of Solid glass micro-sphere (SGM) filled epoxy composites. SGMs of different sizes are embedded in epoxy resin to fabricate composites by hand-layup technique. The composites for various SGM content ranging from 0 to about 35 vol % are thus fabricated and the effective thermal conductivities (k_{eff}) of the composites are estimated. The theoretical values are then compared with k_{eff} values obtained from the experiment. This study shows that the incorporation of SGM results in an improvement in thermal insulation capability of the polymer. Further, the influence of size and content of SGMs in the extent of reduction of k_{eff} was studied. Also, the effect of void content on improving insulation capability of the composites was analysed.

Keywords: *Composites; Solid Glass Micro-spheres; Thermal Conductivity; Insulation*

1. Introduction

Many parts of the country experience large changes in temperature from season to season. The demand for structurally stable, cost effective and light-weight insulation materials is therefore increasing day- by- day but the conventional engineering materials are unable to meet the requirements like high strength, low density and low conductivity. Foamed plastic is commonly used for thermal insulation. However, due to its poor mechanical properties, its application is considerably limited [1]. Therefore, there has been a focus to fabricate a kind of light, porous material with better mechanical strength and good thermal insulation properties. Against this backdrop, polymer composites are emerging as a class of promising engineering insulation materials.

Epoxy is chosen as the polymer to be used as the matrix material for this work [2]. Today, nearly 80% of the world die attach market for plastic encapsulated devices is solvent-less conductive epoxies. A few reasons why epoxies are widespread and common in electronic assemblies include: low temperature cure profiles which allow for greater availability of substrates and temperature sensitive components; compliant nature of the polymer; low out-gassing and good thermal stability at elevated temperatures; very good chemical and solvent resistance; and also because epoxies readily accept fillers.

The heat transfer process of porous materials is very complicated, especially for polymer composites. Considerable work has been reported on the subject of heat conductivity in polymers by Hansen and Ho, 1965[3], Choy and Young, 1977 [4], Tavman, 1991 [5] and 1997[6]. Progelhof et al., 1976 [7] was the first to present an exhaustive overview on models and methods for predicting the thermal conductivity of composite systems. It is quite important, therefore, to understand the mechanisms of heat transfer in polymer composites, which are potential insulating materials.

In this context, Solid glass micro-spheres (glass beads) have some advantages as fillers in polymers such as low thermal conductivity, coefficient and density. Also, they have strong filling ability, smooth spherical surface, small and well distributed internal stress in the products and good processibility of the filled materials. So, SGM filled epoxy would be suitable mostly for applications related to insulation [8-11]. In addition, these micro-particles do not generate stress concentration in the interface between the fillers and the matrix owing to their smooth spherical surface. The orientation effects associated with moulding are also minimal. They are preferred as fillers especially when composite properties such as isotropy or low melt viscosity are important. Compared to the investigations carried out on polymer composites filled with metal and ceramic powders, behaviour of glass micro-sphere filled composites has remained a relatively less studied area. This type of composites can be applied in building materials, space flight and aviation industry.

There are only a few published papers on evaluation of effective thermal conductivity of polymer composites filled with glass beads. Liang and Li, 2006 [8] reported on measurement of thermal conductivity of hollow glass-bead-filled polypropylene composites. Recently, they [9] in 2007also made two-dimensional and three-dimensional finite element analysis on the heat transfer and simulated the variation of effective thermal conductivity of hollow glass microsphere filled polymer composites. Liang and Li, 2007 [10] further studied the heat transfer in polymer composites filled with hollow glass micro-spheres and proposed a theoretical model to predict the thermal conductivity of such composite system. Yung et al., 2009 [11] reported on the preparation and properties of hollow glass microsphere-filled epoxy matrix composites. But all these studies are for polymer composites filled with hollow glass spheres and surprisingly, there is no report available on evaluation of effective thermal conductivity of solid glass microsphere filled polymer composites. In view of the above, the present work is undertaken to evaluate theoretically and experimentally the thermal conductivity of epoxy matrix composites filled solid glass micro-spheres. The objectives of this work include fabrication of a new class of composites to further improve the insulating properties of epoxy by the incorporation of solid glass micro-spheres. Epoxy is chosen primarily because it happens to be the most commonly used polymer and because of its low value of thermal conductivity (about 0.363 W/m-K).

The objective of this work is to fabricate different sets of epoxy composites filled with micro-sized SGM and measurement of their physical, mechanical and thermal characteristics, estimation of the effects of fillers and filler size on properties like void content of these composites and exploring the possibility of their use in applications for thermal insulation.

2. Materials and Methods

2.1Processing of the composites

Low temperature curing Epoxy LY 556 resin, used as the matrix material and the hardener (HY951) are mixed in a ratio of 10:1 by weight as recommended. Epoxy is chosen primarily for its low density (1.1 gm/cc) and low value of thermal conductivity (0.363 W/m K). Solid glass micro-spheres (SGM) of three different sizes (100, 200 and 300 micron diameter) are reinforced in the resin to prepare the composites. The dough (epoxy filled with SGM) is then slowly decanted into the glass molds, coated beforehand with wax and a uniform thin film of silicone-releasing agent. The composites are cast in these molds so as to get disc type cylindrical specimens (dia 20 mm, thickness 5 mm). Composites of 8 different compositions (0, 5, 10, 15, 20, 25, 30 and 35 vol % of SGM respectively) using SGMs of each size are made. The castings are left to cure at room temperature for about 24 hours after which the glass molds are broken and samples are released. Unitherm™ Model 2022 is used to measure thermal conductivity of the composites fabricated for this investigation. The test is carried out in accordance with ASTM E-1530 standard.

2.2 Epoxy Composites filled with solid glass micro-spheres

Epoxy composites of 8 different compositions (Table 1) were made for each of the three microsphere sizes (100, 200 and 300 μm). For each composition, the composites were cast in glass molds so as to get both disc type specimens (diameter 10, 20 mm and thickness 5, 10 mm) and rectangular specimens (length 150 mm, width 20 mm, thickness 4 mm).

Table 1 Epoxy composites filled with Solid glass micro - spheres

Composition
Epoxy + 0 vol % SGM
Epoxy + 5 vol % SGM
Epoxy + 10 vol % SGM
Epoxy + 15vol % SGM
Epoxy + 20vol % SGM
Epoxy + 25vol % SGM
Epoxy + 30vol % SGM
Epoxy + 35vol % SGM

Table 2 Comparison of k_{eff} of composites obtained from different models.

SGM content (vol %)	Effective Thermal Conductivity (W/m-K)			
	Maxwell's equation	Lewis Nielsen's equation	Russell's model	Measured Values
0	0.363	0.363	0.363	0.363
5	0.2663	0.3390	0.345	0.298
10	0.3189	0.3163	0.327	0.287
15	0.2982	0.2947	0.237	0.258
20	0.2783	0.2741	0.221	0.231
25	0.2591	0.2545	0.201	0.189
30	0.2407	0.2358	0.187	0.159
35	0.2229	0.2179	0.172	0.136

Effective thermal conductivity (k_{eff}) of epoxy based composites with SGM content ranging from 0 to 35 vol% are estimated and compared by using some of the existing theoretical models .These values are compared with the corresponding measured values of conductivity (Table 2). The k_{eff} values obtained from the Russell's theoretical model that is Rule of mixture Series Model are in good agreement with the experimentally measured values. It is thus observed that the analytical model serves as a very good empirical model for the estimation of effective thermal conductivity for spherical inclusions and that the correlation can very well be used to estimate k_{eff} of composites for a broad range of SGM concentrations.

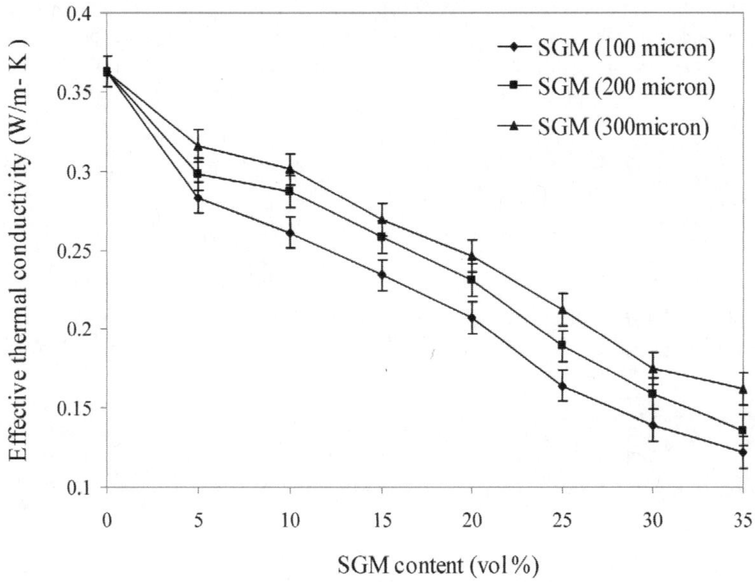

Fig.1:- Variation of effective thermal conductivity with SGM content: Effect of SGM size

Using the theoretical model proposed in this work, the effective thermal conductivity k_{eff} of the composites are calculated. The variation of k_{eff} with the SGM (100 micron) content is shown in Figure 3, which also presents a comparison of the calculated values with the experimentally measured ones. It is seen that the results obtained from the proposed correlation are in good agreement with experimental results. Figure 4 presents the variation of k_{eff} with filler content for different glass micro-sphere sizes and it is evident that the reduction in heat conduction capability is maximum for the SGMs with 100 micron size. It is encouraging to note that the incorporation of SGM results in significant drop in thermal conductivity of epoxy resin. With addition of 35 vol. % of SGM (100 micron), the thermal conductivity decreases by about 66% as compared with neat epoxy resin. It is also noted that, smaller the size of solid glass micro-spheres, higher is the drop in conductivity value of the composites.

2.3. Effect of micro-sphere size on effective thermal conductivity

Thermal conductivities of composites filled with SGMs of three different sizes (100, 200 and 300 μm) are measured to study the effect of micro-sphere size on k_{eff}. The variation of k_{eff} with micro-sphere size is presented in Figure 1.It is evident from the figure that with addition of SGMs, irrespective of the micro-sphere size, there is always a reduction in the heat conduction capability of the composites.

However, it is found that this reduction is maximum in case of SGMs of smallest size i.e. 100 μm. In other words, smaller the size of solid glass micro-spheres, higher is the drop in conductivity of the composites. With addition of 35 vol% of SGM (100 μm), the thermal conductivity decreases by about 66% as compared to that of neat epoxy. At the same time, with the incorporation of same 35 vol% of SGM (300 μm), the drop in thermal conductivity is only about 53%.

3. Physical Characterization

3.1 Density and volume fraction of voids

Density of a composite depends on the relative proportion of matrix and reinforcing materials and it is one of the most important factors for determining the properties of the composites. The theoretical density (ρ_{ct}) of composite materials in terms of weight fractions of different constituents can easily be obtained using the following equation given by Agarwal and Broutman [11].

$$\rho_{ct} = 1 \Big/ \left\{ \left(\frac{w_f}{\rho_f} \right) + \left(\frac{w_m}{\rho_m} \right) \right\} \tag{1}$$

where, w and ρ represent the weight fraction and density respectively. The suffixes f and m stand for the filler and matrix respectively in a composite with single filler. In the present work, since some of the composites are hybrid type i.e. filled with double fillers (SGM and BN), the above expression for density has been modified for them as:

$$\rho_{ct} = 1 \Big/ \left\{ \left(\frac{w_{f_1}}{\rho_{f_1}} \right) + \left(\frac{w_{f_2}}{\rho_{f_2}} \right) + \left(\frac{w_m}{\rho_m} \right) \right\} \tag{2}$$

where, the suffix f_1 and f_2 stand for the first and second filler and m stands for the matrix respectively. The actual density (ρ_{ce}) of the composite can be determined experimentally by the Archimedes principle or water displacement technique (ASTM D 792-91).

The volume fraction of voids (v_v) in the composites is calculated by using the following equation:

$$v_v = (\rho_{ct} - \rho_{ce}) / \rho_{ct} \tag{3}$$

The theoretical and measured densities along with the corresponding volume fraction of voids in the epoxy-SGM composites are presented in Table 3. It may be noted that the composite density values calculated theoretically from weight fractions using Eqn.1 are not equal to the experimentally measured values. This difference is a measure of voids and pores present in the composites. It was found that with the increase in Solid glass micro-sphere content in epoxy resin from 0 to 35 vol%, there was a rise in density of the composite by about 11% although there was a simultaneous increase in the void fraction or porosity from 0.727 % to 2.258% (Table 3).

However, unlike other composites, the void-content in the SGM epoxy composites is desirable as the voids enhance the insulation properties of the composites due to presence of air pockets. It was also observed that smaller sized SGM filled epoxy composites showed less amount of voids than larger particles.

Table 3 Measured and Theoretical densities of the composites
(Epoxy filled with SGM)

Composites	Measured density (gm/cc)	Theoretical density (gm/cc)	Void fraction (%)
Neat Epoxy (hardened)	1.092	1.100	0.727
Epoxy + SGM (5 vol%)	1.107	1.119	1.072
Epoxy + SGM (10 vol%)	1.126	1.140	1.228
Epoxy + SGM (15 vol%)	1.142	1.160	1.551
Epoxy + SGM (20 vol%)	1.158	1.179	1.781
Epoxy + SGM (25 vol%)	1.174	1.200	2.167
Epoxy + SGM (30vol%)	1.193	1.220	2.213
Epoxy + SGM (35 vol%)	1.212	1.240	2.258

3.2 Scanning Electron Microscopy

The micro-structural features of the various particulate filled composite specimens were examined by Scanning Electron Microscope JEOL JSM-6480 LV (Figure 2(a) & 2(b)). The specimens were mounted on stubs with silver paste. To improve the penetration of light and for better surface micrographs, a thin film of platinum is vacuum-evaporated onto the samples before the pictures were taken.

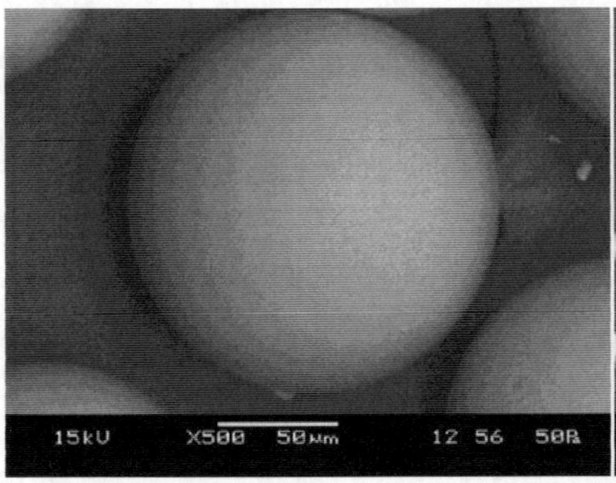

Fig.2 (a):- SEM image of a single glass microsphere used as particulate filler in composite

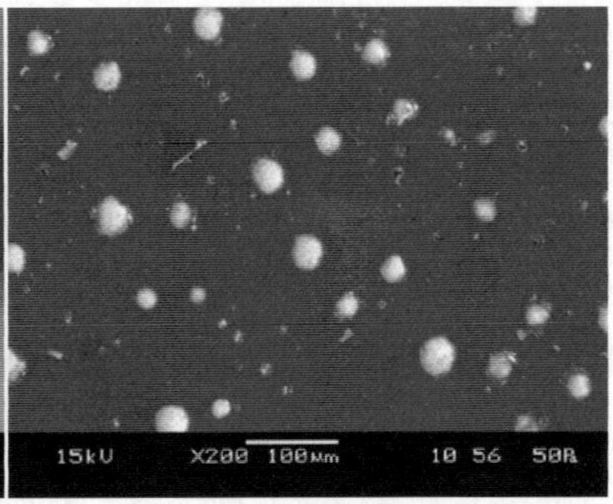

Fig.2 (b) :- SGM-epoxy composite of glass microspheres embedded in epoxy matrix

From the SEM micrographs, it is evident that the shape of SGM particles is spherical and hence they have no stress concentration effects inside the matrix and the distribution of SGMs in the epoxy matrix is uniform. It was also observed that the smaller sized SGMs are more uniformly distributed in the matrix than the larger particles.

4. Conclusion

Successful fabrication of epoxy based composites filled with solid glass micro-spheres by hand-lay-up technique is possible. Incorporation of SGMs results in reduction of thermal conductivity of epoxy resin and there by improves its thermal insulation capability. With addition of 35 vol% of SGM (100 micron size), the thermal conductivity decreases by about 66 % as compared to neat epoxy resin. For same volume fraction of SGM in the composite, the improvement in composite insulation capability is found to be more for smaller SGMs. With light weight and improved insulation capability SGM filled epoxy composite can be used for applications such as electronic packages, insulation board, food container, thermo flasks, building materials, space flight and aviation industry etc.

References

[1] D.Kumlutas et.al., Thermal Conductivity of particle filled polyethylene composite materials, *J. Compos.Sci.Technol.*63(1) (2003) 113-117.

[2] D.Hansen, C.Ho, *J Polym Sci.Part A* 3(2) (1965) 659-670.

[3] C.L.Choy, K.Young, *J Polym Sci.*18(8) (1977) 769-776.

[4] Ismail H.Tavman, Thermal Anisotropy of polymers as a Function of their Molecular orientation, Experimental Heat Transfer, Fluid Mechanics, and Thermodynamics, *Elsevier* 27(1991) 1562-1568.

[5] Ismail.H.Tavman, *J.Powder Technol.* 91(1997) 63-67.

[6] R.C.Progelhof, J.L.Throne, R.R. Ruetsch , Methods of Predicting the Thermal Conductivity of composite Systems, *Polym Eng Sci* 16(9) (1976) 615-625.

[7] J.Z. Liang, F.H. Li, Measurement of thermal conductivity of hollow glass bead filled polypropylene composites, *Polym. Test.* 25(4) (2006) 527-531.

[8] J.Z. Liang, F.H. Li, Simulation of heat transfer in hollow glass-bead- filled polypropylene composites by finite element method, *Polym.Test.* 26(3) (2007) 419-424.

[9] J.Z. Liang, F.H. Li, Heat transfer in polymer composites filled with inorganic hollow micro-spheres:A theoretical model, *Polym. Test.* 26 (2007) 1025-1030.

[10] K.C. Yung ,et al., Preparation and properties of hollow glass microsphere- filled epoxy- matrix composites, *Comp. Sci. Technol.(*69) (2009) 260 -264.

[11] B. D. Agarwal and L. J. Broutman, Analysis and Performance of Fiber Composites, *John Wiley and Sons, New York, NY, USA, 2nd edition*, 1990.

[9] J. Z. Liang , F. H. Li Heat transfer in polymer composites filled with inorganic hollow micro-spheres: A theoretical model *Polym. Test.* 26 (2007)1025-1030.

[10] K. C. Yung et al. Preparation and properties of hollow glass microsphere- filled epoxy- matrix composites *Comp. Sci. Technol.* 69 (2009) 260 -264.

[11] B. D. Agarwal and L .J. Broutman, Analysis and Performance of Fiber Composites, *John Wiley and Sons, New York , NY, USA, 2nd edition*, 1990.

Physical, Mechanical and Water Absorption Behaviour of Coir Fiber Reinforced Epoxy Composites Filled With Al_2O_3 Particulates

Geetanjali Das[*] and Sandhayarani Biswas
Department of Mechanical Engineering, National Institute of Technology, Rourkela-769008, Odisha, India

[*]E-mail: gitabebi@gmail.com

Abstract: The objective of the present work is to study the physical, mechanical and water absorption behaviour of coir fiber reinforced epoxy composites filled with Al_2O_3 particulates. Composites with different compositions were prepared by varying the length of the fiber and content of fiber using hand lay-up technique. The experimental investigation reveals that the properties of composite increases with the incorporation of Al_2O_3 particulates. It is observed that the density of composites increases with increase in fiber content, while a decrease in density is observed with increase in fiber length. The strength properties of the composites increases with the increase in fiber content up to 15 wt.% and 12mm fiber length, however further increase in fiber length and fiber content the value decreases. The maximum tensile strength of 25.71MPa, flexural strength of 29.75MPa and impact strength of 14.76kJ/m^2 is obtained for composites with12 mm fiber length and 15 wt.% of fiber content. The hardness and tensile modulus, on the other hand, increases with increase in fiber length and fiber content. The maximum hardness value of 19.52H_v and tensile modulus of 3.412GPa is obtained for composites with 15mm fiber length and 20 wt.% of fiber content. Finally, morphological analysis is also carried out using scanning electron microscope (SEM) to study the fracture behaviour of the composite samples.

Keywords: Coir fiber, Epoxy, Al_2O_3 particulate, fiber content, fiber length, mechanical properties.

1. Introduction

Over the past two decades, natural fibres have been receiving considerable attention as substitutes for synthetic fibre reinforcements like glass and carbon. Unlike the traditional synthetic fibres these lignocellulosic fibres are able to impart certain benefits to the composites such as low density, low cost, high stiffness, biodegradability, renewability and high degree of flexibility during processing. A great deal of work has been done on the mechanical properties of various natural fibres and how these fibres interact with various thermosets and thermoplastics. Among the different natural fibers, coir is seems to be a promising fiber due to its low cost, high availability, low density, easy production and friendly to environment. Coir consists of cellulosic fibers with hemi-cellulose and lignin as the bonding materials for the fibers. The lignin content in coir fiber is quite high, so the fiber becomes stiffer, tougher and more long-lasting when compared to other natural fibers. Generally, coir is a waste of natural resources and a cause of environment pollution. Hence, research and development efforts have been going on to find out the new areas for coir, along with utilization of coir as reinforcement in polymer composites. Junior et al. [1] studied the tensile behaviour of coir fiber reinforced polyester composites. Monteiro et al. [2] studied the mechanical performance of coir fiber/polyester composites. The mechanical behaviour of coir fiber/polyester composites which revealed the lack of an efficient reinforcement by coir fibers is attributed to their low modulus of elasticity, in comparison with the neat polyester resin. Slate [3] investigated the mechanical properties of coir fiber reinforced cement sand mortar. Ayrilmis et al. [4] studied on coir fiber reinforced polypropylene composite panel for automotive interior applications. This study reported that the coir fiber is a potential reinforcement for thermoplastic composites, specifically for partial replacement of high cost and heavier glass fibers. Mazan et al. [5] studied the viability of coir fiber reinforced composites in sound absorption panel. Among different parameters fiber length and fiber content are the most important factors significantly influencing the properties of composites. Sumaila et al. [6] studied the effect of fiber length on the physical and mechanical properties of random oriented, nonwoven short banana fiber reinforced epoxy composites and concluded that the tensile strength, tensile modulus and percent elongation had their highest values at 15mm fiber length. Jutarat et al. [7] studied the effect of cotton fiber contents and lengths on properties of thermoplastic starch composites prepared from rice and waxy rice starches. Maged et al. [8] investigated the effect of fiber length and concentration on the physical properties of wheat husk fibers rubber composites.

Major constituents in fiber reinforced polymer composite are the reinforcing fibers and a matrix. In addition, particulate fillers can also be used with some polymeric matrices primarily to reduce cost and improve the properties. Hard particulate fillers consisting of ceramic or metal particles are being used to significantly improve the properties of many composite materials. The improved performance of polymers and their composites in industrial and structural applications by the addition of particulate fillers has shown a great promise and so has lately been a subject of considerable interest. Aluminium oxide (Al_2O_3) commonly referred to as alumina, is the most cost effective and widely used material in the family of engineering ceramics and has the potential to be used as filler in various polymer matrices. It is hard, wear-resistant, high strength and stiffness, has excellent dielectric properties, resistance to strong acid and alkali attack at elevated temperatures. With an excellent combination of properties and a reasonable price, it is no surprise that fine grain technical grade alumina has a very wide range of applications.

Although a great deal of work has been done on the coir fiber based polymer composites, however the study on the effect of filler and the fiber parameters such as fiber content and length on the physical, mechanical and water absorption behaviour is limited. To this end, the objective of the present work is to study the effect of fiber content and fiber length on the physical, mechanical and water absorption behaviour of coir fiber reinforced epoxy composites filled with Al_2O_3 particulate.

2. Experimental details

2.1. Composite fabrication

Short coir fiber and Al_2O_3 filler is taken as reinforcement and epoxy as matrix material. The coir fiber is collected from local sources and Al_2O_3 filler is obtained from NICE Ltd., Kolkata, India. The epoxy resin and corresponding hardener is supplied by Ciba Geigy India Ltd. The fabrication of the composite slabs is done by using conventional hand-lay-up technique. The Al_2O_3 filler has average particle size of 80-100 micron and their density is 3.69gm/cm^3. The mould having dimension of $180\times180\times40$ mm^3 is used for composite fabrication. Composites of various compositions were prepared by varying the content of fiber (5wt.%, 10wt.%, 15wt.% and 20wt.%) and length of the fiber (3mm, 6mm, 9mm, 12mm and 15mm) at constant filler loading (10wt%). The cast of each composite material is cured under a load of about 50kg for 24 h before it removed from the mould. Then, specimens of suitable dimension are cut using a diamond cutter for various testing. The designation and composition of composites are listed in the Table1. Fig. 1, 2 and 3 shows the coir fiber, Al_2O_3 filler and the fabricated composite, respectively.

Fig. 1:- Coir fiber

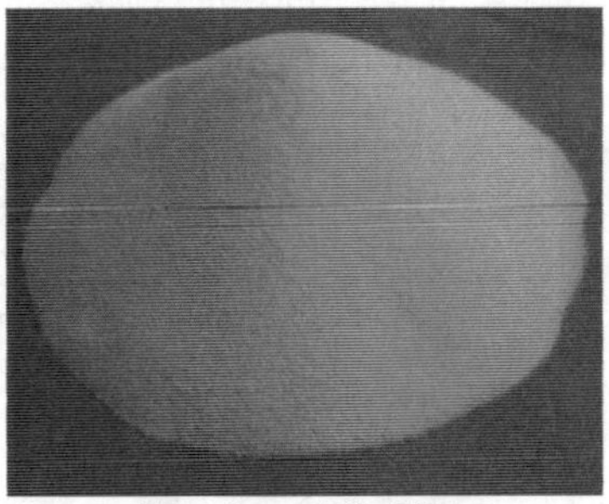

Fig. 2:- Al_2O_3filler

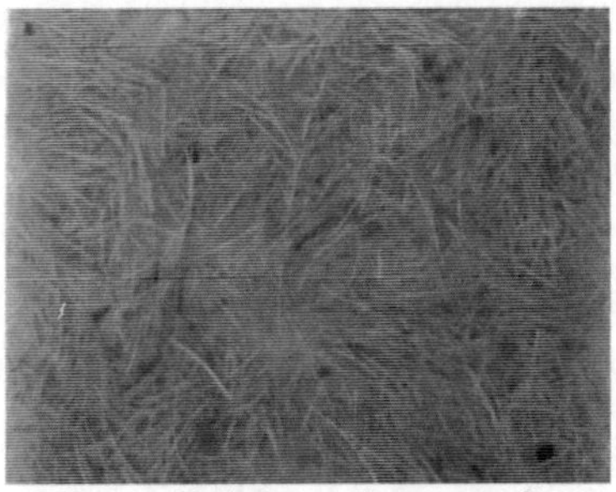

Fig. 3:- Fabricated composites

Table 1: Designation and detailed composition of the composites

Designation	Compositions
S1	Epoxy (85wt%) + Coir Fiber (3mm) (5wt%) + Al_2O_3(10wt%)
S2	Epoxy (85wt%) + Coir Fiber (6mm) (5wt%) + Al_2O_3(10wt%)
S3	Epoxy (85wt%) + Coir Fiber (9mm) (5wt%) + Al_2O_3(10wt%)
S4	Epoxy (85wt%) + Coir Fiber(12mm) (5wt%) + Al_2O_3(10wt%)
S5	Epoxy (85wt%) + Coir Fiber (15mm) (5wt%) + Al_2O_3(10wt%)
S6	Epoxy (80wt%) + Coir Fiber (3mm) (10wt%) +Al_2O_3(10wt%)
S7	Epoxy (80wt%) + Coir Fiber (6mm) (10wt%) + Al_2O_3(10wt%)
S8	Epoxy (80wt%) + Coir Fiber (9mm) (10wt%) + Al_2O_3(10wt%)
S9	Epoxy (80wt%) + Coir Fiber (12mm) (10wt%) + Al_2O_3(10wt%)
S10	Epoxy (80wt%) + Coir Fiber (15mm) (10wt%) + Al_2O_3(10wt%)
S11	Epoxy (75wt%) + Coir Fiber (3mm) (15wt%) + Al_2O_3(10wt%)
S12	Epoxy (75wt%) + Coir Fiber (6mm) (15wt%) + Al_2O_3(10wt%)
S13	Epoxy (75wt%) + Coir Fiber (9mm) (15wt%) + Al_2O_3(10wt%)
S14	Epoxy (75wt%) + Coir Fiber (12mm) (15wt%) + Al_2O_3(10wt%)
S15	Epoxy (75wt%) + Coir Fiber (15mm) (15wt%) + Al_2O_3(10wt%)
S16	Epoxy (70wt%) + Coir Fiber (3mm) (20wt%) + Al_2O_3(10wt%)
S17	Epoxy (70wt%) + Coir Fiber (6mm) (20wt%) + Al_2O_3(10wt%)
S18	Epoxy (70wt%) + Coir Fiber (9mm) (20wt%) + Al_2O_3(10wt%)
S19	Epoxy (70wt%) + Coir Fiber (12mm) (20wt%) + Al_2O_3(10wt%)
S20	Epoxy (70wt%) + Coir Fiber (15mm) (20wt%)+ Al_2O_3(10wt%)

2.2. Physical and mechanical tests

The composites under this study consists of three components namely matrix, fiber and particulate filler. The actual density (ρ_{ce}) of the composite is determined experimentally by simple water immersion technique. However, the theoretical density of composite materials in terms of weight fraction is calculated using the following equation [9]:

$$\rho_{ct} = \frac{1}{(W_f/\rho_f)+(W_m/\rho_m)+(W_p/\rho_p)} \quad (1)$$

Where, W and ρ represent the weight fraction and density respectively. The suffix f, m, p and ct stand for the fiber, matrix, particulate filler and the composite materials respectively. The volume fraction of voids (V_v) in the composites is calculated using the following equation:

$$V_v = \frac{\rho_{ct} - \rho_{ce}}{\rho_{ct}} \quad (2)$$

Micro-hardness measurement is done using a Leitz micro-hardness tester. The tensile test is performed in the universal testing machine Instron 1195 as per ASTM D 3039-76. The three point bend test is conducted on all the composite samples in the universal testing machine Instron 1195. Span length of 40 mm and the cross head speed of 2mm/min are maintained. Impact strength of the composites is evaluated by a low velocity impact tests conducted in an impact tester as per ASTM D 256 test standards. The fractured surfaces after tensile testing are examined using scanning electron microscope (SEM) JEOL JSM-6480LV.

2.3. Water absorption test

The water absorption tests of coir fiber reinforced epoxy composites were done as per ASTM 570 by immersion in distilled water at room temperature. The samples were taken out periodically and after wiping out the water from the surface of the sample weighted immediately using a precise balance machine to find out the content of water absorbed. The specimens were weighed regularly at 24, 48, 72, 96, 120, 144, 168, 192, 216, 240, 264, 288, 312, 336, 360, 384, 408, 432, 456 and 480 hours. The water absorption is calculated by the weight difference. The percentage weight gain of the samples is measured at different time intervals by using the following equation.

$$\text{Water absorption (\%)} = \frac{W_2 - W_1}{W_1} \times 100 \quad (3)$$

Where, W_1 and W_2 are the weight of the dry and wet samples.

3. Results and discussion

3.1. Density

Density is one of the most important factors in determining the properties of composites of a composite material. It mainly depends on the relative proportion of reinforcement and matrix. The void content of composites is the difference between the experimental density and the theoretically density values of composites. Fig. 4 shows the effect of fiber parameters on the density of composites. It is clearly observed from the figure that the density of composites decreases as the fiber length increases from 3mm to 15mm. This is mainly due to the inclusion of long fibers into the composites decreases the packing, which leads to the disruption of fiber distribution and resulting in high void spaces. Apparently, greater void contents yield low density composite. However, the density of composites increases with increase in fiber content. Fig. 5 shows the variation of void content with fiber parameters. It is evident from the figure that the void content in the composites increases with the increase in fiber content. Presence of large amounts of the hydroxyl group in natural fibers makes them polar and hydrophilic in nature. This polar nature also results in high moisture absorption in natural fiber based polymer composites, leading to fiber swelling and voids in the fiber-matrix interface [10].

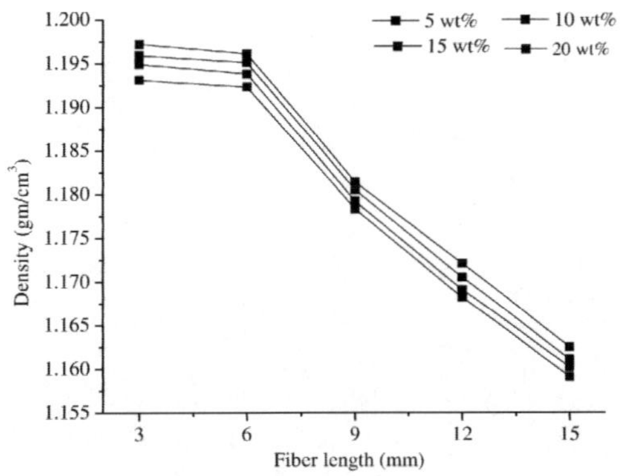

Fig. 4:- Effect of fiber content and length on density of composites

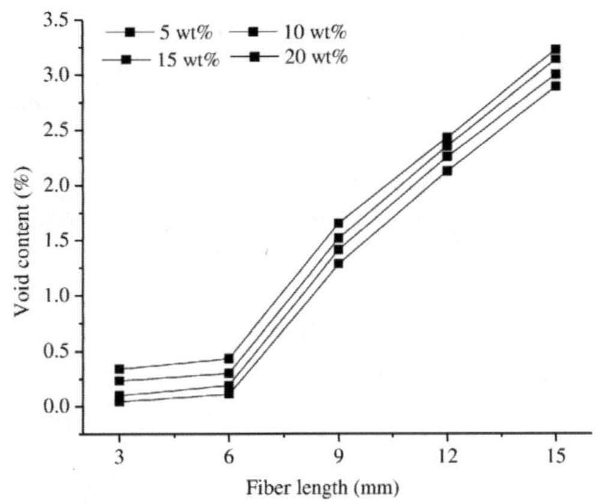

Fig. 5:- Effect of fiber content and length on void of composites

3.2. Tensile strength

The tensile strength of Al_2O_3 filled coir fiber composites is presented in Fig. 6. The tensile strength of composites increases with increase in the fiber content up to 15 wt.%. However, further increase in fiber content the strength decreases. This decrease may be due to the improper adhesion hinders the increase of tensile strength. As the fiber content increases, instead of dispersion the gathering of fibers takes place and the resin cannot wet the fibers due to non-entrance of resin in-between the two adjacent fibers. It is also observed that as the fiber length increases, the tensile strength of composites increases and then decreases irrespective of fiber content. In case of small fiber length, tensile strength is less due to the fact that length may not be sufficient enough for proper load distribution. On the other hand, for the composites of longer fiber length, tensile strength decreases. The reason may be due to the fact that longer fiber may not become compatible with the matrix properly.

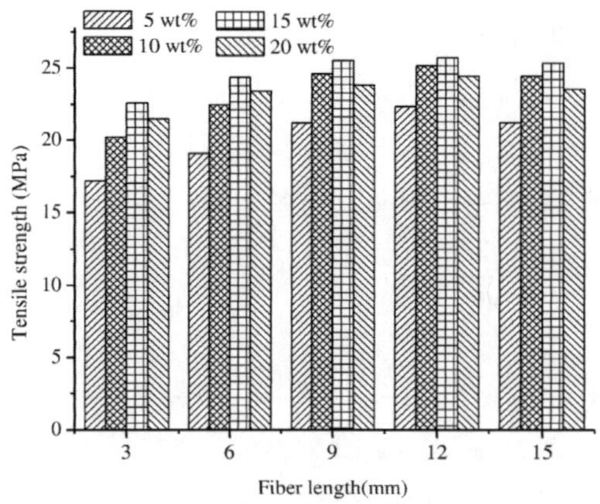

Fig. 6:- Effect of fiber content and length on tensile strength of composites

3.3. Tensile modulus

The effect of fiber parameters on the on the tensile modulus of composites is shown in Fig. 7. It is evident from the figure that the tensile modulus increases with the increase in fiber content. The increase in fiber content results in increased brittleness of the composites; thus stress/strain curve becomes steeper. Thus, as the fiber content increases, the degree of hindrance increases, which in turn increases the stiffness. On the other hand as the fiber length increases, the tensile modulus of the composites also increases.

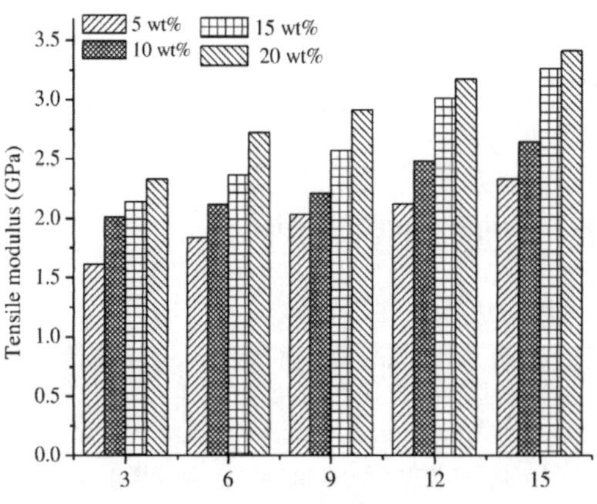

Fig. 7:- Effect of fiber content and length on tensile modulus of composites

3.4. Flexural strength

The effect of fiber content and length on the flexural strength of composites is shown in Fig. 8. It is evident from the figure that the flexural strength increases with increase in fiber content up to 15 wt.%, and then it decreases. It follows the similar trend as tensile behaviour. The reasons for the lower flexural properties at higher fiber content are probably due to the weak fiber-to-fiber interaction, void and poor dispersion of fiber in the matrix. Generally, Al_2O_3 filler offer greater resistance to crack initiation and propagation in the composite. Due to which there is an increase flexural strength of composites with Al_2O_3 filler as compared to without filler.

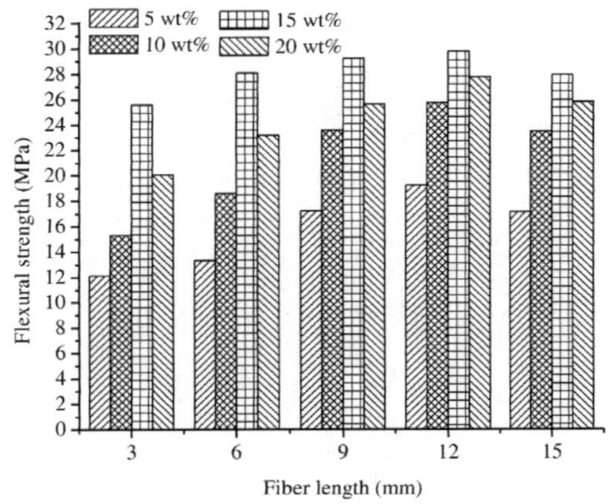

Fig. 8:- Effect of fiber content and length on flexural strength of composites

3.5. *Micro-hardness*

Fig. 9 shows the effect of fiber content and fiber length on the micro-hardness of composites. It is observed from the figure that as the weight percentage of fiber in the composite increases, the hardness of composite also increases. Similarly, as the fiber length increases, the hardness of the composite also increases. Similar trend of increase in hardness of the composites with increase in fiber length has also been reported by the researchers [11].

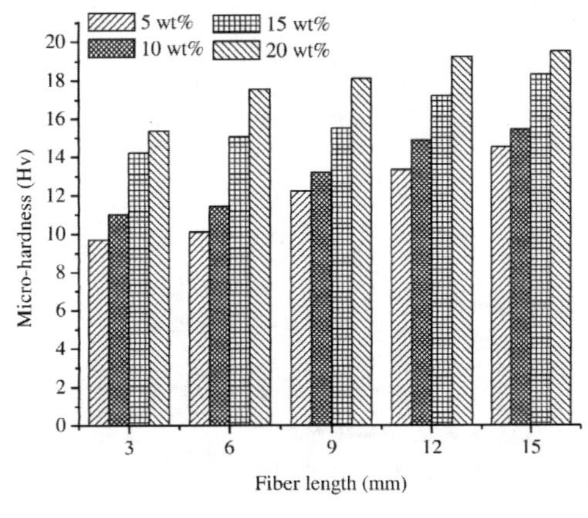

Fig. 9:- Effect of fiber content and length on micro-hardness of composites

3.6. *Impact strength*

Fig. 10 shows the effect of fiber content and fiber length on the impact strength of composites. It is observed that the impact strength increases with the increase in fiber content up to 15 wt.% and further it decreases. It has been reported that high fiber content increases the probability of fiber agglomeration and it stress concentration requiring less energy for crack propagation. The impact strength of all composites increased with fiber content up to 15 wt.%. The reasons are that the fiber is capable of absorbing energy and compression pressure which removes the voids contents in the composites because of appreciative mix-up fiber and matrix. The maximum impact strength of 14.76 kJ/m^2 is obtained for composites with 12 mm fiber length and 15 wt.% fiber content.

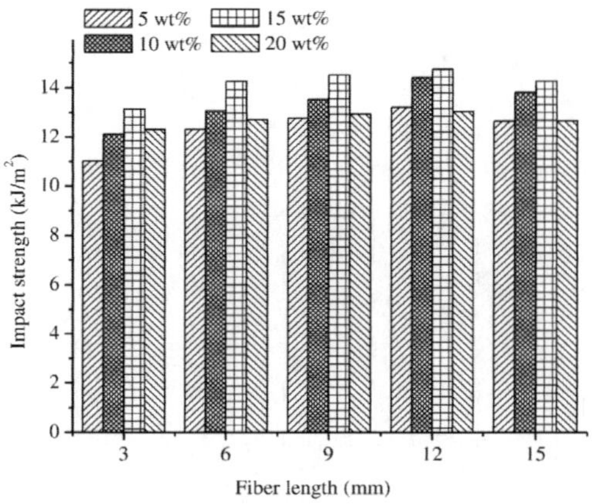

Fig. 10:- Effect of fiber content and length on impact strength of composites

3.7. Water absorption

Fig. 11 shows the effect of fiber parameters on the water absorption of composites with increase in immersion time. It is observed from the figure that the water absorption process is sharp at the beginning and levelled off for some length of time where it approaches to equilibrium. Generally, the rate of water absorption is greatly affected by the composite's density and void content. Longer the fiber, the higher is the water absorption. Similarly, it is also evident from the figure that the rate of water absorption increases with increase in fiber content. Composites with 20 wt.% coir fiber content shows more water absorption rate as compared to 5 wt.% fiber content. The reason may be due to the coir fibers contain more polar hydroxide groups, which result in a high moisture absorption level of natural fiber based polymer composites.

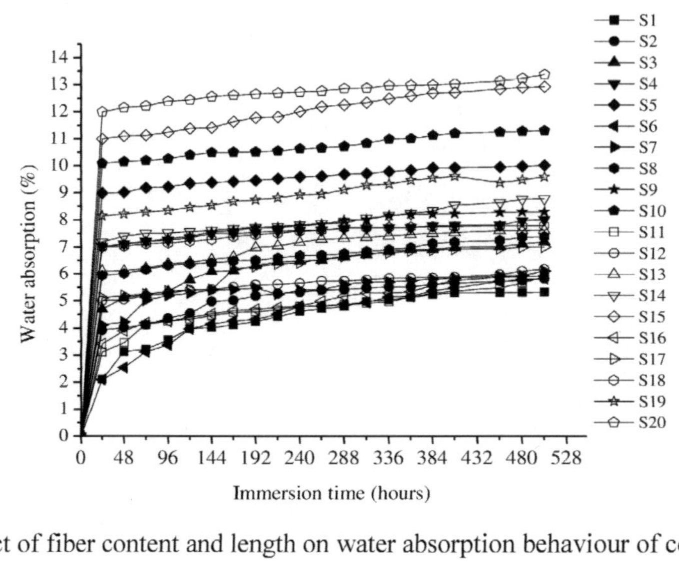

Fig.11:- Effect of fiber content and length on water absorption behaviour of composites

3.8. Surface Morphology

The fracture surfaces study of composite after tensile test is shown in Fig. 12 (a) and (b). From Fig. 12 (a) it is clear that the fibers are detached from the resin surface due to poor interfacial bonding. Pulled-out fibers are clearly visible for composites with 5 wt.% fiber content and 3mm length. However, the composite with 15 wt.% fiber and 12 mm length shows good matrix/fiber adhesion. Only very small fiber pull-out which coated with matrix material is observed as shown in Fig. 12 (b).

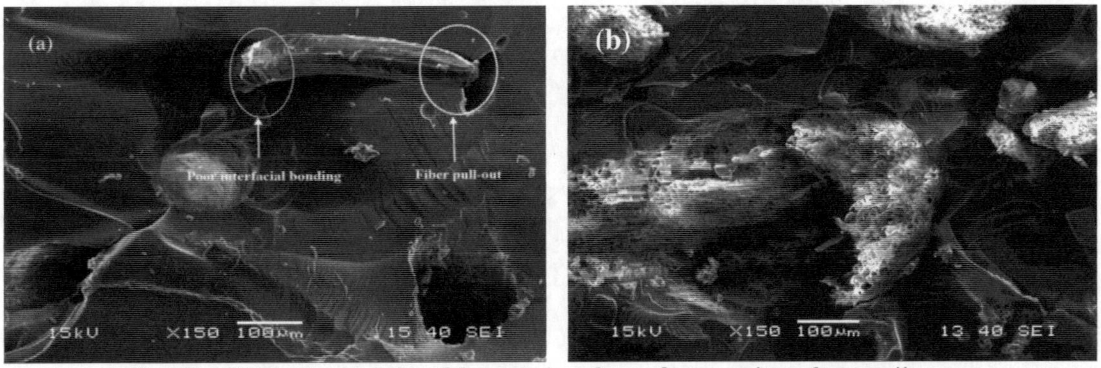

Fig. 12:- SEM micrographs of fractured surface of composites after tensile tests

4. Conclusion

The properties of the coir fiber reinforced epoxy composites filled with Al_2O_3 filler are significantly influenced by the filler and fiber parameters. The void content of the composites increases with the increase in both fiber content and fiber length. The strength properties of composite increases with increase in the fiber content up to 15 wt.% and then decreases. Therefore, the optimum fiber content is found to be 15 wt.% for better mechanical properties. Similarly, 12 mm fiber length is found to be effective in increasing the strength properties of composites. On the other hand, the hardness and tensile modulus of composites increases with increase in both fiber content and fiber length. The improvement in mechanical properties of composites with Al_2O_3 filler is observed as compared to unfilled one. The minimum water absorption is observed for composites with 5 wt.% fiber content at 3 mm fiber length. Fiber pull-out is clearly observed for composites with 5 wt.% fiber content and 3mm length from the SEM micrographs. On the other hand composite with 15 wt.% fiber and 12 mm length shows better matrix/fiber adhesion.

References

[1] H.P.G. Santafe Junior, F.P.D. Lopes, L.L. Costa, S.N. Monteiro, Mechanical properties of tensile tested coir fiber reinforced polyester composites, *Revista Materia* 15 (2010) 113-118.
[2] S.N. Monteiro, L.A.H. Terrones, J.R.M. D'Almeida, Mechanical performance of coir fiber/polyester composites, *Polymer Testing* 27 (2008) 591-595.
[3] F.O. Slate, Coconut Fibers in concrete, *Engineering Journal of Singapore* 3 (1976) 51-54.
[4] N. Ayrilmis, S. Jarusombuti, V. Fueangvivat, P. Bauchongkol, R.H. White, Coir fiber reinforced polypropylene composite panel for automotive interior applications, *Fibers and Polymers* 12 (2011) 919-926.
[5] S. Mahzan, A.M. Ahmad Zaidi, N. Arsat, M.N.M. Hatta, M.I. Ghazali, S. Rasool Mohideen, Study on sound absorption properties of coconut coir fibre reinforced composite with added recycled rubber, *International Journal of Integrated Engineering* 2 (2010) 29-34.
[6] M. Sumaila, I. Amber, M. Bawa, Effect of fiber length on the physical and mechanical properties of random oreinted, nonwoven short banana (musa balbisiana) fibre/epoxy composite, *Asian Journal of Natural & Applied Sciences* 2 (2013) 39-49.
[7] J. Prachayawarakorn, P. Ruttanabus, P. Boonsom, Effect of cotton fiber contents and lengths on properties of thermoplastic starch composites prepared from rice and waxy rice starches, *Journal of Polymer Environmental* 19 (2011) 274-282.

[8] M.S. Sobhy, M.T. Tammam, The influence of fiber length and concentration on the physical properties of wheat husk fibers rubber composites, *International Journal of Polymer Science* 2010 (2010) 1-8.

[9] B.D. Agarwal, L.J. Broutman, Analysis and performance of fiber composites, Second edition, John wiley & Sons, (1990).

[10] H.M. Akil, M.F. Omar, A.A.M. Mazuki, S. Safiee, Z.A.M. Ishak, A.A. Bakar, Kenaf fiber reinforced composites: A review, *Materials and Design* 32 (2011) 4107-4121.

[11] F. Rezaei, R. Yunus, N.A. Ibrahim, E.S. Mahdi, Effect of fiber loading and fiber length on mechanical and thermal properties of short carbon fiber reinforced polypropylene composite, *Malaysian Journal of Analytical Sciences* 11 (2007) 181-188.

Analysis of Effects of Cutting Parameters of Wire Electrical Discharge Machining on Material Removal Rate and Surface Integrity

H. R. Tonday*, **A. M. Tigga**

Department of Manufacturing Engineering, National Institute of Technology Jamshedpur, Jamshedpur-831014, Jharkhand, India

*Email: hulasnifft.13@gmail.com, Phone: +916572374118

Abstract. As wire electrical discharge machining is pioneered as a vigorous, efficient and precise and complex nontraditional machining technique, research is needed in this area for efficient machining. In this paper, the influence of various input factors of wire electrical discharge machining (WEDM) on output variable has been analyzed by using Taguchi technique and analysis of variance. The design of experiments has been done and by applying L8 orthogonal arrays method and experiments have been conducted and collected required data. The objectives of the research are to maximize the material removal rate and to minimize the surface roughness value (Ra). Surface morphology of machined workpiece has been obtained and examined by employing scanning electron microscopy (SEM) technique.

Keywords: wire EDM; Inconel 625; Material Removal Rate (MRR); surface Roughness; scanning electron microscopy (SEM).

1. Introduction

Wire electrical discharge machining process has been established as a crucial non-conventional thermoelectric machining technique to machine the materials which are a conductor of electricity. The mechanism of material removal is same as the common EDM process. The cutting tools are electric sparks, which have been set up and controlled with the help of a generator. Because the wire electrode is not quite touching the workpiece, the electrical charge has produced to a voltage that permits it to cross the gap between electrode and workpiece, and zaps off a tiny bit of the workpiece. The electrolyte not only cools down the cutting zone but also wash away the debris between wire electrode and workpiece to restrict them from being obstacles for the next spark. This process is employed in the aerospace and automotive, die & tool manufacturing industries. It is possible to obtain and measure the desired surface finish on a workpiece during machining in wire EDM process in any intricate shape of the workpiece irrespective of its hardness. As in Wire EDM, wire electrode cannot be reused, therefore, it is difficult to control the wire wastage and minimize the cost of wire.Inconel alloys are difficult to cut materials by the conventional machining process. It, however, is essential to find a productive and competent machining technique for Inconel alloys. For machining electrically conductive materials, wire EDM is pioneered as a vigorous, dynamic and precise technique. Parameters' details and procedures provided in machine manufacturer's manual have not contained enough information for machining new materials and different profiles desired by researchers or production engineer. The sophisticated software adjusts the gap and various parameters in the generator of machine in real time to develop sparkscontrolled carefully. Sparks with less energy develops less heat; whichresults in smaller heat-affected zones. A lot of research is being performed by many researchers for investigating the influence of wire EDM on the material removal rate, surface integrity of nickel alloys regarding recast layer and HAZ.

2. Literature Review

Kanlayasiri K &Boonmung S. [1] [2] came upon that pulse on time and pulse peak current had been important factors affecting the surface roughness during wire electrical discharge machining of DC53 die steel. The surface roughness of workpiece was increased when the pulse on time and pulse peak current was raised. Mao Yong Lin et al. [3] used micro-milling EDM for machining of Inconel 718 and optimized the machining parameters for obtaining desired results of multiple performance parameters by using Grey Taguchi method. Sharma N et al. [4] shown that the pulse on time is an important factor, which affects the cutting speed and dimensional accuracy in wire electrical discharge machining of HSLA steel. Experiments had been carried out on the basis of the central composite rotatable design of response surface methodology. Neeraj Sharma et al. [5] investigated the effect of parameters on metal removal rate for WEDM using HSLA as work-piece. They have observed that metal removal rate and surface roughness increases with increase in pulse on time and peak current. Response Surface methodology (RSM) is used to optimize the process parameter for metal removal rate and surface roughness. Somashekhar K.P. et al. [6] presented the formulation and solution of optimization of various process parameters for the selection of the best control settings on a micro-wire electrical discharge machining process.

Durairaj M. Sudharsun D, et al. [7] summarized the Grey relational theory and Taguchi optimization technique, in order to optimize the cutting parameters in Wire EDM for SS304. The objective of optimization is to attain the minimum kerf width and the best surface quality simultaneously and separately. Ho K H & Newman S T et al. [8] reviewed the research activities done in the field of wire EDM in which optimization of process parameters, monitoring and control of wire EDM process has been reported. NithinArvind et al. [9] optimized performance parameters using Taguchi's experimental design of WEDM with brass wire is fed to the workpiece. The authors predicted the optimal input parameters to maximize MRR and minimize surface finish. Palmers J, Stappen M V et al. [10] compared the white layers formed on workpiece surfaces in grinding and wire EDM in relation to the PVD TIN coating adhesion. They have suggested that white layers could be removed by tempering or annealing of the machined surface.

From the above literatures following statements have been reviewed: Electrode wire tension, wire wear rate and wire breakage and have not been examined appropriately and in many of the research papers. The authors have not performed complex cutting types such as taper cutting, turning, corner cutting, gear cutting etc. Heat distribution and heat transfer in the workpiece during WEDM process have not been explored in any research paper. The least number of researches have been done in which surface roughness is measured by direct contact method and surface morphology is examined by scanning electron microscopy (SEM).

3. Experimental Methodology

Inconel 625 has been used a workpiece material in this research work which known as a difficult to machine material and has the application in aerospace, tool and dies; medical parts manufacturing industries. The shape of the workpiece of Inconel 625 is a rectangular block which has been cut in total eight small pieces of 2 mm thickness. The specification of each piece is 10mm cutting length (l); 15mm cutting thickness (h) and 0.37mm kerf (k) in the rectangular shape as illustrated in figure (1). The chemical composition of Inconel 625 is material shown in the table (1). Electronica MAXICUT wire electrical discharge machine has been used for conducting the experiments and measured the machining time for each workpiece. In this machine, all the axes are servo controlled and can be programmed to follow a CNC code which is fed by the control board.

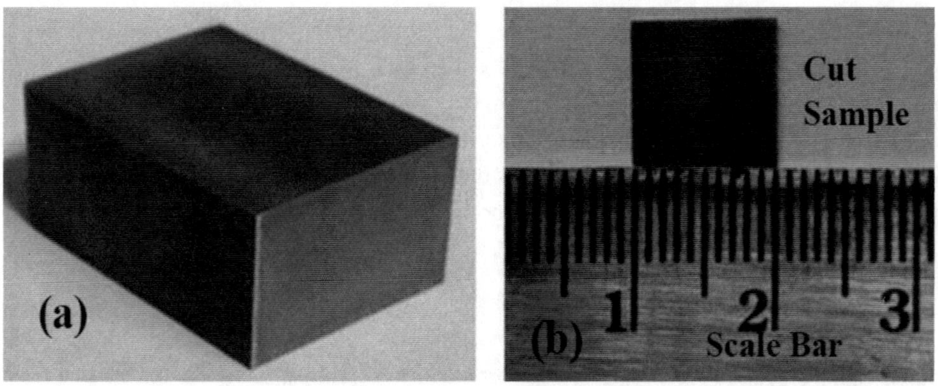

Fig. 1 (a) Inconel 625 workpiece before cutting, (b) Sample after cutting by WEDM.

Table 1. Chemical composition of Inconel 625

Element	Ni	Cr	Mo	Nb	Fe	Al	Ti	C	Mn
Content (%)	58	21.5	09	3.95	5	0.4	0.4	0.1	0.5

In this wire EDM tool, all three axes have an accuracy of 1µm. A brass wire electrode of 0.25mm diameter has been used and taken de-ionized water as the dielectric fluid. Diamond guides usually expensive to replace, but last a long time. The input parameters such spark voltage, pulse on time, and pulse off time and wire tension are considered for analysis with two different levels of each parameter as shown in the table (2) and other parameters remain constant. Surface Roughness has been measured and noted by using Talysurf surface roughness tester instrument of Taylor-Hobson. The design of experiments has been carried out by using L8 orthogonal arrays method and experiments have been conducted and collected the required output data as illustrated in table (3).

Table 2. Input factors and their levels for wire electrical discharge

Sr. No.	Input Parameter	Unit	Level 1	Level 2
1	spark voltage	volts	9	14
2	on time	µs	6	9
3	off time	µs	10	16
4	wire tension	gram-force	1200	1350

Table 3. Experimental design by using L_8 orthogonal arrays and output parameters

Exp. NO.	Spark Voltage	On Time	Off Time	Wire Tension	MRR	Surface Roughness
1	9	6	10	1200	1.968085	1.817
2	9	6	16	1350	**1.554187**	2.15
3	9	9	10	1350	2.120749	1.6
4	9	9	16	1200	1.676737	1.922
5	14	6	10	1350	2.328997	**2.566**
6	14	6	16	1200	1.616662	2.139
7	14	9	10	1200	2.724595	2.232
8	14	9	16	1350	2.494382	2

4. Experimental Results and Analysis

A collection of data is accomplished after cutting the Inconel 625 material by wire EDM. Machining time has been observed and noted after each experiment and material removal rate (MRR) have been calculated by applying formula shown in equation (1). Similarly, Surface roughness of each machined piece has also been measured by utilizing Telysurf instrument. These collected data have been analyzed by using powerful statistical software Minitab 16. In this software, Taguchi method has been considered for analysis of collected values of response parameters.

$$MRR = \frac{l \times h \times k}{t_m} \quad (\text{mm}^3/\text{minute}) \qquad \textbf{(1)}$$

Where, l = cutting length, h = cutting thickness, k= kerf of cutting, t_m = machining time

During statistical analysis and ANOVA analysis, larger is better concept is used for deciding the significant input parameters during analysis of MRR. Smaller is better criterion is considered for deciding the significant input parameters during analysis of surface roughness (Ra).ANOVA for the signal to noise ratio of MRR is shown in the table (4); and ANOVA for signal to noise ratios of surface roughness (Ra) is shown in the table (5). ANOVA has been conducted to identify and determine the significant input factors and their interactions which have the greatest impact on output parameters so that these input parameters can be controlled to obtain desired value of response parameters. Signal to noise (S/N) ratios for each output variable has also been computed. Response tables of the signal to noise ratios of MRR and surface roughness have also been calculated as illustrated in the table (6) and table (7).By using Minitab software, main effects plot and normal probability plot of signal to noise ratios of MRR and surface roughness have been drawn as shown in figure (2) and figure (3) respectively.

Table 4. Analysis of variance for signal to noise ratios of MRR

Source	DF	SS	MS	F	P
Spark Voltage	1	6.9025	6.9025	8.1	0.065
On Time	1	5.1805	5.1805	6.08	0.09
Off Time	1	8.0583	8.0583	9.45	0.054
Wire Tension	1	0.7163	0.7163	0.84	0.427
Residual Error	3	2.5571	0.8524		
Total	7	23.4117			

SS = sum of squares, MS = mean of squares, DF = degree of freedom

Table 5. Analysis of variance for signal to noise ratios of Surface Roughness

source	DF	SS	MS	F	P
spark voltage	1	4.7903	4.79028	3.61	0.154
on time	1	1.8753	1.8753	1.41	0.32
off time	1	0.0338	0.03382	0.03	0.883
wire tension	1	0.0308	0.0308	0.02	0.889
residual error	3	3.986	1.32865		
Total	7	10.7162			

SS = sum of squares, MS = mean of squares, DF = degree of freedom

Table 6. Response table for signal to noise ratios of MRR (Larger is Better)

level	spark voltage	on time	off time	wire tension
1	5.183	5.307	7.115	5.812
2	7.04	6.916	5.108	6.411
Delta	1.858	1.609	2.007	0.598
Rank	2	3	1	4

Table 7. Response table for signal to noise ratios of surface roughness

level	spark voltage	on time	off time	wire tension
1	-5.398	-6.656	-6.107	-6.11
2	-6.946	-5.688	-6.237	-6.234
Delta	1.548	0.968	0.13	0.124
Rank	1	2	3	4

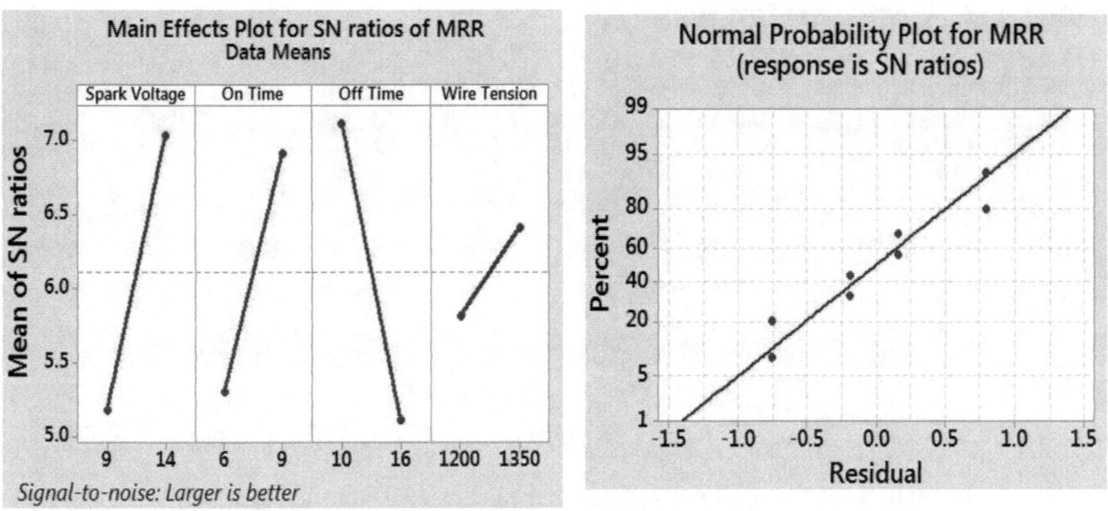

Fig. 2. Main effects plot and Normal probability for SN ratios of MRR

Surface Texture Analysis

JEOL JSM-6390LV Scanning electron microscopy has been used for surface morphology examination. It is a high-performance, low-cost, scanning electron microscope with a high resolution of 3.0nm. The specimen chamber can accommodate a specimen of up to 6-inches in diameter. The surface texture of machined pieces with various experiment numbers are shown in figure (4).SEM analysis of Exp. No- 02 and Exp. No- 07 has been carried out because they possess minimum MRR and Maximum MRR respectively. SEM analysis of Exp. No- 03 and exp. No- 05 has been performed because they concerned with the lower and higher surface roughness (Ra) respectively. It has been revealed that surface texture changes with variation in input parameters. Some material defects such as daub and black spots have been also observed in SEM image of exp. No. (07); and kerf of cutting is also recorded as shown in figure (5).

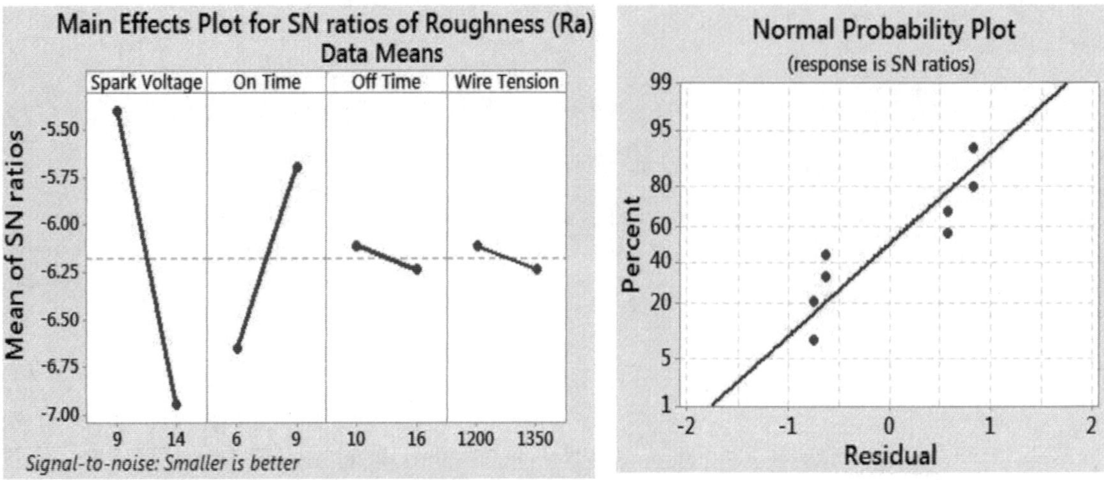

Fig. 3. Main effects plot and normal probability plot for SN ratios of surface roughness

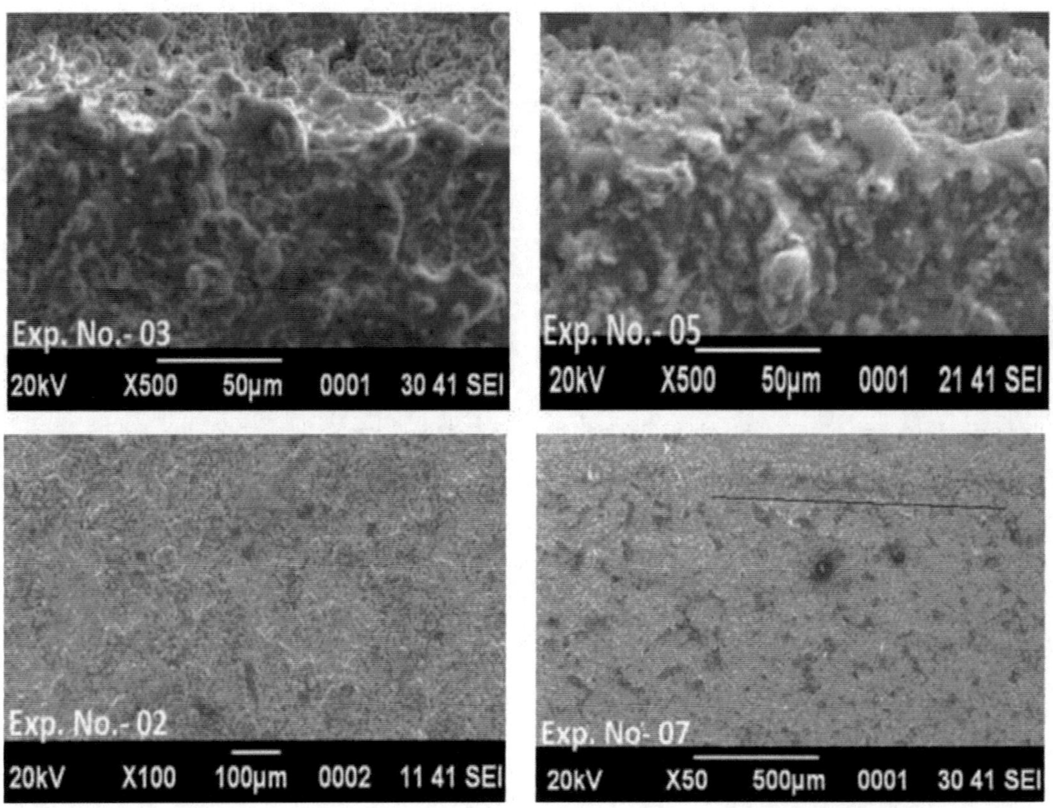

Fig. 4. SEM images of machined pieces with various experiment numbers

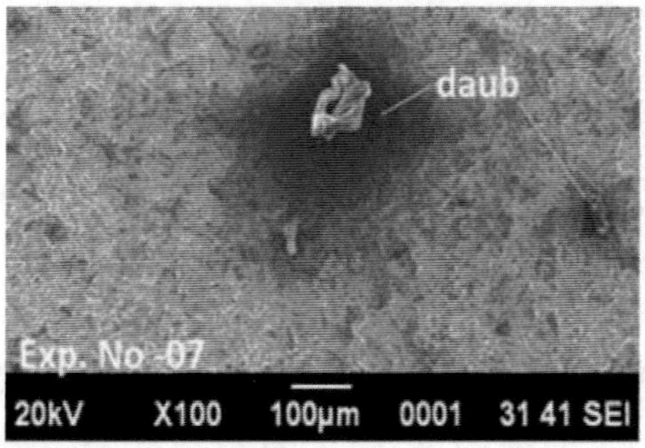

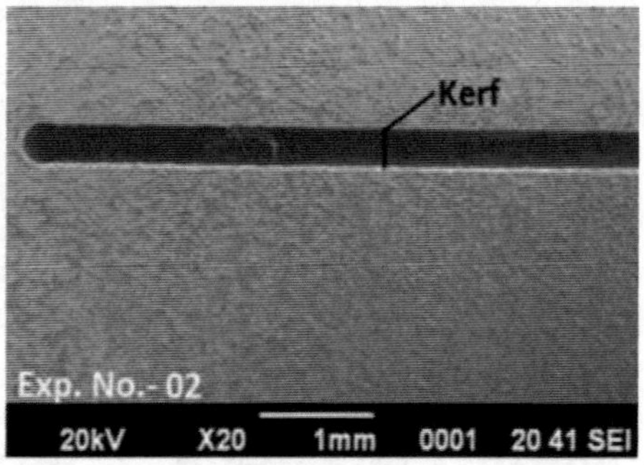

Fig. 5. Exp No -07 with material defect daub, Exp no- 02 representing Kerf of cutting.

Conclusion

There is the following conclusions are made after analysis of data obtained:

(i) Wire tension has no effect on both the output parameters.
(ii) On time has the highest impact on both the output parameters (MRR and surface roughness).
(iii) SEM images revealed that surface morphology of machined piece changes with the variation in input parameters.
(iv) Experiment number 7 has the maximum MRR value as obtained by experiments.
(v) Experiment number 03 has the minimum surface roughness (R_a) value.
(vi) Input parameter off time has the greatest impact on MRR whereas spark voltage has the largest impact on surface roughness (R_a) among all other parameters.

References

[1] K. Kanlayasiri, S. Boonmung, An investigation on effects of wire-EDM machining parameters on surface roughness of newly developed DC53 die steel,*J. Mater. Process. Technol.*,**187–188** (2007) 26–29.

[2] K. Kanlayasiri, S. Boonmung, Effects of wire-EDM machining variables on surface roughness of newly developed DC 53 die steel: Design of experiments and regression model,*J. Mater. Process. Technol.*, **192–193** (2007) 459–464.

[3] M. Y. Lin, C. C. Tsao, C. Y. Hsu, A. H. Chiou, P. C. Huang, and Y. C. Lin, Optimization of micro milling electrical discharge machining of Inconel 718 by Grey-Taguchi method, *Trans. Nonferrous Met. Soc. China (English Ed.)*, **23**(2013) 661–666.

[4] N. Sharma and R. Khanna, Modeling and multiresponse optimization on WEDM for HSLA by RSM, *Int. J. Adv. Manuf. Technol.*, **67** (2013) 2269–2281.

[5] N. Sharma, R. Khanna, R. Gupta, Multi Quality Characteristics Of WEDM Process Parameters With RSM,*Procedia Eng.*, **64** (2013) 710–719.

[6] K. P. Somashekhar, J. Mathew, and N. Ramachandran, A feasibility approach by simulated annealing on optimization of micro-wire electric discharge machining parameters,*Int. J. Adv. Manuf. Technol.*, **61** (2012) 1209–1213.

[7] M. Durairaj, D. Sudharsun, and N. Swamynathan, Analysis of Process Parameters in Wire EDM with Stainless Steel using Single Objective Taguchi Method and Multi Objective Grey Relational Grade,*Procedia Eng.*, **64** (2013) 868–877.

[8] K. H. Ho, S. T. Newman, S. Rahimifard, and R. D. Allen, "State of the art in wire electrical discharge machining (WEDM)," *Int. J. Mach. Tools Manuf.*, **44**(2004) 1247–1259.

[9] Nithin Aravind, S. Sowmyi, Optimization Of Metal Removal Rate And Surface Roughness On Wire EDM Using Taguchi Method,*IEEE-International Conf. Adv. Eng. Sci. Manag.*, (2012) 155–159.

[10] J. Palmers, M. Van Stappen, J. D'Haen, M. D'Olieslaeger, L. M. Stals, G. Uhlig, M. Foller, and E. Haberling, Influence of the presence of white layers formed during grinding and wire-electro-discharge machining on PVD TiN coating adhesion,*Surf. Coatings Technol.*, **74–75** (1995) 162–167.

Free Vibration and Bending Behaviour of CNT Reinforced Composite Plate using Different Shear Deformation Theory

K Mehar[*] and S K Panda

Department of Mechanical Engineering, National Institute of Technology Rourkela. India.

[*]E-mail:kulmanimehar@gmail.com

Abstract. In the present study, the free vibration and the bending behaviour of carbon nanotube reinforced composite plate are computed using three different shear deformation theories under thermal environment. The material properties of carbon nanotube and matrix are assumed to be temperature-dependent, and the extended rule of mixture is used to compute the effectivematerial properties of the composite plate. The convergence and validity of the present modelalso have been checked by computing the wide variety of the numerical example. The applicability of the proposed higher-order models has been highlighted by solving the wide variety of examples for different geometrical and material parameters underelevated thermal environment.The responses are also examined using the simulation model developed in commercial finite element package (ANSYS).

Keywords: CNT, free vibration, bending, thermal environment,FSDT,HSDT.

1. Introduction

From last few decade, many types of researchhave already been completedon the fibre or the laminated composite materials to show the significant improvementof the specific strength and the stiffness behaviour of the advanced structure or the structural components. Recently, carbon nanotubes (CNTs) are used as the reinforcement material in nanocomposites due to their excellent specific strength, specific stiffness, thermal, electrical, and chemical properties. The CNT has got numerous applications in the aerospace, the nuclear plant, automobile and marine structures due to their outstanding properties. The effective material properties of the carbon nanotube reinforced composite (CNTRC) are evaluated using the variousavailable method such as molecular dynamics (MD) simulation, [1-3] representative volume element (RVE) [4-5],the rule of mixture [6-7] and Eshelby–Mori–Tanaka approach [8].In general, the CNTs are classified based on their configurations namely, armchair, zigzag and chiral and their elastic properties varyaccordingly. Ayatollahia *et al.* [9] estimated the nonlinear mechanical properties of the single wall nanotubes under three different set of load (tensile, bending and torsional load) using a molecular model based on the finite element method (FEM) for both armchair and zigzag configuration.Aragh*et al.*[10] computed the free vibration behaviour of randomly oriented CNTRC cylindrical shell based on third-order shear deformation theory (TSDT). Lin and Xiang [11] investigated the free vibration response of FG-CNTRC beam using the p-Ritz method based on the first-order shear deformation theory (FSDT) and TSDT kinematic model with von Karman sense geometric nonlinearity. Shen *et al.*[12] investigated the vibration response of multi-wall carbon nanotube (MWCNT) based biosensor using nonlocal

Timoshenko beam theory(TBT).Ke*et al.*[13] computed the dynamic stability, free vibration and buckling behaviour of functionally graded carbon nanotube reinforced composite (FG-CNTRC) beam using TBT. Alva and Raja [14]examined the effects of the weight percentage and the diameter of the MWCNT on the damping characteristics of the MWCNT/epoxy composite. Jarali*et al.*[15]reported the effect of agglomeration and the volume fractions on the mechanical, thermal, electrical and the moisture properties of the CNT/epoxy composite. Vodenitcharova and Zhang[16] computed the local buckling, and the bending behaviour of the SWCNT reinforced composite plates using Airy'sstress function. Formica *et al.* [17]numericallysolved the free vibration responses of the CNTRC plate using FEM and the effective material properties of the CNTRCplate are computed based on Mori-Tanaka scheme. Zhu *et al.* [18] examined the free vibration and bending analysis of the FG-CNTRC plate using the FSDT kinematics. Yas and Heshmati[19] estimated the free and force vibrational behaviour of functionally graded nanocomposite beams reinforced by randomly oriented straight SWCNTs based on Timoshenko and Euler–Bernoulli beam theories using FEM. Lei *et al.* [20-22] computed the free vibration and the buckling behaviour of the FG-CNTRC cylindrical panels based on the FSDT shell theory using the element-freekp-Ritz method. Alibeigloo and Liew[23] examined the flexural behaviour of FG-CNTRC rectangular plate with all edges simply supported boundary conditions based on the three-dimensional theory of elasticity. Lei *et al.* [24] evaluated the flexural strength and free vibration behaviour of FG-CNTR cylindrical shell panels based on the FSDT. Mehrabadi and Aragh[25] computed the static response of FG-CNTRC cylindrical shell panel using TSDT.

From the above review, weconcluded that considerable amount of work has already been completed on the property evaluation of the CNT and the CNT reinforced composites. In addition to that the structuralanalysis(free vibration, bending and buckling) of the CNT reinforced composites are computed using the FSDT kinematics with and without temperature load. Based on the authors' knowledge no study has been reportedyet on the free vibration and bending behaviour of the CNTRCplate under thermal environmentusing the HSDT kinematic model. The objective of the present work is to compute the free vibration and the bending responses of the CNTRCflat panel numerically using the FSDT and HSDT mid-plane kinematics in conjunction with FEM. The domain is discretized using the suitable isoparametric finite element steps through a nine noded element with nine (model 1), six (model) or ten (model 3) degrees of freedom (DOF) per node. The responses are obtained numerically through a customized computer code developed in MATLAB environment. The validity and the convergence behaviour of the present numerical model have been checked, and the responses are also computed usingthe simulation model developed in ANSYS through ANSYS parametric design language (APDL) code. Finally, the effect of various design parameters (aspect ratios, support conditions, thickness ratios, volume fractions and temperature load) on the free vibration, static, stress and deformation behaviour of the CNTRC plate are highlighted by solving the wide variety of examples.

2. Theory and Formulation

In this analysis, the rectangular configuration of CNTRC plate is considered for the free vibration and the bending analysis as shown in fig. 1. The dimensions of CNTRC plate are the length (a), width (b) and the thickness (h). The CNTs are assumed to be uniformly distributed and alignalong the length of the plate.The displacement field kinematics within the plate is assumed to be based on the HSDT/FSDT, where the in-plane displacements are expanded as cubic/linear functions of the thickness coordinate while the transverse displacement varies either linearly and/or constant through the plate thickness. In this study, three different kinematic models have been utilized for the analysis purpose and discussed in the following lines.

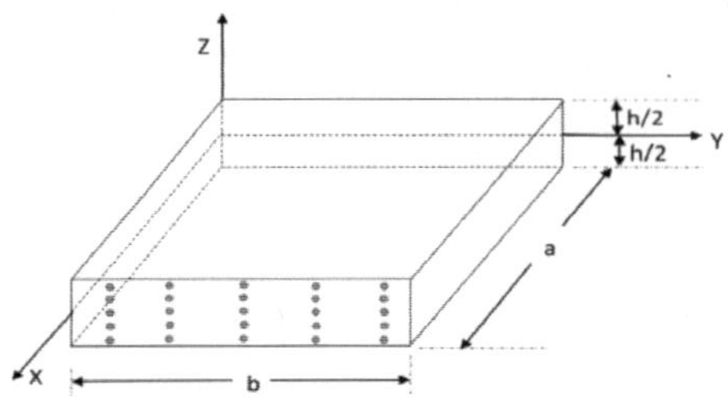

Fig. 1:- Carbon nanotube reinforced composite plate geometry and configuration.

Model 1

As a first step, the mathematical model is developed using the HSDT mid-plane kinematics with nine DOF of the CNTRCs plate as in Kishore *et al.* [26]:

$$\left.\begin{array}{l} u(x,y,z,t)=u_0(x,y,t)+z\varphi_x(x,y,t)+z^2\psi_x(x,y,t)+z^3\theta_x(x,y,t) \\ v(x,y,z,t)=v_0(x,y,t)+z\varphi_y(x,y,t)+z^2\psi_y(x,y,t)+z^3\theta_y(x,y,t) \\ w(x,y,z,t)=w_0(x,y,t) \end{array}\right\} \quad (1)$$

Model 2

Similarly, one another mathematical model is developed using the FSDT mid-plane kinematics with six DOF of the CNTRCs plate as in Szekrenyes[27]:

$$\left.\begin{array}{l} u(x,y,z,t)=u_0(x,y,t)+z\varphi_x(x,y,t) \\ v(x,y,z,t)=v_0(x,y,t)+z\varphi_y(x,y,t) \\ w(x,y,z,t)=w_0(x,y,t)+z\varphi_z(x,y,t) \end{array}\right\} \quad (2)$$

Model 3

Further, the mathematical model is developed using another HSDT mid-plane kinematics with ten DOF of the CNTRCs plate as in Kishore *et al.*[26]:

$$\left.\begin{array}{l} u(x,y,z,t)=u_0(x,y,t)+z\varphi_x(x,y,t)+z^2\psi_x(x,y,t)+z^3\theta_x(x,y,t) \\ v(x,y,z,t)=v_0(x,y,t)+z\varphi_y(x,y,t)+z^2\psi_y(x,y,t)+z^3\theta_y(x,y,t) \\ w(x,y,z,t)=w_0(x,y,t)+z\varphi_z(x,y,t) \end{array}\right\} \quad (3)$$

where, *u, v,* and *w* are the displacement of any point within the plate along X, Y and Z directions, respectively. u_0, v_0, and w_0 are midpoint displacement along X, Y, and Z direction respectively. φ_x and φ_y are the rotation of normal to the mid-plane about the Y and X direction respectively. φ_z, ψ_x, ψ_y, θ_x and θ_y are higher-order terms of Tayler series expansion in the mid plane of the plate.

Again, Eqs.(1), (2) and (3) are rewritten in the matrix form as follows:

$$\{\lambda\}=[f_n]\{\lambda_{0n}\} \quad (4)$$

where, $\{\lambda\}$, $[f_n]$ and $\{\lambda_{0n}\}$ are the displacement field vector of any point, thickness coordinate matrix and displacement field vector within the mid-plane of nth model. The details of each term are provided in the Appendix.

The mid-plane kinematics of the CNTRC plateis evaluated using the following strain-displacement relation:

$$\{\varepsilon\}=\begin{Bmatrix}\varepsilon_{xx}\\ \varepsilon_{yy}\\ \varepsilon_{zz}\\ \gamma_{yz}\\ \gamma_{zx}\\ \gamma_{xy}\end{Bmatrix}=\begin{Bmatrix}\frac{\partial u}{\partial x}\\ \frac{\partial v}{\partial y}\\ \frac{\partial w}{\partial z}\\ \frac{\partial v}{\partial z}+\frac{\partial w}{\partial y}\\ \frac{\partial u}{\partial z}+\frac{\partial w}{\partial x}\\ \frac{\partial u}{\partial y}+\frac{\partial v}{\partial x}\end{Bmatrix}=\begin{Bmatrix}\varepsilon_x^0\\ \varepsilon_y^0\\ \varepsilon_z^0\\ \gamma_{yz}^0\\ \gamma_{zx}^0\\ \gamma_{xy}^0\end{Bmatrix}+z\begin{Bmatrix}\varsigma_x\\ \varsigma_y\\ \varsigma_z\\ \varsigma_{yz}\\ \varsigma_{zx}\\ \varsigma_{xy}\end{Bmatrix}+z^2\begin{Bmatrix}\zeta_x\\ \zeta_y\\ \zeta_z\\ \zeta_{yz}\\ \zeta_{zx}\\ \zeta_{xy}\end{Bmatrix}+z^3\begin{Bmatrix}\xi_x\\ \xi_y\\ \xi_z\\ \xi_{yz}\\ \xi_{zx}\\ \xi_{xy}\end{Bmatrix} \tag{5}$$

or $$\{\varepsilon\}=[T]\{\bar{\varepsilon}\} \tag{6}$$

where,

$$\{\bar{\varepsilon}\}=\left[\varepsilon_x^0\ \varepsilon_y^0\ \varepsilon_z^0\ \varepsilon_{yz}^0\ \varepsilon_{zx}^0\ \varepsilon_{xy}^0\ \varsigma_x\ \varsigma_y\ \varsigma_z\ \varsigma_{yz}\ \varsigma_{zx}\ \varsigma_{xy}\ \zeta_x\ \zeta_y\ \zeta_z\ \zeta_{yz}\ \zeta_{zx}\ \zeta_{xy}\ \xi_x\ \xi_y\ \xi_z\ \xi_{yz}\ \xi_{zx}\ \xi_{xy}\right]^T$$

$$[T]=\begin{bmatrix}1&0&0&0&0&0&z&0&0&0&0&0&z^2&0&0&0&0&0&z^3&0&0&0&0&0\\0&1&0&0&0&0&0&z&0&0&0&0&0&z^2&0&0&0&0&0&z^3&0&0&0&0\\0&0&1&0&0&0&0&0&z&0&0&0&0&0&z^2&0&0&0&0&0&z^3&0&0&0\\0&0&0&1&0&0&0&0&0&z&0&0&0&0&0&z^2&0&0&0&0&0&z^3&0&0\\0&0&0&0&1&0&0&0&0&0&z&0&0&0&0&0&z^2&0&0&0&0&0&z^3&0\\0&0&0&0&0&1&0&0&0&0&0&z&0&0&0&0&0&z^2&0&0&0&0&0&z^3\end{bmatrix}$$

here, $[T]$ and $\{\bar{\varepsilon}\}$ are the thickness coordinate matrix and the mid-plane strain vector, respectively. The general stress-strain relationship of the composite is expressed as follows:

$$\begin{Bmatrix}\sigma_{xx}\\ \sigma_{yy}\\ \tau_{xy}\\ \tau_{zx}\\ \tau_{yz}\end{Bmatrix}=\begin{bmatrix}Q_{11}&Q_{12}&0&0&0\\Q_{21}&Q_{22}&0&0&0\\0&0&Q_{66}&0&0\\0&0&0&Q_{55}&0\\0&0&0&0&Q_{44}\end{bmatrix}\left(\begin{Bmatrix}\varepsilon_{xx}\\ \varepsilon_{yy}\\ \gamma_{xy}\\ \gamma_{zx}\\ \gamma_{yz}\end{Bmatrix}-\begin{Bmatrix}\alpha_{11}\\ \alpha_{22}\\ 0\\ 0\\ 0\end{Bmatrix}\Delta T\right) \tag{7}$$

where, $Q_{11}=Q_{22}=E/(1-\upsilon^2)$, $Q_{12}=Q_{21}=E\upsilon/(1-\upsilon^2)$, $Q_{66}=Q_{55}=Q_{44}=E/2(1+\upsilon)$

In which we assumed that $G_{13}=G_{12}$ and $G_{23}=1.2\times G_{12}$. ΔT is the uniform temperature rise across the panel thickness.Eq. (7) can also be rewritten as

$$\{\sigma\}=[Q]\{\varepsilon-\varepsilon_{th}\} \tag{8}$$

where, [Q] is the reduced stiffness matrix.

Theeffective material properties of CNTRC plate can be computed using the extended rule of mixture as[18]:

$$E_{11} = \eta_1 V_{CNT} E_{11}^{CNT} + V_m E^m \tag{9}$$

$$\frac{\eta_2}{E_{22}} = \frac{V_{CNT}}{E_{22}^{CNT}} + \frac{V_m}{E^m} \tag{10}$$

$$\frac{\eta_3}{G_{12}} = \frac{V_{CNT}}{G_{12}^{CNT}} + \frac{V_m}{G^m} \tag{11}$$

and,
$$V_{CNT} + V_m = 1 \tag{12}$$

where, E_{11}, E_{22},and G_{12} are the effective Young's modulus in the longitudinal and lateral direction and shear modulus of CNTRCplate respectively. Similarly, E_{11}^{CNT}, E_{22}^{CNT} and G_{12}^{CNT} are the properties of the CNT in their corresponding directions. In addition, E^m and G^m are the Young's modulus and shear modulus of the matrix material respectively and η_1, η_2 and η_3 are the effective parameter of CNT in the material directions.

The effective Poisson's ratio (v_{12}) and the density of each constituent material (ρ) areobtained using the formulas as [18]:

$$v_{12} = V_{CNT} v^{CNT} + V_m v^m \tag{13}$$

$$\rho = V_{CNT}\, \rho^{CNT} + V_m \rho^m \tag{14}$$

where, v^{CNT} and v^m are Poisson's ratio of the SWCNT and the matrix material, respectively. Similarly, ρ^{CNT} and ρ^m arethe densities of the CNT and the matrix, respectively.

2.1. *Finite Element Method*

FEM has been widely appreciated numerical tool for the structural analysis with various geometrical and the material complexities. In this present analysis, the displacement fields for different assumed kinematic models are expressed in terms of desired field variables and the models are discretized using suitable FEM steps. The displacement vector $\{\lambda_0\}$ at any point within the mid-surface is given by:

$$\{\lambda_0\} = \sum_{i=1}^{9} N_i \{\lambda_{0i}\} \tag{15}$$

where, $\{\lambda_{0i}\} = \left[u_{0_i}\ v_{0_i}\ w_{0_i}\ \varphi_{x_i}\ \varphi_{y_i}\ \psi_{x_i}\ \psi_{y_i}\ \theta_{x_i}\ \theta_{y_i}\right]^T$, $\{\lambda_{0i}\} = \left[u_{0_i}\ v_{0_i}\ w_{0_i}\ \varphi_{x_i}\ \varphi_{y_i}\ \varphi_{z_i}\right]^T$ and $\{\lambda_{0i}\} = \left[u_{0_i}\ v_{0_i}\ w_{0_i}\ \varphi_{x_i}\ \varphi_{y_i}\ \varphi_{z_i}\ \psi_{x_i}\ \psi_{y_i}\ \theta_{x_i}\ \theta_{y_i}\right]^T$ are the nodal displacement vector formodel 1, model 2 and model 3 respectivelyand N_i is the corresponding interpolating function associated with node '*i*'.

The total strain energy of the CNTRC plate can be expressed as:

$$U = \frac{1}{2}\iint\left[\int_{-h/2}^{+h/2} \{\varepsilon\}^T [Q]\{\varepsilon\}\, dz\right] dxdy \tag{16}$$

Now, Eq. (16) can be rewritten by substituting the strainfromEqs. (6) and conceded as:

$$U = \frac{1}{2}\iint\left(\{\bar{\varepsilon}\}^T [D]\{\bar{\varepsilon}\}\right) dxdy \tag{17}$$

where, $[D]=\int_{-h/2}^{+h/2}[T]^T\left[Q_{ij}\right][T]dz$.

The total work done by externally applied load (p) and in-plane thermal load is given by

$$W=\iint\{\lambda_0\}^T\{p\}dxdy+\iint\left(\{\varepsilon\}^T[Q]\{\varepsilon_{th}\}\right)dxdy$$

$$W=\{\lambda_0\}^T\{F_m\}+\{\lambda_0\}^T\{F_{th}\} \tag{18}$$

where, $\{F_m\}$ and $\{F_{th}\}$ are the mechanical and thermal load vectors, respectively.

The kinetic energy of the CNTRC plate can be expressed as:

$$T=\frac{1}{2}\int_V\rho\{\dot{\lambda}\}^T\{\dot{\lambda}\}dV \tag{19}$$

where, ρ and $\{\dot{\lambda}\}$ are the mass density and the global velocity vector.

Using, Eqs. (3) and (19), the kinetic energy expression of the CNTRC plate with thickness h can be written as:

$$T=\frac{1}{2}\int_A\left(\int_{-h/2}^{+h/2}\{\dot{\lambda}_0\}^T[f]^T\rho[f]\{\dot{\lambda}_0\}dz\right)dA=\frac{1}{2}\int_A\{\dot{\lambda}_0\}^T[m]\{\dot{\lambda}_0\}dA \tag{20}$$

where, $[m]=\int_{-h/2}^{+h/2}[f]^T\rho[f]dz$ is the inertia matrix.

2.2. *ANSYS model and solution steps*

As discussed earlier, a simulation model has been developed in ANSYS using APDL code to compute the desired bending and the free vibration responses of the CNTRC plate under thermal environment. The present model is discretized using eight node serendipity element (SHELL281) with six degrees of freedom at each node from the ANSYS element library. The SHELL281 element is capable of solvingthe thick/thin panel under combined thermo-mechanicalload. The solution procedure used in the ANSYS are as follows:

(i) The first step is to create the desired geometry of the plate with sides a and$b$.
(ii) Assign the material properties.
(iii) Discretize the plate model using SHELL281 element, from ANSYS element library, to obtain the required mesh.
(iv) Then, apply the boundary condition and the load to get the required responses.
(v) The linear eigenvalue vibration problem is solved using Block Lanczos method.

2.3. *Bending analysis*

The final form of the governing equation of the CNTRC plate is obtained using variational principle and conceded as:

$$\delta\Pi=\delta U-\delta W=0 \tag{21}$$

where, δ is the variational symbol and Π is the total potential energy.

IOP Conf. Series: Materials Science and Engineering **115** (2016) 012014 doi:10.1088/1757-899X/115/1/012014

Now, the desired governing equation for the static analysis of the CNTRC plate is obtained by substituting the values of the strain energy and total work done from the Eqs. (17) and (18) in the Eq. (21) and rewritten as:

$$[K_S]\{\lambda_0\} = \{p\} \quad (22)$$

where, $[K_S]$ is the global stiffness matrix.

2.4. Free vibration analysis

The governing equation of free vibrated CNTRC plate is derived using Hamilton's principle and expressed as:

$$\delta\int_{t_1}^{t_2}(T-U)dt = 0 \quad (23)$$

The equilibrium equation for vibrated CNTRC plate can be obtained by substituting the values of the kinetic energy (*T*) and the strain energy (*U*) in Eq. (23) and modified as follows:

$$[M]\{\ddot{\lambda}\} + [K]\{\lambda\} = 0 \quad (24)$$

where, $[M]=[N]^T[m][N]$ is the system mass matrix and $[K]=[B]^T[D][B]$ is the system stiffness matrix.

Again, Eq. (24) is rewritten in the eigenvalue equation form and represented as:

$$\left([K]-\omega^2[M]\right)\Delta = 0 \quad (25)$$

where, ω is the natural frequency and Δ is the corresponding eigenvector.

Table 1. Material properties of (10, 10) SWCNT (*L*= 9.26 nm, *R*=0.68 nm, *h*=0.067 nm, v_{12}^{CNT} = 0.175). [28]

Temperature (K)	E_{11}^{CNT} (TPa)	E_{22}^{CNT} (TPa)	G_{12}^{CNT} (TPa)	α_{11}^{CNT} (10^{-6}/K)	α_{22}^{CNT} (10^{-6}/K)
300	5.6466	7.0800	1.9445	3.4584	5.1682
400	5.5679	6.9814	1.9703	4.1496	5.0905
500	5.5308	6.9348	1.9643	4.5361	5.0189
700	5.4744	6.8641	1.9644	4.6677	4.8943

Table 2. Effective parameter of CNT. [28]

V_{CNT}	η_1	η_2	η_3
0.12	0.137	1.022	0.7185
0.17	0.142	1.626	1.138
0.28	0.141	1.585	1.109

3. Results and Discussions

The free vibration and bending behaviour of the CNTRC plate are investigated using suitable finite element model developed in MATLABenvironment. For the present analysis, PMMAis considered as the matrix phase and the SWCNT as the fibre phase.The material properties of both matrix and fibre are assumed to be temperature-dependent. The properties of PMMA are taken same as[28]ρ^m =1150kg/m^3, v^m = 0.34 ,E^m = (3.52-0.0034*T*) GPa. andα^m = 45(1+0.0005 ΔT) $\times10^{-6}$ in which ΔT= *T* –

T_0 and T_0= 300K (ambient temperature). Similarly, the properties of the armchair (10, 10) configuration of the SWCNTs is taken same as the reference [28] and presented in Table 1. The effectiveness parameters of CNT isalso provided in Table 2and the effective material properties of the composite plate are obtained using the extended rule of mixture.It is true that the extended rule of mixture is one of the appropriate methodfor the analysis of the short fibre reinforced composite structure. Table 3 shows comparison study of the elastic properties computed using the extended rule of mixture and compared with MD simulation results. It is clear from the table that the present results are showing very good agreement and MD simulation is well accepted in materials related analysis due to its accuracy. In the present analysis,the thickness of the composite plate is taken as 2mm throughout if not stated otherwise.

Table 3. Comparison of effective material properties of the CNT/PMMA composite.

V_{CNT}	MD [2]		Extended rule of mixture	
	E_{11} (GPa)	E_{22} (GPa)	E_{11}(GPa)	E_{22} (GPa)
0.12	94.6	2.9	95.021	2.8917
0.17	138.9	4.9	138.38	4.8776
0.28	224.2	5.5	224.72	5.4807

The following sets of support conditions are employed to compute the desired responses:

(a) All edges simply supported condition (SSSS):
$v = w = \varphi_y = \psi_y = \theta_y = 0$ at $x = 0, a$
$u = w = \varphi_x = \psi_x = \theta_x = 0$ at $y = 0, b$

(b) All edges clamped condition (CCCC):
$u = v = w = \varphi_x = \varphi_y = \psi_x = \psi_y = \theta_x = \theta_y = 0$ for both end i.e., $x = 0, a$ and $y = 0, b$.

(c) Combine simply support and clamped condition (SCSC)
$v = w = \varphi_y = \psi_y = \theta_y = 0$ at $x = 0, a$ and
$u = v = w = \varphi_x = \varphi_y = \psi_x = \psi_y = \theta_x = \theta_y = 0$ at $y = 0, b$.

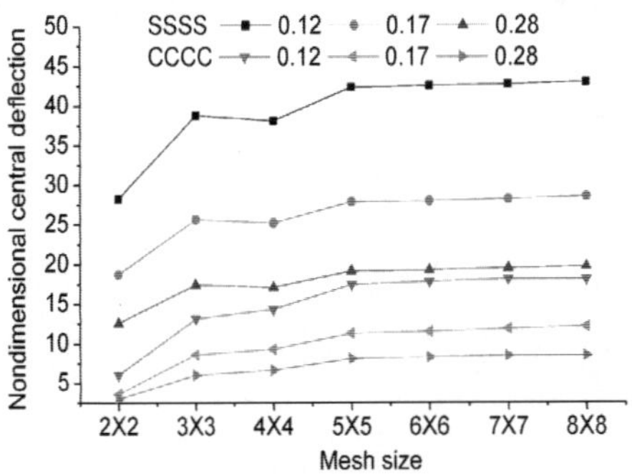

Model 1

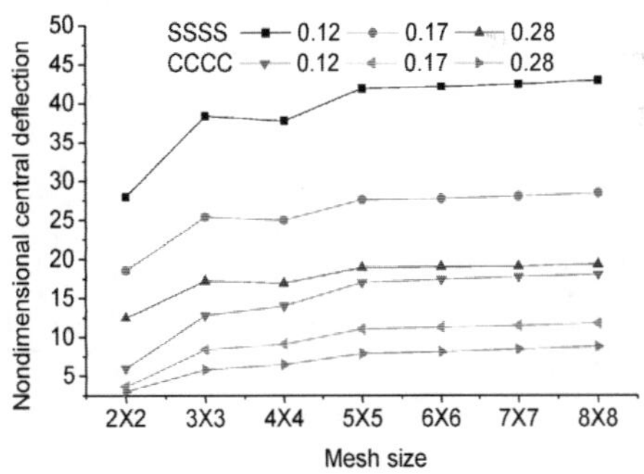

Model 2

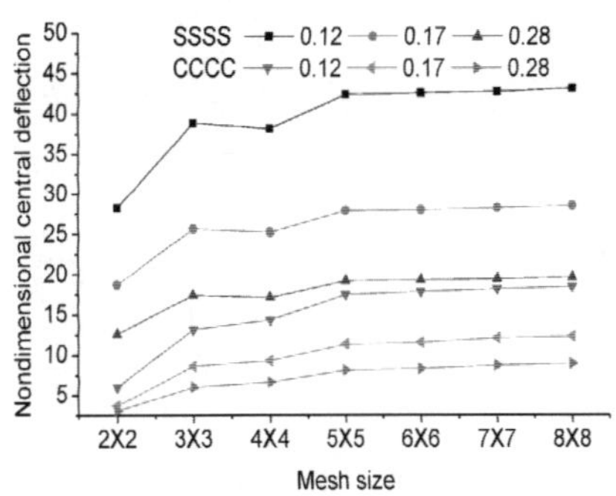

Model 3

Fig. 2:- Convergence behavior of the nondimensional central deflection: (a) Model 1 (b) Model 2 and (c) Model 3.

In this present analysis, the nondimensional form of the central deflection and the frequencies are obtained using the formulae: $\overline{W} = W_{\max} / h$ and $\overline{\omega} = \omega(a^2 / h)\sqrt{\rho_0 / E_0}$ respectively, where E_0, ρ_0 and W_{max} are the Young's modulus and the density of the matrix material at ambient temperature (300K) and the maximum central deflection of the plate, respectively.

3.1. Convergence and validation study

In this section, the convergence behaviour of the present numerical models of the CNTRC plate is computed for different support conditions and volume fractions. The convergence behaviour of the nondimensional central deflection parameter and the nondimensional fundamental frequencies of the CNTRC plate is presented in fig. 2(a)-(c) and fig. 3(a)-(c), respectively. In this example, the responses are computed for two support conditions (SSSS and CCCC) and three volume fraction (V_{CNT} = 0.12, 0.17 and 0.28) at room temperature (300K). The nondimensional central deflections are obtained under point load of 10KN at the center of the plate. It is clearly observed from the figures that the present results are converging well with mesh refinement for all the volume fractions and the support conditions. Therefore, a (6×6) mesh is utilized to compute the further analysis.

Now the proposed finite element model is extended to show the validation behaviour by comparing the responses with the available published literature and the simulation model. The nondimensional bending and vibration responses of the clamped CNTRC plate is computed and presented in Table 4 and Table 5, respectively. From results, it is clear that the presently developed models are good agreement with the previously published results. For the comparison purpose all the geometrical and material properties are taken same as to the Zhu *et al.*[18].

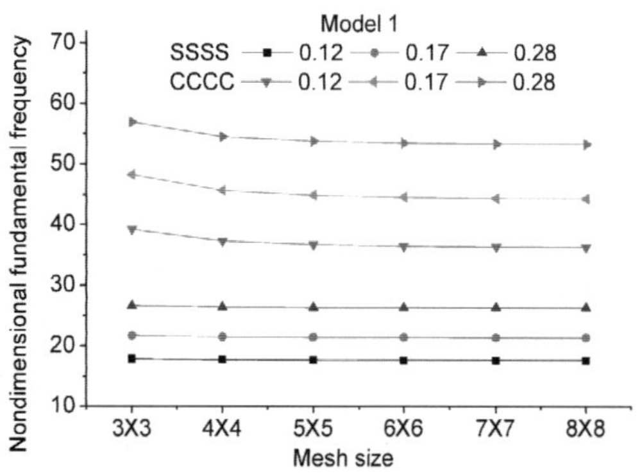

(a) Model 1

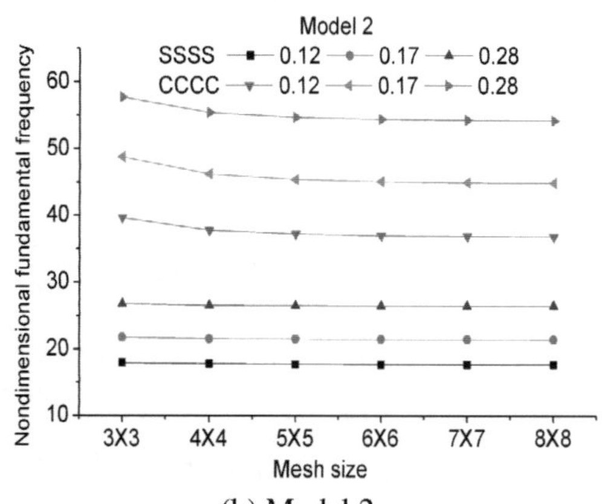

(b) Model 2

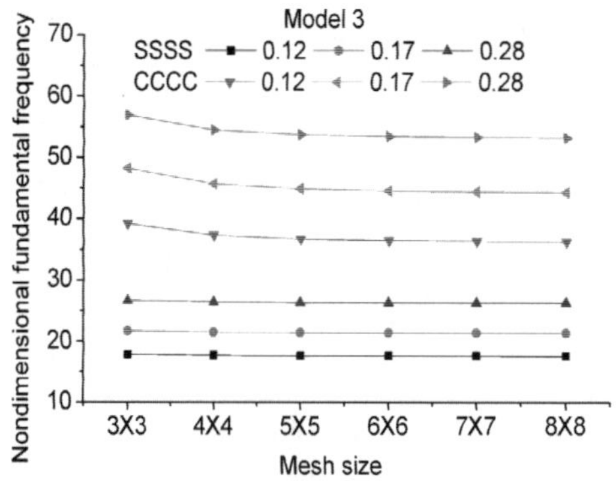

(c) Model 3

Fig. 3:- Convergence behavior of the nondimensional fundamental frequency: (a) Model 1 (b) Model 2 and (c) Model 3.

Table 4. Comparison study of nondimensional central deflection of CNTRC plate (T = 300K and uniformly distributed load = 0.1MPa).

a/h	V_{CNT}	Model 1	Model 2	Model 3	ANSYS	Zhu *et al.* [18]
10	0.11	0.0021	0.0017	0.0021	0.00222	0.0022
	0.14	0.0020	0.0016	0.0020	0.00210	0.0021
	0.17	0.0014	0.0011	0.0013	0.00199	0.0014
20	0.11	0.0134	0.0111	0.0132	0.01335	0.0134
	0.14	0.0118	0.0097	0.0117	0.01190	0.0119
	0.17	0.0085	0.0071	0.0085	0.01090	0.0086

Table 5. Comparison study of nondimensional fundamental frequencies of square CNTRC plate (T =300K).

a/h	V_{CNT}	Model 1	Model 1	Model 2	ANSYS	Zhu *et al.* [18]
10	0.11	18.0904	18.5970	16.5539	17.6385	17.625
	0.14	18.6661	19.2038	17.0819	18.1493	18.127
	0.17	22.5748	23.2008	20.6588	22.0240	22.011
50	0.11	40.0976	37.7016	36.4986	39.6704	39.730
	0.14	43.8828	41.4403	39.9825	43.5893	43.583
	0.17	49.5328	46.5099	45.0802	48.9897	49.074

Table 6. Effect of volume fraction on the nondimensional central deflection.

Support conditions	Load (KN)	V_{CNT}= 0.12			V_{CNT}= 0.17			V_{CNT}= 0.28		
		Model 1	Model 2	Model 3	Model 1	Model 2	Model 3	Model 1	Model 2	Model 3
SSSS	2	8.511	8.430	8.511	5.598	5.550	5.598	3.859	3.482	3.859
	4	17.021	16.859	17.021	11.195	11.101	11.195	7.717	6.963	7.717
	6	25.532	25.289	25.532	16.793	16.651	16.793	11.576	10.445	11.576
	8	34.042	33.719	34.042	22.391	22.202	22.391	15.435	13.927	15.435
	10	42.553	42.148	42.553	27.989	27.752	27.989	19.293	17.409	19.293
CCCC	2	3.568	3.482	3.568	2.308	2.258	2.308	1.661	1.614	1.662
	4	7.137	6.963	7.137	4.616	4.515	4.616	3.323	3.229	3.323
	6	10.705	10.445	10.705	6.924	6.773	6.924	4.984	4.843	4.985
	8	14.274	13.927	14.274	9.232	9.031	9.232	6.646	6.457	6.646
	10	17.842	17.409	17.842	11.540	11.289	11.540	8.307	8.072	8.308
SCSC	2	8.394	8.317	8.394	5.481	5.436	5.481	3.835	3.793	3.835
	4	16.789	16.633	16.789	10.963	10.873	10.963	7.670	7.585	7.670
	6	25.183	24.950	25.183	16.444	16.309	16.444	11.505	11.378	11.505
	8	33.578	33.266	33.578	21.926	21.745	21.926	15.341	15.170	15.341
	10	41.972	41.583	41.972	27.407	27.182	27.407	19.176	18.963	19.176

3.2. Parametric Study

Based on the convergence and comparison study, the developed higher-order CNTRC plate model is extended to compute the responses for the different geometrical and material parameters. The effect of various parameters on the flexural and free vibration behaviouris discussed in detailed. In general the responses are computed for a/h = 50, V_{CNT} =0.12 anda/b = 1 at 300K throughout the analysis if not stated otherwise. For the computational purpose, the thickness of the composite panel is taken as h = 0.002 m and discussed in detailed.

3.2.1. Flexural behaviour

In this section, flexural behaviour of CNTRC panel is investigated for different geometrical and material parameters. The effect of the CNT volume fractions on the bending responses of the square CNTRC plate is examined for three volume fractions (V_{CNT} = 0.12, 0.17 and 0.28), five point loads (p = 2, 4, 6, 8, and 10KN) at the center with three support conditions (SSSS, CCCC and SCSC) and presented in Table6. It is clearly observed from the figures that the nondimensional central deflections are increasing with externally applied load and decreasing with the volume fractions of the CNT.

The thickness ratio of any structural component playsan important role on the stiffness behavior of the structure. The effect of five thickness ratios (a/h = 10, 20, 30, 40 and 50) on the nondimensional central deflections of the square CNTRC plate are examinedfor two volume fraction (V_{CNT} = 0.12 and 0.17) and three support conditions (SSSS, CCCC and SCSC) with point load (p=10KN) at the center point of the composite plateand presented in Table 7. It can be seen that the nondimensional central deflections are increasing with thickness ratios of CNTRC plate.

Table 7. Effect of thickness ratio on the nondimensional central deflection.

		V_{CNT} = 0.12			V_{CNT} = 0.17		
	a/h	Model 1	Model 2	Model 3	Model 1	Model 2	Model 3
SSSS	10	4.9587	4.6417	4.9587	3.0398	2.8504	3.0398
	20	10.0051	9.6305	10.0051	6.3644	6.1432	6.3644
	30	17.9343	17.5405	17.9344	11.6228	11.3918	11.6229
	40	28.7761	28.3745	28.7761	18.8273	18.5925	18.8273
	50	42.5531	42.1482	42.5531	27.9887	27.7522	27.9887
CCCC	10	4.0394	3.7729	4.0394	2.4268	2.2644	2.4269
	20	6.2431	5.8627	6.2431	3.8452	3.6192	3.8453
	30	9.2030	8.7840	9.2030	5.7963	5.5505	5.7963
	40	13.0553	12.6239	13.0553	8.3543	8.1031	8.3543
	50	17.8422	17.4087	17.8423	11.5402	11.2887	11.5403
SCSC	10	4.7813	4.4959	4.7814	2.9159	2.7464	2.9160
	20	9.8152	9.4615	9.8152	6.2033	5.9967	6.2033
	30	17.6576	17.2812	17.6577	11.3652	11.1467	11.3652
	40	28.3673	27.9817	28.3674	18.4284	18.2053	18.4284
	50	41.9723	41.5829	41.9724	27.4070	27.1818	27.4071

The CNT is well known for its good thermal property, and it becomes more significant when the structure is exposed to the elevated temperature field. In this present example, the CNT properties are assumed to be temperature-dependent, and, therefore, environmental temperature plays a significant role on the flexural strength of the CNTRC plate. The nondimensional central deflections of the square CNTRC plate are analyzed for three different temperature loading (T = 300K, 500K and 700K), three support conditions (SSSS, CCCC and SCSC) by varying the value of point load andpresented in Table 7. It is observed that the nondimensional central deflections of the CNTRC plate are increasing as the temperature load increase.

The deformation behaviour of the CNTRC plate is computed for model 1 and model 2 with three support condition (SSSS, CCCC and SCSC) and presented in fig. 4 (a)-(f). The responses are computed by setting a/h = 50, V_{CNT} = 0.12 and p= 10KN at room temperature (T = 300K). It is clearly observed that the deformation of the plate is maximum and minimum for the SSSS and the CCCC support conditions, respectively and the responses are within the expected line.

Table8. Effect of temperature on the nondimensional central deflection.

Support Conditions	Load (KN)	Temperature (K)								
		300			500			700		
		Model 1	Model 2	Model 3	Model 1	Model 2	Model 3	Model 1	Model 2	Model 3
SSSS	2	8.5106	8.4296	8.5106	9.0049	8.9099	9.0049	9.5673	9.4526	9.5673
	4	17.0212	16.8593	17.0212	18.0097	17.8197	18.0097	19.1346	18.9053	19.1346
	6	25.5318	25.2889	25.5319	27.0146	26.7296	27.0146	28.7018	28.3579	28.7019
	8	34.0425	33.7185	34.0425	36.0194	35.6395	36.0195	38.2691	37.8106	38.2691
	10	42.5531	42.1482	42.5531	45.0243	44.5493	45.0243	47.8364	47.2632	47.8364
CCCC	2	3.5685	3.4817	3.5685	3.8354	3.7331	3.8354	4.1621	4.0381	4.1621
	4	7.1369	6.9635	7.1369	7.6708	7.4663	7.6708	8.3243	8.0761	8.3243
	6	10.7054	10.4452	10.7054	11.5062	11.1994	11.5062	12.4864	12.1142	12.4864
	8	14.2738	13.9270	14.2738	15.3416	14.9326	15.3416	16.6486	16.1522	16.6486
	10	17.8423	17.4087	17.8423	19.1770	18.6657	19.1770	20.8107	20.1903	20.8107
SCSC	2	8.3945	8.3166	8.3945	8.9255	8.8334	8.9255	9.5240	9.4118	9.5240
	4	16.7889	16.6332	16.7889	17.8510	17.6668	17.8510	19.0480	18.8236	19.0480
	6	25.1834	24.9497	25.1834	26.7765	26.5001	26.7765	28.5720	28.2354	28.5720
	8	33.5779	33.2663	33.5779	35.7020	35.3335	35.7020	38.0960	37.6472	38.0960
	10	41.9723	41.5829	41.9724	44.6275	44.1669	44.6275	47.6200	47.0590	47.6200

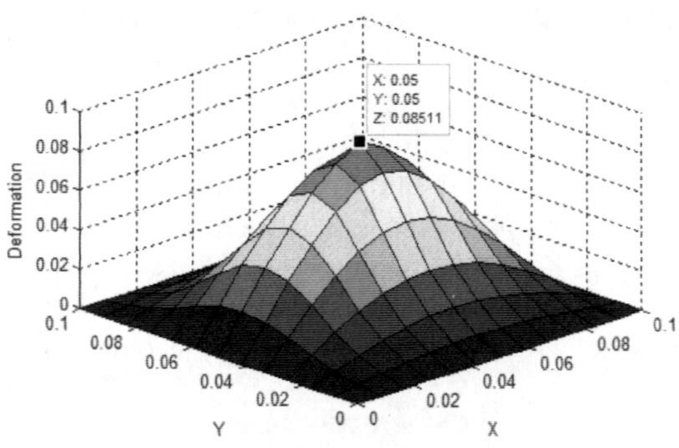

(a) Model 1, SSSS

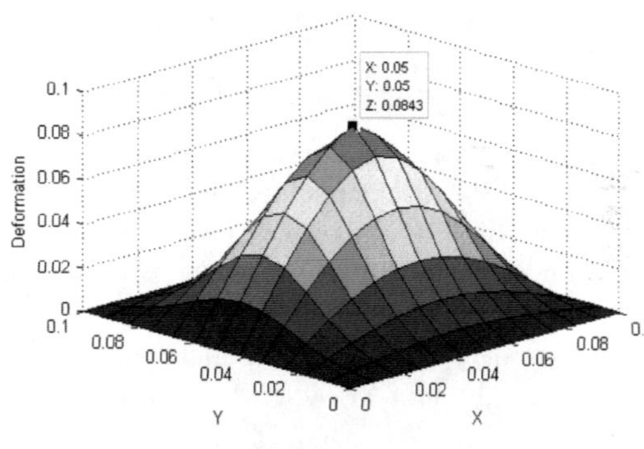

(b) Model 2, SSSS

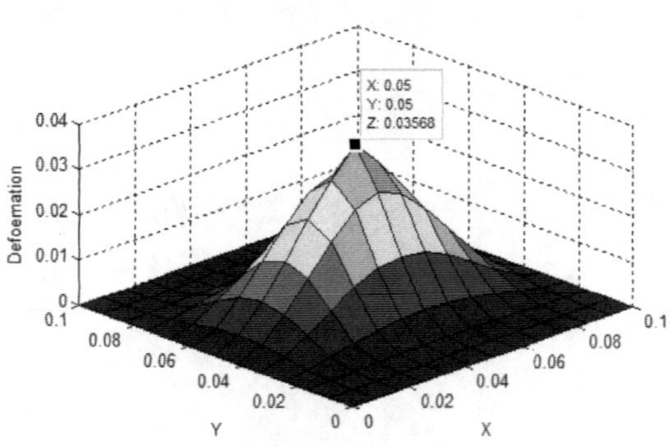

(c) Model 1, CCCC

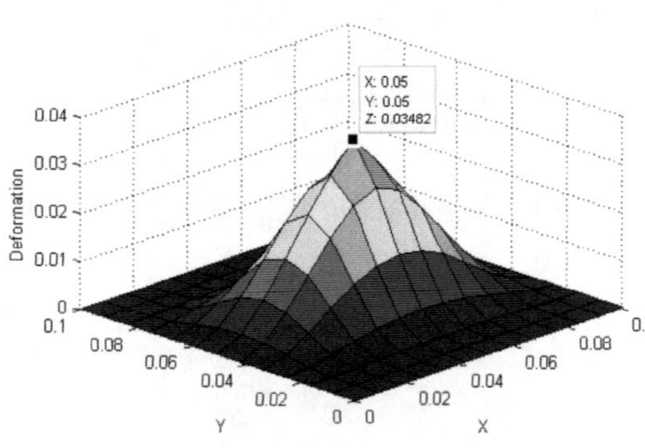

(d) Model 2, CCCC

Fig. 4:- Deformed configuration of CNTRC panel: (a) Model 1, SSSS, (b) Model 2, SSSS, (c) Model 1, CCCC and (d) Model 2, CCCC.

3.2.2. Vibration and modal analysis

In this section, the effect of different geometrical and material parameters on the free vibration responses are examined using the developed three micromechanical finite element model of the CNTRC plate. In general, the responses are examined for T = 300K V_{CNT} = 0.12 and a/h = 50 throughout the analysis if not stated otherwise.

It is well known that the volume fraction of the CNT plays a significant role in stiffness response of the CNTRC plate. The effect of volume fractions (V_{CNT} = 0.12, 0.17 and 0.28) of the CNT on the nondimensional fundamental frequencies of the square CNTRC plate is analyzed usinga/h =50 under two temperature (T = 300K and 400K) load. The nondimensional fundamental frequency of the CNTRC plate is computed and presented in Table9 for three support conditions (SSSS, CCCC and SCSC). It is clearly observed that the nondimensional fundamental frequencies are increasing with the volume fractions of the CNT and decreasing with the temperature.

Table 9. Effect of volume fraction on the nondimensional fundamental frequency.

	V_{CNT}	T=300K			T=400K		
		Model 1	Model 2	Model 3	Model 1	Model 2	Model 3
SSSS	0.12	17.6564	17.7262	17.6563	17.3654	17.4427	17.3651
	0.17	21.3910	21.4649	21.3910	21.0425	21.1245	21.0425
	0.28	26.3164	26.4439	26.3163	25.9105	26.0522	25.9104
CCCC	0.12	36.4265	36.9554	36.4264	35.51	36.0831	35.5101
	0.17	44.5219	45.0966	44.5245	43.4643	44.0877	43.4651
	0.28	53.4669	54.3944	53.4666	52.0619	53.062	52.0617
SCSC	0.12	18.9113	18.9813	18.9125	18.4697	18.5471	18.4698
	0.17	23.1087	23.1871	23.1118	22.5583	22.6404	22.5601
	0.28	27.8613	27.9886	27.8616	27.2702	27.4090	27.2683

In this present analysis, the material properties of the CNTand the matrix are assumed to be temperature-dependent,therefore,the temperature field effect on the vibration behaviour of the CNTRC plate is investigated. Fig.5(a) and (b) shows the nondimensional fundamental frequency of the square CNTRC plate under fourdifferent temperature loads (T=300K, 400K, 500K and 700K), two support conditions (SSSS and CCCC), and three volume fractions of CNT (V_{CNT}). It is observed that the nondimensional fundamental frequency responses of the CNTRC plate decrease as the temperature load increases for all type support conditions.

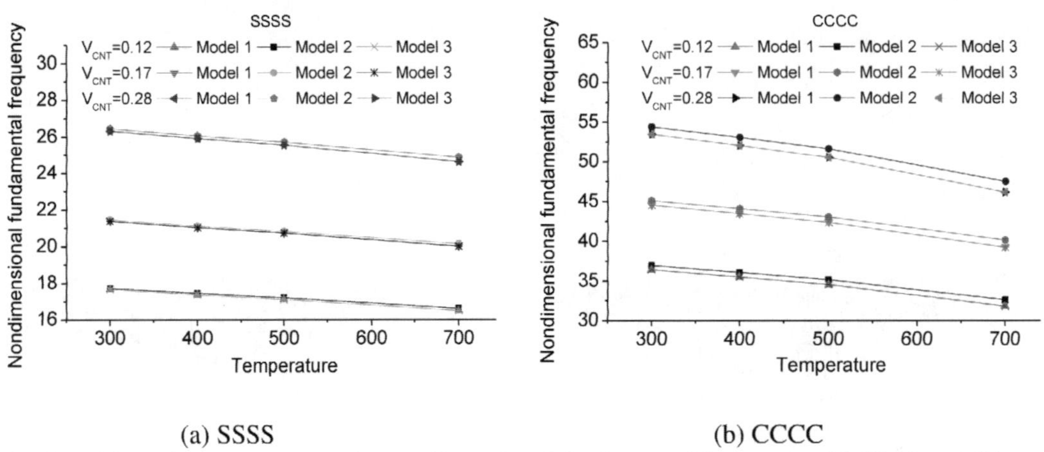

(a) SSSS (b) CCCC

Fig. 5:- Effect of temperature on the nondimensional fundamental frequency: (a) SSSS and (b) CCCC.

In this example, the effect of thickness ratios on the nondimensional fundamental frequency of CNTRC square plate is examined. The responses are computed for five thickness ratios (a/h = 10, 20, 30, 40 and 50) and two support conditions (SSSS and CCCC) under two uniform thermal loads (T = 300K and 700K) with V_{CNT} =0.12 and plotted in fig.6 (a) and (b).It is observed that the nondimensional fundamental frequencyis increasedwith the volume fractions of the CNT. Moreover, the nondimensional fundamental frequency increase sharply with increase the thickness ratio but the rate of increase reduces for thehigher value of thickness ratio.

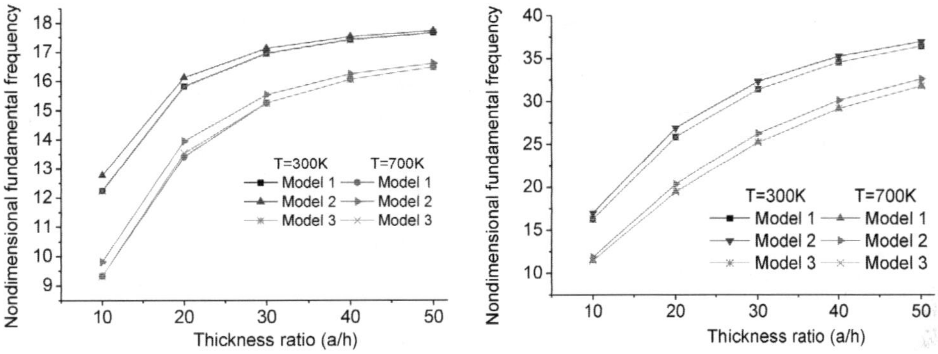

(a) SSSS (b) CCCC

Fig. 6:- Effect of thickness ratio (a/h) on the nondimensional fundamental frequency: (a) SSSS and (b) CCCC.

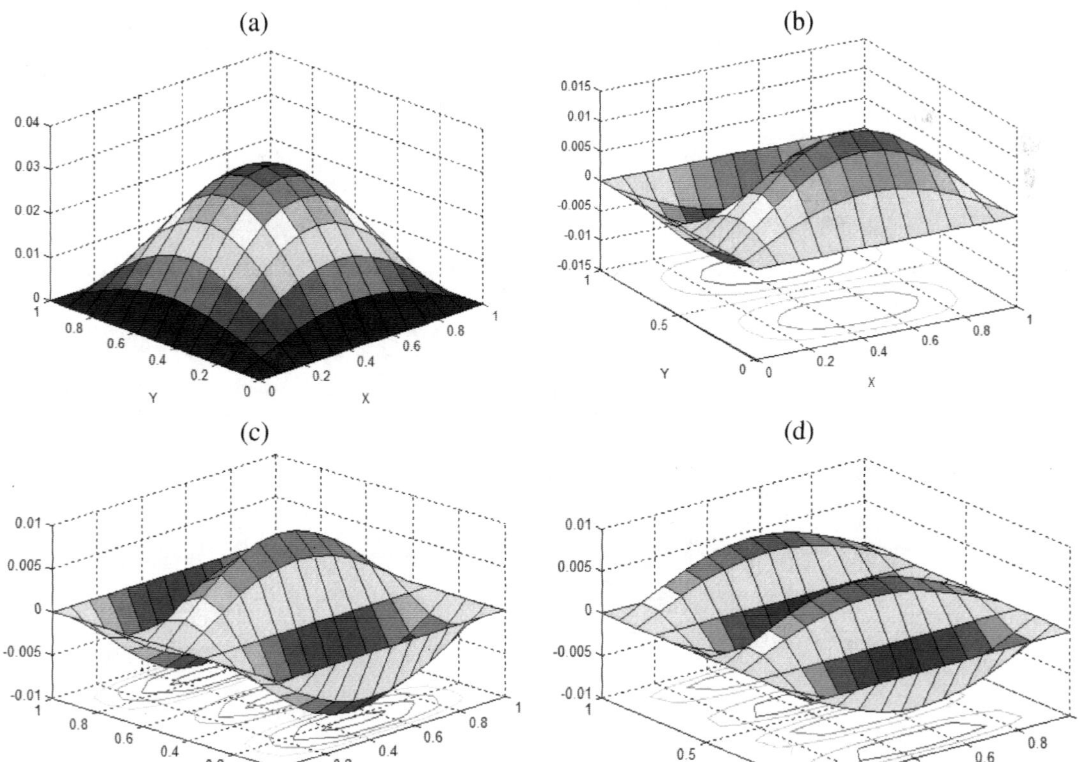

Fig.7:- Free vibration mode shapes of a simply supported uniformly distributed CNTRC square plate (Model-1): (a) 1st Mode (b) 2nd Mode (c) 3rd Mode (d) 4th Mode.

Fig. 7(a)-(d) represent the four mode shapes of the square clamped CNTRC plate (Model 1) at ambient temperature (T = 300K) by setting the geometrical parameters as a/h = 50 and V_{CNT} = 0.12. It is well known that the mode shapes are not giving any numerical value rather it provides the direction of vibration and the present responses are following the expected line.

4. Conclusions

In the present article, the free vibration and bending responses of the SWCNT reinforced composite plate are computed using the FSDT and HSDT mid-plane kinematics in conjunction with the FEM under uniform thermal environment. To achieve the realistic behaviour, the material properties of both (CNT and matrix) are assumedto be temperature dependent.The CNTs are uniformly distributed with respect to the dimension of the plate, and the effective material properties of the composite are computed using the extended rule of mixture. The desired governing equation of bending and free vibration are obtained using the suitable FEM steps. The responses are computed numerically using the suitableFEM steps with the help of ninenodedisoparametricelements with nine (model 1), six (model 2)and ten (model 3) DOFseachnode. The present models are showing good convergence rate with mesh refinement. The comparison study also indicates the validity of the present first and higher order models. The models arealso validated by comparing the responses of the simulation model developed in ANSYS using APDL code. The effect of the different geometrical and material parameters on the free vibration and bending under thermal environment are computed.The following conclusions are drawn from the detailed parametric study.

(i) The convergence and comparison study shows the accuracy of the present FSDT and HSDT models with and without temperature load.

(ii) The nondimensional central deflections are increasing with the rise of the mechanical load, temperature load and the thickness ratio. However, the deflections are showing a reverse trend for the volume fraction.

(iii) The nondimensional fundamental frequency is increasing with the increasein the volume fractionand the thickness ratios. However, the nondimensional frequency is showing a reverse trend for the temperature load.

(iv) The results indicated that the FSDT model (Model 2) is the stiffest configuration as compared to the HSDT models (Model 1 and Model 3).

Appendix

$$\{\lambda\}=\{u\ v\ w\}^T,\{\lambda_{01}\}=\left[u_0\ v_0\ w_0\ \varphi_x\ \varphi_y\ \psi_x\ \psi_y\ \theta_x\ \theta_y\right]^T,\{\lambda_{02}\}=\left[u_0\ v_0\ w_0\ \varphi_x\ \varphi_y\ \varphi_z\right]^T$$

$$\{\lambda_{03}\}=\left[u_0\ v_0\ w_0\ \varphi_x\ \varphi_y\ \varphi_z\ \psi_x\ \psi_y\ \theta_x\ \theta_y\right]^T$$

$$[f_1]=\begin{bmatrix}1&0&0&z&0&z^2&0&z^3&0\\0&1&0&0&z&0&z^2&0&z^3\\0&0&1&0&0&0&0&0&0\end{bmatrix},\ [f_2]=\begin{bmatrix}1&0&0&z&0&0\\0&1&0&0&z&0\\0&0&1&0&0&z\end{bmatrix},$$

$$[f_3]=\begin{bmatrix}1&0&0&z&0&0&z^2&0&z^3&0\\0&1&0&0&z&0&0&z^2&0&z^3\\0&0&1&0&0&z&0&0&0&0\end{bmatrix}$$

References

[1] Moradi-Dastjerdi R, Pourasghar A, ForoutanM and Bidram M 2014*J Compos Mater***48**1901-1913.
[2] Han Y, Elliott J 2007*Comput Mater Sci***39**315–23.
[3] Shen H S and Zhu Z H 2012*Eur J Mech A-Solid***35**10-21.
[4] Liu Y J and Chen X L 2003 *Mech Mater***35**69-81.
[5] Chen X L and Liu Y J 2004*Compos Mater Sci***29**1-11.
[6] Alibeigloo A 2013*Comp Struct***95**612-22.
[7] Lin Fand Xiang Y 2014*Appl Math Model***38**3741–3754.
[8] LiewK M, LeiZ X,Yu J LandZhangL W2014 *Comput Methods ApplMechEngrg***268**1-17.
[9] AyatollahiaM R, ShadlouS and ShokriehM M 2011*Compos Struct***93**2250-2259.
[10] Aragh B S, FarahaniE B and Barati A H N 2012*Math Mech Solids***18** 264-284.
[11] Lin F and Xiang Y 2014*International Journal of Structural Stability and Dynamics* DOI:10.1142/S0219455413500569
[12] Shen Z B, Sheng L P and Li X F 2012*Physica E***44** 1169-1175.
[13] Ke L L, Yang J and Kitipornchai S 2013*Mechanics of Advanced Materials and Structures***22**28-37.
[14] AlvaA and Raja S 2014*Mechanics of Advanced Materials and Structures***21**197-206
[15] Jarali S, PatilS F and Pilli S C 2015*Mechanics of Advanced Materials and Structure***22** 428-439.
[16] VodenitcharovaT and Zhang L C 2006*Int J Solids Struct***43**3006-3024.
[17] Formica G, LacarbonaraWand Alessi R 2010 *J Sound Vib***329**1875-1889.
[18] Zhu Z H, Lei Z X and Liew K M 2012 *Compos Struct***94**1450-1460.
[19] YasM H and Heshmati M 2012*Appl Math Model***36**1371-1394.
[20] Lei Z X, Yu J L and Liew K M 2013 *International Journal of Materials Science and Engineering***1**36-40.
[21] Lei Z X, LiewK M and Yu J L 2013*Compos Struct***106**128-138.
[22] Lei Z X, LiewK M and Yu J L 2013*Compos Struct***98**160-168.
[23] AlibeiglooAand Liew K M 2013 *ComposStruct***106** 873-881.
[24] Lei Z X, Zhang L W, LiewK M and Yu J L 2014*Compos Struct***113**328-338.
[25] MehrabadiS J and Aragh B S 2014*Thin WallStruct***80**130-41.
[26] Kishore M D V H, Singh B N and Pandit M K 2011*AerospSciTechnol***15** 224-235.
[27] Szekrenyes A 2014 *J Compos Mater***48** 1441-1457.
[28] Shen H S and Xiang Y 2013*EngStruct***56** 698-708.

Effect of post-deposition annealing on the structural and electrical properties of RF sputtered hafnium oxide thin films

K C Das[1], S P Ghosh[1], N Tripathy[1], G Bose[2] and J P Kar[1*]

[1]Department of Physics and Astronomy, National Institute of Technology, Rourkela, India 769008

[2]FST, IFHE university, Hyderabad, India 501203

E-mail: karjp@nitrkl.ac.in

Abstract. Hafnium oxide films were deposited on silicon substrates by RF sputtering at room temperature. Post-deposition rapid thermal annealing of the sputtered HfO_2 films was carried out in the temperature range of 400°C to 800°C in oxygen ambient. The structural properties ware studied by X-ray diffraction (XRD), where the enhancement in the crystallinity of HfO_2 ($\bar{1}11$) orientation was observed. The Capacitance –Voltage (C–V) and Current density –Voltage (J–V) characteristics of the annealed dielectric film were investigated employing Al/HfO_2/Si Metal Oxide Semiconductor (MOS) capacitor structure. The flatband voltage (V_{fb}) and oxide charge density (Q_{ox}) were extracted from the high frequency (1 MHz) C–V curve. Leakage current was found to be minimum for the annealing temperature of 600°C.

1. Introduction

In the past few decades, continuously scaling down of silicon dioxide (SiO_2) gate dielectrics has facilitated increased-numbers of transistors per chip with enhanced circuit functionality and performance at low costs in complementary metal-oxide semiconductor (CMOS) technologies as predicted by Moore's law [1]. However, conventional SiO_2 gate dielectric (low-k) has reached the maximum operational limit during the downscaling of the CMOS devices. In order to overcome these problem, high-k dielectric materials such as Ta_2O_5, TiO_2, ZrO_2 and HfO_2 are currently under active consideration. Among them hafnium oxide thin film has long been recognized as most suitable high dielectric material owing to its relatively high dielectric constant, wide bandgap (5.68 eV), compatibility with polysilicon gate electrodes, low density of interface states and good electrical reliability [2–4]. HfO_2 is also a potential candidate for several other applications such as gas sensing devices, tunnel junctions, laser damage-resistance optical coatings and thin film capacitors [5]. The growth process of gate oxide film is a very important issue and received considerable interest in order to get high quality reliable high-k gate dielectric thin films for MOS structures. Various methods such as pulsed laser deposition, chemical vapor deposition, e-beam evaporation and radio frequency (RF) magnetron sputtering have been used to achieve good quality high-k thin films. Among them, RF sputtering, a physical vapor deposition (PVD) based technique combined with plasma, has several advantages such as low thermal budget, ease of handling, less consumption of source materials, nontoxic nature and possibility of multi-target processing.

The characteristics of MOS devices are strongly dependent on the electrical properties of the film. Not only the deposition techniques, but also the post deposition heat treatments have the ability to modify electrical properties of the thin films. Furthermore, if these devices will be monolithically integrated with integrated circuits (IC) signal processing chip, the as-deposited HfO_2 film has to go through a high temperature process. Therefore, it is utmost important to evaluate HfO_2 film properties with post deposition annealing. In this work, a systematic study on the structural and electrical properties of HfO_2 thin films with various post deposition annealing temperature are investigated.

2. Experimental work

Hafnium oxide thin films were deposited at room temperature by RF magnetron sputtering technique on p-type silicon [(100), 1-10 Ohm.cm] wafers using pure hafnium target (99.995%) in presence of high purity argon and oxygen gases. A schematic layout of experimental set up is shown in the figure 1. The process parameters such as RF power, operating pressure, argon/oxygen flow ratio for the deposition of HfO_2 were kept at 150 W, 8×10^{-3} mbar and 3:2 respectively. The thickness of HfO_2 films was kept around 20 nm. The post-deposition annealing of the films was performed at 400°C, 600°C and 800°C in O_2 ambient using Rapid Thermal Annealing (RTA) system.

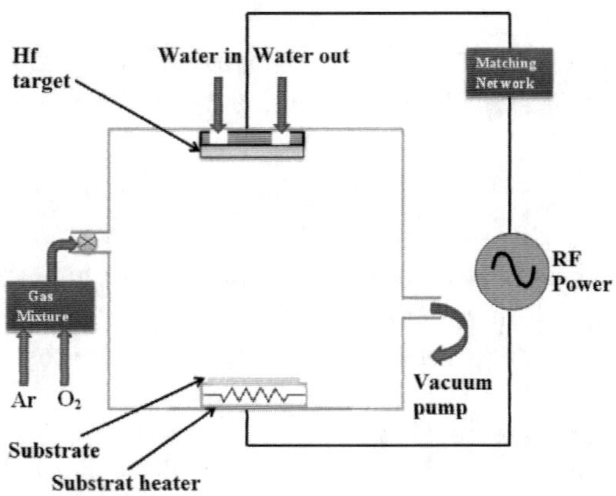

Figure 1. Schematic layout of experimantal set up.

The structural properties of HfO_2 thin films were investigated by X-ray diffraction system (Rigaku ultima IV). The MOS capacitors were fabricated by thermal evaporation technique with Al as top and bottom electrodes. The capacitance–voltage (C–V) and leakage current density–voltage (J–V) measurements for the Al/HfO_2/Si MOS structures were performed using an Agilent E4980A precision LCR meter and Keithley 6487 Picoammeter/Voltage source, respectively.

3. Results and discussion

3.1. Structural properties

Figure 2. represents the XRD plot of HfO_2 dielectric thin films with various annealing temperatures. A weak and broad $(\bar{1}11)$ diffraction peak is observed for as deposited sample, which corresponds to the mixed structure of amorphous and crystalline phase. The intensity of HfO_2 $(\bar{1}11)$ peak is increased with annealing temperature, which indicates the increase in crystallinity. This may be due to the orientation of the crystallites in a preferred direction after receiving the adequate amount of thermal energy. In addition, RTA treatment at 400°C or higher gives rise to the evolution of (002), ($\bar{1}$02), (300), (221) and (130) peaks, which corresponds to the monoclinic crystal structures of the HfO_2 dielectric thin films [6]. Figure 3 shows the FESEM image of HfO_2 thin film annealed at 600°C. A crack free uniform film with homogeneous microstructure is observed .

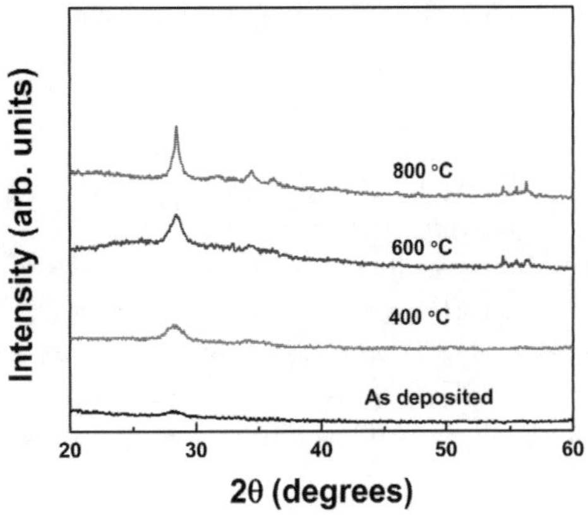

Figure 2. X-ray diffraction patterns of HfO_2 thin films with various RTA temperatures.

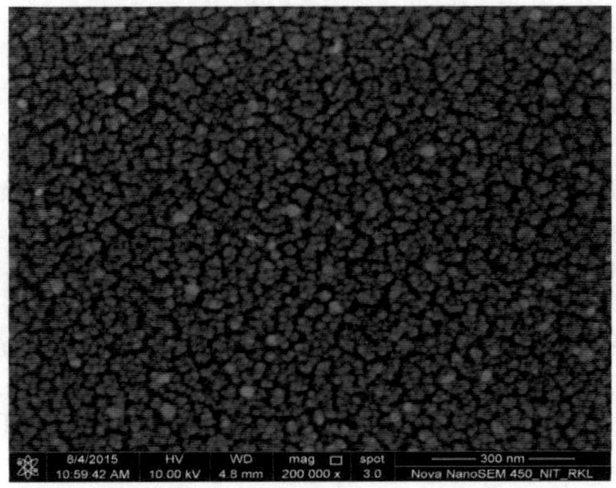

Figure 3. FESEM image of HfO_2 thin film annealed at 600°C.

3.2. Electrical properties

3.2.1. Capacitance –Voltage measurement: The normalized high frequency (1 MHz) C–V curves of annealed HfO_2 thin film is shown in figure 4. Three regions of the C-V graph, namely accumulation, depletion and inversion are observed formed due to the variation in carrier concentrations at the interface with bias voltage. A negative shift in the C–V curve of HfO_2 film is observed for the film annealed at 400°C. This indicates the positive charges generated in the film and an increase in magnitude of the density of oxide traps (Q_{ox}) (0.38×10^{12}–1.67×10^{12}

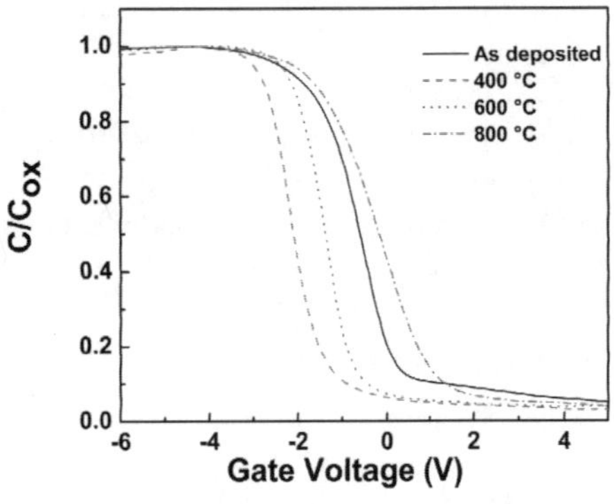

Figure 4. Capacitance –Voltage characteristics of HfO_2 thin film with various RTA temperatures.

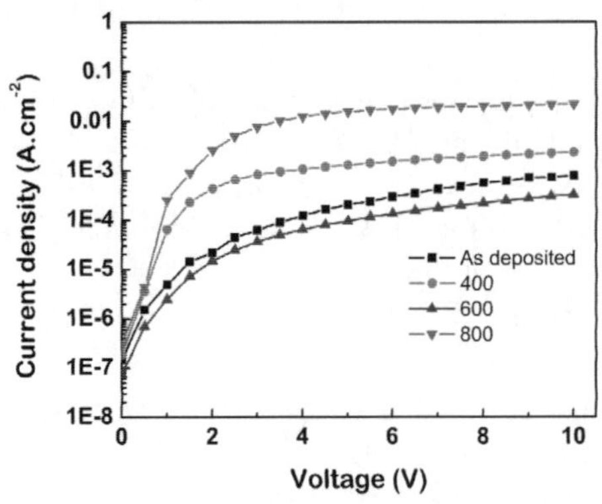

Figure 5. Current density –Voltage characteristics of HfO_2 thin film with various RTA temperatures.

cm^{-2}), which is determined by the $Q_{ox} = C_{acc}(\phi_{ms} - V_{fb})/q$ [7], where C_{acc} is accumulation capacitance per unit area, ϕ_{ms} metal-semiconductor work function, V_{fb} is flatband voltage, q is electron charge. RTA treatment of the film at 600°C and above shows a positive shift of the C–V curve. Thus indicates the reduction in oxide charge density of the film annealed at 600°C in comparision to annealed at 400°C. The magnitude of trapped oxide density (Q_{ox}) for 600°C and 800°C annealing temperatures are found to be 4.56×10^{11} cm^{-2} and 1.48×10^{12} cm^{-2} respectively. The Q_{ox} is found to be minimum for the RTA treatment of 600°C.

3.2.2. Current density –Voltage measurement: Figure 5. shows the current density–Voltage (J–V) characteristics of HfO_2 thin films with various annealed temperature in oxygen ambient. At the lower bias, there is a rapid increase in the current and afterwards it becomes quasisaturated. It is observed that the leakage current increases with the increase in annealing temperature from as-deposited sample to 400°C and lower leakage current was observed for annealed at 600°C. This indicates that the thermal annealing reduces the number of the defects (or oxide traps) and increases the packing density or decrease in defect density in the film, causing the leakage current to decrease[8].

4. Conclusion

RF sputtered HfO_2 films were subjected to rapid thermal annealing over a variety of temperatures ranging from 400°C to 800°C in oxygen ambient. The as-deposited hafnium oxide thin films has shown a mixed structure of amorphous and crystalline phase. The increase in HfO_2 ($\bar{1}11$) intensity is observed with the increase in annealing temperature. The oxide trap density was found to be a strongly dependent on the annealing temperature, where the Q_{ox} significantly reduced to 4.56×10^{11} cm^{-2} at 600°C. The HfO_2 thin films, annealed at 600°C has shown minimum leakage current, which may be due to the increase in packing density or decrease in defects in the films.

Acknowledgments

This work was supported by Science and Engineering Research Board (SERB), DST project (SR/FTP/PS-099/2012) under Fast Track Scheme for Young Scientist. K .C .Das thankfully acknowledge Human Resource Development Ministry, India for research fellowship.

References

[1] Moore G 1975 Progress in digital integrated electronics *Electron Devices Meeting, 1975 International* vol 21 pp 11–13
[2] Lee B H, Kang L, Nieh R, Qi W J and Lee J C 2000 *Appl. Phys. Lett.* **76** 1926
[3] Will G D, Wallace R M and Anthony J M 2001 *J. Appl. Phys.* **89** 5243
[4] Tirmali P, Khairnar A G, Joshi B N and Mahajan A 2011 *Solid-State Electron.* **62** 44
[5] Kaupo Kukli M R, Keinonen J and Leskela M 2002 *Chem. Vap. Deposition* **8** 199
[6] Lin C L, Chou M Y, Kang T K and Wu S C 2011 *Microelectronic Engineering* **88** 950
[7] Nicolian E 1982 *Metal Oxide Semiconductor Physics and Technology* (New York: John Wiley and Sons) p 420
[8] Lin C L, Chou M Y, Kang T K and Wu S C 2011 *Microelectronic Engineering* **88** 950

Stress Ratio Effect on Ratcheting Behavior of AISI 4340 Steel

K Divya Bharathi[*] and K Dutta
Department of Metallurgical and Materials Engineering, National Institute of Technology, Rourkela- 769008, India

[*]E-mail: k.divya.212@gmail.com

Abstract. Ratcheting is known as accumulation of plastic strain during asymmetric cyclic loading of metallic materials under non-zero mean stress. This phenomenon reduces fatigue life of engineering materials and thus limits the life prediction capacity of Coffin-Manson relationship. This study intends to investigate the ratcheting behavior in AISI 4340 steel which is mainly used for designing of railway wheel sets, axles, shafts, aircraft components and other machinery parts. The effect of stress ratio on the ratcheting behaviour in both annealed and normalised conditions were investigated for investigated steel. Ratcheting tests were done at different stress ratios of -0.4, -0.6 and -0.8. The results showed that the material responds to hardening behavior and nature of strain accumulation is dependent on the magnitude of stress ratio. The post ratcheted samples showed increase in tensile strength and hardness which increases with increasing stress ratio and these variations in tensile properties are correlated with the induced cyclic hardening.

1. Introduction

Ratcheting is a secondary phenomenon of cyclic plasticity that can accelerate fatigue damage or even act as the failure mechanism itself [1]. It occurs during asymmetric cyclic loading, in presence of positive or negative mean stress and thus plastic strain gets accumulated during each cycle in the direction of the applied mean stress. Accumulation of ratcheting strain can substantially reduce the fatigue life of a component [2-7]. In practice, machine components can come across both ratcheting and fatigue simultaneously and therefore failure of a component can occur due to fatigue, ratcheting or due to combination of both the effects. So it will be a-priori to understand effect of ratcheting of a material before designing for a particular application to define its proper fatigue life and to avoid catastrophic failure.

AISI 4340 steel is a kind of medium carbon steel which is traditionally used in aerospace applications (e.g. crank shafts and landing gears) and in automotive industries (e.g. connecting rods and axles); the steel is known to be normalising heat treatment before being designed to any component [8]. One cannot rule out the presence of ratcheting deformation during applications of such machine components. Generally, in mechanical design, loading condition is usually presented by maximum stress and stress ratio, responding to the maximum load and cyclic features respectively. Therefore the maximum stress and stress ratio are chosen as control parameters to describe the stress features. Currently, only few reports exist on stress ratio effect and almost negligible amount of reports exist that focus ratcheting and fatigue interaction of such steel [9-10]. Hence it is of utmost importance to investigate the ratcheting behavior at different stress ratios of such steel.

Therefore, in the present work, effect of stress ratio on ratcheting behavior of AISI 4340 steel with different heat treated (annealed and normalised) conditions has been studied. Post ratcheting tensile and hardness tests also carried out to correlate the properties with ratcheting effect on investigated steel.

2. Experimental details

2.1 Material selection and specimen specification

The sample selected for the current investigation was a medium carbon steel designated as AISI 4340. The steel was available in the form of rods of 18 mm diameter. The chemical composition of the selected material was assessed using optical emission spectrometer as in wt%: C-0.35, Ni-1.52, Cr-1.44, Mo-0.18, Mn-0.69, Si-0.31, S-0.02, P-0.03 and balance Fe. The as-received steel rods were subjected to stress relief annealing and normalising at a temperature of 750°C and 870°C respectively with soaking time of 2 h. Specimens with approximate height of 10 mm with cross section were cut from the annealed materials for microstructural characterizations and determination of hardness. The microstructural constituents of the investigated steel have been examined using an optical microscope (Carl Zeiss, model no.: 3327000403, Germany) connected to an image analyzer (Software: Axio vision release 4.8 S.P 3) and a series of representative photographs have been recorded. The microstructure of the investigated steel exhibited larger grains for annealed samples and smaller grains for normalized ones. Using Vickers hardness tester (LECO LV400, US), the hardness has been measured at a load of 20kgf with dwelling time of 15s. Tensile and fatigue test specimens have been fabricated as per ASTM standards E8M [11] and E-606 [12] respectively from heat treated rods.

2.2 Ratcheting test

Ratcheting tests were carried out at room temperature at a constant stress rate of 50MPa/s and stress ratios of -0.4, -0.6 and -0.8. The maximum stress selected as 80% of ultimate tensile strength i.e., 800 MPa and 1300 MPa for annealed and normalised specimens respectively. All these details are listed in Table 1.

Table 1: Text matrix for ratcheting tests at varying stress ratios

S. No.	Stress ratio (R)	Annealed		Normalised	
		σ_a (MPa)	σ_m (MPa)	σ_a (MPa)	σ_m (MPa)
1	-0.4	560	240	910	390
2	-0.6	640	160	1040	260
3	-0.8	720	80	1170	130

During each test, the load extension as well as the actuator displacement data has been continuously recorded by using the attached software to the computer. It was aimed to acquire at least 250 data points per cycle during test. All fatigue tests have been done up to 100 cycles and then post ratcheting tensile test were done for further analysis on these samples. Post ratcheting tensile test has been done for ratcheted specimens and post ratcheting hardness tests were done by cutting the transverse sections of the gauge portion of the ratcheted samples with approximate height of 10 mm, after cutting the portion containing the fracture surface.

3. Results and discussion

3.1 Study of mechanical properties

A comparison of the hardness values of annealed and normalised samples indicates that average hardness of the normalised samples was 321 VHN which was higher as compared to the annealed samples with average hardness of 228 VHN. The normalised samples are being subjected to faster cooling rate showed increase in its hardness value. These results are in accordance with a few published reports [8], hence it can be stated that the heat treatment of investigated steel was proper.

3.2 Tensile test

Typical engineering stress-strain diagrams for the investigated steel for annealed and normalised specimens are illustrated in Figure 1(a) and (b). Both annealed and normalised specimens showed continuous yielding behavior from elastic to plastic region and hence their yield strength is estimated by 0.2% strain off-set procedure, as suggested in ASTM standard E8M [11]. The tensile properties of the annealed and normalised samples are given in Table.2. All the values which are tabulated were almost near to standard values of this particular selected steel as mentioned in ASM hand book [8]. Hence it can be stated that the heat treatment and specimen design of investigated steel were proper. The main focus to perform this test is that to calculate stress amplitude and mean stress from the ultimate tensile stress which were used in fatigue test and to compare the properties with standard values of this steel.

Estimating the stress values from the above stress strain plots for annealed and normalised specimens and comparing the results, the yield strength and ultimate tensile strength of the annealed specimen were found to be 616 MPa and 906 MPa which are less as compared to the normalised specimen which were 1300 MPa and 1467 MPa respectively. This shows that the normalised specimen has more strength compared to annealed specimen because of finer grain size induced due to air cooling as expected. The uniform elongation (ε_u) of annealed specimen found as 8.86% and for normalised specimen was 4.62%. The total elongation (ε_t) of annealed specimen and normalised specimen was 16.11% and 11.94% respectively. From this it can be concluded that the normalised specimen being stronger and less ductile and less elongation showed as compared to annealed specimen.

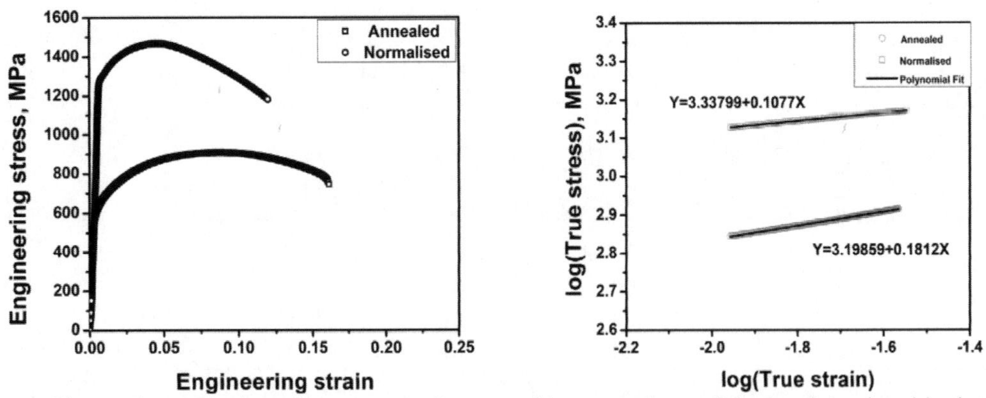

Figure 1: (a) Engineering stress strain curve (b) comparison of the log (σ) – log (ε) plots for annealed and normalised samples of AISI 4340 steel.

The strain hardening exponent (n) of the AISI 4340 steel was estimated by calculating the true stress (σ) and true strain (ε) values from the engineering stress and engineering strain respectively. The log(true stress) vs. log (true strain) plots in the strain range of 1.68 to 4.13 for annealed, 1.11 to 2.88 for normalised samples result into straight lines as shown in Figure 2(b). The strain hardening exponent values were calculated by using Hollomon equation $\sigma = K\varepsilon^n$, where K is strength coefficient. The values of n are summarized in the Table 2.

Table 2: Tensile properties of annealed and normalised samples of AISI 4340 steel.

Strength	Annealed	Normalised
Yield Strength (MPa)	616	1300
Ultimate tensile strength (MPa)	906	1467
Strain Hardening Exponent (n)	0.18	0.10
Uniform elongation (ε_u) %	8.86	4.62
Total elongation (ε_t) %	16.11	11.94

3.3 Ratcheting by varying stress ratio

The results of cyclic tests conducted up to 100 cycles under various stress ratios in annealed and normalised conditions are presented and discussed in this section. Figure 2 (a) and (b) show the hysteresis loops of first, 50th and 100th cycles at stress ratio of R = -0.4 for both annealed and normalised specimens respectively. During stress controlled fatigue, the accompanying variation in strain range with progression of cycles is due to the change in hardening or softening response of the material.

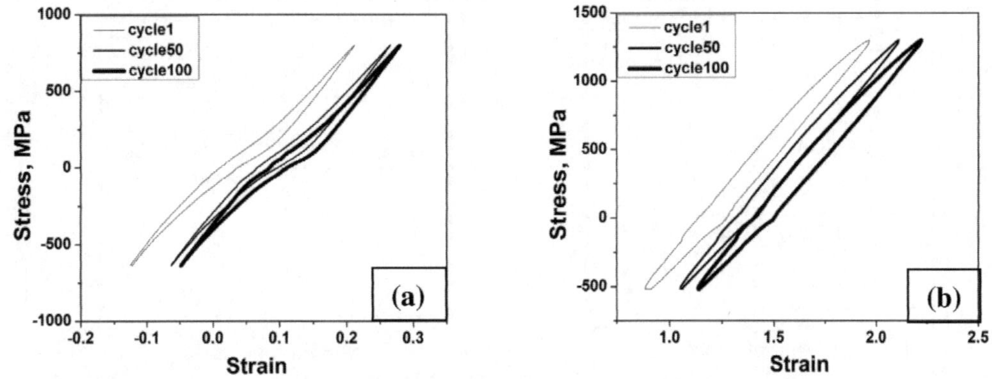

Figure 2: Typical hysteresis loops generated during ratcheting tests at R= -0.4 for (a) annealed (b) normalised specimens

Kang et al. [14] explained that the ratcheting strain increases gradually and stress–strain hysteresis loops become wider, if there is cyclic softening feature of the material. As one can visualize that, if the hysteresis loop area increases, a material shows cyclic softening behavior and if the hysteresis loop area decreases there exists cyclic hardening [15-18]. This hardening/softening feature of materials depends greatly on the different heat treatments experienced [15]. From Figure 2, it is clear that hysteresis loops shift towards more strain direction during ratcheting deformation, which in turn induces plastic strain to the material. Strain accumulation is calculated by taking the average of minimum and maximum strain in particular cycle. The loop area decreased from first cycle to last cycle i.e., the strain accumulation is decreased from first cycle to last cycle. This indicates clearly that the selected material shows cyclic hardening behavior.

Yoshida et al. [18] reported that the strain accumulation for R = -0.4 is slightly larger than the R = 0.8, 0.0, -0.8 and -1.0. But the difference in the magnitude of strain accumulation is not so large between stress ratios of others. Kang et al. [14] explained that the most significant ratcheting occurs at the stress ratio of 0.889, even if the increasing stress ratio also results in a weaker ratcheting behavior in 42CrMo steel. Figure 3 (a) and (b) depict the variations in ratcheting strain with number of cycles at different stress ratios in both annealed and normalised conditions. As the stress ratio increases ratcheting strain also increases in both annealed and normalised samples and the magnitude of strain accumulation in ratcheting of R = -0.4 is slightly higher than that in case of R= -0.6 and -0.8. Larger strain accumulation at a given number of stress cycles is found at stress rate of R= -0.4. All these details summarized in Table 3. These features are partially similar to those obtained by Yoshida et al. [18], for SUS304 stainless steel but totally different from 42CrMo steel [14]. This fact causes increased amount of dislocation generation when stress ratio is increased to higher level. The increase in strain accumulation with increasing stress ratio can be described as a consequence of increasing dislocation density, at higher stress ratio levels.

Table.3: Ratcheting strain variation with respect to stress ratio

S. No.	Stress ratio (R)	Ratcheting strain%	
		Annealed	Normalised
1	- 0.4	1.16	1.0
2	- 0.6	0.62	0.94
3	- 0.8	0.40	0.81

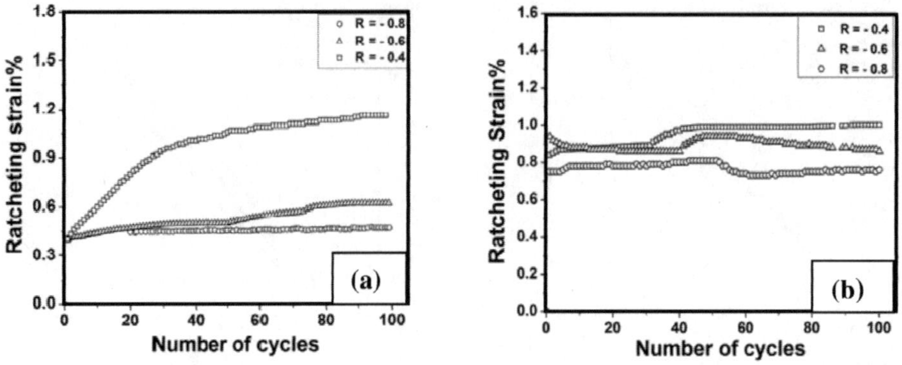

Figure 3: Effect of stress ratio on ratcheting strain at R= -0.4, -0.6 and -0.8 in (a) annealed (b) normalised specimens

Xia et al. [20] observed that the rate of accumulation of ratcheting strain i.e., the increment of ratcheting strain in each cycle decreases gradually with the number of cycles due to its cyclic hardening feature in ASTM A-516 Gr.70 steel. Chen et al. [21] observed that the ratcheting strain increases in high-nitrogen steel X13CrMnMoN18-14-3 whereas its rate decreases continuously with increasing number of cycles. Dutta et al. [22] observed that rapid accumulation of ratcheting strain in the initial few cycles followed by attainment of a steady state value in ratcheting rate are the characteristic features of the asymmetric cyclic deformation behavior of IF steel. All these studies indicate that strain accumulation takes a saturation plateau after few cycles of loading. In this study also, the nature of attainment of steady state has been examined. Figure 4 (a) and (b) depict the variations in ratcheting strain rate effect with number of cycles at different stress ratios in both annealed and normalised conditions respectively. From Figure 4, it can be seen that the rate of strain accumulation decreases continuously up to initial few cycles after that it reaches saturation level in both annealed and normalised conditions. It can also be stated that there is no much variation in ratcheting strain rate at different stress ratios. One can note that the saturation level for annealed steel is about 30 cycles while that for normalized steel is about 15 cycles.

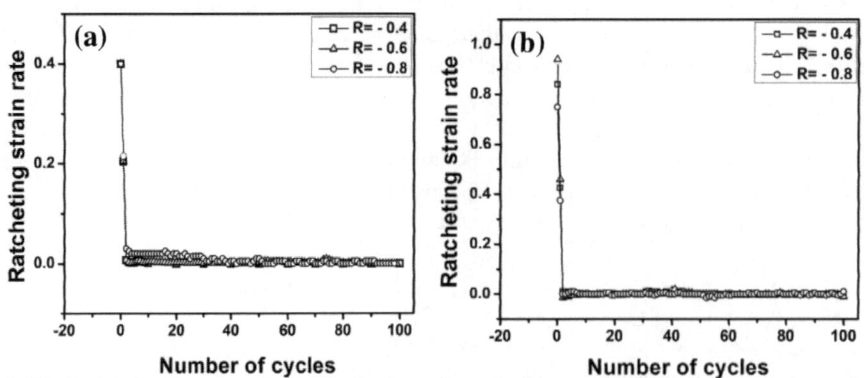

Figure 4: Variation in the rate of accumulation of ratcheting strain with increasing number of cycles for (a) annealed (b) normalised AISI 4340 steel at different stress ratio.

3.4 Post ratcheting tensile

To understand the effect of previous strain accumulation on tensile properties of ratcheted specimens; tensile tests were carried out on a series of specimens after 100 cycles of ratcheting, subjected to varying stress ratio conditions. Mahato et al. [23] observed that post ratcheting yield and ultimate tensile strength of copper increased as compared to unratcheted samples and these are decreased with decreasing stress ratio. Dutta et al. [19, 22] reported that both yield strength and ultimate tensile strength of IF steel and Aluminum alloy increase as compared to unratcheted values. Figure 5 (a) and (b) show the post ratcheting tensile plots for both the heat treated conditions. All the details related to these tests are summarized in Table 4. The results clearly showed that yield strength and ultimate tensile strength increased as also shown in Figure 5. Further, the results indicate that the strength values are higher at stress ratio R= -0.4 than at R= -0.6 and -0.8 in both annealed and normalized samples. The strain hardening exponent values for post ratcheted tensile samples increased compared to unratcheted one in both annealed and normalised conditions. This fact indicates that the strength of the ratcheted steel is governed by increased strain hardening [22]. From the results it can be concluded that post ratcheting yield strength and ultimate tensile strength of selected steel increased as compared to unratcheted samples and these are decreased with decreasing stress ratio.

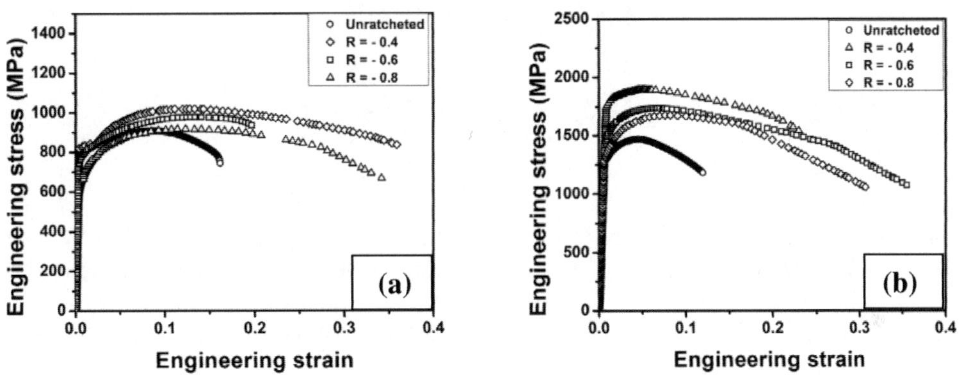

Fig.5: Post ratcheting tensile stress-strain plots for (a) annealed and (b) normalised samples of investigated steel.

Table.4: Post ratcheting tensile values for annealed and normalised samples at different stress ratios

Properties	Stress ratio	Yield strength (MPa)	Ultimate tensile strength (MPa)	Strain hardening exponent (n)
	Unratcheted	616	906	0.18
	-0.4	854	1019	0.18
Annealed	-0.6	833	978	0.17
	-0.8	769	916	0.15
	Unratcheted	1300	1467	0.10
	-0.4	1380	1895	0.10
Normalised	-0.6	1621	1737	0.11
	-0.8	1504	1675	0.13

3.5 Post ratcheting hardness

To assess the extent of deformation during cyclic loading, post-ratcheting hardness tests were also carried out on ratcheted samples. In order to this, Vickers hardness tests were done on a series of specimens after post ratcheting tensile, subjected to varying stress ratio conditions and the results were

Table 5. Post ratcheting hardness values for annealed and normalised samples at different stress ratios

Stress ratio	**Annealed**		**Normalised**	
	HV_{20}	**Average**	**HV_{20}**	**Average**
	229		323	
Unratcheted	226	228	321	321
	228		320	
	329		462	
-0.4	330	331	463	462
	333		459	
	301		447	
-0.6	304	302	442	444
	302		443	
	252		427	
-0.8	253	253	424	424
	255		421	

listed in Table 5. It can be noticed that hardness values of the investigated steel specimens increased after ratcheting deformation. Bhattacharyya et al. [24], in their rolling cycle fatigue experiments, have reported hardness increase can be a result of cyclic hardening, which takes place due to continuous plastic strain accumulation (via ratcheting). Hence in this investigation also increase in hardness can be considered as due to effect of strain hardening due to deformation. It is clear from the table that hardness varies with stress ratio i.e., as the stress ratio decreases, hardness increases. From the results it can be concluded that at more negative stress ratio, hardness becomes more in the investigate steel.

4.Conclusions

The obtained results and their pertinent analyses related to fatigue-ratcheting interaction behavior of AISI 4340 steel at room temperature assist to infer:

1. Accumulation of ratcheting strain increased with increasing stress ratio in both annealed and normalised samples of AISI 4340 steel. Maximum accumulation of ratcheting strain (1.16% for annealed and 1.02% for normalised) was observed at R = -0.4.
2. Rate of strain accumulation is decreased from first cycle to last cycle. This indicates clearly that the selected material shows cyclic hardening behavior. The increase in strain accumulation can be explained with increased dislocation densities in the ratcheted samples.
3. Post ratcheting yield strength, tensile strength and hardness of investigated steel increased as compared to unratcheted samples and these increased with increasing stress ratio. The fact can be attributed to the previous cyclic hardening during ratcheting tests.

References

[1] A.V. Farahani, Theoretical and Applied Fracture Mechanics 73 (2014) 152–160.
[2] S. Vishnuvardhan, G. Raghava, P. Gandhi, M. Saravanan, D.M. Pukazhendhi, S. Goyal, P. Arora, S.K. Gupta, Procedia Engineering 2 (2010) 2209–2218.
[3] Z. Xia, D. Kujawski and F. Ellyin, International Journal of Fatigue 18 (1996) 335–341.
[4] S. Kwofie, International Journal of Fatigue 23 (2001) 829–836.
[5] G. Kang, Q. Gao, X. Yang, Mechanical Materials, 34 (2002) 521–531.
[6] X.M. Chen, Y.C. Lin, J. Chen, Journal of Alloys and Compounds 579 (2013) 540–548.
[7] X. Chen, R. Jiao, K.S. Kim, International Journal of Plastics, 21 (2005) 161–184.
[8] ASM Handbook volume 1 properties and selection irons steels and high permance alloys (1990) 1124–1131.
[9] Y. Xiong, Q. Yu, Y. Jiang, International Journal of Plasticity 53 (2014) 107–124.
[10] G. Kang, Y. Liu, Material Science and Engineering A, 472 (2008) 258–268.
[11] E 8M-12, 'Annual Book of ASTM Standards', Standard test method for tension testing of metallic materials (Metric), 03.01; 2012, West Conshohocken, PA.
[12] E606/E606M-12, Standard Test Method for Strain-Controlled Fatigue Testing, ASTM International, 2012 West Conshohocken, PA.
[13] Y. Zhang , C.L. Hu, Z. Zhao, A.P. Li, X.L. Xu, W.B. Shi, Materials and Design 44 (2013) 612–621.
[14] G. Kang, Q. Gao, X. Yang, Mechanical Materials, 34 (2002) 521–531.
[15] G. Kang, Y. Liu, Z. Li, Material Science and Engineering A, 435–436 (2006) 396–404.
[16] M. Wen, H. Li, D. Yu, G. Chen , X. Chen, Materials and Design 46 (2013) 426–434.
[17] Y. Liu, G. Kang, Q. Gao, International Journal of Fatigue, 30 (2008) 1065–1073.
[18] F. Yoshida, International Journal of Pressure Vessels and Piping 44 (1990) 207–223.
[19] K. Dutta, K.K. Ray, Materials Science and Engineering A 540 (2013) 127–135.
[20] Z. Xia, D. Kujawski and F. Ellyin, International Journal of Fatigue 18 (1996) 335–341.
[21] G. Chen, S.C. Shan, X. Chen, H. Yuan, Computational Materials Science 46 (2009) 572–578.
[22] K. Dutta, K.K. Ray, Materials Science and Engineering A 540 (2012) 30–37.
[23] J. K. Mahato, P. Sarathi, A. Sarkar, A. Kundu, P. C. Chakraborthi, Procedia Materials Science 5 (2014) 1358–1367.
[24] A. Bhattacharyya, G. Subhash, N. Arakere, International Journal of Fatigue, 59 (2014) 102–113.

Effect of loading rate on tensile properties and failure behavior of glass fibre/epoxy composite

K K Mahato*, M Biswal, D K Rathore, R K Prusty, K Dutta, B C Ray

Composite Materials Group, Metallurgical and Materials Engineering Department, National Institute of Technology, Rourkela-769008, India

*Email: kishorepce@gmail.com

Abstract. Fibre reinforced polymeric (FRP) composite materials are subjected to different range of loading rates during their service life. Present investigation is focused on to study the effects of variation of loading rates on mechanical behavior and various dominating failure modes of these potential materials when subjected to tensile loading. The results revealed that on the variation of loading rates the ultimate tensile strength varies but the tensile modulus is mostly unaffected. Furthermore, the strain to failure is also increasing with increase in loading rates. Different failure patterns of glass/epoxy composite tested at 1, 10,100, 500 and 1000 mm/min loading rates are identified. Scanning electron micrographs shows various dominating failures modes in the glass/epoxy composite.

1. Introduction

Combining with two or more constituents different from physical and mechanical properties, composite materials can obtain a desirable characteristic. Nowadays, the fibre reinforced polymeric (FRPs) composites are one among the highest used structural materials in the globe. Glass reinforced polymer matrix composite material has already evolved to be a kind of important engineering material due to its light- weight feature. Their superior favourable properties include high strength to weight ratio, high specific stiffness, low density, high corrosion resistance, high endurance limit, high fatigue resistance over traditional metallic materials. The FRP composites discovers a wide range of application in several fields such as automotive, aerospace, marine, sporting goods, low temperature application such as cryogenic fuel tanks, cryogenic fuel delivery lines, cryogenic wind tunnels and parts of the cryogenic side of turbo-pumps due to their easy usage, better mechanical properties and lower fabrication cost [1].

During the fabrication, in-service life and storage the FRP composite materials are exposed to various loading conditions ranging from quasi-static to dynamic loading. In many structural applications FRP's are subjected to high energy and high velocity dynamic loadings that can produce multi-axial dynamic states of stress [2]. FRP composites are anisotropic materials, so the damages due to these stresses are complex phenomenon involving various failure mechanisms (frequently interactive) in micro scale. Numerous methods have been suggested and discussed, including fracture mechanics, damage mechanics, nonlinear viscoplastic constitutive modelling, and macroscopic (global) failure criteria. For quasi-static loading conditions the latter are the commonly followed method in design and analysis of composite structures. Under certain biaxial states of stress and under dynamic loading conditions available failure criteria and design guidelines are still not promising and fully reliable [3]. These loadings are typically highly transient and the material and structural response occurs over very short (dynamic) time scales (of the order of milliseconds or microseconds).
Furthermore, in these composites the reasons for adding the fibres are often rather complex; for example, improvements may be sought in tensile, fatigue, creep, wear, fracture toughness, thermal

stability. They are vulnerable to keep their performance effectively even in harsh space environments despite of having a lot of positive aspects of utilizing FRP's in space structures.

Type of fiber	Tensile strength (MPa)	Young's Modulus (GPa)	Density (g/cm^3)	Thermal Exp. Co-efficient (mm/°C)
E-glass	2000	80	2.58	5.4
Epoxy	85	3.5	1.2	1.5-10

Table 1. Mechanical Properties of E-glass and epoxy.

The present investigation deals with the loading rate sensitivity of glass/epoxy composite at 1, 10, 100, 500 and 1000 loading rates at room temperature (30°C). The tensile behaviour of glass/epoxy composite was also evaluated in terms of ultimate tensile strength, modulus and strain at break.

2. Materials and Methods

2.1 Testing Materials

The current experimental investigation consists of fabrication of glass/epoxy composite and Diglycidyl Ether of Bisphenol A (DGEBA) based epoxy resin was used as matrix and triethylene tetra amine (TETA) was used as hardener. The woven fabric E-glass fibres used in the experimental program was supplied by Saint Gobin industries. Epoxy resin and hardener were obtained from Atul Industries Ltd, Gujarat having the trade name of Lapox L-12 and K-6 respectively. The epoxy to hardener weight ratio was taken as 10:1 as per the supplier's recommended standard.

2.2 Preparation of glass/epoxy laminates

The glass/epoxy composite laminated was fabricated with 10 layers of woven fabric E-glass fibres having 60:40 weight fraction of glass fibres and epoxy respectively. The fabrication was done using hand layup method. The glass/epoxy laminates were cured in hot press compression moulding machine at 60°C temperature for 20 minutes at a pressure of 5 kgf/cm^2. Then the glass/epoxy laminates were dried in an oven at 60°C for 2 hours to remove moisture and volatile substances. As per the ASTM D3039 standard the samples were cut using diamond cutter as shown in figure 1(a). After that the samples were post cured in an oven at 140°C for 6 hours [4].

2.3 Testing Method of glass/epoxy composite

2.3.1 Tensile Test

The test was performed with tension fixture in Instron 8862 Universal testing machine (UTM) shown in figure 1(b) in accordance to the ASTM D3039 standard samples to evaluate the tensile properties. The samples were tested at room temperature with different loading rates viz. 1, 10, 100, 500 and 1000 mm/min.

2.3.2 *Scanning Electron Microscopy (SEM) Analysis.* The fractured surfaces of the tested specimens were further analysed using scanning electron microscope (SEM) for determining the various dominating modes failure mechanisms using JEOL-JSM 6480 LVSEM at 20KV.

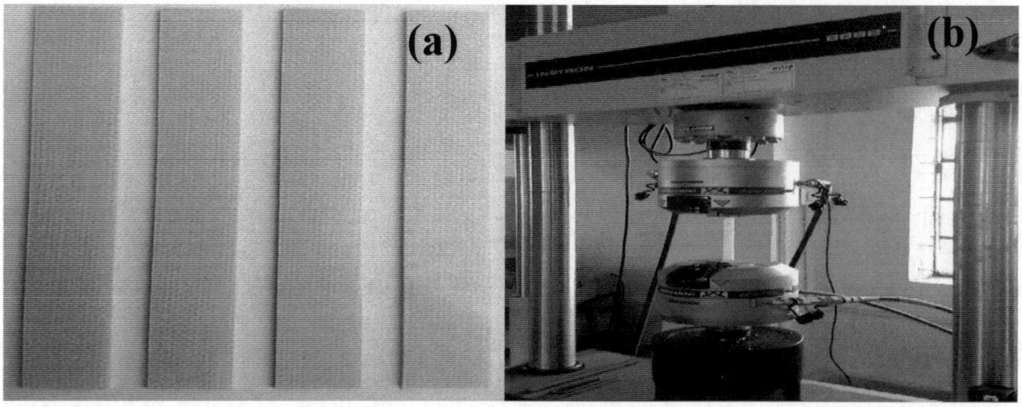

Figure 1. (a) Glass/epoxy standard tensile samples and (b) Instron 8862 UTM with tension fixture.

3. Results and Discussions

3.1 Effect of loading rates on tensile behaviour of Glass/Epoxy composite

The samples were tested at room temperature (30°C) with different loading rates viz. 1, 10, 100, 500 and 1000 mm/min.

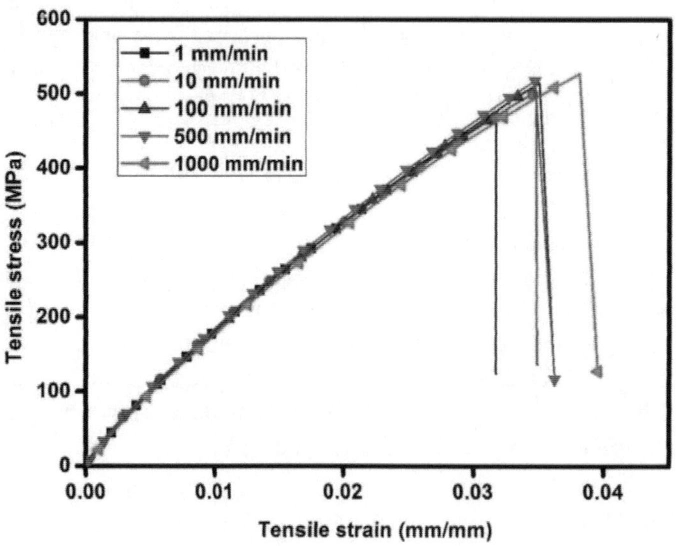

Figure 2. Tensile stress Vs Tensile strain curve with 1, 10, 100, 500 and 1000 loading rates tested at room temperature (30°C).

Figure 2 shows the curve between tensile stress and tensile strain. It is evident from the figure that with increase in loading rate i.e. from 1 mm/min to 1000 mm/min the value of tensile stress is also increasing.

The UTS Vs Loading rate curve is shown in figure 3 and it is evident from the figure that the tensile strength of the glass/epoxy composite increases with increase in the loading rate. Its shows the load carrying capacity of the glass/epoxy composite is better at higher loading rate. It is due to the better cohesive bonding between the fibre/matrix interfaces and also due to proper post curing between fibre/matrix.

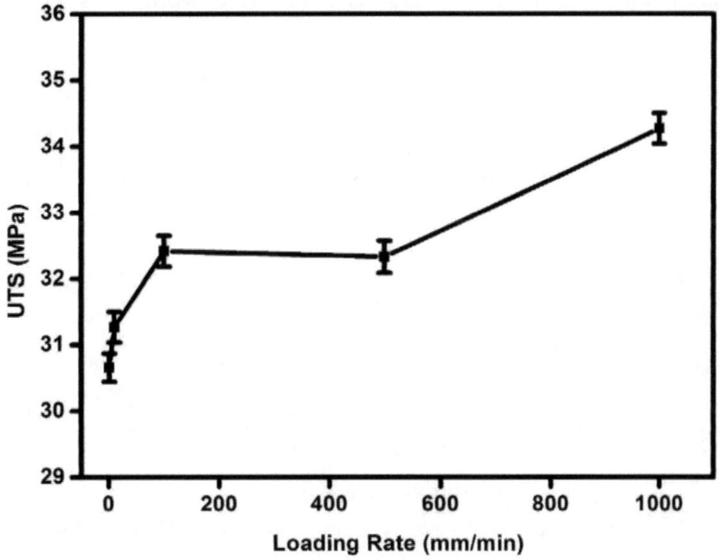

Figure 3. UTS Vs Loading rate curve with 1, 10, 100, 500 and 1000 mm/min loading rates tested at room temperature (30°C).

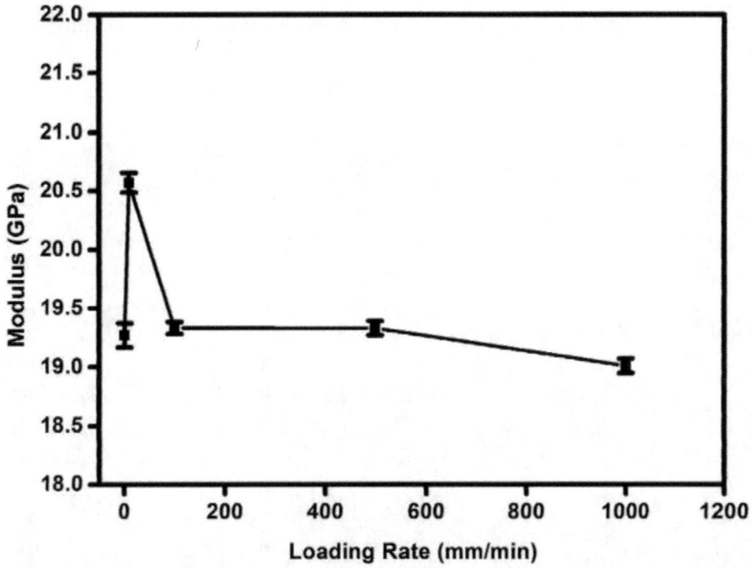

Figure 4. Modulus Vs Loading rate curve with 1, 10, 100, 500 and 1000 mm/min loading rates tested at room temperature (30°C).

Figure 4 shows the curve between Modulus Vs Loading rate and it reveals that at 10 mm/ min the modulus of the composite is higher and after that with increase in the loading rate the modulus of glass/epoxy composite decreases and least modulus was observed at 1000 mm/min. But the modulus value is found to be unaffected with loading rate. Therefore in glass/epoxy composite the modulus value is found to be loading rate is in-sensitive.

The Strain at break Vs Loading rate is shown in figure 5 and it is evident from the figure that as load carrying capacity of the composite system increases the strain value of the glass/epoxy composite is also increasing.

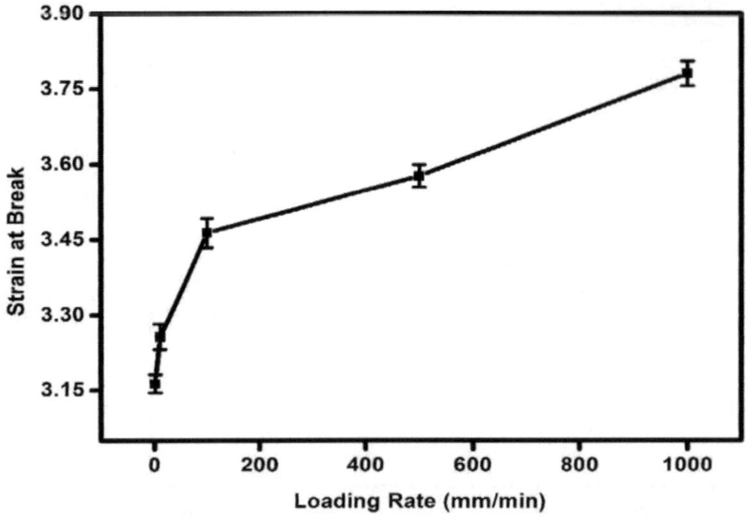

Figure 5. Strain at break Vs Loading rate curve with 1, 10, 100, 500 and 1000 mm/min loading rates tested at room temperature (30°C).

3.2 *Fractography Analysis*

After the tensile testing, fractured samples were analysed using scanning electron microscope (SEM). The Figure 6(a) and 6(b) shows scanning electron micrographs of glass/epoxy composite tested at 1 mm/min loading rate at room temperature. The various dominating modes of failures modes obtained in the composite are matrix fracture, and delaminations of fibres and matrix.

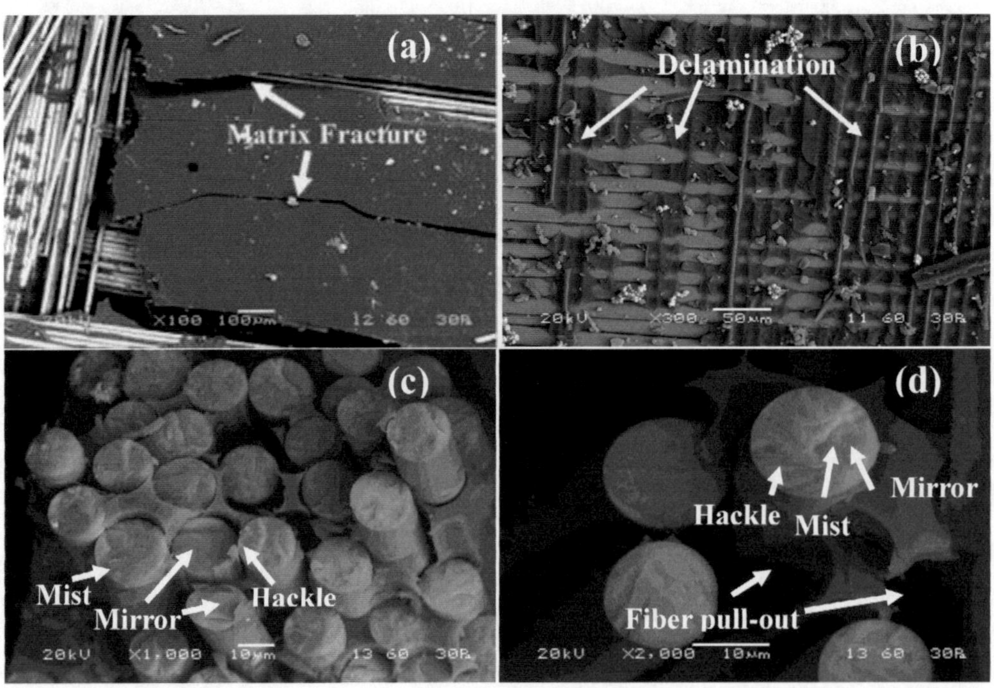

Figure 6. Scanning electron micrographs of glass/epoxy composite tested at (a) 1mm/min, (b) 1mm/min (c) 1000 mm/min (d) 1000 mm/min loading rates respectively.

The Figure 6(c) and 6(d) shows scanning electron micrographs of glass/epoxy composite tested at 1000 mm/min loading rate at room temperature (30°C). The various dominating modes of failures modes obtained in the composite are mirror, mist and hackle and fibre pullout.

3.3 Failure patterns of glass/epoxy composite

Figure 7 shows the bulk images of glass/epoxy composite with different failure pattern tested at (a) 1mm/min, (b) 10 mm/min (c) 100 mm/min (d) 500 mm/min and (e) 1000 mm/min loading rates. At lower loading rates the initiation of crack generates at the middle portion of the specimens. But at higher loading rates multiple numbers of cracks are generating throughout the samples i.e. at middle portions as well as at the tab portions.

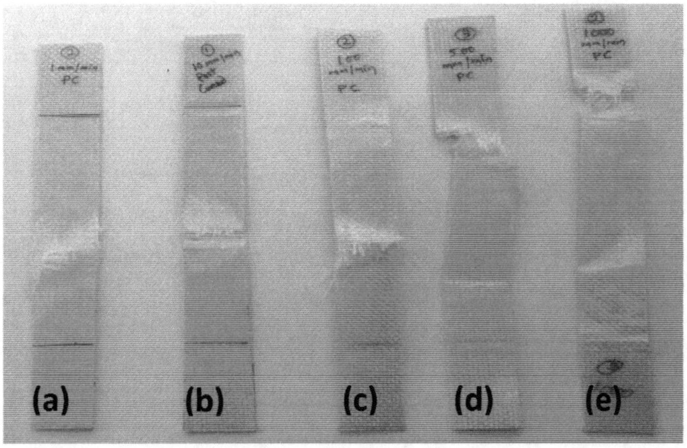

Figure 7. Different failure patterns of glass/epoxy composite tested at (a) 1mm/min, (b) 10 mm/min (c) 100 mm/min (d) 500 mm/min and 1000 mm/min loading rates.

4. Conclusions

The investigation on the effect of tensile properties and failure behavior of glass/epoxy composite comprises the following conclusions

- The glass/epoxy composite is found to be loading rate sensitive. As the loading rate increases the ultimate tensile strength of the composite is also increasing.
- The tensile modulus of the composite is found to be loading rate in-sensitive.
- The strain to failure of the glass/epoxy composite is also increasing as the loading rate increases.
- At bulk level different failure patterns of the glass/epoxy composite is observed. At lower loading rates failure generates from the middle portion of the glass/epoxy composite, but at higher loading rates it is throughout the glass/epoxy composite.

References

[1] Ray B C and Rathore D 2014, Durability and integrity studies of environmentally conditioned interfaces in fibrous polymeric composites: Critical concepts and comments. *Adv. Colloid Interface Sci.* 209, 68–83

[2] Sethi S, Rathore D K and Ray B C 2015, Effects of temperature and loading speed on interface-dominated strength in fibre/polymer composites: An evaluation for in-situ environment. *Mater design*, 65, 617-626

[3] Shokrieh M M and Omidi M J 2009, Tension behavior of unidirectional glass/epoxy composites under different strain rate, *Compos. Struct.* 88, 595–601

[4] Kumar D S, Shukla M J, Mahato K K, Rathore D K, Prusty R K, and Ray B C 2015, Effect of post-curing on thermal and mechanical behavior of GFRP composites, *IOP Conference Series: Materials Science and Engineering*, 75, 01 2012

Ratcheting fatigue behaviour of Al-7075 T6 alloy: Influence of stress parameters

Lala Amarnath, Antara Bhattacharjee and K. Dutta

Department of Metallurgical and Materials Engineering, National Institute of Technology, Rourkela-769008, India

Email- antarabhattacharjee91@gmail.com

Abstract. The use of aluminium and aluminium based alloys are increasing rapidly on account of its high formability, good thermal and electrical conductivity, high strength and lightness. Aluminium alloys are extensively used in aerospace, automobile, marine and space research industries and are also put into structural applications where chances of fatigue damage cannot be ruled out. In the current work, it is intended to study the ratcheting fatigue behavior of 7075-T6 aluminium alloy at room temperature. This Al alloy is potentially used in aviation, marine and automotive components as well as in bicycle parts, rock mounting equipment and parts of ammunition where there is every chance of failure of the parts due to deformation caused by ratcheting. Ratcheting is the process of accruement of plastic stain produced when a component is subjected to asymmetric cyclic loading under the influence of low cycle fatigue. To accomplish the requirements of the projected research, stress-controlled cyclic loading experiments were done using a ±250 kN servo-hydraulic universal testing machine (Instron: 8800R). The effect of stress parameters such as mean stress and stress amplitude were investigated on the ratcheting behavior of the selected aluminium alloy. It was observed that, ratcheting strain increased with increase in the value of stress amplitude at any constant mean stress while a saturation in strain accumulation attained in the investigated material after around 10-20 cycles, under all test conditions. The analyses of hysteresis loop generated during cyclic loading indicate that the material exhibits cyclic hardening in the initial fifty cycles which gets softened in further loading up to about 70-80 cycles and finally attains a steady state. The increase in the ratcheting strain value with stress parameters happens owing to increased deformation domain during cycling. The cyclic hardening accompanied by softening is correlated with characteristic precipitation features of the investigated Al 7075 alloy.

1. Introduction

Aluminium is being used widely due to its high strength, lightness, high formability and good thermal and electrical conductivity. The metal and its alloys are used in automobile, aerospace, marine and space research industries and are also put into structural applications. The current investigation is done using AA7075-T6, an alloy of high strength and used for aircraft components and structural applications. The upper and lower receivers and extension tubes of high precision M16 rifles used for military applications are particularly made from this alloy. The 7XXX series alloy is a heat treatable

alloy series with Zinc as the major alloying element. The Zn (4-6 % of weight) and Mg (1-3 % of weight) present increases the strength of the alloy by forming strengthening precipitates but they also reduce the corrosion resistance. T6 refers to the process of solution heat treatment followed by artificial ageing. This improves the mechanical properties of the alloy. Solution heat treatment of the alloy was done at 470°C for 1 hour, followed by rapid quenching in water and finally artificial aging at 120°C for 24 hours, which was followed by air cooling up to room temperature [1].
When used as aircraft components, the structures are subjected to cyclic loading during landing and take-off and therefore the chief mode of failure is fatigue. Many research works has been done to investigate and thus prevent the fatigue failure of the engineering components. Fatigue can be of two type high cycle and low cycle fatigue. One particular type of low cycle fatigue is ratcheting. Ratcheting is the process of strain accumulation during asymmetric cyclic loading of metallic parts under application of non-zero mean stress at different stress amplitudes [2]. It deteriorates the performance of components as a result of collective effect of fatigue damage and buildup of permanent ratcheting strain, which leads to further development of fatigue damage by continuous thinning out of the components' cross-sectional area, and their combined effect leads to premature failure of the material [3]. About in the year 1911, Bairstow [4] introduce a new phenomenon calling it cyclic creep phenomenon and started research on it. In this research, he studied effect of positive mean stress on the nature of strain accumulation under uniaxial cyclic stressing in steel. Thereafter in the decade of 1950s, came a burst of investigations to understand axial strain accumulation due to asymmetric uniaxial cyclic loading at elevated temperature [5-7] as well as room temperature [8-10]. Hassan and Kyriakides [11] examined ratcheting behavior of consistently hardenable and softenable materials under uniaxial cyclic loading and concluded that material qualities incorporate cyclic hardening/softening. Later in 1996, the effect of mean stress and ratcheting on the fatigue life of steel was described by Xia et al. [12]. After 2000 onwards numerous investigations have been done on various materials viz. metals, compounds, polymers, etc. to understand their ratcheting behavior. Kang et al. [13] in 2002 reported uniaxial cyclic ratcheting and properties of plastic flow of SS304 at the room as well as elevated temperature. Development of dislocation pattern and internal stresses during the cyclic creep process was reported by Gaudin and Feaugas [14]. Kang et al. reported that softening behavior material during cyclic loading increases the ratcheting strain with number of cycles and hardening behavior of material during cyclic loading decreases the ratcheting strain with increases the number of cycle [15].
However as per the knowledge of the current investigator, very limited research has been done to study the ratcheting fatigue behavior of Al-7075 T6 alloy. In the current study, ratcheting fatigue behavior of Al-7075 T6 alloy is investigated under different combinations of stress amplitude and stress mean. Experimental observation, which exhibits material's cyclic hardening in the initial fifty cycles which gets softened in further loading up to about 70-80 cycles and finally attains a steady state, has been discussed.

2. Experimental Procedures

2.1. Material and Heat Treatment

The aluminium 7075 alloy was received in the form of rods of 16 mm diameter. The chemical composition of the alloy is given in Table 1. Solution heat treatment of the alloy was done at 470°C for 1 hour, followed by rapid quenching in water and finally artificial aging at 120°C for 24 hours, which was followed by air cooling up to room temperature [1].

Table 1: Chemical composition of the investigated Al-7075 alloy

Element	Zn	Mg	Cu	Fe	Cr	Si	Mn	Al
Wt.%	5.203	2.163	1.856	0.261	0.256	0.206	0.087	Balance

2.2. Microscopic examinations and Hardness measurements

The cylindrical samples of 16 mm diameter with an approximate height of 10 mm were cut from the as received and heat treated materials for performing microstructural characterization and determining Vickers hardness. The samples were electro-polished using stainless steel rod as cathode and sample as anode with phosphoric acid as electrolyte, preheated to 70°C for 10 minutes at 100 mA/cm^2 current density and 12 V voltage. After washing the electro-polished sample with distilled water, these were etched using freshly prepared Keller's reagent [2 ml HF (1%), 3ml HCl (1.5%), 5 ml HNO_3 (2.5%) and 190 ml H_2O (95%)]. The microstructures of the investigated materials were examined using an optical microscope (Model: ZEISS Axiocam ERc 5s) connected to an image analyzer and a series of representative photographs were recorded. The hardness measurements of aluminium alloy were made at indentation load of 5 kgf and 10 kgf. These tests were carried out using a Vickers hardness testing machine (Leco, Model: LV 700, Michigan, USA) for a dwell time of 10 s.

2.3. Tensile and Fatigue Test

Al-7075 T6 samples were mechanically tested for tensile and fatigue properties. Round bar specimens, for tensile and fatigue tests having a gauge lengths of 25mm and 13 mm with diameters of 6 mm and 7 mm respectively were fabricated as per ASTM standards E8M and E-606 respectively.
These tests were done on servo-hydraulic universal testing machine (Model: INSTRON 8800R). All tensile tests were carried out at a crosshead speed of 1mm/min. Samples for ratcheting tests were with very well surface finish. All these tests were done at a constant stress rate of 50 MPa/s. Whole tests carried on varying mean stress (σ_m) and stress amplitude (σ_a). A summary of the test conditions is shown in Table 2. Fractographic studies were carried out by using scanning electron microscope (Model: JEOL-JSM 6480LV).

Table 2: Stress values for ratcheting tests.

Sl. No.	σ_m /MPa	σ_a /MPa
1	55	25 35 45
2	65	25 35 45
3	75	25 35 45

3. Results and Discussion

As described above, the as received Al-7075 samples were subjected to T6 heat treatment. Optical microstructure of as received alloy is presented. Nature of hardness, tensile and ratcheting behavior of the sample after this heat treatment is presented below in a detail. All tests were performed at room temperature.

3.1 Microstructural analysis and hardness

Figure 1 reveals the microstructure of as received Al-7075 alloy. It consists of elongated pancake shaped grains of uniform size. The average grain was calculated using the

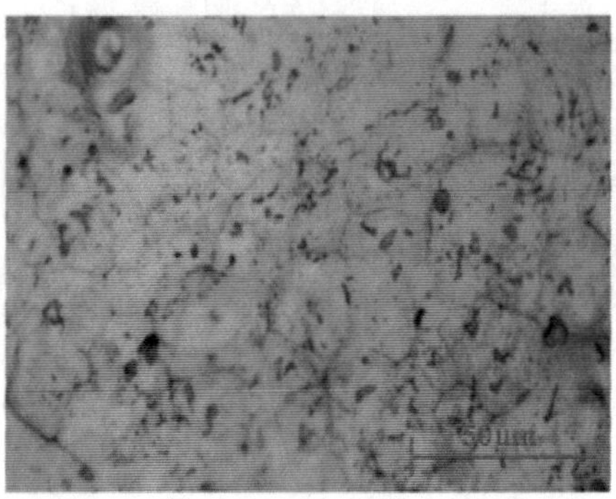

Figure 1: Optical microstructure of 7075 alloy

linear intercept method and was found to be 11.76 ± 1.11 μm. The average hardness values for 5kg load applied of as received and heat-treated alloy are 92 Hv and 150 Hv respectively and for 10 kg load applied are 95 Hv and 151 Hv respectively. The values of hardness for 5 kg load are graphed in Figure 2. The hardness values for as received alloy was observed to be lower than that of the T6 alloy at both loads. In T6 condition, a continuous uniform grain structure was observed (Fig.1). Kumar et al. concluded that there exists Al and Cu particle rich continuous fine grain structure in the alloy. In association, precipitates like $MgZn_2$ at the grain boundaries helps to increase the hardness value. The temperatures, time and cooling rate are important factors of aging treatment, which changes the properties of the alloy. Kumar et al. studied effect of post weld heat treatment on the hardness of AA7075. They reported that hardness value of the alloy was 170 VHN for T6 heat treated condition. The composition of the current alloy is similar with Kumar et al. and the hardness values also close to their results [16].

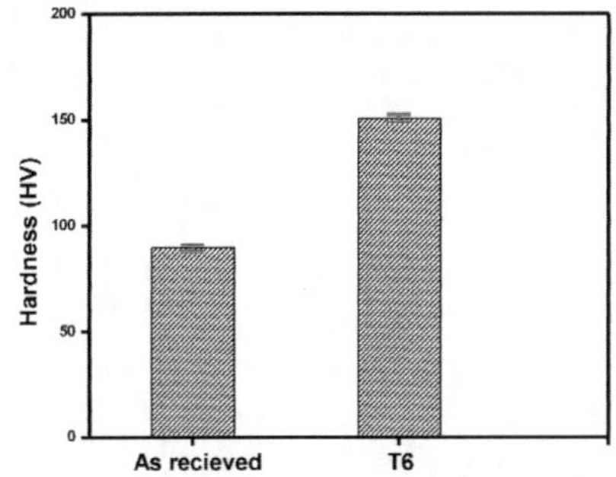

Figure 2: Hardness values of the as received as well as T6 samples.

3.2 Tensile properties

The obtained tensile properties of the material are summarized in Table 3. These data were analyzed to calculate the value of yield strength, ultimate tensile strength (UTS), percentage uniform elongation ($\%\varepsilon_u$) and percentage total elongation ($\%\varepsilon_t$) of alloy. Typical engineering stress-strain curve is plotted which is shown in Figure 3.

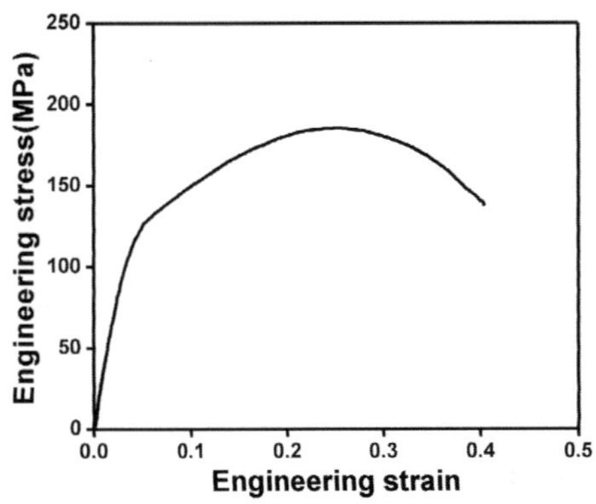

Figure 3: Engineering stress strain plot of Al-7075 T6 alloy

Table 3: Tensile properties of the selected Al-7075 T6 alloy

YS (MPa)	TS (Mpa)	%UE	%TE
70	188	25.6	40.3

YS = yield strength, TS = tensile strength, UE = uniform elongation, TE = total elongation

The fracture surfaces of broken tensile specimens were examined using field emission scanning electron microscope (FESEM, Model: Nova Nano SEM/ FEI). Typical fractographs obtained from the FESEM are shown in Figure 4. The fractographs reveals dimples on the surface, the typical nature of fracture surface of a ductile material, as expected. The dimple fracture in aluminium alloy was observed due to presence of second phase particles in it and there is de-cohesion of the particle – matrix inter phase. The fracture surface also indicates presence of some inclusion-like particles.

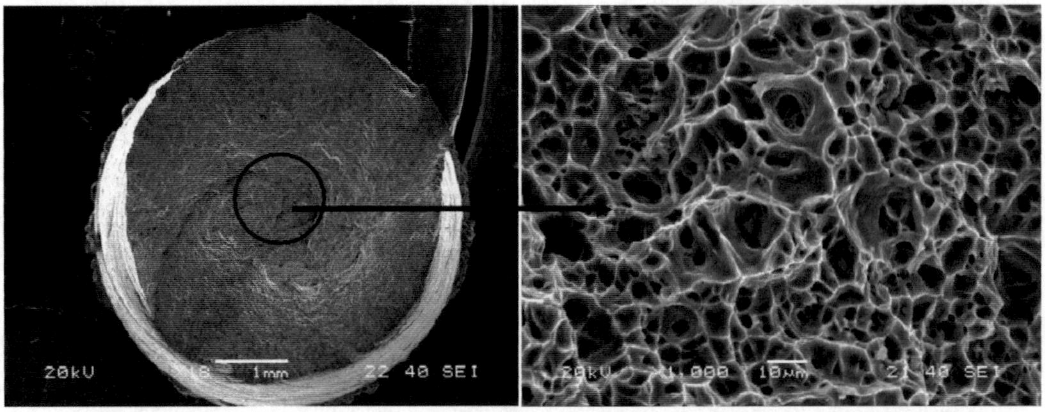

Figure 4: SEM micrographs of the fracture surface of the broken tensile sample

3.3 Ratcheting behaviour

In this section, the uniaxial ratcheting behaviour of the investigated alloy is discussed. The tests were carried out under different combinations of mean stress (σ_m) and stress amplitude (σ_a). Earlier investigations on ratcheting suggest that strain accumulation due to ratcheting attains a saturation value in the range of 50 to 100 cycles, after which the rate of accumulation of strain varies negligibly even up to failure of the material [17]. Keeping this feature into mind, all ratcheting tests in the current investigation were done up to 100 cycles, to understand the trend of strain accumulation in the investigated alloy under tension-tension mode.

3.3.1 Uniaxial ratcheting behavior with varying stress amplitude and at constant mean stress

The ratcheting strain accumulated in the alloy under various stress amplitude is shown in Figure 5. It was observed that at constant mean stress with increasing stress amplitude, accumulation of ratcheting strain increases.

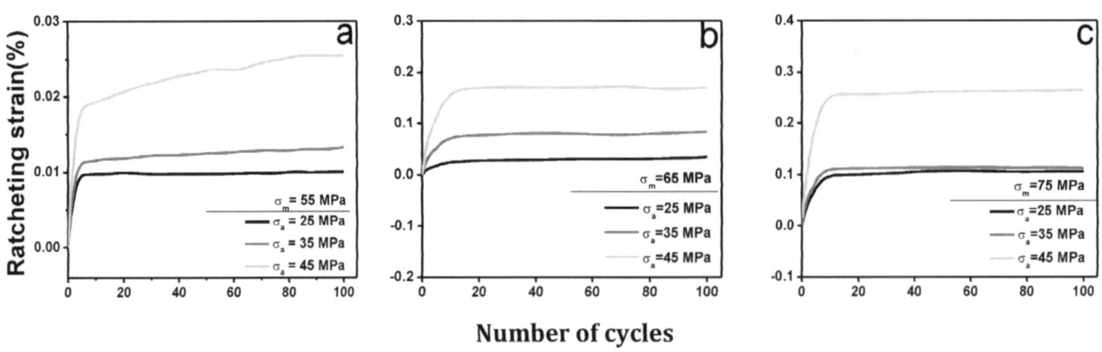

Figure 5: Typical ratcheting strain versus no. of cycle of the investigated aluminium 7075-T6 alloy: (a) at constant σ_m = 55 MPa with varying σ_a = 25 MPa, 35 MPa and 45 MPa.(b) at constant σ_m = 65 MPa with varying σ_a = 25 MPa, 35 MPa and 45 MPa.(c) at constant σ_m = 75 MPa with varying σ_a = 25 MPa, 35 MPa and 45 MPa.

It was also deduced from the graphs that the saturation of strain accumulation occurs in the alloy after about 20 cycles for all the test conditions. Initially the accumulation of ratcheting strain is sharp up to 10 cycles, which slowed down after that and finally attainment of steady state takes place after 20 cycles, for all combinations of mean stresses and stress amplitudes. The increase in strain accumulation with increase in stress amplitude can be expressed as the variation in the height of the hysteresis loops which corresponds to the deformation zone. From the graph it is clear that with increase in stress amplitude the height of the hysteresis loop increases. This fact causes more strain in a particular cycle during the cyclic loading event. On reverse loading, a part of the gathered strain is recovered while a part of strain retains in the sample as to add in the measure of ratcheting strain. Considering the phenomenon of dislocation generation, more amount of dislocations generate due to increased maximum stress during loading with higher stress amplitude.

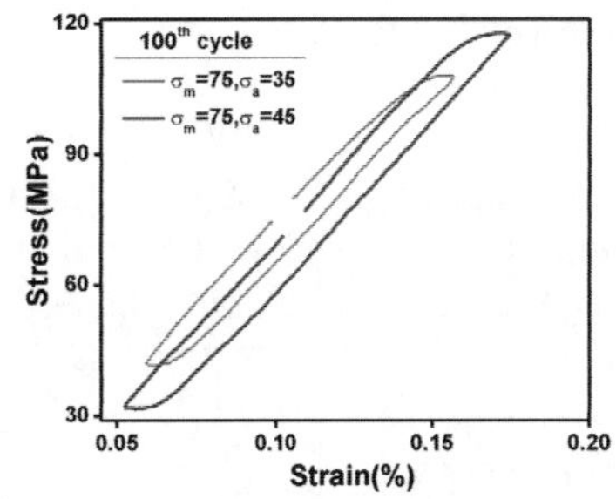

Figure 6: Typical hysteresis loop of 100th cycle at σ_m =75 and σ_a =35, 45 MPa

3.3.2 Uniaxial ratcheting behavior with varying mean stress and at constant stress amplitude

The ratcheting strain accumulated in the alloy under various stress mean is shown in Figure 7. It was observed that at constant stress amplitude with increasing mean stress, accumulation of ratcheting strain increases significantly as comparison to constant mean stress with varying stress amplitude.

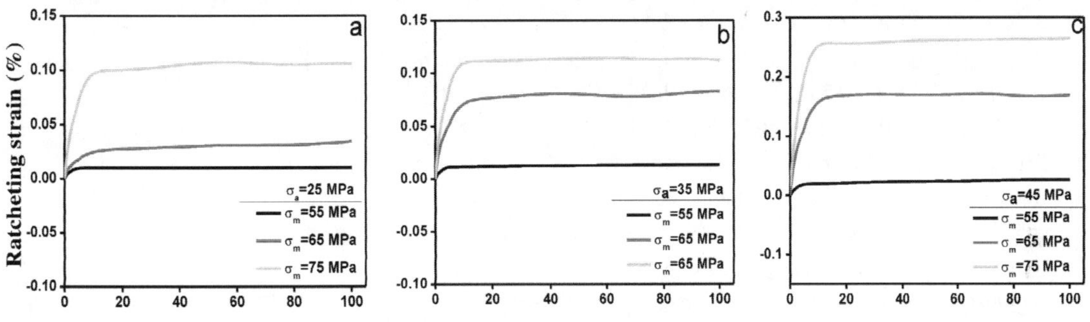

Figure 7: Typical ratcheting strain versus no. of cycle of the investigated aluminium 7075-T6 alloy: (a) at constant σ_a = 25 MPa with varying σ_m = 55 MPa, 65 MPa and 75 MPa,(b) at constant σ_a = 35 MPa with varying σ_m = 55 MPa, 65 MPa and 75 MPa,(c) at constant σ_a = 35 MPa with varying σ_m = 55 MPa, 65 MPa and 75 MPa

4. Conclusions

Uniaxial ratcheting behavior of Al-7075 T6 alloy has been examined in this investigation. The following conclusions can be drawn from the experimental outcomes and their pertinent discussion.

(1) Strain accumulation due to ratcheting increases with increase in mean stress and/or stress amplitude for either of these is constant. This increase in strain accumulation occurs due to enhanced deformation zone during cyclic loading operations with higher stress amplitude, which thereby increases the remnant dislocation density.
(2) The analyses of hysteresis loops produced during cyclic loading indicate that the alloy exhibits cyclic hardening in the first fifty cycles. However it gradually softens on further cycling up to the

range of 70-80 cycles and lastly attains a steady state. This characteristic of softening in the range of 70-80 cycles is considered to be attributed to the formation of Al-Cu type precipitates in cyclic loading.

Acknowledgements

The authors would like to thank the Department of Science and Technology, New Delhi, India for providing financial support (SERB/F/1499/ 2014-2015) for the research. The authors also gratefully acknowledge Prof. D. Chakrabarti of Department of Metallurgical and Materials Engineering, Indian Institute of Technology, Kharagpur for the conduction of fatigue tests.

References

[1] Feng L J, Wei P Z, Xing L C, Qiang J Z, Jing C W and Qiao Z Z 2008 *Transactions of Nonferrous Metals Society China* **18** 755–762
[2] Dutta K and Ray K K 2012 *Materials Science and Engineering A* **540** 30-37
[3] Paul S K, Sivaprasad S, Dhar S and Tarafder S 2011 *Materials Science and Engineering A* **528** 4874–4882
[4] Bairstow L 1911 *Philosophical Transaction of Royal Society* **210** 359–386
[5] Lazan B J 1949 *American Society for Testing of Materials Proceedings* **49** 757 787
[6] Kenedy A J 1956 *International Conference on Fatigue of Metals* 401–410
[7] Meleka A H and Evershed A V 1960 *Journal of Institute of Metallurgists* **88** 411–414
[8] Coffin L F 1960 *ASME Journal of Basic Engineering* **82** 671–682
[9] Benham P P 1961 *Journal of Institute of Metals* **89** 328–339
[10] Benham P P and Ford H 1961 *Journal of Mechanical Engineering and Sciences* **3** 119–132
[11] Hassan T and Kyriakides S 1994 *International Journal of Plasticity* **10** 149–184
[12] Xia Z, Kujawski D and Ellyin F 1996 *International Journal of Fatigue* **18** 335–341
[13] Kang G, Gao Q and Yang X 2002 *Mechanics of Materials* **34** 145–159
[14] Gaudin C and Feaugas X 2004 *Acta Materialia* **52** 3097–3110
[15] Kang G Z, Li Y G, Zhang J, Sun Y F and Gao Q 2005 *Theoretical and Applied Fracture Mechanics* **43** 199–209
[16] Kumar P V, Reddy G M and Rao K S 2015 *Defence Technology* **20** 1–8
[17] Kreethi R, Verma P and Dutta K 2015 *Transactions of Indian Institute of Metals* **68(2)** 229–237

Deformation of $Ni_{20}W_{20}Cu_{20}Fe_{20}Mo_{20}$ high entropy alloy for tensile followed by compressive and compressive followed by tensile loading: A molecular dynamics simulation based study

Md. Meraj, S. Pal *

Computational Materials Engineering Group, Department of Metallurgical and Materials Engineering, National Institute of Technology Rourkela, Rourkela-769008, Odisha, India

*E-mail: pals@nitrkl.ac.in, Phone: +91 -661 – 2462573, Fax No. +91 -661 – 2462550

Abstract. In this paper, molecular dynamics (MD) simulations based study on deformation behavior during uniaxial tension followed by compression and compression followed by tension after 0.6 pre-strain for Ni20W20Cu20Fe20Mo20 high entropy alloy (20 at. % each element) single crystals has been reported. This MD simulation is carried out at strain rate of 108 s-1 and at the temperature of -10oC. The influence of observed nano twin on deformation behaviour for such two types of loading process (i.e. tensile followed by compressive and compressive followed by tensile) has been investigated thoroughly. It is found that the dominant deformation mechanism is twin for tensile forward loading in Ni20W20Cu20Fe20Mo20 high entropy alloy single crystal, whereas atomic diffusion is the dominating factor for deformation behaviour in compressive reverse loading direction of high entropy alloy.

Keywords: Molecular Dynamics; High entropy alloy; Twin; Atomic diffusion

1. Introduction

An equi-molar solid solution of five or more principal metallic elements having high entropy of mixing is known as High-entropy alloys (HEA) [1]. These kind of multi-component alloys are usually comprised of metallic element with a concentration between 5 and 35 at. % each. According to experimental findings reported in literature, higher mixing entropy of these alloys leads to the formation of solid solution phase having similar structure and consequently restricts formation of different phases [1]. HEAs have some advantageous properties like high strength/hardness, outstanding wear resistance, exceptional high-temperature strength, good structural stability, good corrosion and oxidation resistance. Moreover, these alloys are highly irradiation resistant. During particle irradiation, the high atomic-level stress facilitates amorphorization followed by local melting and re-crystallization. This process leaves fewer defects in HEAs as compared to conventional alloys which makes HEAs applicable for high temperature applications like nuclear sectors [2]. HEAs can also be used as a coating material for anti-sticky molds, and material having diffusion barrier application because they have amorphous structure and low roughness. HEA has also potential application as brazing filler, for welding cemented carbide and steel [3]. The HEAs have better fatigue properties in comparison with conventional alloys, such as steels, titanium alloys, and even advanced bulk metallic glasses [4]. The recent discovery and development of high-entropy alloys has opened the door to introduce customization in stacking fault energy of materials which in turn to help to design material having high mechanical strength. High-entropy alloys (HEAs) can have either high strength or high ductility, but achievement of both at same time is still facing a serious challenge [1]. The inferior cast ability and compositional segregation of HEAs also limit their technological applications. To tackle these problems, here we have proposed a novel strategy to design HEAs using molecular dynamics simulation. Experimental studies on the deformation mechanism is difficult as it has numerous drawbacks such as gripping of the samples, rotation of the samples, expensive setups to maintain the high precision equipment/facilities and uncertainty in nano-scale measurements [5-8]. Classical molecular dynamics simulations (MD) have been emerged as an effective simulation method for studying the behaviour and underlying mechanism of deformation at atomic level [9-11]. According to simulation findings reported in literature, micro-structures and stress on the atomic level are important to appreciate the mechanical behaviors and relevant mechanisms by molecular dynamics [12]. According to author knowledge, tensile followed by compressive and compressive followed by tensile of HEAs are not still reported. A nano scale simulation using MD has been performed here to study deformation behavior of HEA ($Ni_{20}W_{20}Cu_{20}Fe_{20}Mo_{20}$ high entropy alloy (20 at. % each element)) at low temperature during tension followed by compression and compression followed by tension loading. In addition, twinning and amorphization along with their features and interactions are also investigated.

2. Computational details

Classical MD simulation deformation studies of high entropy single crystals alloy have been performed using Large Scale Atomic/Molecular Massively Parallel Simulator (LAMMPS, Sandia National Laboratory, USA) [13]. Atomic positions and visualization of snapshots are done using open visualization tool software OVITO [14, 15]. In the present study we used EAM (embedded atom method) potential developed by Zhou et al. [16] to model the interactions between Ni, W, Fe, Cu, and Mo atoms. The expression for total energy of the crystal is given below

$$E=\frac{1}{2}\sum_{i,j,i\neq j}\phi_{ij}(r_{ij})+\sum_{i}F_i(\rho_i) \tag{1}$$

where ϕ_{ij} represents the pair energy between atoms i and j separated by a distance r_{ij}, and F_i is the embedding energy associated with embedding an atom i into a local site with an electron density ρ_i.

In the present study single crystal Ni sample is prepared by creating a perfect FCC crystal of Ni with lattice constant of 3.52 Å and Ni-W-Fe-Cu-Mo alloys are created by randomly replacing Ni atoms of the single crystal by W, Fe, Cu, Mo atoms (20 at. % each). The samples are cylindrical in shape with 100 Å diameter and 100Å in length as shown in **Fig. 1** and comprises ~72300 atoms. After the initial atomic configuration, the samples are relaxed by the conjugate gradient algorithm through local energy minimum for 100 ps at temperature of -10 °C. The samples are subjected to uniaxial tensile-compressive and compressive-tensile loading separately along Y-direction [010] under periodic boundary condition along loading direction simulated at -10 °C and at strain rate of 108 s^{-1}. Simulations are performed under NVT ensemble, integrated by velocity-Verlet algorithm with a time-step of 0.002 ps.

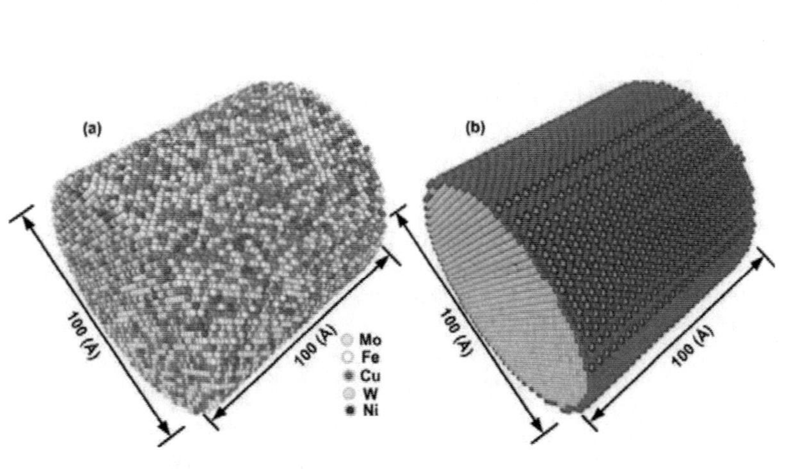

Fig.1: Atomic configuration 3-D snapshots of HEA sample (a) initial structure and (b) CSP of initial structure

3. Results and discussion

3.1. Sress-strain behaviour

The stress-strain curves for high entropy alloy single crystal under uniaxial loading (tension and compression) simulated at -10 °C temperature and strain rate of 108 s^{-1}, are shown in **Fig. 2**. Four different stages, i.e. a linear elastic region, a small plastic region, and again a linear elastic region followed by plastic deformation are observed in both tensile and compressive stress-strain curves for high entropy alloy single crystal.

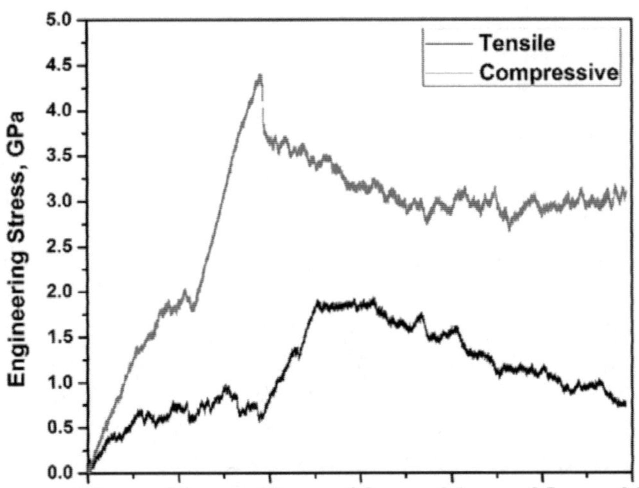

Fig.2: Engineering stress-strain curves for high entropy alloy single crystal under uniaxial loading.

The tensile stress of maximum value is ~1.9 GPa corresponding to strain ~0.25. On other hand, the compressive stress of maximum value is ~4.4 GPa corresponding to strain ~0.19. The small plastic portion of the tensile and compressive curve are due to twin formation. After small plastic deformation followed by linear elastic portion of the tensile and compressive curve are due to strain hardening. The stress-strain curves of high entropy alloy single crystals for tensile followed by compressive loading after ~0.60 pre strain and the pictorial representation of corresponding atomic snapshots colored according to CSP (Centro symmetric parameter) values at different position of deformation path have been presented in **Fig. 3.**

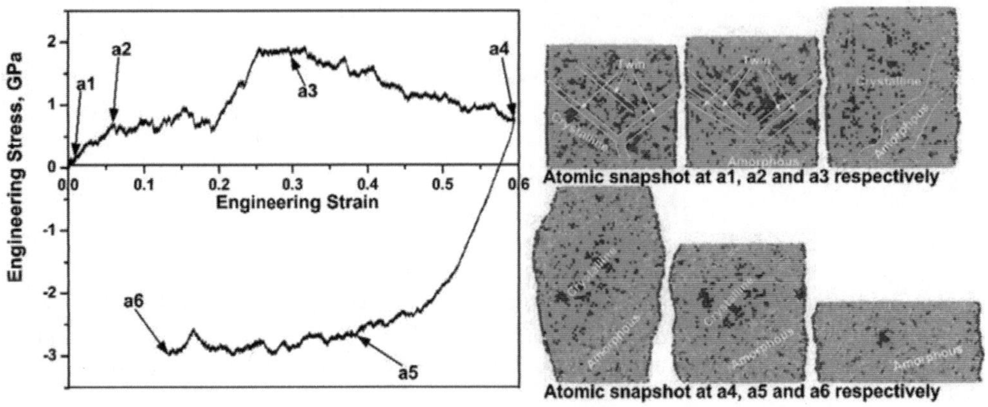

Fig.3: Engineering stress-strain curves of high entropy alloy single crystals for tensile followed by compressive loading after ~0.60 pre strain with the atomic snapshots marked on corresponding point of stress-strain curve.

It is revealed from **Fig. 3** that the presence of plastic deformation feature is presented of forward loading direction and twin are observed at a1 and a2 point of stress-strain curve during tensile part of the loading. On the other hand, crystalline and amorphous structure are observed during compressive part of the loading reversed after 0.6 strain. The stress-strain curves of high entropy alloy single crystals for compressive followed by tensile loading after ~0.60 pre strain and The pictorial representation of corresponding atomic snapshots colored according to CSP values at different position of deformation path have been presented in **Fig. 4.** It is revealed from Fig. 4 that the presence of plastic deformation feature is presented of forward loading direction and twin are observed at a1 point of stress-strain curve during tensile part of the loading. On the other hand, crystalline and amorphous structure is observed during compressive part of the loading reversed after 0.6 strain. Linear elastic region and serrated plastic region are observed. Further, reverse loading at different direction after which yield stress more in compressive followed by tensile type of loading with respect to tensile followed by compressive type of loading of high entropy alloy. The generation of twins are observed mostly during forward loading (i.e. tensile part of tensile followed by compressive and compressive part of compressive followed by tensile) and subsequently these twins are found to act as potential sources of amorphous formation. This phenomenon is not reported in available literature for single crystal during tensile followed compressive type of loading and compressive followed tensile type of loading. Twins are found to be inclined at ~45° to the loading direction.

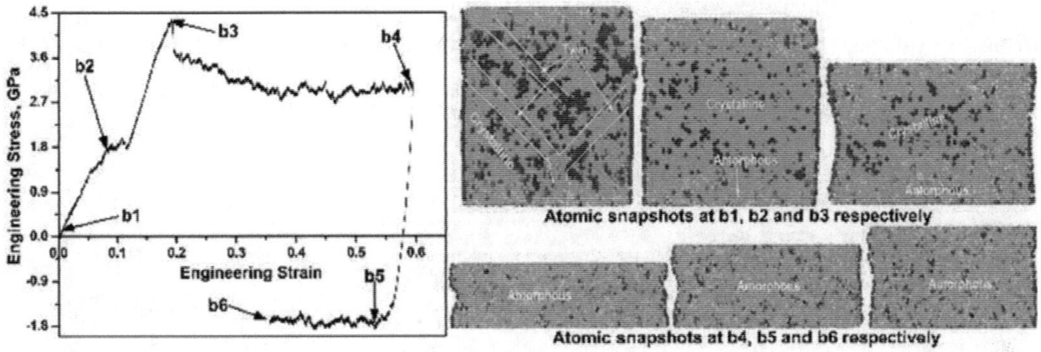

Fig.4: Engineering stress-strain curves of high entropy alloy single crystals for compressive followed by tensile loading after ~0.60 pre strain with the atomic snapshots marked on corresponding point of stress-strain curve.

3.2. Radial distribution function (RDF)
The RDF patterns at different loading direction such as tension followed by compression and compression followed by tension loading after 0.60 pre-strain for high entropy alloy single crystals have been presented in **Fig. 5 (a)** and **5 (b)** respectively. Sharp peaks observed at different strains indicate that there is no amorphization of the high entropy alloy single crystal during the forward or reverse loading. The stress-strain curves of high entropy alloy single crystal exhibits linear elastic behavior and serrated plastic behavior. This is due to the transformation from crystalline to partial amorphous structure which is also evident in RDF plots provided in **Fig. 5.**

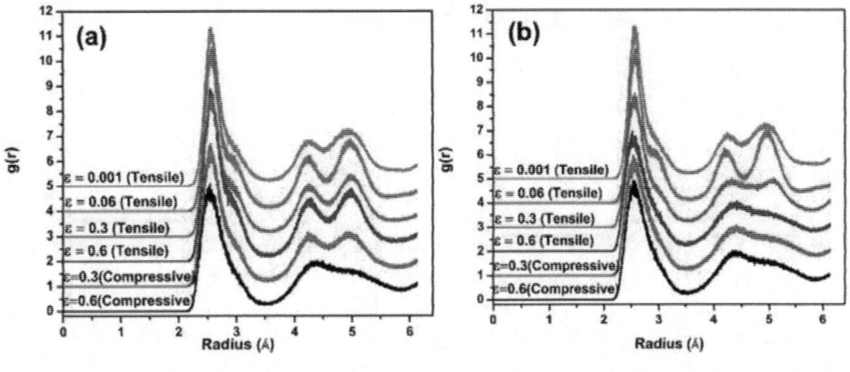

Fig.5: The radial distribution functions at different strains during (a) tensile followed by compressive and (b) compressive followed by tensile loading after ~0.60 pre strain.

3.3. Stress Distribution

Von Mises' stress contours plots representing development of stress at vertical cross section of all the specimens at different strain of forward and reverse loading have been shown in **Fig. 6.** The stress distributions at different strain are observed to be almost homogeneous throughout the specimen and slight localization of stress field near the surface is identified at the point of reverse loading. There is no significant polarization of stress field in high entropy alloy single crystals at different strain of forward and reverse loading as evident in **Fig. 6.**

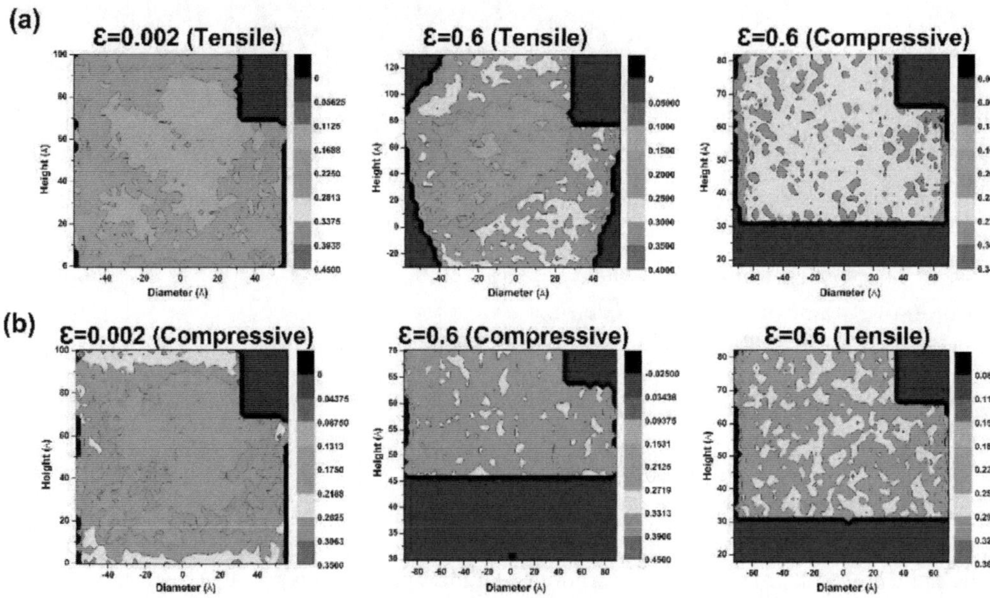

Fig.6: Stress distribution at different strain (a) 0.002 (tensile), 0.6 (tensile) and 0.6 (compressive) during tensile followed by compressive and (b) 0.002 (compressive), 0.6 (compressive) and 0.6 (tensile) during compressive followed by tensile loading of high entropy alloy single crystals,

4. Conclusions

Molecular dynamics simulations have been carried out for $Ni_{20}W_{20}Cu_{20}Fe_{20}Mo_{20}$ high entropy alloy (20 at. % each element) single crystals with EAM potential for studying the underlying deformation mechanism during tension followed by compression and compression followed by tension. This MD simulation reveals that plastic deformation in $Ni_{20}W_{20}Cu_{20}Fe_{20}Mo_{20}$ high entropy alloy (20 at. % each element) single crystals due to the twin and atomic diffusion. Formation of high entropy alloy single crystal causes reduction in crystallinity significantly and this leads to decrease in dependence of deformation process on dislocation controlled mechanisms. Twin dominated deformation mechanism prevails during forward loading. Deformation of high entropy alloy during reverse loading is controlled by atomic diffusion. Stress distribution throughout the system almost homogenous.

Acknowledgements
The authors want to thank the computer centre of National Institute of Technology Rourkela for providing high performance computing facility (HPCF).

References

[1] M. H. Tsai, J. W. Yeh, High-entropy alloys: a critical review. Materials Research Letters, 2, (2014) 107-123.
[2] T. Egami, W. Guo, P.D. Rack, T. Nagase, Irradiation resistance of multi-component alloys. Metall. Mater. Trans. A 45 (2013) 180-183.
[3] Q.Y. Zhai, J.F. Xu, (2011), China Patent CN101554686B.
[4] M. A. Hemphill, T. Yuan, G. Y. Wang, J. W. Yeh, C. W. Tsai, A. Chuang, P. K. Liaw, Fatigue behavior of Al 0.5 CoCrCuFeNi high entropy alloys. ActaMaterialia, 60 (2012) 5723-5734.
[5] Y. Lu, C. Peng, Y. Ganesan, J. Y. Huang and J. Lou, Quantitative in situ TEM tensile testing of an individual nickel nanowire, Nanotechnology 22 (2011) 355702-355708.
[6] C. H. Lin, H. Ni, X. Wang, M. Chang, Y. J. Chao, J. R. Deka and X. Li, In situ Nanomechanical Characterization of Single-Crystalline Boron Nanowires by Buckling, Small 6 (2010) 927-931.
[7] J. Y. Huang, H. Zheng, S. X. Mao, Q. Li and G. T. Wang, In situ nanomechanics of GaN nanowires, Nano letters 11 (2011) 1618-1622.
[8] M. T. McDowell, A. M. Leach and K. Gall, Bending and tensile deformation of metallic nanowires, Modelling and Simulation in Materials Science and engineering 16 (2008) 045003.
[9] M. Widom, W. P. Huhn, S. Maiti, W. Steurer, Hybrid Monte Carlo/molecular dynamics simulation of a refractory metal high entropy alloy. Metallurgical and Materials Transactions A, 45 (2014) 196-200.
[10] S. W.Kao, J. W. Yeh, T. S. Chin, Rapidly solidified structure of alloys with up to eight equal-molar elements-a simulation by molecular dynamics. Journal of Physics: Condensed Matter, 20 (2008) 145214.
[11] M. C. Gao, D. E. Alman, Searching for next single-phase high-entropy alloy compositions. Entropy, 15 (2013) 4504-4519.
[12] H. Diao, L. J. Santodonato, Z. Tang, T. Egami, P. K. Liaw, Local structures of high-entropy alloys (HEAs) on atomic scales: An overview. JOM, 67 (2015) 2321-2325.
[13] S. Plimpton, Fast parallel algorithms for short-range molecular dynamics, Journal of computational physics 117 (1995) 1-19.
[14] A. Stukowski, Visualization and analysis of atomistic simulation data with OVITO-the Open Visualization Tool, Modelling and Simulation in Materials Science and Engineering 18 (2010) 015012.
[15] A. Stukowski, Structure identification methods for atomistic simulations of crystalline materials, Modelling and Simulation in Materials Science and Engineering 20 (2012) 045021.
[16] X. W. Zhou, R. A. Johnson and H. N. G. Wadley, Misfit-energy-increasing dislocations in vapor-deposited CoFe/NiFe multilayers, Physical Review B 69 (2004) 1441131-14411310.

A Novel Technique for Making Cold Briquettes for Charging in Blast Furnace

M K Mohanty[1], S Mishra[1], B Mishra[2], S Sarkar[1] and S K Samal[1]

[1]Department of Metallurgical and Materials Engineering, National Institute of Technology Rourkela, Rourkela 769008, India
[2]Dalmia Institute of scientific & Industrial Research, Rajgangpur, Odisha, India

E-mail: manojamohanty@gmail.com

Abstract. Different metallurgical wastes are generated during pyro processing of iron ore, which is used for making sponge iron or hot metal and for producing steel. Apart from these wastes, coke fines are generated during the coke making, and iron ore fines are generated during mining of iron ore. Although iron ore fines are used for making pellet after beneficiation still, it generates a huge quantity of iron ore waste during beneficiation with comparatively lower iron content. In the present study, briquettes are made by a stiff extrusion process from metallurgical waste like iron ore fines and coke fines with the addition of Portland cement as a binder and clay as a rheology modifier. Physical properties of the briquettes are evaluated, and reducibility of the briquettes is studied in comparison to lumpy iron ore. Phase analysis and microstructural analysis of the briquettes and lumpy iron ore are carried out after firing at different temperatures in the simulated blast furnace condition. Physical and mineralogical properties are correlated with the reducibility of the briquettes and lumpy iron ore. Briquettes made by a stiff extrusion process show a better mechanical strength fired at a different temperature to take the load of burden and better reducibility than lumpy iron ore. The briquettes after self-curing are charged to a 23 mt3 blast furnace which shows encouraging results.

1. Introduction

Large quantities of useful iron bearing wastes are generated at different stages of iron and steel making processes. Apart from these wastes large quantity of iron ore fines are generated during mining, crushing and grinding. Most of them are in fine fraction and difficult to use even for making sinter use as a feed material for the blast furnace. Now a day's pelletization is getting popular which uses iron ore fine and becomes a feed material after induration. But different iron-bearing waste generated could not be used for pallet making due to different metallurgical reasons.

Another method of agglomeration is briquetting, which consists of three basic technologies i.e. vibro-pressing, roller pressing and the stiff extrusion. The stiff-extrusion technique which was mastered by J. C. Steel & Sons (USA), had its 1st industrial application in the year 1993 [1]. In this cold briquetting technology various steel plant by-products can be easily agglomerated using very few amount of binder material and also cost effective than other agglomerating processes, due to their less processing steps.In this advanced technology homogenizing mixture of raw materials with the moisture content of 12-16% passes through the vacuum chamber and then pushed through the die holes by pressure 3-3.5 Mpa [2]. The industrial stiff-extrusion process lines shown in figure 1.

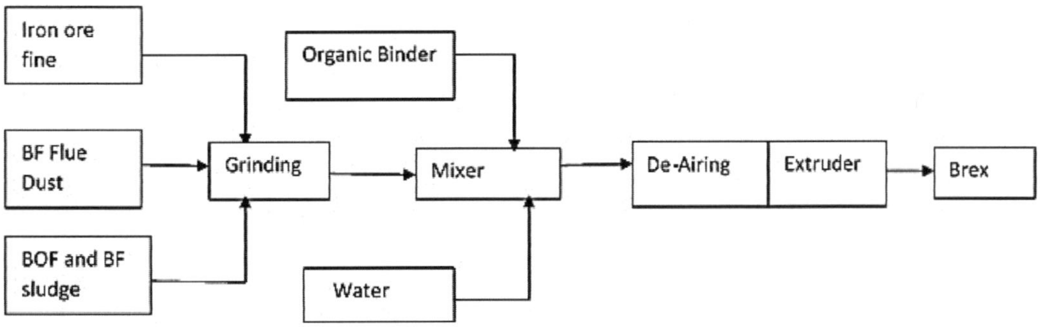

Figure 1. Schematic diagram of the extruded briquette preparation process.

During the passing of homogenizing mixture of raw materials through the vacuum chamber of the extruder, approximately 90% of compressed air is removed [3]. This causes the high mechanical strength of extruded briquette, even after the outlet of the extruder. A Steele 25 extruder used in the present work for briquetting is a combination of Pug sealer and extruder as shown in the figure 2. The Pug Sealer sits on top of the extruder having high pressure screw mixer that mixes the materials intensively. The vacuum inside the vacuum chamber is created by a vacuum pump. The mixture passes through the vacuum chamber and drops down at the blades of the extruder's auger. The vacuum is maintained throughout the working volume of the extruder. Then the material mixture is pressed through the extrusion dies under high pressure and vacuum to generate cold briquettes.

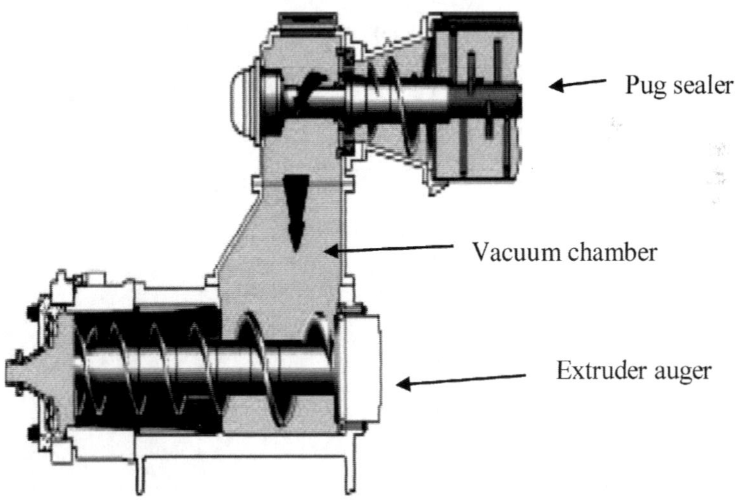

Figure 2. Sectional view of extruder vacuum chamber.

2. Experimental Methodology

2.1 Materials

The materials used in this work are iron ore fines, BF dust, LD sludge, sludge and mill scale. The waste materials were collected from mining area and different steel plants. These materials were evaluated for their chemical properties. Their chemical analyses are given in table 1. Phase analysis is done by X-ray diffractometer at DISIR, Rajgangpur. Phase analysis data are given in table 2. Chemical analysis of Portland cement and bentonite are given in table 3.

Table 1. Chemical analysis (wt. %) of Metallurgical waste.

Constituents	Iron ore fines (wt. %)	BF dust (wt. %)	LD sludge (wt. %)
Fe_2O_3/FeO	78.5	51.5	87.5
SiO_2	5.6	6.3	0.6
CaO	-	4.9	9.5
MgO	-	0.2	1.2
Al_2O_3	5.4	5.1	0.3
TiO_2	0.8	-	-
C	-	30.5	-
Fe_T	53.5	35.6	64

Table 2. Phase analysis by XRD.

Constituents	Major phases	Minor phases
Iron ore fine	Hematite	Goethite, Gibbsite, Kaolin, Pyroxene
BF dust	Magnetite, Hematite, Graphite	Wustite, quartz, Dicalcium aluminate
LD sludge	Magnetite, Wustite	Wollastonite, calcite

Table 3. Chemical analysis (wt. %) of Portland cement and Bentonite.

Constituents	Portland cement	Bentonite
CaO	63.2	0.8
SiO2	20.5	58.4
Al_2O_3	4.5	12.6
Fe_2O_3	4.2	10.6
MgO	2.1	0.3
K_2O+Na_2O	0.6	4.4
SO_3	2.6	-
LOI	1.5	9.4

The raw materials were taken according to their weight percentage, as given in table 4 and mixed properly. The mixture was passed through the homogenizing mixer and the vacuum chamber. Where, vacuum maintained was $0.5x10^{-3}$Bar. The briquette formed was extruded from the die at a pressure of 100 kg/cm^2.After that, the brex were cured in an open air.

Table 4. Mixing percentage of raw materials

Briquette Sample No	LD sludge (Wt. %)	BF dust (Wt. %)	Iron ore fine (Wt. %)	Binder (Wt. %)	Plasticizer (Wt. %)
1	47.28	28.3	18.9	4.7	0.9

The chemical compositions of the raw briquette and lumpy iron ore are shown in table 5. XRD analysis of raw iron ore lump shown in Fig 3 and for raw briquette shown in figure 4.

Table 5. Chemical composition of Briquette and iron ore.

Constituents	Iron ore (Wt. %)	Briquette (Wt. %)
CaO	0.2	8.4
SiO_2	2.21	4.1
Al_2O_3	3.11	2.35
Fe_2O_3	88.53	68.53
MgO	-	1.4
K_2O+Na_2O	0.2	0.2
TiO_2	0.22	0.23
LOI	5.82	15.02

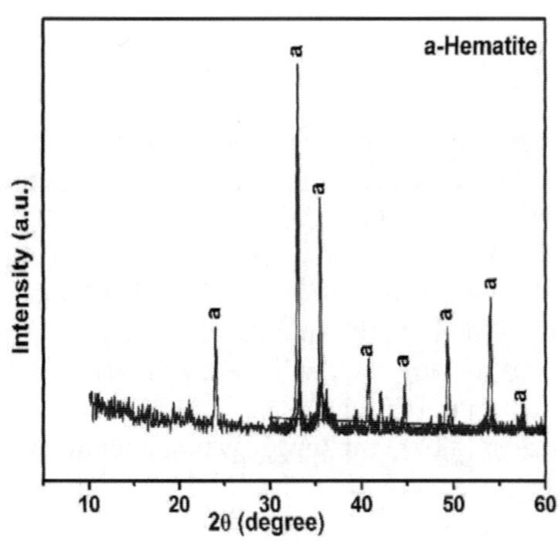

Figure 3. XRD analysis of raw iron ore lump.

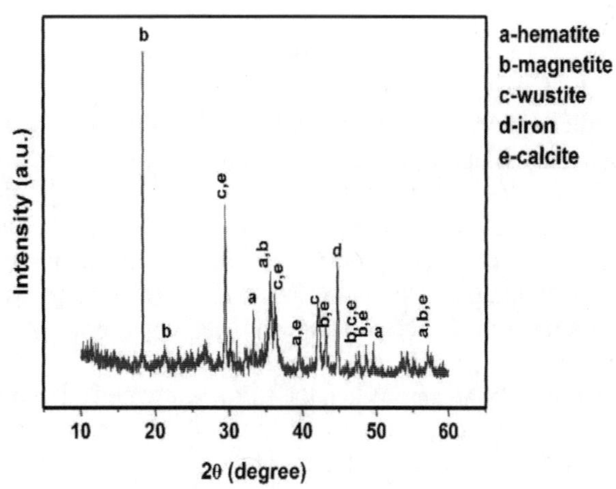

Figure 4. XRD analysis of raw briquette.

2.2 Mechanical strength of briquette and lumpy iron ore

The mechanical properties of briquette and iron ore are given in table 6. The cold compressive strength of briquette after natural curing at different days are shown in figure 5. For getting the mechanical properties which are shown in table 6, the briquette and iron ore lumps were rotated in tumbler drum at

25RPM for 200 revolutions. Then it passed through 6.3 mm sieve and a 0.6 mm sieve. The retained weight percentage of the samples on 6.3 mm size sieve give the tumbler index. Abrasion index is calculated by taking the weight percentage of the fines passed through 0.6 mm size sieve.

Table 6. Mechanical properties.

Sample	Tumbler Index	Abrasion index
Briquette	96	1.8
Iron ore	82	2.9

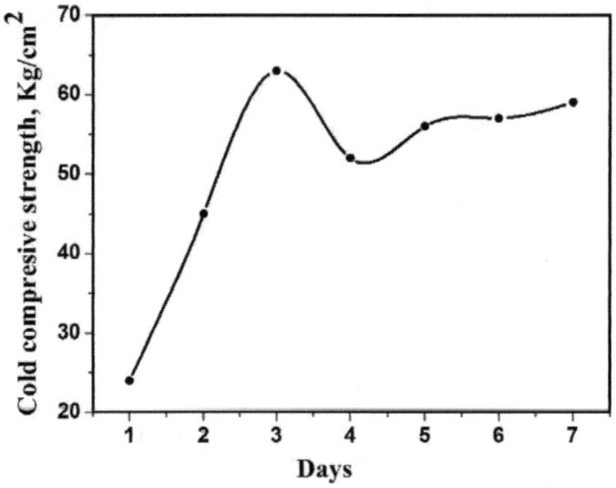

Figure 5. Cold compression strength (CCS) after natural curing.

The cold compressive strength of cured briquettes was measured by Tonipact3000. The strength shows a max on completion of third say followed by a decrease on the fourth day and then showed an increasing trend up to the studied range. Such behavior is due to the creation of the coagulation structure in the cement-bentonite-water system, leading to modifications in the properties of the binder. Reduced strength, observed after the third day, can be explained by the destruction of structures.

2.3 Reduction test of Briquette

A hollow alumina tube having a length 1000 mm, outside diameter 85 mm and inside diameter 75mm was taken to carry out the reduction test. The weights of the briquettes and iron ore were taken before and after the test. Briquettes and iron ore were kept in a graphite crucible and the crucible was then placed inside the alumina tube covered by coke breeze having size 1-5mm throughout the height engulfing the graphite crucible completely. The thickness of the coke layer was 25mm. Then it is blocked with insulating blocks. Water cooled metallic clamps were fitted on both the sides of alumina tube. The vacuum pump connected in one of the clamps along with air supply connected with a flow meter. Before heating, the inside chamber of the alumina tube was evacuated up to 0.5 mbar. The furnace was heated at a rate of 10°C/min. When the furnace attained a temperature of 800°C, air supply was started at a rate of 10 liters /min till the completion of the reduction test. The process repeated for all the temperatures from 1000°C to 1500°C. Both briquettes and iron ore were fired from 1000°C to 1500°C with a soaking period of 2 hours. Photograph of the apparatus is given in figure 6 and figure 7.

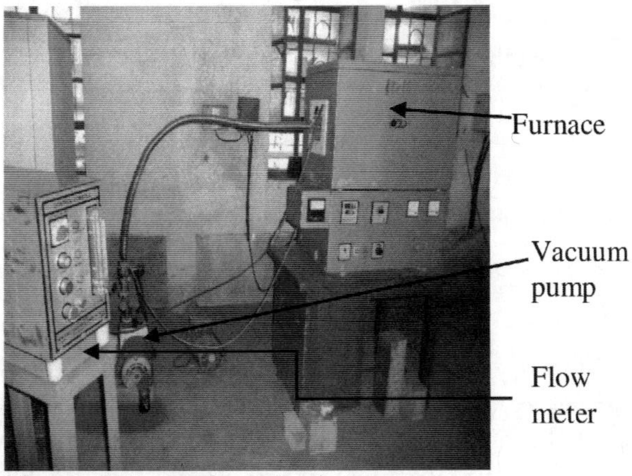

Figure 6. Set-up for reduction test

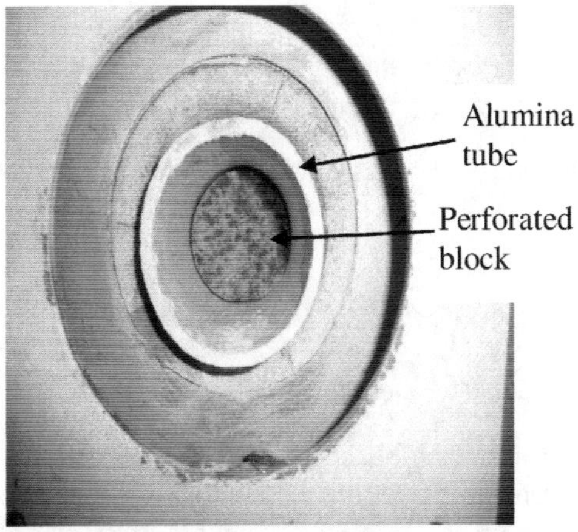

Figure 7. Crucible & alumina tube setup.

3. Results and discussion

3.1 Effect of Temperature on the Reduction of Briquettes and Iron ore lump

The degree of reduction and metallization were calculated. The results of metallic iron, total iron, the degree of reduction and % metallization for iron ore lump are given in table 7. The results of metallic iron, total iron, the degree of reduction and % metallization for briquette are given in table 8. The degree of reduction of briquettes and iron ore lumps was determined by using the formula given in equation (1).

$$\text{Degree of reduction}\ (\%) = \left(\frac{W_i - W_f}{W_o}\right) \times 100 \qquad (1)$$

Where W_i = the mass of sample before firing. W_f = the mass of the sample after firing. W_0 = the total mass of oxygen present in every individual mixture in the form of Fe_2O_3, Fe_3O_4 & FeO.

The degree of metallization of briquettes and iron ore lumps was determined by using the formula given in equation (2)

$$\text{Degree of metallization}\ (\%) = \left(\frac{\text{Metallic Fe}}{\text{Total Fe}}\right) \times 100 \qquad (2)$$

Table 7. Degree of reduction & Metallization for Iron ore.

Temperature (^{0}C)	Fe_M (%)	Fe_T (%)	Metalization (%)	Degree of reduction (%)
1000/2 hr	2.16	67.48	3.20	35
1100/2 hr	12.69	69.52	18.25	52
1200/2 hr	17.65	70.25	25.10	61
1300/2 hr	30.33	73.3	41.30	65
1400/2 hr	53.75	80.79	66.53	82
1500/2 hr	84.25	90.4	93.20	93

Table 8. Degree of reduction & Metallization for Briquette.

Temperature	Fe_M	Fe_T	Metalization (%)	Degree of reduction (%)
1000/2 hr	14.20	60.5	23.5	48
1100/2 hr	20.96	61.7	33.97	62
1200/2 hr	30.36	63.7	47.66	72
1300/2 hr	44.74	65.6	68.2	80
1400/2 hr	47.04	67.46	69.7	88
1500/2 hr	79.22	80.87	97.95	99

Table 7 and 8 show that briquettes have a higher degree of reduction than iron ore lump since the reducibility of lumpy iron ore dependent on lump size [4]. The reason for this is evidently related with the presence of the carbon in the BF sludge which is highly reactive and helps for gasification reaction at a comparatively lower temperature where iron ore fines work as a catalyst. From the mineralogical point of view iron ore after heat treatment >1100°C forms iron silicate and iron aluminium silicate which has a melting point around 1200°C. This melting temperature goes down to 1150°C with 1-2% lime. This envelopes the grain and seals the pores resulting in retardation of reducibility. Also, reducibility of iron silicate or iron aluminium silicate is very poor and requires very high temperature. In the case of briquettes, it forms calcium ferrite. Even BF dust also contains some calcium ferrite. The melting temperature of calcium ferrite is high which is >1400°C facilitating the pores which help in better reduction. Apart from this reducibility of calcium ferrite is even better than Magnetite. So to say a reason for higher reducibility of briquette appears to be due to their higher porosity values. Swelling index of briquette at different temperatures obtained are shown in table 9. The swelling indices of briquettes were determined by using the formula given in equation (3).

$$\text{Swelling index (\%)} = \left(\frac{V_f - V_i}{V_i} \right) \times 100 \tag{3}$$

Where V_f = initial volume of the briquette. V_i = final volume of the briquette.

Table 9. Swelling index at different firing temperatures of briquette

Firing Temperature (^{0}C)	Swelling Index (%)
1000^0	19
1100^0	15
1200^0	13
1300^0	9
1400^0	4

The observed values of swelling index of the briquette are less than 20.This indicating normal swelling at each of the studied firing temperatures. Table 9 also indicates that the value of swelling index decreased drastically with a rise in firing temperature. This is expected to be due to the sintering of iron particles. The sintering effect restricts the growth of fibers responsible for swelling [5].

3.2 X-ray Analysis of Reduced Sample

Phase analysis was done by X-ray diffractometer. X-ray Diffraction patterns of iron ore fired at 1300°C, 1400°C & 1500^{0}C are shown in figure 8.

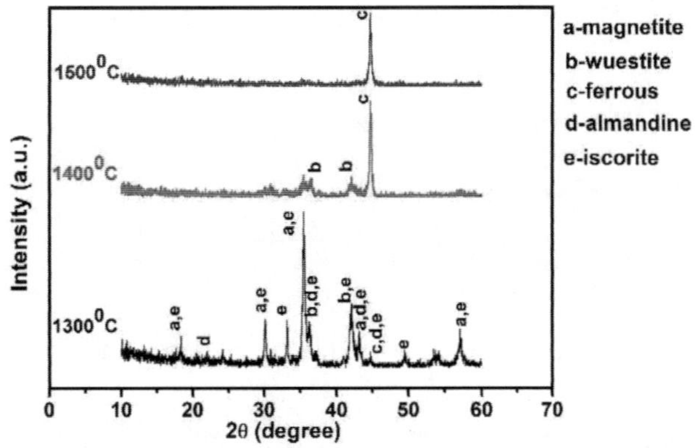

Figure 8. XRD analysis of iron ore fired at 1300°C, 1400 °C, and 1500 °C for 2hr.

At 1300^{0}C, the presence of magnetite and wustite as major phases with ferrous as a minor phase confirms that hematite had been reduced to the next level in the reduction process. At 1400^{0}C, the presence of ferrous as a major phase along with wustite as a minor phase confirms the complete reduction of magnetite to wustite and iron, for which degree of metallization had been enhanced to a greater extent. XRD pattern of reduced iron ore at 1500^{0}C shows that wustite was completely reduced to metallic iron and for which highest degree of metallization was achieved.

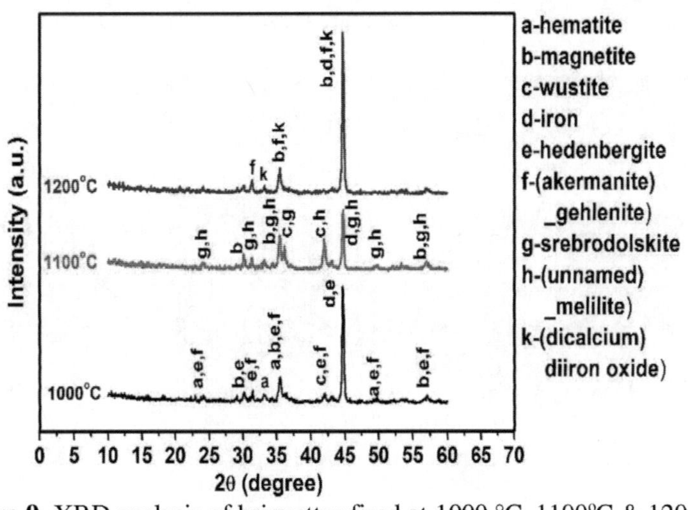

Figure 9. XRD analysis of briquettes fired at 1000 °C, 1100°C & 1200 °C.

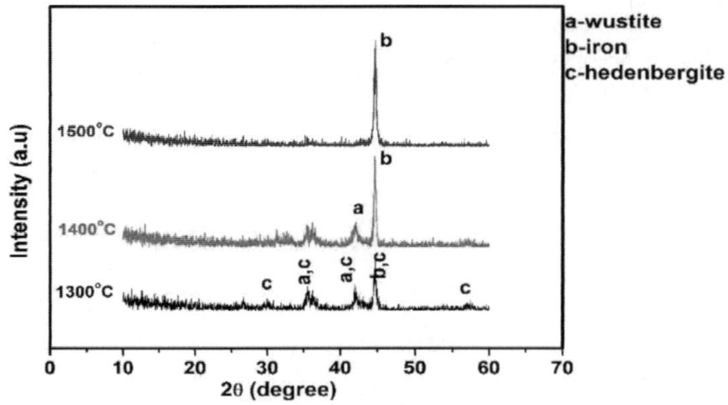

Figure 10. XRD analysis of briquettes fired at 1300 °C, 1400°C & 1500 °C.

XRD pattern of reduced briquettes fired at 1000°C, 1100°C and 1200°C are given in figure 9. XRD pattern of reduced briquettes fired at 1300°C, 1400°C and 1500°C are given in Figure 10. At 1300°C wustite and ferrous phases are present, whereas, only ferrous phase found in the diffraction pattern of briquette reduced at 1500°C. The presence of wustite and iron at 1400° lowered the degree of metallization value as compared to that of briquette reduced at 1500° C.

3.3 Microscopic Analyses

Microscopic evaluation of iron ore and briquette fired at the different temperatures were evaluated on the polished section under the reflected light in a universal microscope. Optical micrograph of iron ore fired at 1200°C is given in figure 11. Photomicrograph of briquette fired at 1200°C is given in figure 12.

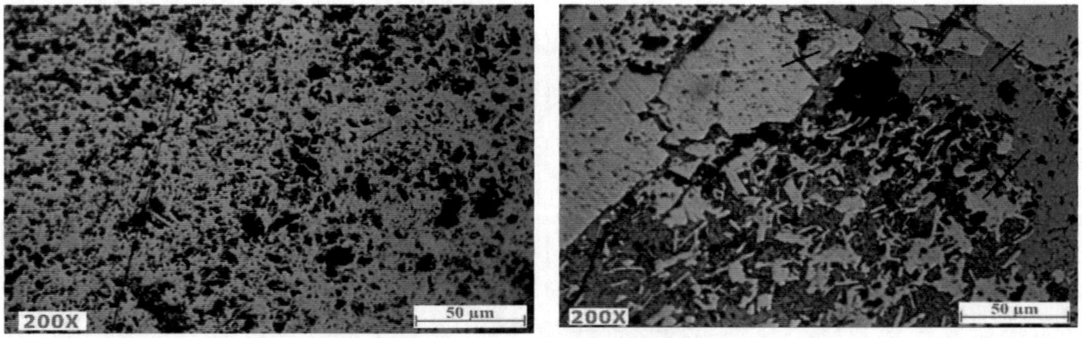

1- Hematite, 2- Magnetite, 3- Wustite, 4- Glass

Figure 11. Optical microscopy of the reduced iron ore lump at 1200°C in reducing atmosphere for 2hr.

From below microscopic images of reduced briquette at 1200°C shows the presence of Dicalcium ferrite, which is responsible for faster reduction. At temperature 1400°C in figure 13, one can see the structure of the metal surrounded by the silicates and ferrites. Thus, the metallization of the surface layer of the agglomerated product helps to keep the integrity during its reduction with the creation of the metallic phase supported by the silicate and a ferrite matrix [6]. Blast furnace operating parameters of a 23 M^3 blast furnace, given in table 10 where the charging % of extruded briquette varied w.r.t lumpy iron ore.

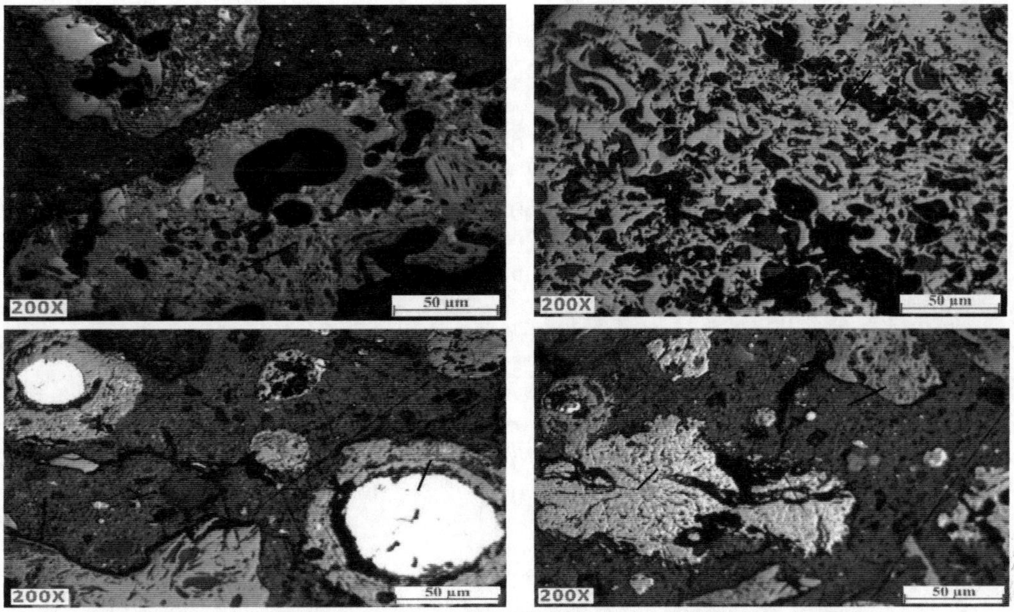

1- Hematite, 2- Magnetite, 3- Wustite, 4- Metallic Fe, 5- Dicalcium ferrite

Figure 12. Optical microscopy of the reduced briquette at 1200^{0}C in reducing atmosphere for 2hr.

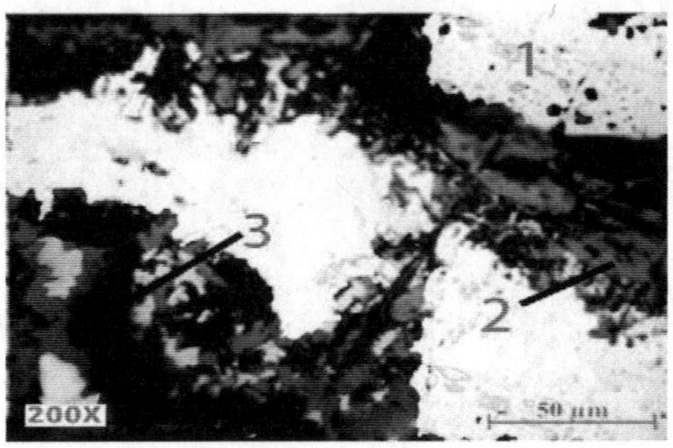

1- Metal, 2- Ferrite, 3- Silicate

Figure 13. Optical microscopy of the reduced briquette at 1400^{0}C in reducing atmosphere for 2hr.

Table 10. Blast Furnace operation parameters.

BF operation parameters	100% iron ore	80% Briquettes	100% Briquettes
Feed Material	Iron ore, Limestone, Dolomite, sponge iron	Briquettes, Iron ore, Dolomite, Mn ore	Briquettes, Iron ore, Dolomite, Mn ore
Fe content of the charge (%)	58	50.5	46
Blast temp in °C	950	950	975
Tuyere pressure (kg/cm2)	0.45-0.5	0.38-0.44	0.38-0.44
Hot metal Temp in °C	1410-1430	1400-1420	1400-1420
Slag temp in °C	1430-1450	1430-1440	1430-1440
Coke cons/TLM in kg	680	530	490
Slag volume in kg/TLM	360	420	450

Hot metal chemistry (%)			
Si			
Mn			
C	1-1.5	0.7-0.9	0.5-0.7
S	0.3	0.3-0.4	0.3-0.4
P	3.9-4	3.8-4	3.8-4
Fe	0.05-0.06	0.04-0.05	0.03-0.035
	0.09-0.11	0.09-0.11	0.09-0.1
	93-93.5	93-93.5	93-93.5
Slag chemistry (%)			
CaO	34.86	33.12	34.43
SiO_2	31.98	31.13	31.17
Al_2O_3	19.16	17.98	17.3
MgO	9.46	9.65	9.99
	0.3	0.25	0.31
TiO_2	1.2	1.16	0.67
FeO			

4. Conclusions

- Briquettes made by stiff extrusion principle are a novel technique for making briquettes which do not require any heat treatment. Moreover, it is utilized all the waste which is difficult to use at present.
- The briquettes retain its integrity up to the softening till it enters the cohesive zone where it gets melted.
- Briquettes made by this technique shows better reducibility than iron ore. Because major silicate phases after firing iron ore at different temperature are iron silicate and iron aluminium silicate which envelopes the grain and retard reduction. In the case of briquettes formation of ferrite which is easier to reduce than iron silicate or iron aluminium silicate.
- 100% briquettes indicate minimum reductant consumption despite the slag volume is high.
- Burden with 100% briquettes indicates improved hot metal quality as the silicon content is much lower.
- Utilization of waste helps to preserve the environment by utilizing the waste and less generation of greenhouse gas.

References

[1] Dalmia Y K ., Kurunov I F ., Steele R B . and Bizhanov A M . 2012 Production of a new generation of briquettes and their use in blast-furnace smelting *Metallurgist* **56** 164–8.

[2] Bizhanov A, Pavlov A, Chadaeva O and Dalmia Y 2014 Received on December 18, 2013; accepted on February 24, 2014) **54** 1450–2.

[3] Durov N M, Nushtaev D V and Ryzhov S A 2013 Study of the mechanical strength of Brex. Part 2 **56** 489–93.

[4] Esin O A and Gel'd P V Physical Chemistry of Pyrometallurgical Processes (Metallurgizdat, Sverdlovsk, 1962) *Part I.*

[5] Kumar M and Patel S K 2009 Assessment of Reduction Behavior of Hematite Iron Ore Pellets in Coal Fines for Application in Sponge Ironmaking *Miner. Process. Extr. Metall. Rev.* **30** 240–59.

[6] Kurunov I, Malysheva T Y and Bol'shakova O 2007 A study of the phase composition of iron-ore briquets to assess their behavior in a blast furnace *Metallurgist* **51** 548–57.

Ultrasonic spot welding of dissimilar materials: characterization of welded joints and parametric optimization

M P Satpathy[1] **and S K Sahoo**[1]

[1]Department of Mechanical Engineering, National Institute of Technology Rourkela, Rourkela 769008, India

E-mail: mantraofficial@gmail.com

Abstract. Material joining is one of the key manufacturing processes used to assemble metallic and non-metallic parts for several applications. But the industries are facing many difficulties in joining of thin sheets of dissimilar metals by the conventional welding process because of their differences in chemical composition, physical and mechanical properties. Thus, ultrasonic welding is a solid state joining process used for joining of small elements in microelectronics industries. In this process, acoustic horn and booster are the important assets. The accuracy and strength of the welding depend mainly on their geometry. This proposed work deals with the design and modelling of an acoustic stepped sonotrode with booster using finite element analysis (FEA). From this analysis, the actual length of the horn is obtained by gradually decreasing its theoretical length. The quality of the weld is reckoned by its weld strength and the combinations of different process parameters. These are examined using the principal components coupled with grey relational analysis approach which is showing good agreement between the predicted values with experimental results. Fractographic examination of weld zone and hardness are also used to explore the weld quality.

1. Introduction

Now-a-days there is a necessity of miniaturization, ultra lightweight and ultra-precise parts. So many researchers are looking forward to various innovative micro welding techniques, in which metals and non-metals are united together for numerous industrial applications. The major area of is the automobile industry, which enables high production rate with the use of high-performance materials. The applications are now frequently increasing because of the requirement of top-quality accuracy and safety, which can't be obtained in traditional welding processes. In the dissimilar metal welding, two or more metals are joined together under some particular conditions. But at the same time, it is difficult to join dissimilar metals because of their differences in composition, physical and mechanical properties. However, good welds have not been noticed by the fusion welding process so far. To overcome this difficulty, a solid-state joining process born, in which the bonding between the materials can be created due to the friction, and it happens below the melting temperature of parent materials. Ultrasonic metal welding (USMW) is one of these processes.

USMW is a solid-state joining process in which the joint is created between the workpieces by the application of high-frequency ultrasonic waves under pressure. The ultrasonic welding stack is comprised of six subsystems (i) generator (ii) piezoelectric transducer (iii) booster (iv) horn/sonotrode (v) pneumatic cylinder [1], (vi) anvil. The acoustic horn and booster design is a vital part of the ultrasonic welding process and the efficiency mainly depends on its design correctness. The traditional methods for the design of an acoustic horn are based on empirical relationships which include the equilibrium of a small element under elastic action and inertia forces. But, this will take a lot of time and even failed, when the complicated shape is considered. To overcome this, FEM can be used as an alternative method. Sherrit et al. [2] described a new type of horn designs which are superior to other normal horn designs from a design point of view. Generally resonance frequency depends on the length of the horn and amplitude depends on the different cross-sectional area. Amin et al. [3] did their investigation on various designs of the horn, which can affect ultrasonic cutting. They stated in the different design concept of horns by computer added design procedure and used finite element analysis to validate it. Nad [4] conducted his ultrasonic machining experiment using different horns like the cylindrical, tapered and exponential type. Siddiq and Ghassemieh[5] investigated that, it is possible to join thin layers of titanium alloy with aluminium alloy by the ultrasonic welding process at low temperature, pressure, and energy consumption. The solid state welding could be applied to join nonferrous materials like aluminium with ferrous materials like stainless steel (Tsujino et al. [6]). Sooriyamoorthy et al. [7] attempted to join metal with ceramic, and it was also successful of welding of Al with Al_2O_3. Optimization of process parameters is the major step in the Taguchi method. But one major limitation of this philosophy is that it is failed while solving the multi-response optimization problem. So, to overcome this drawback, Kumar et al. [8] proposed desirability function approach in combination with Taguchi philosophy. There are also some studies that joined aluminum with copper using ultrasonic spot welding process [9]. Zhao et al. [10] reported the formation of Al_4Cu_9intermetallic layer which is main cause of joint failure. Thus, it helps a researcher to find out the possible causes of weldment failure.

Designing of the horn for the ultrasonic welding is really a challenging task because the high-frequency vibration should precisely occur at the welding tip and the amplitude should be maximized at the welding tip. In this study, a horn with the booster is designed using FEA and performing experiments with it. The ultrasonic welding is carried out using Al and brass metals, and the effect of different process parameters on weld strength is studied. The principal component analysis is used to convert the correlated responses to uncorrelated one, and grey based Taguchi method is employed to find out the optimal parameter condition. The quality of the weld is identified by the metallographic studies.

2. Experimental Methodology

2.1 Ultrasonic horn design

In ultrasonic welding, the asymmetrical horn mostly works with the longitudinal mode phenomena. The governing equation of the horn can be derived by considering a free-free vibration in a non-uniform bar.

$$\frac{\partial^2 \xi}{\partial x^2} + \frac{1}{S} \times \frac{\partial S}{\partial x} \times \frac{\partial \xi}{\partial x} + \frac{\omega^2}{C^2} \times \xi = 0 \tag{1}$$

where ξ = Amplitude in μm, S = Cross-sectional area in mm^2, ω = Angular velocity in rad/sec and C = acoustic velocity in m/sec passed through the material. This velocity can be found out by

$$C = \left(\frac{E}{\rho}\right)^{1/2} \tag{2}$$

The length of the horn can be obtained by solving the Webster horn equation as shown in equation (1).

$$C = \lambda f \tag{3}$$

where λ = wavelength of ultrasound waves.

The length of the horn is the half of the wavelength of waves passing through it. Mathematically, it can be represented as

$$L = \frac{\lambda}{2} = \frac{C}{2f} \tag{4}$$

The performance of the ultrasonic system is measured by its amplification ratio/gain. It represents the capability of the system to weld high thickness materials. The dimensions of the input and output ends are fixed by the size of the booster and the specific weld tip size. The input and output diameter of the horn is 44 and 35 mm and for the booster they are 55 and 44 mm respectively. The line diagram of the horn with the booster is presented in figure 1. Mathematically the amplification ratio is represented as

$$\frac{\xi_1}{\xi_2} = \left(\frac{S_2}{S_1}\right) = \left(\frac{d}{D}\right)^2 \quad (5)$$

where S_1, S_2 = Input and output area of the horn. D, d = Input and output diameter of the horn.

The most important factor of the horn and booster to get the accurate length. Because an incorrect design may produce infinite resistance with other elements and can damage to the system. High fatigue strength, hardness and low acoustic losses should be the criteria for the selection of any materials for the horn. If the hardness of the material is higher than the horn, then it will wear very fast. In this study, D2 steel is used as the horn material, and the titanium is used as booster material. These two materials have excellent acoustic properties and are shown in table 1.

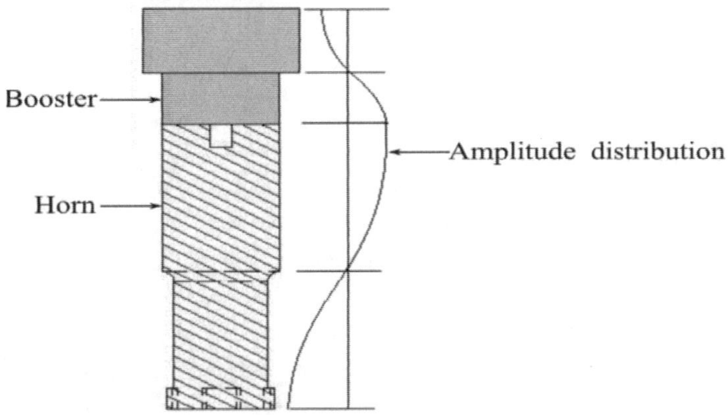

Figure 1. Schematic diagram of horn and booster.

Table 1. Mechanical properties of D2 steel and titanium.

Property	D2 steel	Titanium
Young's modulus (E) (GPa)	210	116
Poison's ratio (v)	0.30	0.34
Density (ρ) (kg/m3)	7700	4500
Velocity of sound (C) (m/Sec)	5222.32	5077.18

Generally analytical techniques fail when there are a complicated contour shape and sudden increase of stress and amplitude in the system. There is two finite element analysis used for this study (i) Modal analysis, (ii) Harmonic analysis. Modal analysis has been performed to find out the natural frequency of the horn and booster without considering any load, and 20018 Hz is obtained as natural frequency. It represents the rigidity of the structure. By accurate design, it will be resonating with other parts of the machine, and this can be achieved by adjusting the length. Harmonic analysis is done to study the response of the horn under periodic loading. This is otherwise also called as dynamic loading. In this analysis, the modal frequency and the amplitude are given as the input. Then the amplitude magnification and stress concentration are obtained at the output end. From figure2, it is clearly observed that less displacement is observed at the nodal plane and high displacement of 70 μm is at the output end. Stress is also more at the nodal plane due to a sudden change in cross-sectional area and shown in figure 3.

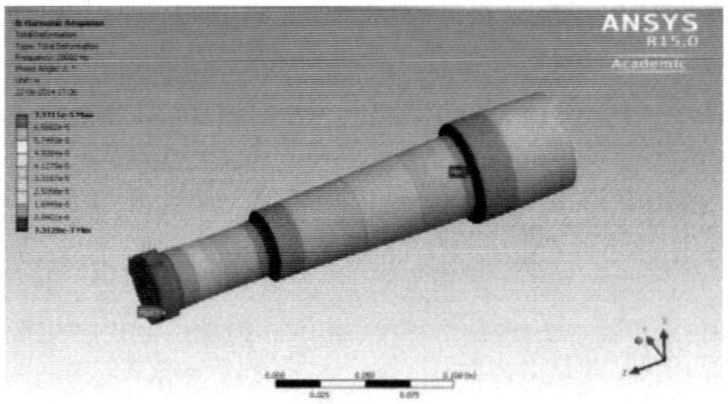

Figure 2. Total deformation from harmonic analysis

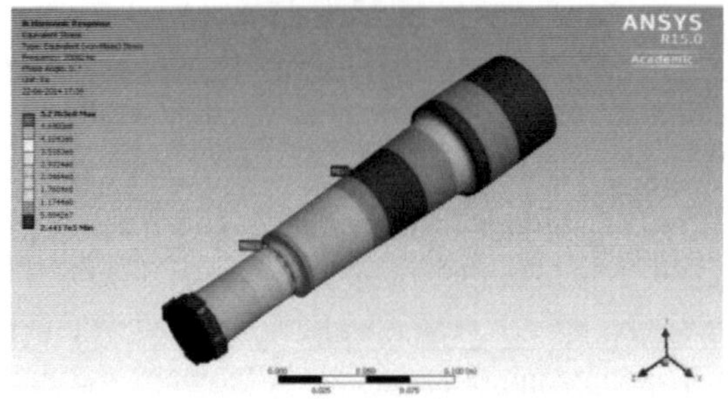

Figure 3. Stress distribution from harmonic analysis

2.2 Experimental setup

During this investigation, the joining has been done between two materials Al and Cu alloys i.e. A1100 and CuZn37 of 0.2-mm thin sheet. Before welding, all the materials were rubbed with P800 waterproof SiC emery paper to remove the oxide layer. All the contaminants and grease were removed from the top surface by using acetone. The size of the sheet is 20×80 mm with an overlap of 20 mm as shown in figure 4. The A1100 sheet is placed on the top of CuZn37 (brass) sheet. All the spot welding experiments have been performed with a Telsonic® lateral drive welding machine at a constant power of 3 kW and a frequency of 20.0 kHz. Figure 5 shows the complete setup of welding machine. The flat welding tip and a rigid anvil with serrations were implemented for the experiments. A universal testing machine was used to find out the weld strength with a displacement rate of 1 mm/min.

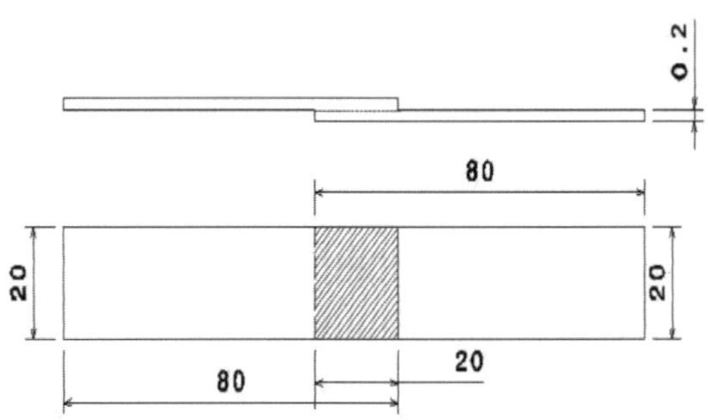

Figure 4. Schematic diagram of specimen.

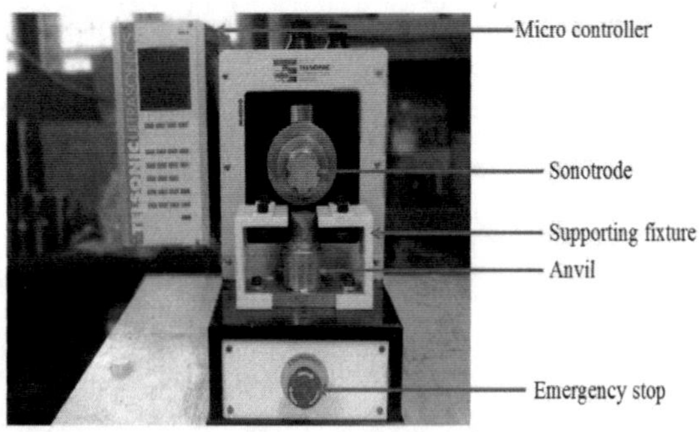

Figure 5. Ultrasonic welding machine.

2.3 Domain of experiment

A lot of screening tests conducted prior to this study and based on that six process parameters were selected as the design variables. Taguchi's L27 orthogonal array design with 5 replicates for each test condition was applied to determine the optimal results from a finite set of data. At the optimum condition of the input factors, micro-hardness and microstructural analysis are performed. Based on the parameter settings available in the system and the idea of the screen test; the domain of experiment has been chosen. The design of experiments is given in table 2.

Table 2. Process parameters with their levels.

Parameters	Unit	Terms	Level 1	Level 2	Level 3
Amplitude	μm	A	54	60	68
Weld pressure	bar	WP	2	3	4
Weld time	Sec	WT	0.2	0.4	0.6

3. Data analysis and optimization

Experimental data have been normalized first to eliminate the ignorance of some factors. There are three different types of normalization procedure like LB (lower-the-better), HB (higher-the-better) and NB (nominal-the-best). For all the responses like tensile shear strength, T-peel strength and weld area, HB criteria have been chosen. After normalization, the correlation between three responses has been checked by using Pearson coefficients, and it is found that the correlation exists among the responses. These responses are converted into uncorrelated one with the help of Principal component analysis (PCA) and are shown in table 3.

Table 3. Principal component analysis results.

	Ψ_1	Ψ_2	Ψ_3
Eigenvalue	2.546	0.389	0.064
	0.585	-0.518	-0.624
Eigenvector	0.608	-0.229	0.760
	0.537	0.824	-0.181
AP	0.849	0.13	0.022
CAP	0.849	0.978	1.000

The individual grey relational coefficients of the three principal components and consequently overall grey relational grade have been calculated using equation (6). The results are shown in table 4.

$$\gamma_i = \frac{1}{n}\sum_{k=1}^{n} \varsigma_i(k) \tag{6}$$

Table 4. Grey analysis results.

Sl no.	Individual grey coefficients			Overall grey relational grade
	G1	G2	G3	γ_i
1	0.3609	0.8206	0.7020	0.6278
2	0.3609	0.8200	0.7005	0.6271
3	0.3608	0.8208	0.7028	0.6281
4	0.7206	0.9995	0.8289	0.8497
5	0.7204	0.9988	0.8260	0.8484
6	0.7207	0.9998	0.8296	0.8500
7	0.6168	0.6349	0.5785	0.6100
8	0.6164	0.6343	0.5817	0.6108
9	0.6168	0.6348	0.5792	0.6103
10	0.4655	0.4714	0.6161	0.5177
11	0.4656	0.4716	0.6146	0.5173
12	0.4655	0.4714	0.6170	0.5180
13	0.9995	0.8136	0.9937	0.9356
14	0.9999	0.8133	0.9991	0.9374
15	0.9993	0.8138	0.9904	0.9345
16	0.4850	0.5041	0.3339	0.4410
17	0.4848	0.5040	0.3348	0.4412
18	0.4849	0.5040	0.3343	0.4410
19	0.5626	0.3544	0.9744	0.6305

20	0.5628	0.3545	0.9690	0.6288
21	0.5627	0.3545	0.9703	0.6292
22	0.6778	0.7158	0.3373	0.5770
23	0.6775	0.7153	0.3380	0.5769
24	0.6780	0.7160	0.3369	0.5770
25	0.8985	0.9518	0.4136	0.7546
26	0.8985	0.9519	0.4140	0.7548
27	0.8987	0.9520	0.4131	0.7546

From the overall grey relational grade value, the main effects plot can be obtained. All the input variables are varied in 3 levels. For each parameter, the over desirability values of a particular level are averaged. Then the difference between the highest and lowest value gives the rank of each parameter. The highest rank signifies that the weld strength is greatly affected by it. All these values are given in Table5, and main effects plot is showing in figure6.

Table 5. Main effects on overall grey relational grade.

Factors	Levels			Difference	Rank	Optimum level
	1	2	3			
Amplitude	0.696	0.592	0.549	0.147	2	A1
Weld pressure	0.632	0.787	0.707	0.155	1	WP2
Weld time	0.654	0.602	0.725	0.123	3	WT3

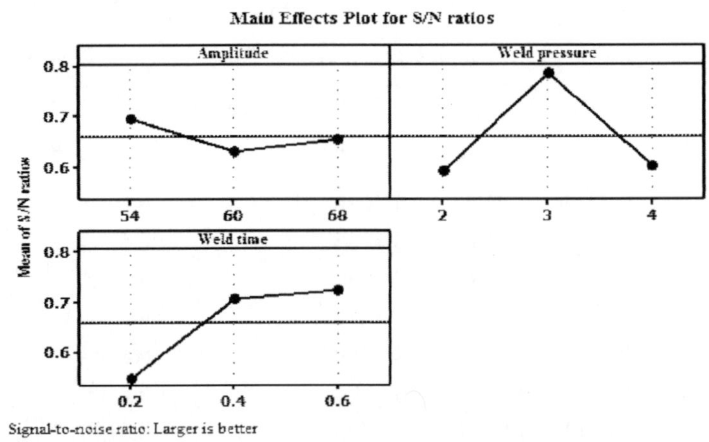

Figure 6. Main effects plot for the overall grey relational grade.

4. Results and discussion

4.1 FEM results for the welding stack

From the dynamic analysis, it has been observed that the input amplitude is modified by the booster, and again it is also amplified by the horn to create a good weld between the sheets. According to the design of both elements, at each stage the amplitude is increased by 1.5 times. Thus, at the end of horn nearly 70 μm amplitude is obtained which is just 2.25 times of the input amplitude. This value is quite close to the theoretical value. Also at the nodal position of the booster and horn, i.e. at 60 mm and 183 mm, the amplitude becomes zero. These variations in amplitude with the length of welding stack are given figure 7. Likewise, figure 8 shows the Von Mises stress distribution along the length of the horn. It increases with the gradual decrease in the dimension, and it is maximum at the suddenly changing section. Thus, a stress of 117.90 MPa and 324.63 MPa are observed for booster and horn correspondingly. The 4.5 mm radius of curvature is also applied at that section to lower the stress concentration. These results are then validated by FEM analysis and are given in Table 6. From the data shown in the table, signifies that there are 2.4% and 4.76 % errors in the length calculation of both systems. Similarly, 5% and 4.45 % errors are obtained while calculating the magnification ratio.

Table 6. Comparison between theoretical and FEM analyzed horn and booster.

Sl No.	Factors	Theoretical calculated horn	Theoretical calculated booster	FEM Analysed horn	FEM Analysed booster	% error in horn	% error in booster
1	Frequency (kHz)	20	20	20	20	0	0
2	Total length (mm)	130	127	126	120	2.4	4.76
3	Diameter ratio	1.25:1	1.25:1	1.25:1	1.25:1	0	0
4	Amplification factor	1.5	1.5	1.58	1.57	5	4.45

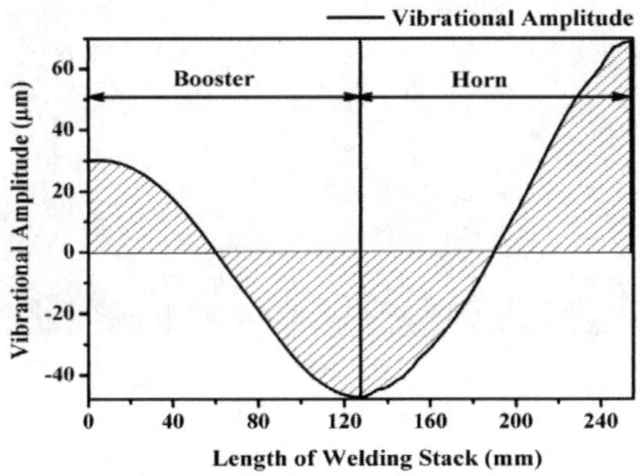

Figure 7. Amplitude variation with length.

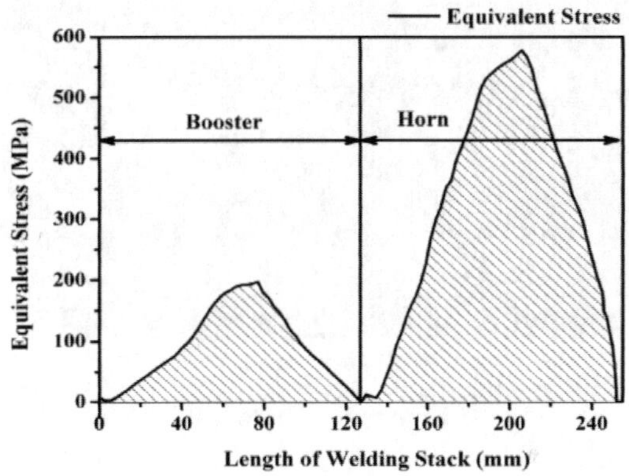

Figure 8. Stress distribution with length.

4.2 Confirmatory test results

Based on table 5, the rank of each process parameter is given. The highest rank parameter contributes most of the responses. So weld pressure is the most important parameter. The optimal condition of parameters is A1WP2WT3. On this condition, high weld strength was obtained. The predicted values for the above optimal condition are given by the PCA based grey Taguchi analysis. The lap joints were performed with the predicted parameters and corresponding S/N ratio is also found out. It is then compared with the predicted parameter results. Table 7 shows the results of the confirmatory experiment. It is observed that the confirmatory test results show a fine agreement with predicted results as the percentage of error is 0.816%.

Table 7. Confirmatory test results.

	Optimal setting	
	Prediction	Experiment
Factors level	A1WP2WT3	A1WP2WT3
S/N ratio	-4.738	-4.777
Overall grey relational grade	0.5795	0.5768
Error between S/N ratio	0.816%	

4.3 Metallographic studies

After the tensile test, a scanning electron microscopy (SEM) is employed to observe the fracture surfaces of the weld coupons. Figure 9 shows the microscopic features such as the plastic flow of the metal in the weld zone of Al-Brass specimens at 3 bar pressure, 0.6 Sec welding time and 54 μm vibration amplitude. The fracture surface of the Al side is shown in figure 9 (a). It is observed that due to the lower hardness of Al than brass, it is severely deformed, and thus less brass is located on it. But as the brass is harder than Al, so it undergoes less deformation, and there is no improvement of weld area which affects the weld quality. It is represented in figure 9 (b). In the meantime, in the weld zone, the hardness of both the materials increases due to increase in temperature, and it is the major cause of strain hardening. For this purpose, Vaiseshika microhardness tester with a load of 100gm and holding time of 15 sec was used and figure 10 shows the microhardness of the material in the parent zone as well as in the weld zone.

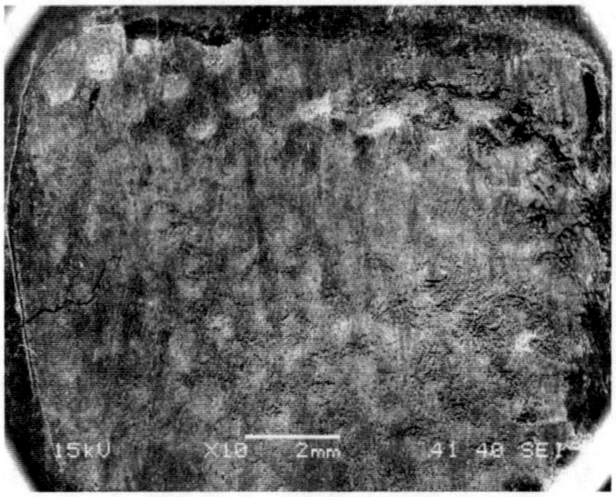

(a)

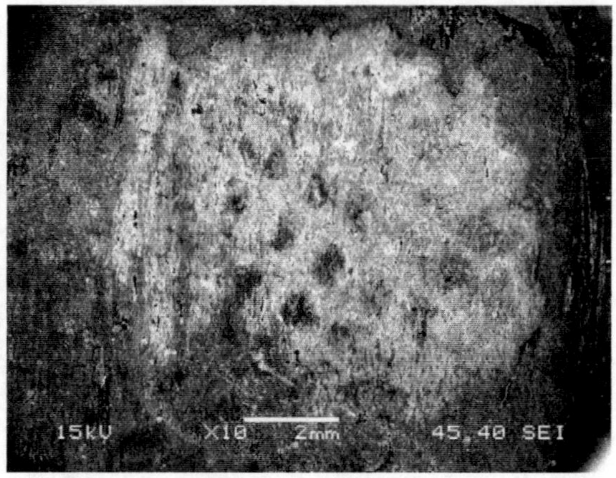

(b)

Figure 9. Fractographic study of weld coupons (a) Al side, (b) brass side

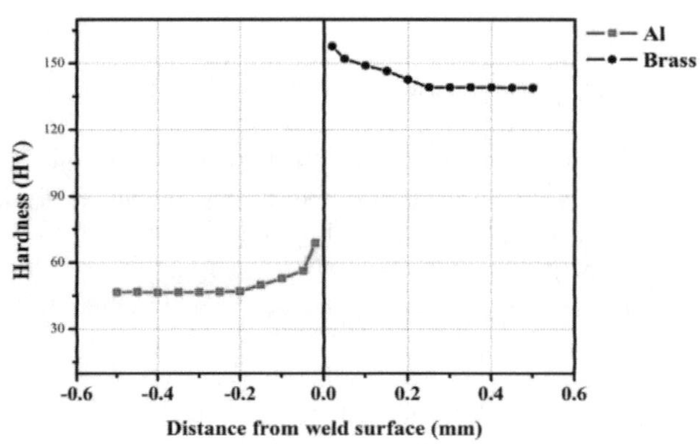

Figure 10. Microhardness variation with length.

5. Conclusions

- The modal analysis has been done to find out the natural frequency of the horn as 20007 Hz, which is quite close to the operating frequency of 20 kHz. These are obtained by adjusting the length of the horn and booster to 126 mm and 120 mm respectively. 2.4 % and 4.76 % errors are noticed while adjusting it.
- The harmonic analysis also has been performed to find out the amplitude and stress distribution in the horn. The magnification factors obtained from this analysis are 1.58 and

1.57. These values are very close to magnification given by the commercial horn. 5 % and 4.45 % errors are found out during this analysis. This is due to the limitations of the FEM or measurement errors in the real system. To encounter the danger of stress concentration, the radius of curvature of 4.5 mm is given to the corners.

- In addition to it, the multi-objective optimization problem has been solved by finding an optimal parametric combination in which appreciable weld strength is obtained. PCA has been implemented to eliminate the correlation between the responses, and grey based Taguchi method is used to find out the optimal sequence of input parameters.
- The confirmatory test also shows an error of 0.816%, which shows a good agreement with the predicted results. From the weld quality studies, the mechanism of plastic deformation and the effect of hardness due to the temperature of the welding zone have been uncovered effectively.

References

[1] Bakavos D and Prangnell P B 2010 Mechanisms of joint and microstructure formation in high power ultrasonic spot welding 6111 aluminium automotive sheet *Mat. Sci.Eng. A* **527**(23): 6320-6334.

[2] Sherrit S,Askins SA and Gradziol M 2002 Novel horn designs for ultrasonic/sonic cleaning, welding, soldering, cutting and drilling *Proceedings of the SPIE Smart Structures Conference* **4701** 353.

[3] Amin S G, Ahmed MH M and Youssef H A 1995 Computer-aided design of acoustic horns for ultrasonic machining using finite-element analysis *J. of Mat. Proc. Tech.* **55**(3-4): 254-60.

[4] Nad M. 2010 Ultrasonic horn design for ultrasonic machining technologies *Applied and Computational Mechanics* **4**: 79-88.

[5] Siddiq A and Ghassemieh E 2011Fibre embedding in aluminium alloy 3003 using ultrasonic consolidation process-thermo-mechanical analyses *Int J. Adv Manuf Technol.* **54**(9-12): 997-1009.

[6] Tsujino J, Hidai K, Hasegawa A, Kanai R, Matsuura H, Matsushima K and Ueoka T 2002 Ultrasonic butt welding of aluminum, aluminum alloy and stainless steel plate specimens *Ultrasonics* **40**(1-8): 371-374.

[7] SooriyamoorthyE, John SP andKalakkath P 2011 Experimental studies on optimization of process parameters and finite element analysis of temperature and stress distribution on joining of Al-Al and Al-Al_2O_3 using ultrasonic welding *Int J.Adv Manuf. Technol.* **55**(5-8): 631-640.

[8] Kumar P, Barua PB, Gaindhar J L2000 Quality optimization (multi-characteristics) through taguchi's technique and utility concept *Qual. Reliab Eng Int.* **16**(6): 475-85.

[9] Matsuoka S-, Imai H. 2009 Direct welding of different metals used ultrasonic vibration. *J Mater Process Technol* **209**(2): 954-60.

[10] Zhao YY, Li D, Zhang YS. 2013 Effect of welding energy on interface zone of Al-Cu ultrasonic welded joint. *Sci TechnolWeld Joining* **18**(4): 354-60.

Fabrication of high-k dielectric Calcium Copper Titanate (CCTO) target by solid state route

N Tripathy[1], K C Das[1], S P Ghosh[1], G Bose[2] and J P Kar[1*]
[1]Department of Physics and Astronomy, National Institute of Technology, Rourkela, India 769008
[2]FST, IFHE University, Hyderabad, India 501203

E-mail: karjp@nitrkl.ac.in, Phone : +916612462732

Abstract. $CaCu_3Ti_4O_{12}$ (CCTO) ceramic pellet of 10mm diameter has been synthesized by adopting solid state route. The structural and morphological characterization of the ceramics sample was carried out by X–ray diffraction (XRD) and scanning electron microscope (SEM), respectively. XRD pattern revealed the CCTO phase formation, where as SEM micrograph shows the sample consisting of well defined grain and grain boundaries. The room temperature dielectric constant of the sample was found to be ~ 5000 at 1kHz. After successful preparation of CCTO pellet, a 2 inch diameter CCTO sputtering target is also fabricated in order to deposit CCTO thin films for microelectronic applications.

Keywords: Ceramic, X-ray diffraction, Dielectrics, Electron Microscopy

1. Introduction

In recent years, development of high-k dielectric thin films for microelectronics has drawn tremendous attention in order to replace SiO_2. The colossal dielectric constant (CDC) phenomenon in $CaCu_3Ti_4O_{12}$ (CCTO) has drawn a great deal of interest. It possesses a distorted cubic perovskite (ABO_3) structure with lattice parameter of 7.391Å. It also shows a negligible temperature dependence on dielectric constant [1, 2]. The origin of the giant dielectric constant in CCTO is not well understood yet. The origin of high dielectric constant in CCTO may be due to stoichiometric variation [3], oxygen vacancies [4] and/or presence of secondary phases which forms layered structures which can be explained by Internal Barrier Layer Capacitance (IBLC) [5]. For fabrication of microelectronic devices, the synthesis of high quality CCTO thin films is highly required. CCTO thin films can be deposited by different techniques notably PLD [6], Sol-gel route [8], CVD, RF sputtering [9] etc. Out of them RF sputtering have some advantages over other deposition techniques such as low cost, low processing temperature, good compositional homogeneity, and large area deposition [10]. For sputtering, a high pure CCTO target is required. In this report, an attempt has been done to fabricate 2 inch diameter target for deposition of high-k thin film.

2. Materials and Methods

Stoichiometric amount of $CaCO_3$, CuO, TiO_2 were mixed thoroughly in acetone medium by using ball milling in a sealed plastic bottle with Zirconium oxide beads for 12 hour. The powder was calcined at different temperature varied from 900°C to 1050°C for 10 hours. The final product was grinded into fine powder with Agate mortar. Formation of single phase CCTO was confirmed by XRD studies. The powder was mixed with 3wt. % polyvinyl alcohol (PVA) solution and was pressed into pellets of 10mm diameter and 1.5mm thickness using uniaxial compression with pressure 20MPa. Finally the pellet was sintered in air ambient at 1100°C for 8 hours.

The structural and morphological properties were studied by X-Ray diffractometer (Rigaku ultima IV) and scanning electron microscope (JEOL JSM- 7001F). For electrical measurements, the sample surfaces were polished and silver paint was applied on both the surfaces. Electrical properties have been investigated using a LCR meter (Wayne Kerr 6500 B). In order to fabricate 2 inch diameter and 3mm thickness sputtering target, the calcined powder was mixed with 5wt.% PVA solution and was pressed into a pellet of 2 inch dia and 3mm thickness using uniaxial compression with pressure 40MPa. Finally the sample was sintered at 1100°C for 8 hours.

3. Results and discussion

Fig. 1. shows the XRD pattern of CCTO powder calcined for 10 hours for the temperatures ranging from 900°C to 1050°C. The single phase CCTO was obtained for the calcination temperature of 1050°C. XRD pattern matches well with body centered cubic perovskite CCTO (ICDD card no – 751149). The diffraction pattern reveals the formation of polycrystalline sample. Fig. 2 shows the SEM micrograph of CCTO pellet sintered at 1100°C for 8 hour. Smaller grains were observed at the grain junctions, which might be due to decomposition of partial CCTO during sintering at high temperature.

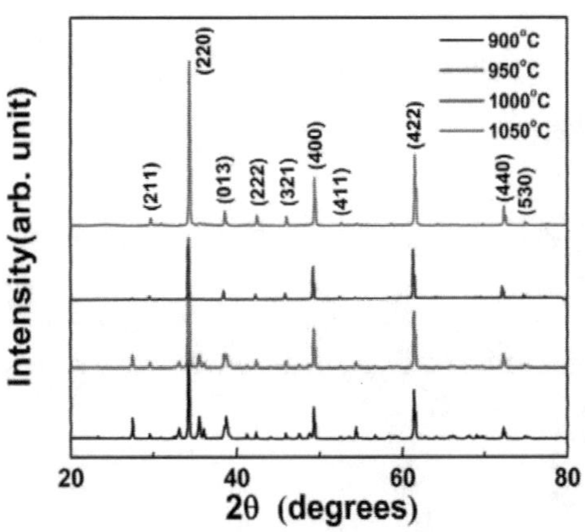

Fig. 1:- X-ray diffraction patterns of CCTO powder calcined at different temperature

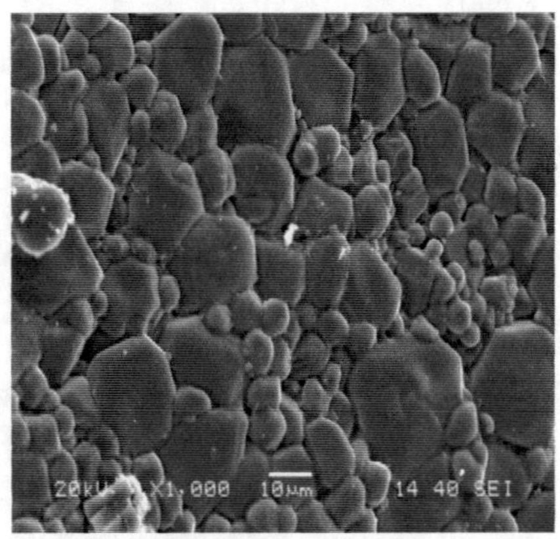

Fig. 2:- SEM micrograph of CCTO pellet

Fig. 3. shows the frequency dependence of dielectric constant in a temperature range of 30°C - 310°C. From figure 3, the dielectric constant of CCTO increased rapidly below 1kHz and in the frequency range of 1kHz to 100kHz it is nearly constant. The dielectric constant at 310°C is very high with a value of 16000 at 100Hz. The ε_r value decreased with an increase in frequency, but for a particular frequency the dielectric constant increased with increase in temperature. This can be attributed to the contribution of charge accumulation at the interface. The high permittivity value of CCTO ceramic is due to the presence of large grains and grain boundaries, which supports internal barrier layer capacitance (IBLC) mechanism [11]. Fig. 4 shows the frequency dependence of dielectric loss (tanδ) in the temperature ranging from 30°C to 310°C. The value of tanδ for CCTO ceramic at 1kHz at room temperature is about 0.14. The value of tan δ is nearly frequency independent at high frequencies (10kHz -1MHz). There is a sharp increase in dielectric loss at 100Hz with increase in temperature. This may be due to exponential increase in conductivity with temperature [12].

A 2 inch diameter and 3mm thick sputtering target was fabricated from calcined CCTO powder.

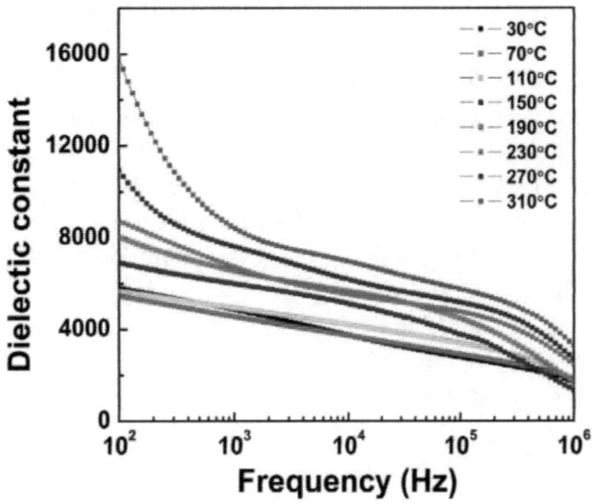

Fig. 3:- Frequency dependence of dielectric constant of CCTO pellet at different temperature

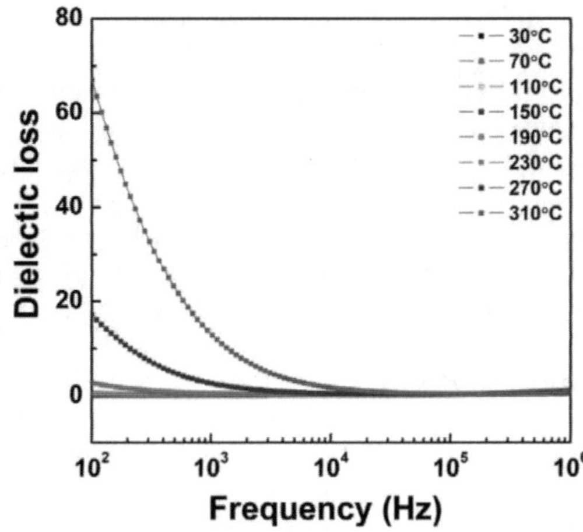

Fig. 4:- Frequency dependence of dielectric loss of CCTO pellet at different temperature

Fig. 5:- Optical image of 2 inch diameter fabricated CCTO sputtering target

4. Conclusions

Perovskite $CaCu_3Ti_4O_{12}$ ceramic sample was synthesized by solid-state reaction method. XRD pattern confirms the CCTO phase formation at calcinations temperature of 1050°C. SEM results depict formation of microstructures with well defined grain and grain boundaries. The sample has shown a dielectric constant of around 5000 at 1kHz frequency with loss factor of 0.14. The high permittivity value is related to capacitance of the grain boundaries which is consistent with the IBLC model.

5. Acknowledgement

This work was supported by Science and Engineering Research Board (SERB), DST project (SR/FTP/PS-099/2012) under Fast Track Scheme for Young Scientist.

6. References

[1] M. A. Subramanian, D. Li, N. Duan, B. A. Reisner, A. W. Sleight, High Dielectric Constant in $ACu_3Ti_4O_{12}$ and $ACu_3Ti_3FeO_{12}$ phases, *Journal of Solid State Chemistry* 151 (2000) 323-325.

[2] A. P. Ramirez, M. A. Subramanian, M. Gardel, G. Blumberg, D. Li, T. Vogt, S. M. Shapiro, Giant dielectric constant response in a copper-titanate, *Solid State Communications* 115 (2000) 217-220.

[3] S. F. Shao, J. L. Zhang, P. Zheng, C. L. Wang, Effect of Cu-stoichiometry on the dielectric and electric properties in $CaCu_3Ti_4O_{12}$ ceramics, *Solid State Communications* 142 (2007) 281-286.

[4] J. Li, A. W. Sleight, M. A. Subramanian, Evidence for internal resistive barriers in a crystal of the giant dielectric constant material: $CaCu_3Ti_4O_{12}$, *Solid State Communications* 135 (2005) 260–263.

[5] M. J. Pan, B. A. Bender, A Bimodal Grain Size Model for Predicting the Dielectric Constant of Calcium Copper Titanate Ceramics, *Journal of American Ceramic Society* 88 (2005) 2611–2614.

[6] L. Fang, M. R. Shen, W. Cao, Effects of post anneal conditions on the dielectric properties of $CaCu_3Ti_4O_{12}$ thin films prepared on Pt/Ti/SiO_2/Si substrates, *Journal of Applied Physics* 95 (2004) 6483–6485.

[7] D. C. Sinclair, T. B. Adams, F. D. Morrison, A. R. West, Microstructure and dielectric properties of pulsed-laser-deposited $CaCu_3Ti_4O_{12}$ thin films on $LaNiO_3$ buffered Pt/Ti/SiO2/Si substrates , *Applied Physics Letters* 80 (2002) 2153–2155.

[8] J. Liu, R. W. Smith, W. N. Mei, Synthesis of the Giant Dielectric Constant Material $CaCu_3Ti_4O_{12}$ by Wet-Chemistry Methods, *Chemistry of Materials* 19 (2007) 6020–6024.

[9] C. R. Foschini, R. Tararam, A. Z. Simões, M. Cilense, E. Longo, J. A. Varela, $CaCu_3Ti_4O_{12}$ thin films with non-linear resistivity deposited by RF-sputtering, *Journal of Alloys and Compounds* 574 (2013) 604-608.

[10] W. Lu, L. Feng, G. Cao, Z. Jiao, Preparation of $CaCu_3Ti_4O_{12}$ thin films by chemical solution deposition, *Journal of Material Science* 39 (2004) 3523-3524.

[11] K. L. W. Chen, Y. F. Liu, X. M. Bao Peng Lu, J. S. Zhu, Investigation of the Size Effect on the Giant Dielectric Constant of $CaCu_3Ti_4O_{12}$ Ceramic, *Chinese Physics Letter* 21 (2004) 1815–1818.

[12] B. S. Prakash, K.V. R. Varma, Influence of sintering conditions and doping on the dielectric relaxation originating from the surface layer effects in $CaCu_3Ti_4O_{12}$ ceramics, *Journal of Physics and Chemistry of Solids* 68 (2007) 490–502.

Development of Cu-E-Glass Fiber Composites by Powder Metallurgy Route

Pallabi Bhuyan*, Harspreet Singh, Lailesh Kumar, Nidhi Sharma, DeepankarPanda, DeepanshuVerma, Syed NasmulAlam

Department of Metallurgical and Materials Engineering, National Instituteof Technology, Rourkela-769008, Orissa, India

*Email:pallabibhuyan92@gmail.com, Phone: +919437678813

Abstract: Cu-E glass fiber composites were developed with different vol. % of E-glass fiber (10, 20, 30 and 40 vol. %) by powder metallurgy route. Both as-received Cu and nanostructured Cu developed by milling as-received Cu powder for 20 h were used to develop various Cu-E-glass fiber composites. The effect of using as-received Cu powder and nanostructured Cu powder on the properties of the various Cu-E-glass fiber composites was analysed. The samples were sintered at 900°C for 1 h in inert atmosphere. The results show good bonding between the matrix and the reinforcement and there is homogeneous distribution of the reinforcement in the matrix. . The hardness of the Cu-E-glass fiber composites was found to increase from 0.8GPa to 2.7GPa with increase in vol. % of the glass fiber in case of unmilled and from 1.2GPa to 2.9GPa for the milled Cu-E-glass fiber composites. The as-milled Cu-E-glass fiber composites shows better densification and sinterability compared to the unmilled Cu-E-glass fiber composites

Keywords: Wear, Hardness, Density, Reinforcement.

1. Introduction

Metal matrix composites (MMCs) have generated a wide interest in research community because of its high strength, stiffness and fracture toughness. Metal matrix composites are capable of resisting elevated temperature. Most of the metals and alloys could be used as matrices and the reinforcements have to be stable over a range of temperature and should be non-reactive. For high strength applications the matrices require reinforcements having high modulus. The strength to weight ratios of resulting composites can be higher than most of the alloys. Several factors such as melting point, physical and mechanical properties of the composites at various temperatures determine the service temperature of the composites. Composites have been developed with great success by the use of fiber reinforcements in the case of polymer matrix materials. However, due to the limitation of the use of polymer matrices at high temperature there has been an increase in research in the area of metallic matrix composites [1-3].

Over the last several decades, there has been considerable interest in the use of Cu-based metal matrix composites (MMCs). For many applications pure Cu cannot be used because of its low strength. Therefore, improvement in the properties of Cu has become essential for its applications in cutting-edge technological applications. The improvement of mechanical properties of Cu is important for its use in a larger number of fields. Composite materials offer several applications in the field of building aerospace applications, automotive, ship building industry as it has different advantages over other conventional metals. Metal matrix composite (MMCs) is combines both metallic properties such as toughness & ductility and ceramic properties such as modulus and high strength which possess shear strength provide service at high temperatures. Cu has a density of 8.96 gm/cc. Cu has a melting point of 1085°C and it has a high thermal conductivity (401 $W.m^{-1}.K^{-1}$) and used as a structural material for cooling. It also has high electrical conductivity (5.96×10^{7}S/m). Cu has a Young's modulus of 130 GPa and its tensile strength is 210 MPa. In order to increase its high temperature properties different reinforcements are being used.Glass fibers are among the most versatile industrial materials known today. They are readily produced from raw materials, which are available in virtually unlimited supply. They exhibit useful bulk properties such as hardness, low susceptibility to moisture absorption, resistance to chemical attack, thermal stability and high melting point, as well as desirable fiber properties such as strength, flexibility, and stiffness. Glass fibers are used in the manufacture of structural composites, printed circuit boards and a wide range of special-purpose products. Glass fibers fall into two categories, low-cost general-purpose fibers and premium special-purpose fibers. Over 90% of all glass fibers are general-purpose products. However, several challenges stand in the way of the development of metal matrix composites reinforced with glass fibers such as higher processing temperatures, fiber/matrix bonding, and the ability to produce desired geometries. The final properties of a composite are determined by the fiber content, matrix material, fiber material, fiber orientation, fiber length and by the distribution in the matrix [4-7].

Here Cu-10, 20, 30 and 40 vol. % E-glass fiber composites have been developed by powder metallurgy route using both as received Cu powder and nanostructured Cu powder developed by mechanical milling of the as-received Cu powder in a high-energy ball mill for 20 h. The effect of nanostructured Cu on sinterability and densification of the composites was analysed. The microstructures of the various composites wereanalysed. The hardness and density of the various composites were determined and the nature of fracture in these composites wasanalysed.

2. Materials and Methods

Elemental Cu powder having purity > 99 % and average particle size of ~ 400 mesh was procured from LobaChemie. E-galss fibers were procured from Saint Gobain. Green pellets of unmilled Cu and as-milled Cu for 20 h with different volume fractions (10, 20, 30 and 40%) of E-glass fiber were developed by blending the constituents and cold compaction of the samples under a load of 665 MPa in a uniaxial cold compaction machine. Nanostructured Cu powder was developed by milling the Cu powder in a high-energy Fritsch P-5 planetary ball mill. The vials and balls were made of hardened chrome steel. Toluene was used as a process control agent. The ball to powder mass ratio was maintained at 10:1. Samples were collected at an interval of 5 h of milling for analysis of the milled powder. Sintering of the samples was done in tubular furnace at 900°C for 1 h in Ar atmosphere. Hardness of the various composites was determined using a Leco Vickers microhardness tester. Densities of the composites were determined using Archimedes' principle. A JEOL JSM-6480LV scanning electron microscope (SEM) equipped with an INCAPentaFET-x3 X-ray microanalysis system was used. A high resolution transmission electron microscope (HRTEM) JEM – 2100 JEOL with point to point resolution 0.194 nm was used for analysing the milled powder. A Phillips *PANalytical*PW 3040/00 XRD was also used to characterize the milled powder.

3. Result and Discussion

Elemental Cu powder was milled in a high-energy planetary ball mill in order to synthesize nanocrystalline Cu. Mechanical milling is a very effective processing technique for the preparation of nanocrystalline metallic and ceramic powders. It is a solid-state powder processing technique which involves repeated welding, fracturing, and rewelding of powder particles in a high-energy ball mill. Due to the heavy plastic deformation of the particles during milling there was reduction in crystallite size and increase in strain. In a high-energy planetary ball mill, the grinding balls and the powder in the vial are acted upon by centrifugal forces due to the rotation of the vials about its own axis. The supporting disc rotates in the opposite direction.The impact of the balls on the powder leads to the formation of nanostructured powder. The strain in the powder also rises due to milling. The reduction in crystallite size and increase in strain leads to the broadening of the x-ray diffraction peaks as can be seen in Figs.1(a, b).

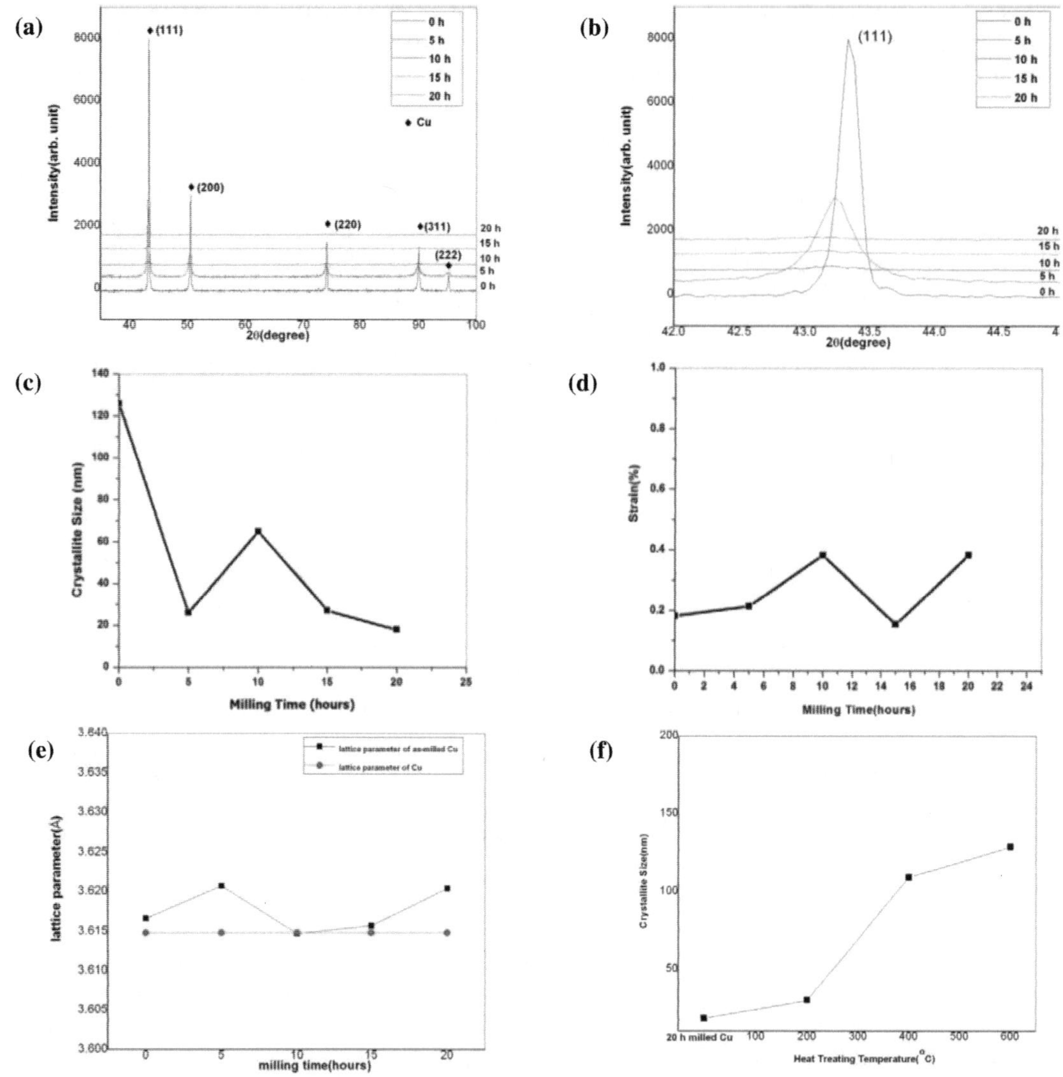

Fig.1 :- (a, b) X -ray diffraction plots of Cu milled for various periods of time and variation of (c) crystallite size (d) r.m.s. strain and (e) lattice parameter of Cu with milling time (f) Variation of crystallite size of the 20 h milled powder with het treatment temperature.

The x-ray diffraction of Cu powder milled for various durations of time are shown in Fig.1 (a). The various peaks in the x-ray diffraction pattern could be indexed to the different crystallographic planes of Cu. No other peaks could be detected in the x-ray pattern suggesting that there is no contamination from the milling media. Fig.1 (b) presents the (111) peak of unmilled Cu and Cu milled for various periods of time. With increase in milling time broadening of the (111) peak of Cu takes place and its intensity reduces. The (111) peak of Cu at 43.3° after 5 h of milling shows shift towards the lower 2θ angle. The shift of the peak towards the lower 2θ angle is possibly due to the increase in lattice parameter of Cu after 5 h of milling. The lattice parameter variation of Cu with milling time in Fig.1 (e) also presents slight increase with milling time. Beyond 5 h of milling the (111) peak of Cu shows significant broadening as can be seen in Fig. 1(b). Voigt's method was used for calculating the crystallite size of the milled Cu powder from the x-ray diffraction plot using the most intense (111) peak. After 20 h of milling the crystallite size of Cu was found to be 18 nm (Fig.1(c)). There is a gradual drop in the crystallite size with milling time. From Fig. 1(c) it is evident that Cu could be reduced to nanometric dimension within 5 h of milling. Cold welding was also evident between 5 to 10 h of milling which led to the increase in crystallite size. There is a gradual increase in the lattice strain due to severe deformation of the milled powder and strain reaches a maximum value after 20 h of milling when the crystallite size is smallest (Fig.1 (d)). Fig. 1(f) presents the growth in crystallite size of Cu when the 20 h milled Cu powder is heat treated at various temperatures for 1 h. It shows that the crystallite size of Cu grows to beyond 100 nm after heat treatment at 400°C for 1 h. Therefore at the sintering temperature of 900°C chosen for the deployment of the composites here the crystallite size of Cu is above 100 nm. However the fine size of the milled Cu powder could lead to better sinterability and densification of the composites. This is why the 20 h as-milled Cu powder has been also used for the development of the composites to see the effect of fine sized Cu powder on the properties of the composites [8-11].

Fig.2:- SEM images of (a) unmilled Cu, Cu milled for (b) 5 h (c) 10 h (d) 15 h (e) 20 h

Figs.2 (a-e) presents the SEM images of Cu powder milled for various periods of time. Due to the ductile nature of Cu it was possible to flatten the Cu particles. The Cu particles which were initially spherical in shape had a flake like shapeafter 5 h of milling. After 15 h of milling they had a characteristic layered microstructure with very fine lamellar spacing (Fig. 2(d)). Thickening of the plates due to coldwelding between the plates leads to increase in the average particle size of the powder which was seen after 10 h of milling (refer Fig. 1(c)).

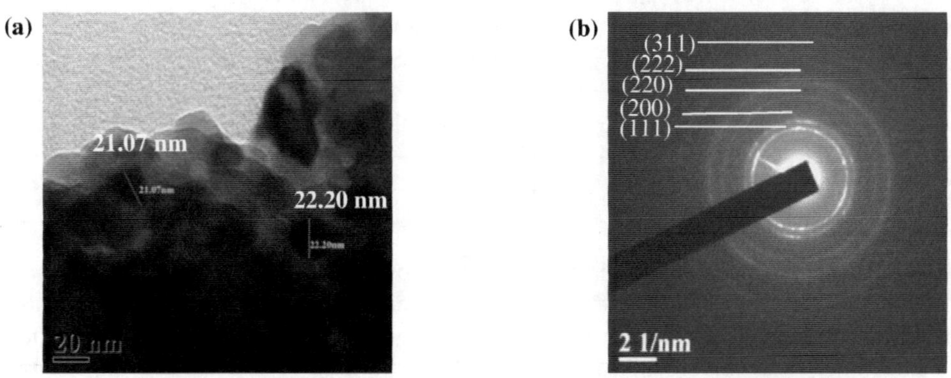

Fig.3 :- (a) HRTEM images and (b) SAD pattern of 20 h milled Cu

The HRTEM image of the 20 h milled Cu powder in Fig.3 (a) indicates that powder has been reduced to nanometric dimension after high-energy milling for 20 h. The SAD pattern in Fig.3 (b) shows complete ring pattern which suggests that the 20 h milled Cu powder has nanometric dimension. The ring patterns could be indexed to the various crystallographic planes of Cu. The particle size analysis of the Cu powder milled for various periods of time was done by particle size analyser. Figs.4 (a, b) shows the particle size analysis for unmilled Cu and 20 h milled Cu powder. It should be noted that size distribution of both the unmilled Cu and the 20 h milled Cu is bimodal in nature showing peaks at less than 1 μm and higher than 10 μm. There is presence of very fine Cu powder particles having size less than 1 μm in both the unmilled Cu powder as well as the 20 h milled Cu powder although the volume fraction of this fine sized powder is very low in both the unmilled and the 20 h milled Cu powder.

(a)

(b)

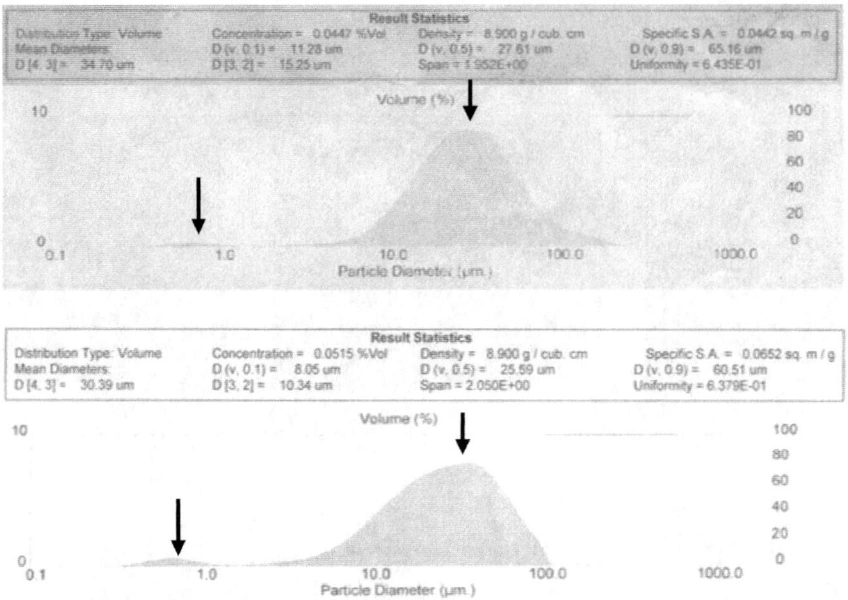

Fig.4:- Particle size analysis for (a) unmilled Cu and (b) 20h milled Cu powder

The average size of the particles was found to be 25.59 μm after 20 h of milling. Table 1 shows the particle size analysis of pure elemental Cu after various periods of milling time. The results show that there is a sudden increase in particle size after 5 h of milling. This is possibly due to the cold welding effect as Cu is soft and ductile material. Beyond 5 h of milling the size of the particles gets reduced due to fracture of the particles which leads to size reduction. The lowest particle size of 25.59 μm is attained after 20 h of milling [12-14].

Table 1:- Particle size analysis of pure Cu at different milling time

Milling Time (hours)	Average Particle Size, D(v,0.5) (μm)
0	27.61
5	57.36
10	35.72
15	25.91
20	25.59

E-Glass fibers are the most widely used glass fibers as reinforcement in the composites. E-glass fibers are alumina-borosilicate glass with less than 1 wt. % alkali oxides and are mainly used for glass-reinforced plastics. E-glass or electrical grade glass was originally developed for use as insulators for electrical wiring which was later found to have excellent fiber forming capabilities and is now used widely as the reinforcing phase in composites and is commonly known as fiber glass. E-glass fibers exhibit useful bulk properties such as hardness (6000 MPa), dimensional stability, resistance to chemical attack and strength. Its density is 2.58 gm/cc. Its tensile strength is 3500 MPa and its Young's modulus is 85 GPa. It has a compressive strength of 5000 MPa [15].

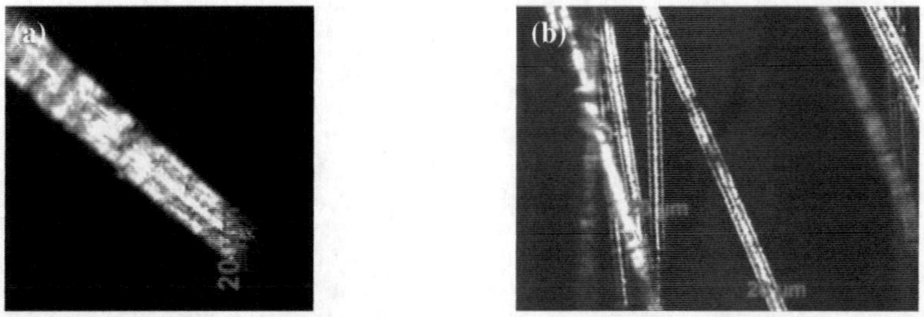

Fig.5 (a, b):- Optical images of E-glass fiber used in the composites

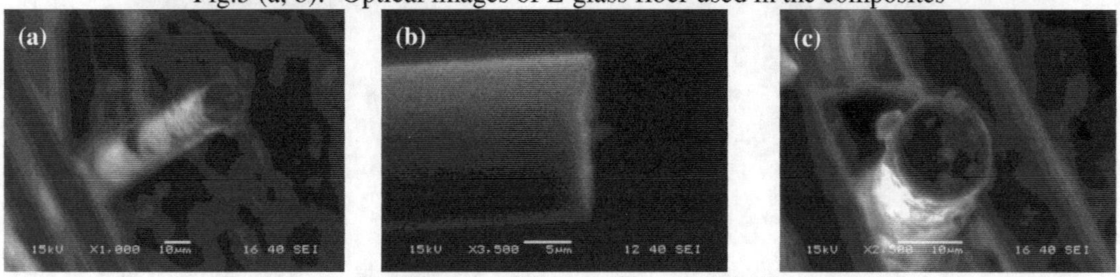

Fig.6 (a-c):- SEM image of E-glass fiber used in the composites

Fig.5 and Fig.6 shows the optical and SEM images respectively of the E-glass fiber used as reinforcement in the Cu-E-glass fiber composites. The diameter of the fibers are ~20 μm. Tensile test of the fiber was carried out in order to find out the mechanical properties of the fibers. The fiber shows brittle fracture with elongation to failure of 4.134 %. The elongation to failure for the E-glass fiber is very low as compared to that of ductile metals. E-glass fiber is highly brittle and shows almost flat fracture surface as can be seen from the SEM images in Fig 6. The E-glass fiber shows 100% recovery when stressed below the point of rupture. The tensile strength of a single E-glass fiber was found to be ~2224 MPa.

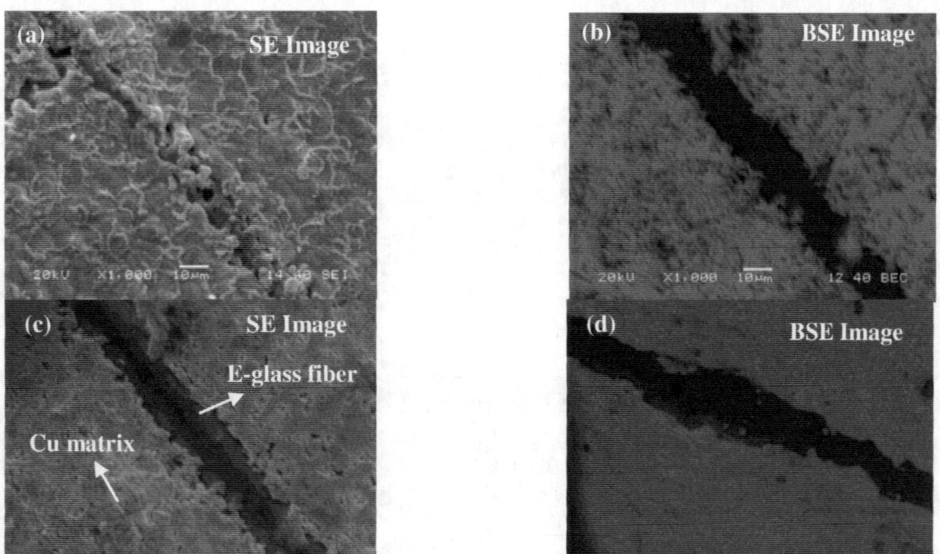

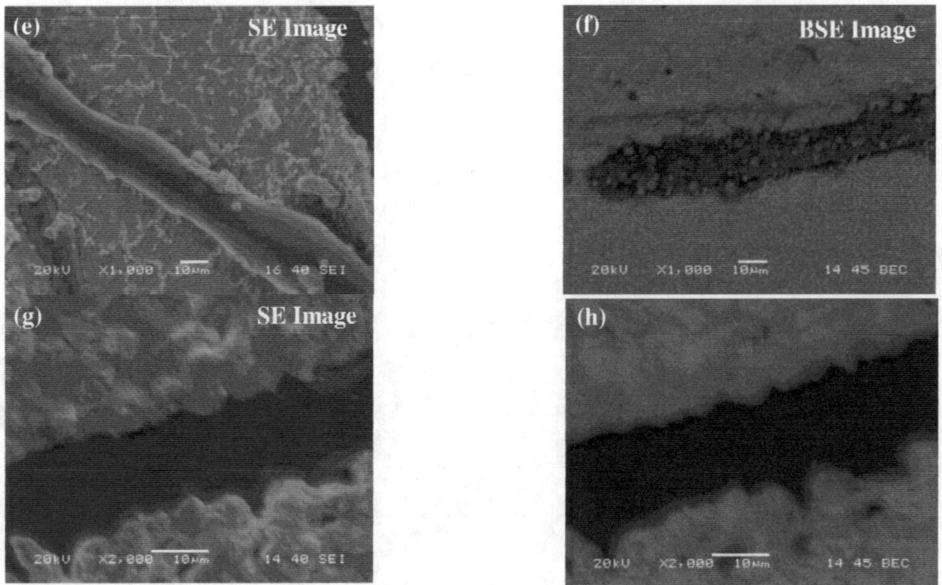

Fig. 8:- SEM image of (a, b) unmilled Cu-10 vol. % E-glass fiber (c, d) unmilled Cu-20 vol. % E-glass fiber (e, f) unmilled Cu-30 vol. % E-glass fiber (g, h) unmilled Cu-40 vol. % E-glass fiber composites sintered at 900°C for 1h

Fig.8 presents the microstructure of various sintered unmilled Cu-E-glass fiber composites developed containing different vol. % of E-glass fiber sintered at 900°C for 1 h in Ar atmosphere. From the microstructural analysis it is evident that the glass fibers are randomly oriented in the Cu matrix and the glass fibers shows very good bonding with the Cu matrix. Back scattered SEM images also show very good wettability of Cu with E-glass fiber. A homogeneous distribution of the reinforcement in the Cu matrix could also be seen. Very few pores could be seen in the Cu matrix as well as the interface of the glass fiber and the Cu matrix [16-18].

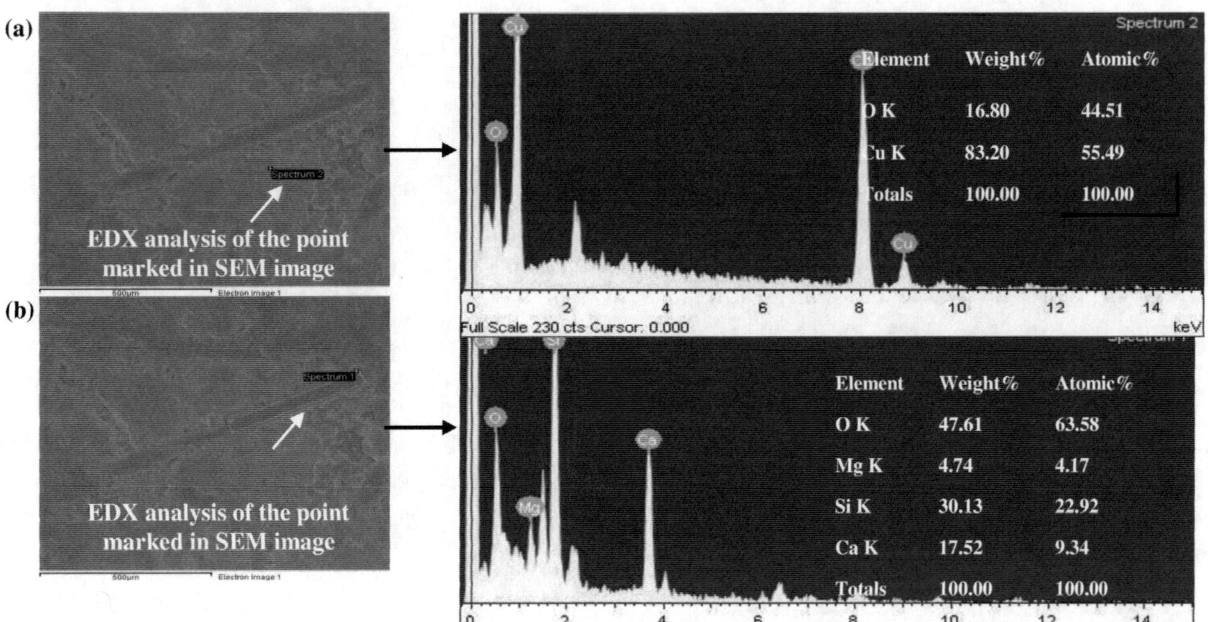

Fig.9 (a, b):- EDX analyses of sintered unmilled Cu-40vol. % composite

Fig.9 presents the EDX of both the reinforcement and the matrix phase in unmilled Cu-40 vol. % composite. The EDX analysis of the Cu matrix in Fig. 9(a) presents the presence of oxygen in the matrix. This is possibly due to the unavoidable oxygen present in the furnace during sintering. The EDX analysis of the E-glass fiber in Fig.9 (b) indicates presence of element like Mg (4.17 at. %), Ca (9.34 at. %), Si (22.92 at. %) and O (63.58 at. %).

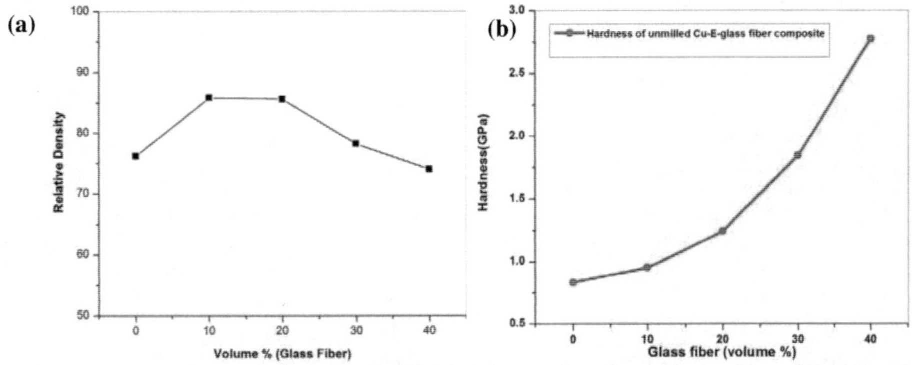

Fig.10:- (a) Variation of (a) relative density (b) hardness of various sintered unmilled Cu-E-glass fiber composites

Fig.10 (a) presents the variation of relative density of unmilled Cu-E-glass fiber composite containing different vol. % reinforcement. The relative density increases initially with increase in the content of E-glass fiber in the composite but it showed slight decreases when higher vol. % of glass fiber was added. It should be noted that a high vol. % of glass fiber could lead to lower theoretical density of the Cu-E-glass fiber composite as compared to that of pure Cu as the density of the E-glass fiber is 2.58 gm/cc and that of Cu is 8.96 gm/cc. Highest relative density of 85.8 % could be achieved in the case Cu-10 vol. % E-glass fiber composite. Fig.10 (b) shows the variation of hardness with the addition of E-glass fiber in the unmilled Cu-E-glass fiber composites. The hardness of the Cu-E-glass fiber composites increases with the increase in vol. % of glass fiber in the composite. This is due to the strengthening of the Cu matrix by the glass fiber. The fracture surfaces of the various composites fractured by impact were also analysed under SEM. The fracture surface of unmilled Cu-10 vol. % E-glass fiber composite in Fig.11 (a) presents the presence of dimples in the Cu matrix suggesting ductile fracture of the Cu matrix. E-glass fibers could be seen embedded in the Cu matrix.

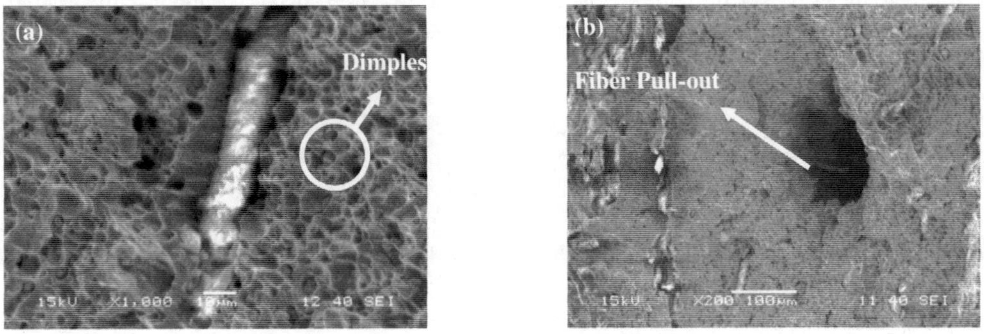

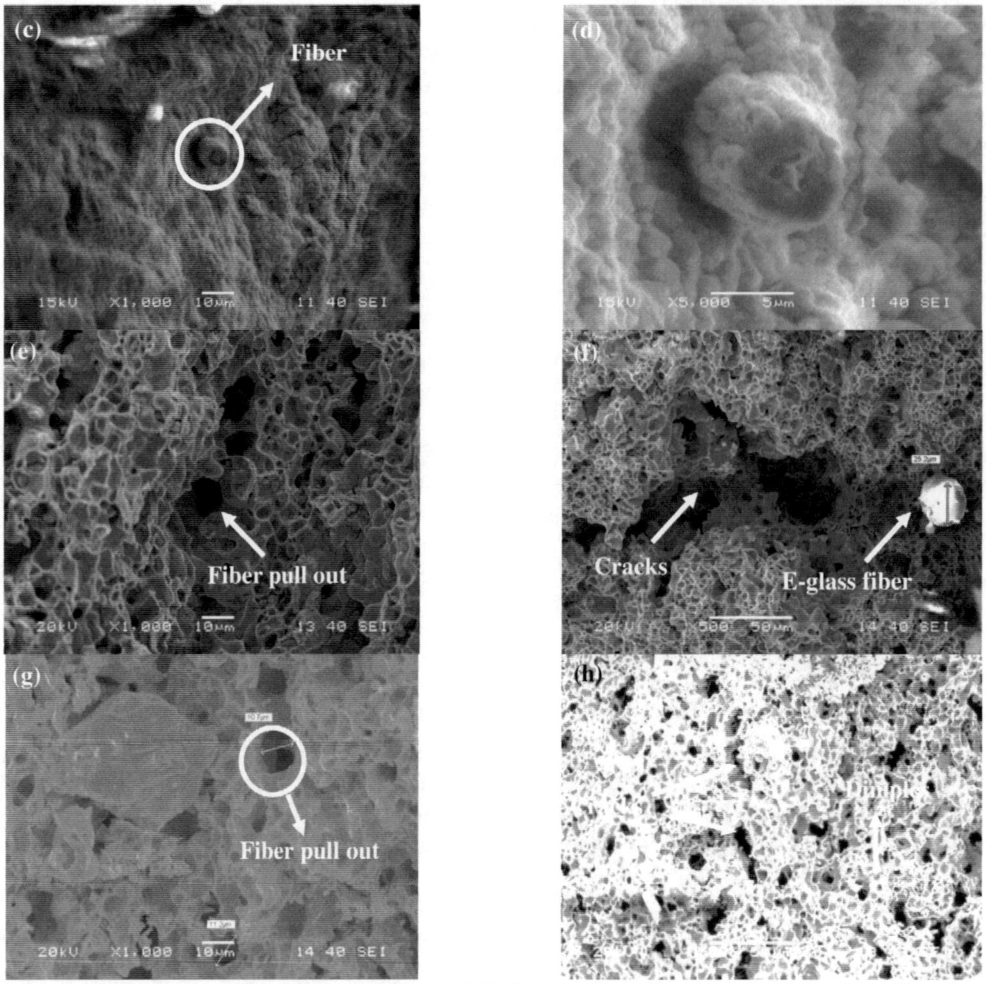

Fig.11:- SEM image of the fracture surface of (a, b) unmilled Cu- 10 vol. % E-glass fiber (c, d) unmilled Cu- 20 vol. % E-glass fiber (e, f) unmilled Cu- 30 vol. % E-glass fiber (g, h) unmilled Cu- 40 vol. % E –glass fiber composite

The SEM image in Fig.11 (b) presents fiber pull out in the fractured surface of the unmilled Cu-10 vol. % E-glass fiber composite. This is possibly due to the lower load bearing capacity as the vol. % of the glass fiber in the composite is very low. Higher stress on each fiber leads to pull-out of the fiber. The stress on the fiber also leads to plastic deformation of the Cu matrix around the fiber. The fiber embedded in the Cu matrix tries to restrict the plastic deformation as is visible from the SEM image. Figs.11(c, d) are the SEM images of the fracture surfaces of unmilled Cu-20 vol. % E-glass fiber composite. From the SEM images it is clear that the nature of fracture in the Cu matrix is ductile whereas the E-glass fiber undergoes brittle fracture. The fractured E-glass fibers have Cu adhered to its surface indicating good wettability and bonding between the Cu matrix and the glass fiber. Figs.11 (e, f) are the SEM images of the fracture surfaces of unmilled Cu-30 vol.% of E-glass fiber composite. The SEM images show ductule nature of fratcure in the Cu matrix and brittle fracture of the glass fibers. The SEM images show pull-out of the glass fiber. The SEM images also show the prences of higher amount of pores in the sample. It can be seen form the variation of relative density of various sintered unmilled Cu-E-glass fiber composites in Fig. 4.10(a) that the realtive density of the unmilled Cu- 30 vol. % E-glass fiber composite (78.2 %) and Cu- 40 vol. % E-glass fiber composite (74 %) are less than the relative density of unmilled Cu- 20 vol. % E-glass fiber composite (85.8%). Addition of E-glass fiber beyond 20 vol. % has led to a large decrease in the relative density of the composite suggesting the presence of larger volume fraction of pores in the composites containing higher vol. % of E-glass fibers. Addition of glass fiber beyond 20 vol. % leads to poor sinterability and densification of the composites.The large volume of pores has led to the crack in these composites during the tensile test. The SEM images of Cu 30, 40 vol. % E-glass fiber composites in Figs. 11(e, f) and Figs. 11(g, h) show a large number of crack. Figs.11 (g, h) show the fracture surface of unmilled Cu-40 vol.% of E-glass fiber composite. The Cu matrix shows dimples on its fracture surface which suggests the ductile fracture. Several cracks can be seen clearly in the Cu matrix [19-21].

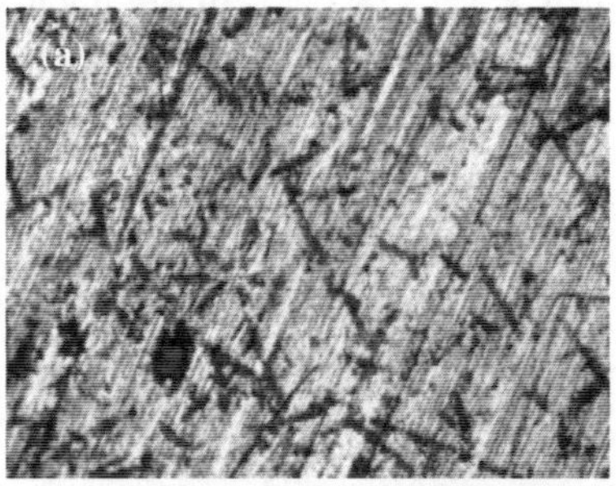

10 vol. % E-glass fiber

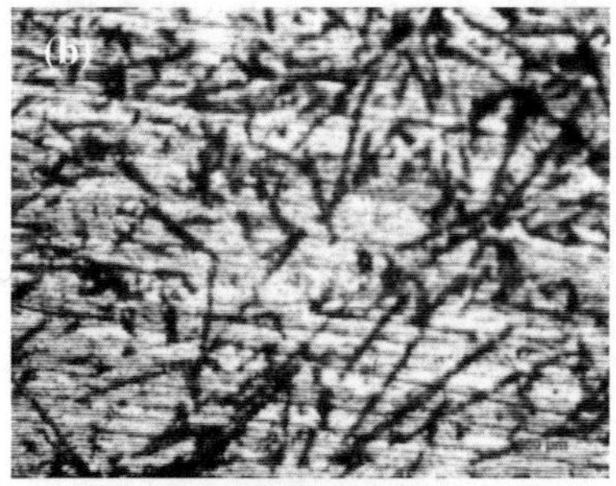

20 vol. % E-glass fiber

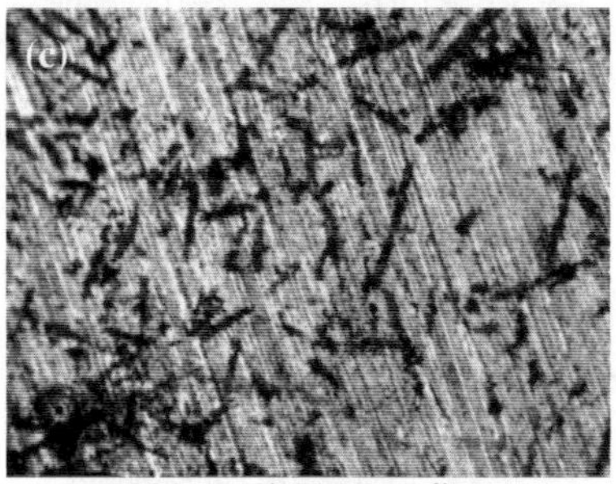

30 vol. % E-glass fiber

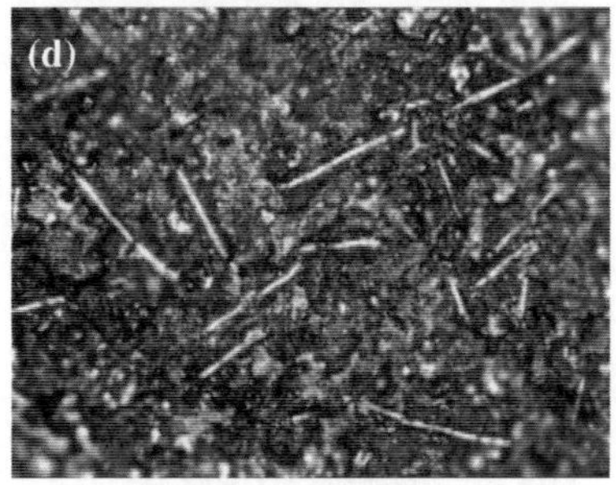

40 vol. % E-glass fiber

Fig.12 (a-d):- Optical micrographs of as-milled Cu-E-glass fiber composites containing different vol. % of E-glass fiber as reinforcement (10, 20, 30 and 40 vol. %).

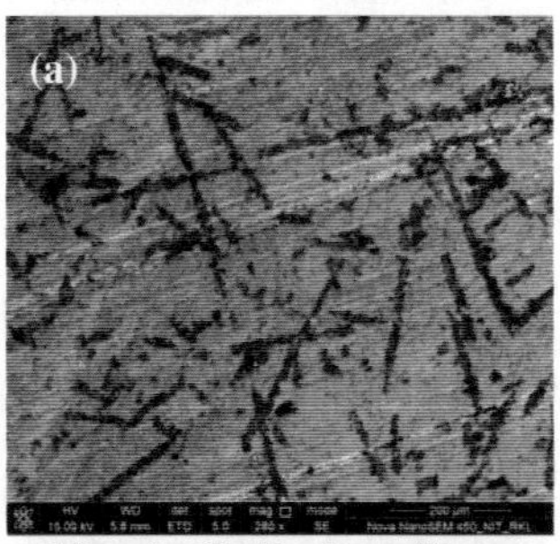

10 vol. % E-glass fiber

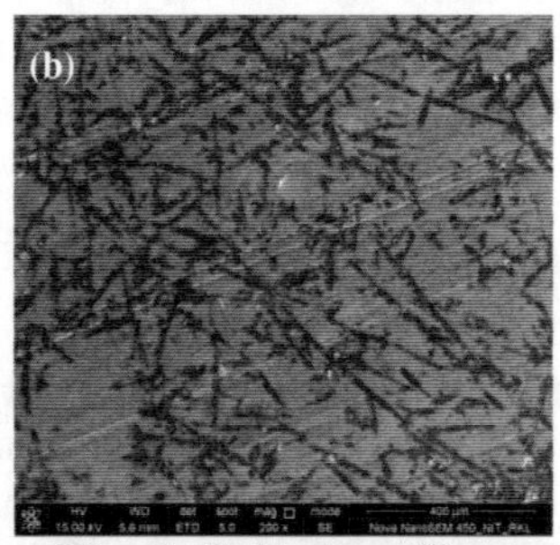

20 vol. % E-glass fiber

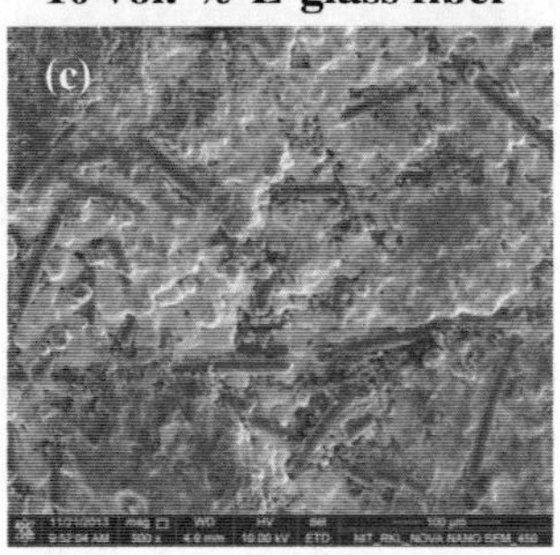

30 vol. % E-glass fiber

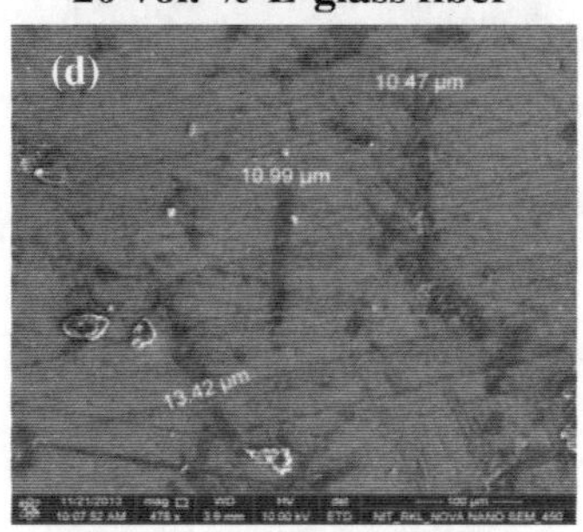

40 vol. % E-glass fiber

Fig.13:- (a-d) SEM images of as- milled Cu-E-glass fiber composites containing different vol. % of E-glass fiber as reinforcement (10, 20, 30 and 40 vol. %).

From Figs.12 (a-d) it is evident that there is a homogenous dispersion of glass fibers throughout the Cu matrix. It is clearly seen that fibers are embedded in the Cu matrix. SEM images for as-milled Cu-E-glass fiber composites in Figs.13 (a-d) also show the random distribution of reinforcements in the Cu matrix. The SEM images show good wettability of Cu and the E-glass fiber. There is strong adhesion between the Cu matrix and the glass fibers. Comparing the SEM images in Fig.8 of unmilled Cu-E-glass fiber composites and the SEM images in Fig.13 of as-milled Cu-E-glass fiber composites the glass fibers seem more embedded in the Cu matrix and there seems to be a better sinterability in the case of milled Cu-E glass fiber composites. This is also evident from the comparison of the relative densities of the unmilled Cu-E-glass fiber composites and the as-milled Cu-E-glass fiber composites shown in Fig. 10(a) and Fig. 14(a) respectively. The as-milled Cu-E-glass fiber composites show higher relative densities as compared to the unmilled Cu-E-glass fiber composites.

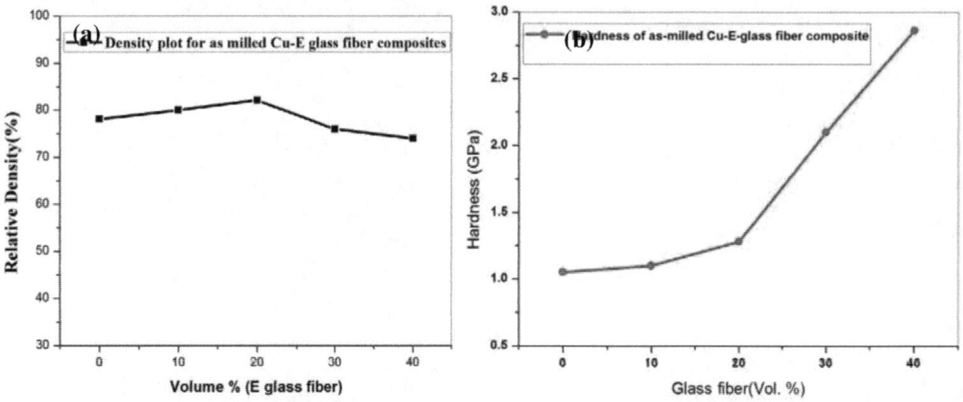

Fig 14:- Variation of (a) relative density (b) hardness of various sintered as-milled Cu-E-glass fiber composites

In as-milled Cu-E-glass fiber composites 20 h milled Cu powder has been used for the development of the composites. The 20 h milled Cu powder is nanocrystalline in nature and as a result has led to better sinterability and densification as shown in Fig.14 (a). Cu has a melting temperature of 1083°C. Nanocrystalline Cu has higher surface area per unit volume and this could reduce the melting point of nanocrystalline Cu. This is why as-milled Cu possibly softens at the sintering temperature of 900°C and results in better sinterability and densification compared to unmilled Cu-E-glass fiber composites. Fig.14 (b) presents variation of microhardness of the various composites with the addition of glass fibers in the as-milled Cu-E-glass fiber composites. The hardness of the composites shows a gradual increase with increase in the content of E-glass fiber in the composites. Comparing the hardness of the unmilled Cu-E-glass fiber composites and as-milled Cu-E-glass fiber composites in Fig.10 (b) and 14(b) respectively it can be seen that the as-milled Cu-E-glass fiber composites possess better hardness. This is due to the better densification of the as-milled Cu-E-glass fiber composites during sintering at 900°C for 1 h.

The SEM images of the fracture surface of as-milled Cu-E-glass fiber composites fractured by impact are shown in Fig.15. The fracture surface in Figs.15 (a, b) shows dimples in the Cu matrix which suggests the ductile nature of fracture of the Cu matrix. It can also be seen that fibers are embedded inside the Cu matrix and has good binding with the matrix. The SEM image shown in Fig.15 (b) clearly shows the brittle fracture of E-glass fiber. The fracture surface of the glass fibers are flat and suggest the brittle nature of their fracture. The glass fibers are well embedded d in the Cu matrix and there is good wettbility and adhesion between the Cu matrix and the glass fibers.

IOP Conf. Series: Materials Science and Engineering **115** (2016) 012023 doi:10.1088/1757-899X/115/1/012023

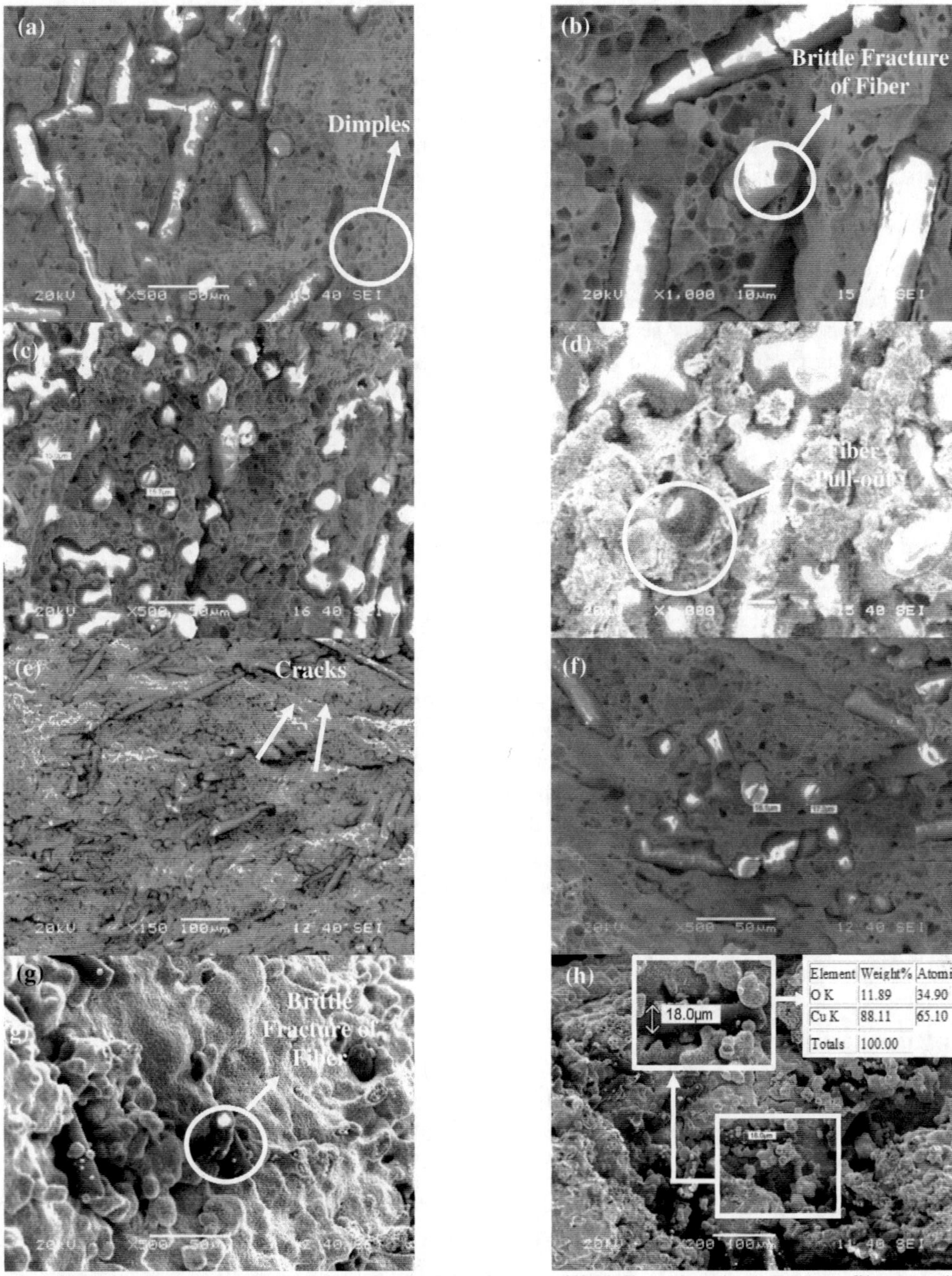

Fig.15:- SEM image of the fracture surface of (a, b) as-milled Cu- 10 vol. % E-glass fiber (c, d) as-milled Cu- 20 vol. % E glass fiber (e, f) as-milled Cu- 30 vol. % E- glass fiber (g, h) as-milled Cu- 40 vol. % E- glass fiber composites. Inset images in Fig. (h) show the magnified image of a section of the SEM image and the EDX of agglomerated Cu particle

Figs. 15 (c, d) presents SEM image of as-milled Cu-20 vol.% E-glass fiber composites. The fibers and the Cu matrix show good bonding beteen them and the fibers are firmly held in the Cu matrix. There is a random distribution of fibers inside the Cu matrix. Fiber pull-out can be seen in Fig.15 (d). Figs.15 (e, f) presents the SEM images of the fracture surface of as-milled Cu-30 vol.% E-glass fiber composite. The fibers show good interfacial integrity with the Cu matrix. Cracks could be seen in the Cu matrix. Comparision of the SEM images in Fig.11 and Fig.15 shows that in the case of as-milled Cu-E-glass fiber composite there are fewer numbers of pores and cracks in the Cu matrix as compared to the unmilled Cu-E-glass fiber composite. This is possibly due to better sinterability and densification in the case of as-milled Cu composite as compared to the unmilled Cu composite. The fracture surfaces of the unmilled Cu-E-glass fiber composites show a large number of pores. Plastic deformation in the Cu matrix due to the stress on the fiber during could be seen in the unmilled Cu-E-glass fiber composite (refer Fig. 11(a)). However, such plastic deformations could not be seen in the as-milled Cu-E-glass fiber composites. This suggests stronger bonding between the glass fibers and the Cu matrix in the case of as-milled Cu-E-glass fiber composites. This is due to the better sinterability and densification of the as-milled Cu-E-glass fiber composites. Comparison of the relative densities of unmilled Cu-E-glass fiber composites and as-milled Cu-E-glass fiber composites in Fig. 10 (a) and Fig. 14(a) clearly shows higher relative densities in the case of as-milled Cu-E-glass fiber composites. Figs.15(g, h) presents the SEM image of the fracture surface of as-milled Cu-40 vol.% E-glass fiber composite. The SEM images clearly show Cu adhered to the E-glass fiber surface. The brittle fracture of the glass fiber could be clearly seen in Fig.15 (g). In the case of as-milled Cu- 40 vol. % E- glass fiber composite a large number of Cu rich spherical particles could be seen (refer Fig. 15(h)). This is possibly due to the agglomeration of finer Cu particles which tend to come together to form larger particles and there by reducing the surface area per unit volume. This was not seen in the case of unmilled Cu composites (Fig.11). This is possibly due to the Gibbs-Thomson effect which predicts that nanoparticles of radius r dissolve at lower electrochemical potentials than bulk materials and thereby decreasing the surface area per unit volume (1/r). The Gibbs-Thomson effect is seen due to the reduction of local chemical potential caused by nanoscale curvature [22-24].

4. Conclusion

1. Milling of elemental Cu powder for 20 h led to the formation of nanostructured Cu. The crystallite size of Cu after 20 h of milling was found to be around 18 nm. There is a gradual increase in the lattice strain with milling time due to severe deformation of the milled powder and the strain reaches a maximum after 20 h of milling when the crystallite size is smallest.
2. The hardness values of Cu-E-glass fibre composites increase with increase in vol. % of E-glass fiber in the composite both in the case of milled and unmilled Cu-E-glass fiber composites. The maximum hardness of 2.86 GPa was found in the case of as-milled Cu-40 vol. % E-glass fiber composite.
3. As-milled Cu-E-glass fiber composites shows better densification and sinterability compared to the unmilled Cu-E-glass fiber composites due to the finer size of Cu particles in the milled Cu powder. Better densification and sinterability has also led to higher hardness in the case of milled Cu-E-glass fiber composites.

Acknowledgement

We would like to thank the staff in the SEM and XRD laboratories at the National Institute of Technology-Rourkela, India. The authors would also like to thank the Central Research Facility, Indian Institute of Technology Kharagpur, for allowing them to analyze the samples in the HRTEM laboratory.

References

[1] B. Harris, Engineering Composite Materials,London, IOM (1999).[2] D. Hull and T. W. Clyne, An Introduction to Composite Materials,New York, Cambridge University Press (1996).[3] F. L. Matthews and R. D. Rawlings, Composite Materials: Engineering and Science,Boca Raton, CRC (2006).
[4] S.-M. Choi and H. Awaji, Nanocomposites-a new material design concept,*Sci. Technol. Adv. Mater.*6 (2005) 2-10.
[5] K. U. Kainer, Basics of Metal Matrix Composites,in*Metal Matrix Composites*, K. U. Kainer, Ed. Wiley-VCH Verlag GmbH & Co. KGaA, (2006) 1-54
[6] D. B. Miracle,Metal matrix composites – From science to technological significance,*Compos. Sci. Technol.* 65 (2005) 2526-2540.

[7]K. KChawla ,Composite Materials: Science and Engineering,Springer Science & Business Media, New York (2012).

[8] C. Suryanarayana, E. Ivanov, and V. V. Boldyrev,The science and technology of mechanical alloying,*Mater.Sci. Eng.* A304-306 (2001) 151-158.

[9] C. Suryanarayana,Mechanical alloying and milling,*Prog. Mater. Sci.,* 46 (2001) 1-184.

[10] D. Balzar and H. Ledbetter, Voigt-function modeling in Fourier analysis of size- and strain-broadened X-ray diffraction peaks,*J. Appl. Crystallogr.*, 26 (1993) 97-103.

[11] D. Balzar, N. Audebrand, M. R. Daymond, A. Fitch, A. Hewat, J. I. Langford, A. Le Bail, D. Louër, O. Masson, C. N. McCowan, N. C. Popa, P. W. Stephens, and B. H. Toby, Size-strain line-broadening analysis of the ceria round-robin sample,*J. Appl. Crystallogr.,*
37 (2004) 911-924.

[12] C. C. Koch, Materials Synthesis by Mechanical Alloying,Annu. *Rev. Mater. Sci.,* 19(1989)121-143.

[13] B. S. Murty, M. Mohan Rao, and S. Ranganathan, Milling maps and amorphization during mechanical alloying,*Acta Metall. Mater.,* 43 (1995) 2443-2450.

[14] T. Raghu, R. Sundaresan, P. Ramakrishnan, and T. R. Rama Mohan, Synthesis of nanocrystalline copper–tungsten alloys by mechanical alloying,*Mater.Sci. Eng.A*304- 306 (2001) 438-441.

[15] F. T. Wallenberger, Commercial and Experimental Glass Fibers, in *Fiberglass and Glass Technology,* F. T. Wallenberger and P. A. Bingham, Eds., US, Springer, (2010) 3-90.

[16] G. Zak, M. N. Sela, V. Yevko, C. B. Park, and B. Benhabib,Layered-Manufacturing of Fiber-Reinforced Composites,*J. Manuf. Sci. Eng.,* 121 (1999) 448-456.

[17] C. L. Schutte, Environmental durability of glass-fiber composites, *Mater. Sci. Eng. RRep.,*13 (1994) 265-323.

[18] T. W. Chou, A. Kelly, and A. Okura, Fibre-reinforced metal-matrix composites,*Composites*, 16 (1985) 187-206.

[19] J. R. Brockenbrough, S. Suresh, and H. A. Wienecke, Deformation of metal-matrix composites with continuous fibers: geometrical effects of fiber distribution and shape,*Acta Metall. Mater.,* 39 (1991) 735-752.

[20] B. Budiansky, J. W. Hutchinson, and A. G. Evans, Matrix fracture in fiber-reinforced ceramics,*J. Mech. Phys. Solids*, 34 (1986) 167-189.

[21] J. E. Bolander, S. Choi, and S. R. Duddukuri, Fracture of fiber-reinforced cement composites: effects of fiber dispersion,*Int. J. Fract.,*154 (2008) 73-86.

[22] V. Srinivasa, V. Shivakumar, V. Nayaka, S. Jagadeeshaiaih, M. Seethram, R. Shenoy, and A. Nafidi, Fracture morphology of carbon fiber reinforced plastic composite laminates,*Mater. Res.,* 13 (2010) 417-424.

[23] Q. Du, M. Perez, W. J. Poole, and M. Wells, Numerical integration of the Gibbs–Thomson equation for multicomponent systems,*Scr. Mater.,*66 (2012) 419-422.

[24] I. McCue, J. Snyder, X. Li, Q. Chen, K. Sieradzki, and J. Erlebacher, Apparent Inverse Gibbs-Thomson Effect in DealloyedNanoporous Nanoparticles,*Phys. Rev. Lett.*, 108(2012) 225-503 .

High Temperature Mechanical Behavior of Aluminum-$Cu_{50}Zr_{50}$Metallic Glass Interface

P Gupta[*] and N Yedla

Computational Materials Engineering Group, Department of Metallurgical and Materials Engineering, National Institute of Technology Rourkela, India-769008

*E-mail: 514mm1005@nitrkl.ac.in, phone: +9109862826625

Abstract. Molecular dynamics (MD) simulations are carried out to determine interface strength between aluminum (metal) and $Cu_{50}Zr_{50}$ (metallic glass) at temperature of 500 K and at strain rate of 10^8 s^{-1}. Simulation box of size 100 Å (x) × 110 Å (y) × 50 Å (z) is used for the above studies. At first Al–$Cu_{50}Zr_{50}$ crystalline interface model is built with the base layer-Al of 50 Å and the top layer-$Cu_{50}Zr_{50}$ of 55 Å along y–direction. Later $Cu_{50}Zr_{50}$ metallic glass is obtained by quenching at a cooling rate of 4 × 10^{12} Ks^{-1}. NPT ensemble is used in metallic glass preparation simulation. The interface model is then equilibrated at 300 K for 500 ps to relieve the internal stresses. EAM (Embedded Atom Method) potential is used for modelling the interaction between Al-Cu-Zr atoms. The interface strength of Al–$Cu_{50}Zr_{50}$ model interface is determined by applying load in the directions normal (mode-I) and parallel (mode-II) to the interface. NVT ensemble is used for the deformation studies. In mode-I for perfect and cracked interface, the interface fractures in the Al-region via necking. Sticking of the Al-atoms to the metallic glass is observed in both the loading conditions. Also, multiple voids are nucleated at the interface.

Keywords: Interface; Molecular Dynamics: High Temperature

1. Introduction

Among the advanced materials for aviation, defense and automobile applications, Al-based metal Matrix composites (MMC) are of extraordinary mechanical properties, including low density, high elastic modulus, strength, good fatigue and good wear resistance [1–3]. The main drive behind the advancement of MMC is the likelihood to tailor their properties to meet particular requirements, which renders this sort of material remarkable in correlation with general engineering materials [4]. In common MMCs, the most broadly utilized reinforcement are ceramic, for example, Al_2O_3 or SiC, as fibers, flakes or particulates [5]. In any case, the interfaces between the reinforcement and the metal matrix are generally not good, which brings about very permeable (porous) materials and poor wettability which reduces the mechanical properties and also reduces prolonged consumption life of the MMCs. To take care of this issue, metallic glasses have been proposed as a novel reinforcement in MMCs [6–9], considering the metastable/amorphous nature of the metallic glass materials. The metallic glass reinforcement, containing practically metallic components, are accepted to be better with the metal matrix and to bring about better interface [10–13]. They are viewed as reasonable for use as reinforcement in light metal with moderately low melting temperature, for example, Al and Mg metals matrix. Mechanical properties of the composite are influence by the interface between reinforcement and the matrix. Moreover when the size of the reinforcement is reduced to the nano level, remarkable improvement in the mechanical properties like high elastic strain and large yield stress are observed [12] due to high interface strength. Deformation mechanisms such as Dislocation glide, twinning are impossible or too expensive to experimentally observe because experimental equipment are very sensitive to factors like crystal defects, impurities, lattice mismatch etc. [14]. To study the interface properties at nano scale, computer simulation like molecular dynamics (MD) offers a good alternative. In present study, we aim at investigating the deformation behavior of the interface between the aluminum and $Cu_{50}Zr_{50}$ metallic glass subjected to mode-I and mode-II loading at temperature of 500 K and strain rate of $10^8 s^{-1}$.

2. Simulation method and model

We performed MD simulation on LAMMPS (Large Scale Atomic/Molecular Massively Parallel Simulator) [15] platform. The interaction between the Al-Cu-Zr atoms during mode-I and mode-II deformation is used by EAM potential developed by Zhou et al. [16]. As per the EAM, the total energy E of the crystal is expressed as below

$$E=\frac{1}{2}\sum_{i,j,i\neq j}\phi_{ij}(r_{ij})+\sum_{i}F_i(\rho_i)\text{------------------}(1)$$

where, the pair energy represented by (ϕ_{ij}) between the atoms i and j separated by a (r_{ij}) distance, and the embedding energy (F_i) associated with embedding an atom i into a local site with an electron density (ρ_i).The following equation calculate the electron density

$$\rho_i=\sum_{j,j\neq i}f_j(r_{ij})\text{------------------------}(2)$$

where, the electron density represented by ($f_j(r_{ij})$) at the site of atom i arising from atom j at (r_{ij}) distance.

$$f(r)=\frac{f_e\exp[-\beta(r/r_e-1)]}{1+(r/r_e-\lambda)^{20}}\text{--------------------}(3)$$

where, f_e represents the scaling constant and for pure metals its value is 1, r_e represents the equilibrium nearest neighbour distance and λ is additional parameter for the cutoff. The generalized elemental pair potential equation is given by [16]

$$\phi(r)=\frac{A\exp[-\alpha(r/r_e-1)]}{1+(r/r_e-k)^{20}}-\frac{B\exp[-\beta(r/r_e-1)]}{1+(r/r_e-\lambda)^{20}}\text{------------}(4)$$

Where, A, B, α, β and re are represented as fitted parameters, and k and λ are two additional parameter for cutoff. The pair potential for a and b species are given below:

$$\phi^{ab}(r)=\frac{1}{2}\left[\frac{f^b(r)}{f^a(r)}\phi^{aa}(r)+\frac{f^a(r)}{f^b(r)}\phi^{bb}(r)\right]\text{--------------}(5)$$

The potential stated above is found to well fit to fundamental material properties, for example, lattice constants, elastic constants, bulk modulus, vacancy development energies and it predicts accurately the heat of solution of alloy. For instance, Meng et al. [17] effectively utilized Zhou potential parameters to study phase transitions in FeCo and FeNi nanoparticles. The EAM potential parameters utilized for Al, Cu and Zr are shown in Table 1.

Table 1 EAM potential parameters

	Al	**Cu**	**Zr**
r_e	2.863924	2.556162	3.199978
f_e	1.403115	1.554485	2.230909
ρ_e	20.418205	21.175871	30.879991
ρ_s	23.19574	21.175395	30.879991
A	6.613165	8.127620	8.559190
B	3.527021	4.334731	4.564902
A	0.134873	0.396620	0.424667
B	0.365551	0.548085	0.640054
K	0.379846	0.308782	0.5
Λ	0.759692	0.756515	1.0
F_{n0}	-2.807602	-2.170269	-4.485793
F_{n1}	-0.301435	-0.263788	-0.293129
F_{n2}	1.258562	1.088878	0.990148
F_{n3}	-1.247604	-0.817603	-3.202516
F_0	-2.83	-2.19	-4.51
F_1	0	0	0
F_2	0.622245	0.561830	0.928602
F_3	-2.488244	-2.100595	-0.981870
H	0.785902	0.310490	0.597133
F_e	-2.824528	-2.186568	-4.509025

3. Simulation of interface

Simulation box represented in Fig. 1 with dimension 100 Å (x) × 110 Å (y) × 50 Å (z) is used for the prediction of interface strength of Al (metal)-$Cu_{50}Zr_{50}$-(metallic glass). At first Al–$Cu_{50}Zr_{50}$ crystalline model is constructed with the bottom layer (Al) of 50 Å (Al-atoms represented as red colour) and the top layer of 60 Å ($Cu_{50}Zr_{50}$) in height along y–direction (Cu-atoms and Zr-atoms are represented as blue and yellow in colour respectively). NPT ensemble is used in metallic glass preparation simulation. Isothermal-isobaric (npt) ensembles which updates the position and velocity for atoms in the group each timestep. Thereafter, a cooling rate of 4×10^{12} K/s is used to obtain $Cu_{50}Zr_{50}$ metallic glass by rapid cooling. To relieve the stresses interface model is equilibrated at 500 K for 500 ps. A defect (void) interface model is obtained by deleting atoms in the interface region of 20Å diameter.

Mode-I and mode-II loading is used to determine the fracture strength of Al–$Cu_{50}Zr_{50}$ (perfect and defect) model interface. For mode-I, loading direction is y-axis [0 1 0] while for mod-II it is z-axis [0 0 1]. Periodic boundary condition is used along z-direction for mode-II loading while non-periodic boundary condition in all the directions for mode-I. All the simulations are carried out at timestep of 0.002 ps, strain rate of $10^8 s^{-1}$ and temperature of 500 K. The perfect and defect interface model is allowed for full separation under tensile load.

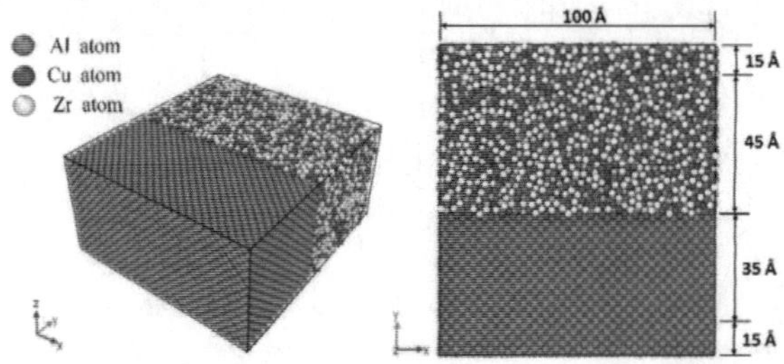

Figure 1: Atomic configuration snap shot of the Al (metal)-$Cu_{50}Zr_{50}$ (metallic glass) perfect interface model.

4. Results and discussion

The simulation results of the perfect interface and cracked interface simulated at temperature of 500 K and deformed at strain rate of 10^8 s^{-1} for mode-I and mode-II will be presented here. The engineering strain is calculated by comparing the boundary distance from the reference configuration of the undeformed structure in both the x and y -directions. Similar methodology was also used in the studies of Dandekar and Shin [18], and Zhou et al [19] Figs. 2(a) and 2(b) shows the stress-strain relation of the interface simulated at 500 K for strain rate of 10^8 s^{-1} under mode-I and Figs. 2(c) and 2(d) for mode-II conditions for perfect and cracked sample respectively. The regions on the stress-strain curve where changes occur in the interface have been numbered (1-4). The corresponding atomic position snap shot are shown in Figs. 3(a-d). The centro-symmetry parameter [20] is used to observe plastic deformation mechanism features and are shown in the Figs. 4(a-d).

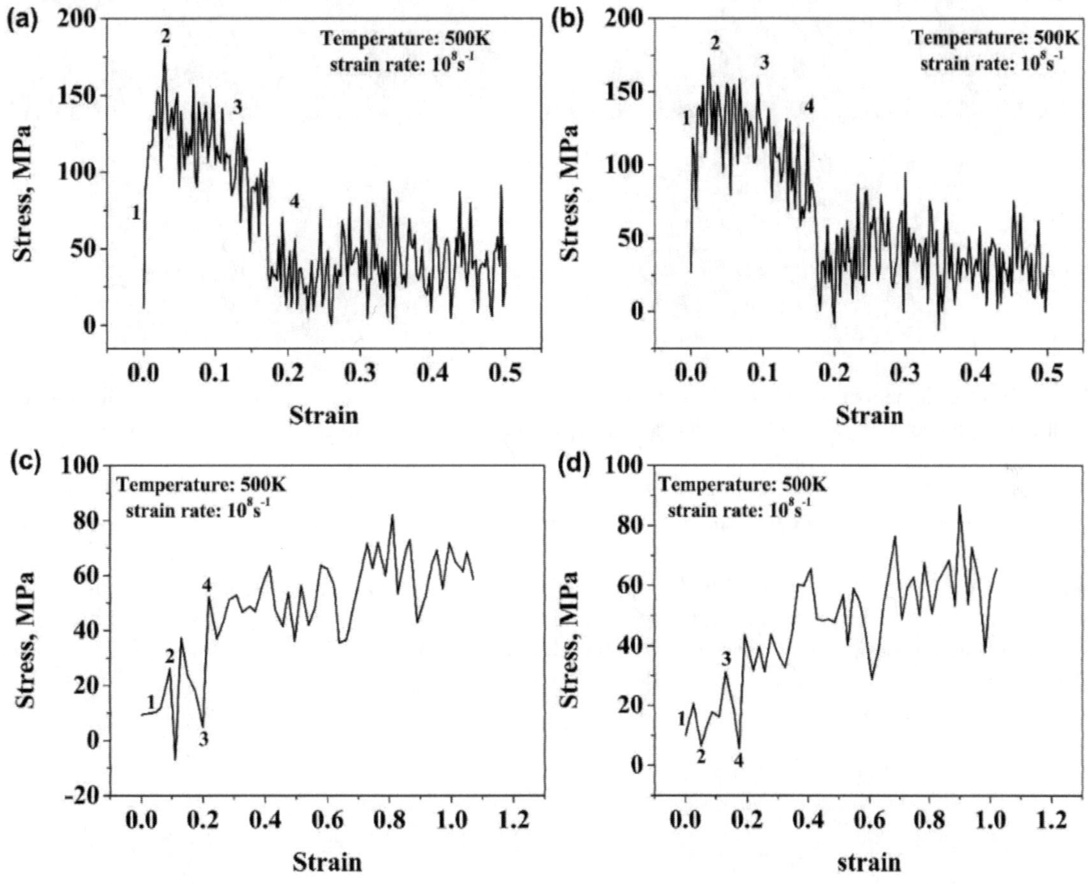

Figure 2: Stress-strain plot of the interface subjected to strain rate of $10^8 s^{-1}$ at temperature 500 K for mode-I loading: (a) perfect sample (b) defect sample with crack; for mode-II loading: (c) perfect sample and (d) defect sample with crack.

Simulated stress-strain features of the interface at temperature of 500 K and strain rate of 10^8 s^{-1} from its beginning stage to final stage of fracture are shown in Fig. 2(a-d). The disfigurement of the interface (glass and crystalline locale) starts with versatile distortion from its beginning state to the first yield state (Fig. 2a) ~121.9 MPa relates to strain of 0.019 in mode-I deformation and reaches to maximum stress of 181.94 MPa. When crack is introduced in the interface (Fig. 2(b)), yield strength and

IOP Conf. Series: Materials Science and Engineering **115** (2016) 012024 doi:10.1088/1757-899X/115/1/012024

the maximum strength decreases. The stress-strain features are more serrated due to vibration of atoms about the mean positions at such high temperatures [21].

Fig. 3a and Fig. 3b shows the atomic position snapshots of the interface without and with void at the centre of the crystalline (aluminum) and metallic glass ($Cu_{50}Zr_{50}$) interface simulated at temperature of 500 K and deformed strain rate of $10^8 s^{-1}$ under mode-I condition. It can be seen that with progress of deformation amorphization of the crystalline region of the interface occurs along with nucleation of voids. In the interface with crack, its closure occurs as strain increases. The load bearing capacity decreased in the presence of void.

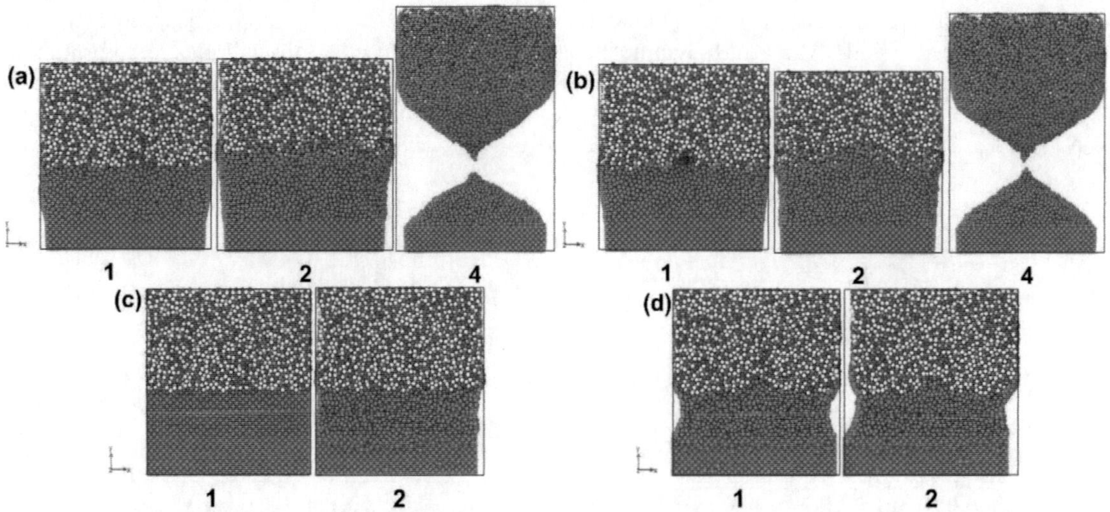

Figure 3: Atomic snapshot of the interface under strain rate of $10^8 s^{-1}$ at temperature 500 K for mode-I loading (a) perfect sample and (b) with crack sample and for mode-II loading (c) perfect sample and (d) with crack sample respectively.

Figs. 4a-d show centro-symmetry parameter of the perfect and defect interface under mode-I and mide-II loading conditions. It can be clearly seen that defects are nucleated in the crystalline region of the interface. Corresponding to mode-I of Fig. 2 (a) at point 1, two Shockley partials (Fig. 5) with burgers vector 1/6<112> are observed. At higher strains dislocations are not observed due to amorphization of the crystalline region at high temperature, as the atoms randomized for amorphization and dislocation dissolves [24].

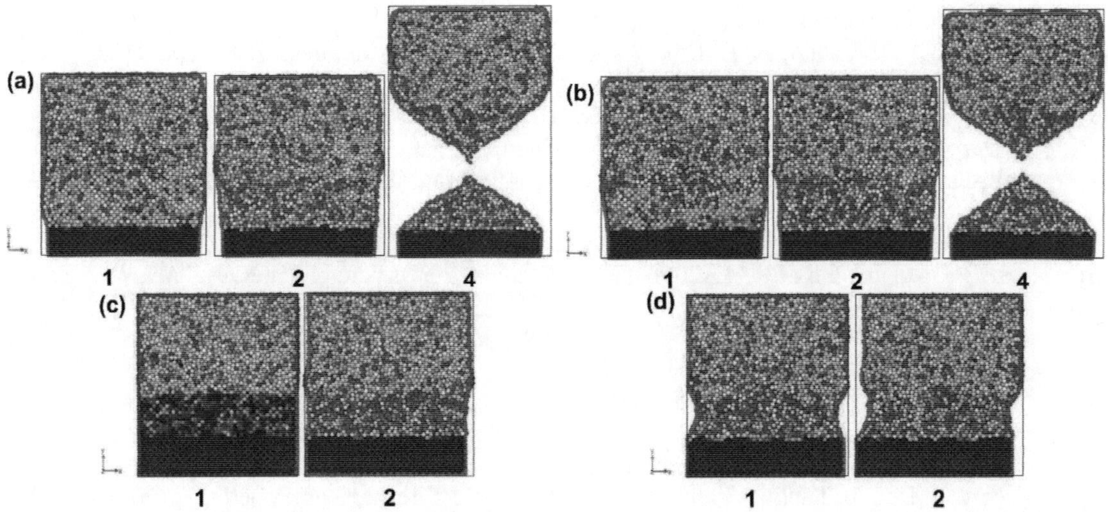

Figure 4: Centrosymmetry Parameter snapshot of the interface under strain rate of $10^8 s^{-1}$ at temperature 500 K for mode-I loading (a) perfect sample and (b) with crack sample and for mode-II loading (c) perfect sample and (d) with crack sample respectively.

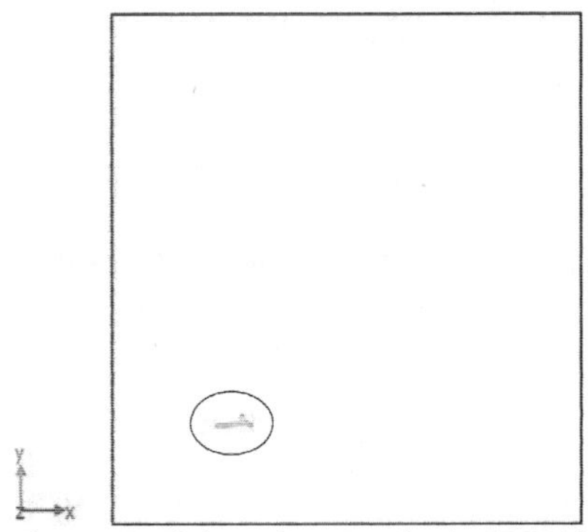

Figure 5 Dislocation Analysis (DXA)

5. Conclusions

- The stress-strain curve is more serrated due to deformation carried out at high temperature.
- Yield strength and maximum strength is high for perfect interface.
- During mode-I loading, the fracture occurs by necking in the aluminium region of the interface.
- Void closure occurs during progress of deformation in the cracked interface.
- Adhesion of the Al atoms at the interface has been observed for mode-I and mode-II loading.
- Centro-symmetry parameter (CSP) revel amorphization of the interface in the early stages of the deformation due to high temperature.
- Vacancies are observed in the crystalline region of the interface.
- Formation of dislocations at high temperature is very low due to amorphization.

References

[1] Kainer, K. U., 2006, Metal Matrix Composites: Custom-made Materials for Automotive and

Aerospace Engineering, *John Wiley & Sons*.

[2] Starke, E. A., and Staley, J. T., 1996, "Application of modern aluminum alloys to aircraft," *Prog. Aerosp. Sci.*, **32**(2-3), pp. 131–172.

[3] Wang, W. H. H., Dong, C., and Shek, C. H. H., 2004, "Bulk metallic glasses," *Mater. Sci. Eng. R Reports*, **44**(2–3), pp. 45–89.

[4] Callister, W. D., and Rethwisch, D. G., 2010, Materials Science and Engineering: An Introduction, *8th ed. Hoboken, NJ: Wiley*.

[5] Chawla, K.K., 2006. Metal matrix composites. *Wiley-VCH Verlag GmbH & Co. KGaA*.

[6] ASHBY, M., and GREER, A., 2006, "Metallic glasses as structural materials," *Scr. Mater.*, **54**, pp. 321–326.

[7] Johnson, W. L., Inoue, A., and Liu, C. T., 1999, Bulk Metallic Glasses:, *Cambridge University Press*.

[8] Scudino, S., Liu, G., Prashanth, K. G., Bartusch, B., Surreddi, K. B., Murty, B. S., and Eckert, J., 2009, "Mechanical properties of Al-based metal matrix composites reinforced with Zr-based glassy particles produced by powder metallurgy," *Acta Mater.*, **57**(6), pp. 2029–2039.

[9] Eckert, J., Das, J., Pauly, S., and Duhamel, C., 2007, "Mechanical properties of bulk metallic glasses and composites," *J. Mater. Res.*, **22**(02), pp. 285–301.

[10] Rajan, T. P. D., Pillai, R. M., and Pai, B. C., 1998, "Reinforcement coatings and interfaces in aluminium metal matrix composites," *J. Mater. Sci.*, **33**, pp. 3491–3503.

[11] Lee, M., 2004, "Fabrication of Ni–Nb–Ta metallic glass reinforced Al-based alloy matrix composites by infiltration casting process," *Scr. Mater.*, **50**(11), pp. 1367–1371.

[12] Yu, P., Kim, K. B., Das, J., Baier, F., Xu, W., and Eckert, J., 2006, "Fabrication and mechanical properties of Ni–Nb metallic glass particle-reinforced Al-based metal matrix composite," *Scr. Mater.*, **54**(8), pp. 1445–1450.

[13] Inoue, A., 2000, "Stabilization of metallic supercooled liquid and bulk amorphous alloys," *Acta Mater.*, **48**(1), pp. 279–306.

[14] Galan, U., and Sodano, H. A., 2012, "Molecular dynamics prediction of interfacial strength and validation through atomic force microscopy," *Appl. Phys. Lett.*, **101**(15), p. 151603.

[15] Plimpton, S., 1995, "Fast parallel algorithms for short-range molecular dynamics," *J. Comput. Phys.*, **117**(1), pp. 1–19.

[16] Zhou, X. W., Johnson, R. A., and Wadley, H. N. G., 2004, "Misfit-energy-increasing dislocations in vapor-deposited CoFe/NiFe multilayers," *Phys. Rev. B*, **69**(14), p. 144113.

[17] Meng, L., Peng, X., and Zhang, K., 2012, "Structural phase transitions of FeCo and FeNi nanoparticles: A molecular dynamics study," *J. Appl. Phys.*, **111**(2), p. 024303.

[18] Dandekar, C. R., and Shin, Y. C., 2011, "Molecular dynamics based cohesive zone law for describing Al-SiC interface mechanics," *Compos. Part A Appl. Sci. Manuf.*, **42**(4), pp. 355–363.

[19] Zhou, X. W., Zimmerman, J. A., Reedy, E. D., and Moody, N. R., 2008, "Molecular dynamics simulation based cohesive surface representation of mixed mode fracture," *Mech. Mater.*, **40**(10), pp. 832–845.

[20] Stukowski, A., 2010, "Visualization and analysis of atomistic simulation data with OVITO–the Open Visualization Tool," *Model. Simul. Mater. Sci. Eng.*, **18**(1), p. 15012.

[21] Yedla, N., Meraj, M., Gupta, P., Sarat, V., Kabi, A. J., and Pal, S., 2015, "The effect of nano-void on deformation behaviour of Al-Cu intermetallic thin film compounds," *Metall. Res. Technol.*, **112**(5), p. 505.

[24] Huang, S., Zhang, S., Belytschko, T., Terdalkar, S. S., & Zhu, T., 2009, "Mechanics of nanocrack: fracture, dislocation emission, and amorphization". *J. Mech Phy Solids*, **57**(5), pp. 840-850.

Characterisation of Suspension Precipitated Nanocrystalline Hydroxyapatite Powders

P K Mallik[*], P.K.Swain, S.C Patnaik
Department of Metallurgical and Materials Engineering, IGIT Sarang, Odisha India

E-mail: prafulla.mallik@gmail.com

Abstract: Hydroxyapatite (HA) is a well-known biomaterial for coating on femoral implants, filling of dental cavity and scaffold for tissue replacement. Hydroxyapatite possess limited load bearing capacity due to their brittleness. In this paper, the synthesis of nanocrystalline hydroxyapatite powders was prepared by dissolving calcium oxide in phosphoric acid, followed by addition of ammonia liquor in a beaker. The prepared solution was stirred by using magnetic stirrer operated at temperature of 80°C for an hour. This leads to the formation of hydroxyapatite precipitate. The precipitate was dried in oven for overnight at 100°C. The dried agglomerated precipitate was calcined at 800°C in conventional furnace for an hour. The influence of calcium oxide concentration and pH on the resulting precipitates was studied using BET, XRD and SEM. As result, a well-defined sub-rounded morphology of powders size of ~41 nm was obtained with a salt concentration of 0.02 M. Finally, it can be concluded that small changes in the reaction conditions led to large changes in final size, shape and degree of aggregation of the hydroxyapatite particles.

Keywords: Nanocrystalline; HA; Suspension precipitation; Morphology; Biomaterial

1. Introduction

In last two decades, hydroxyapatite has been used as synthetic biomaterials for orthopaedic research due to its special property of being highly bioactive as well as its ability to form direct interfacial bond with surrounding tissue, leading to better osseointegration [1-2]. Because of the insufficient electrical conductivity and fracture toughness, the use of HA as load bearing hard tissue replacement is rather limited. However, the use of conventional process to fabricate of HA ceramics results in poor microstructure and properties due to decomposition of HA [3]. It was also reported that oxyapatite ($Ca_{10}O(PO_4)_6$) is highly reactive at a temperature lower than 800°C, even under cooling in vacuum (10^{-4} -10^{-6} torr) and forms a stable compound of oxyhydroxyapatite ($Ca_{10} (PO_4)_6O_{0.75} (OH)_{0.5}V_{0.75}$, where V is a neutral vacancy in the OH site at room temperature. However, it decomposes into a mixture of tri- and tetra- calcium phosphates at temperature higher than 1050°C [4]. From the above discussion, it is more important to have a Ca/P ratio of 1.67. If the stoichiometric ratio is maintained, then HA should be stable up to temperatures of 1300°C without phase changes [5]. Therefore, HA-based composites have been developed by using different additives to improve the mechanical reliability [6-8]. However, these additives do not only tend to decrease the biocompatibility and bioactivity of pure HA, but also decreases the decomposition temperature of monolithic HA during conventional sintering. In this research, our main objective is to synthesis the nanocrystalline hydroxyapatite powders were made by dissolving calcium oxide in phosphoric acid and heating at 80°C for an hour to address the above issues.

2. Experimental procedure

2.1 Hydroxyapatite (HA) powder synthesis

Hydroxyapatite, $Ca_{10} (PO_4)_6(OH)_2$ is synthesized using widely used method of suspension-precipitation route [9]. The powders used were CaO (SD Fine-Chem) and H_3PO_4 (Merck). Initially, CaO was dispersed in distilled water at a concentration of 18.6 gm/l. The dispersed solution was kept on a hot plate (~80°C) with electromagnetic stirrer for continuous stirring of the dispersed medium. By keeping the Ca/P ratio as desired (1.67), an equivalent amount of H_3PO_4 solution with 0.17M was added drop wise in the dispersed CaO solution. The above set up was held at 80°C and stirred for 3-4 hours to result completion of reaction. Thereafter, concentrated NH_4 (Qualikems) was added drop wise to bring the pH of solution to 10. (For making 25gm of HA, about 5-10 ml of NH_4 was needed). After the completion of the reaction, the product was allowed to precipitate by keeping it at room temperature for 24 hours and was collected by filtering the solution. The collected precipitate was dried. The lump was crushed into powder in an agate mortar and subsequently calcined at 800°C for 2 hours.

2.2 Microstructural Characterization

Detailed investigation of microstructure of powders was performed using Scanning Electron Microscope (SEM, FEI Quanta 200 and JEOL JSM-6330F) attached with EDS (Energy Dispersive X-ray Spectroscopy) to determine the compositional analysis. X-ray diffraction (XRD) of powders was used to characterise the phase assemblage and in particular to asses the phase purity. The powder diffraction patterns of powders was measured on a (Brucker Xpert diffractometer using a CuKα line (wavelength 1.54 Å, 40 kV, 40 mA, 0.5° divergence slit, and 0.5° antiscatter slit). The XRD data were recorded using a step size of 0.02° with a scan speed of 0.5° per minute. The recorded diffraction patterns were compared to powder diffraction files from the JCPDS database. Quantitative estimations of phase composition were made using the semi-quantitative feature in the XPert High score Plus (PANalytical, the Netherlands, v2.2b) software.

2.3 Powders size distribution using BET analysis

The distribution of powder particle sizes was measured using a Mastersizer 2000 (Malvern Instruments). To prepare samples for measurement, approximately 0.5 g of powder was ultra-sonicated in 800 mL of water for 2 min. The measurements were made over the size range of 0.02 – 2000 μm. The reported values are the means of three measurements taken for each sample.

3. Results and discussions

3.1 Microstructural analysis

X-ray diffraction (XRD) analysis is shown in the Fig.1 indicates all peaks are related to the hydroxyapatite as data obtained from JCPCDF 09-0432. There is no trace of CaO and phosphate present in the XRD pattern. It is suggested that high purity nano crystalline hydroxyapatite (HA) powders can produced by the suspension precipitation technique, which is not possible in any other methods.

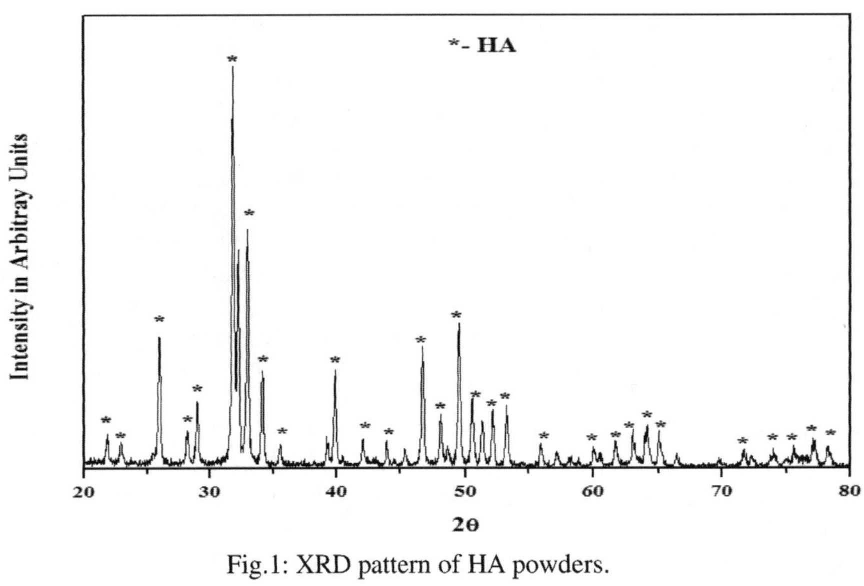

Fig.1: XRD pattern of HA powders.

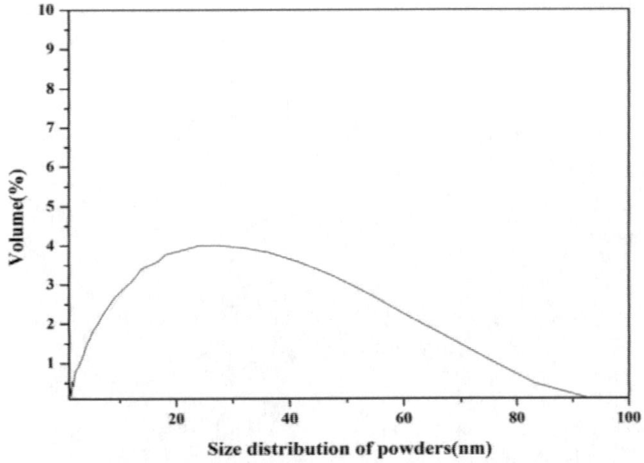

Fig. 2: Size distribution of hydroxyapatite powders.

The presence of agglomerated powders of HA is revealed by the bimodal size distribution, as shown in Fig.2. Secondary electron SEM (SE-SEM) images of HA powders are presented in Fig.3 (a & b). The images reveal considerable agglomeration between the particles, which increases the size of the crystalline powders.

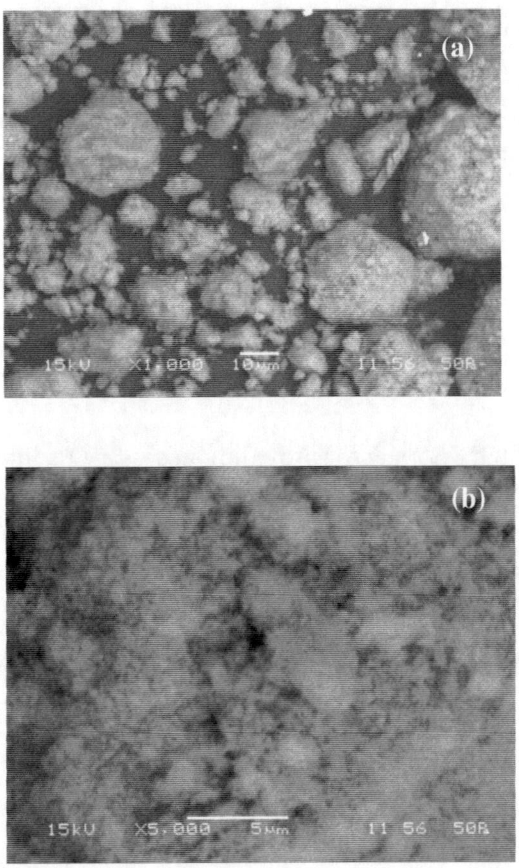

Fig. 3 (a & b): SEM images show the morphology of hydroxyapatite powders.

In the Energy-dispersive X-ray spectroscopy (Fig.4), compositional analysis of powders reveals the presence of Ca, O and P. While the larger peaks of O that of the comparatively smaller peaks reveal peaks of Ca and P. From the table 1, it can observed that stoichiometry ratio of Ca/P is approximately 1.67, which indicates the stability of hydroxyapatite will be enhanced in further processing of the powders. It is clearly indicated that the purity of HA powders without any impurities can be synthesised using the suspension precipitation technique.

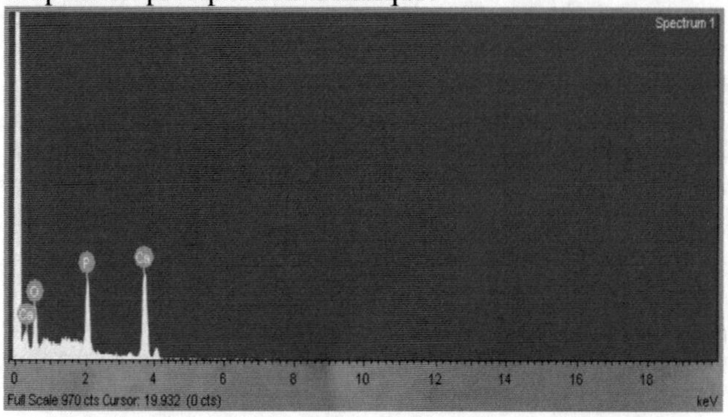

Fig. 4: EDS analysis of hydroxyapatite powders.

Table 1: Shows EDS analysis of the atomic % of Ca, P and O present in the powders.

Element	App	Intensity	Weight%	Weight%	Atomic%
	Conc.	Corrn.		Sigma	
O K	11.46	0.5645	44.99	2.91	65.08
P K	12.05	1.4386	18.59	1.40	13.89
Ca K	16.63	1.0136	36.41	2.14	21.02
Totals			100.00		

4. Conclusion

From the above study, the following conclusion would be derived that the XRD, SEM and BET analysis indicates the nano crystalline of hydroxyapatite (HA) powders could be produced by controlling the temperature and pH of the suspension. EDS analysis indicates composition of Hydroxyapatite (HA) is in the stoichiometry ratio of Ca/P is ~1.65.

Acknowledgement

The authors thank the IGIT Sarang for providing necessary facilities.

References

[1] Hench L.L1998 *J.Am.Ceram.Soc.* **81(7)** 1705.

[2] Ducheyne P, Qiu Q 1999 *Biomaterials* **20** 2287.

[3] Jarcho M Dolen C.H, Thomas M. B, Bobick J, Kay J. F, and Doremus R. H. 1976 *J. Mater. Sci.* **11** 2027.

[4] Trombe J. C, Montel G 1978 *Journal of Inorganic and Nuclear Chemistry* **40(1)**15.

[5] Fang Y,Agrawal D. K Roy D. M, Roy R 1995 *Materials Letters*, **23(1-3)**147.

[6] Nath S, Biswas K, Wang K, Bordia R. K, Basu B 2010 *Journal of the American Ceramic Society* **93 (6)** 1639.

[7] Ignjatović N, Savić V, Najman S, Plavšić M,Uskoković D2001 *Biomaterials* **22 (6)** 571.

[8] Goller G, Demirkıran H, Oktar F. N, Demirkesen E 2001 *Ceramics international* **29 (6)**721.

[9] Santos M. H, M. Oliveira L. P. F. Souza, Mansur H. S, Vasconcelos W. L 2004 *Materials Research* **7(4)** 625.

IOP Conf. Series: Materials Science and Engineering **115** (2016) 012026 doi:10.1088/1757-899X/115/1/012026

Fabrication and Characterization of novel $W_{80}Ni_{10}Nb_{10}$ alloy produced by mechanical alloying

R. Saxena, A. Patra[1], S. K. Karak, A. Pattanaik, S. C. Mishra

Department of Metallurgical and Materials Engineering, National Institute of Technology, Rourkela-769008, Orissa, India

Corresponding author Email : patraa@nitrkl.ac.in/anspat.met@gmail.com

Abstract: Nanostructured tungsten (W) based alloy with nominal composition of $W_{80}Ni_{10}Nb_{10}$ (in wt. %) was synthesized by mechanical alloying of elemental powders of tungsten (W), nickel (Ni), niobium (Nb) in a high energy planetary ball-mill for 20 h using chrome steel as grinding media and toluene as process control agent followed by compaction at 500 MPa pressure for 5 mins and sintering at 1500°C for 2 h in Ar atmosphere. The phase evolution and the microstructure of the milled powder and consolidated product were investigated by X-ray diffraction (XRD), Scanning electron microscopy (SEM) and Transmission electron microscopy (TEM). The crystallite size of W in $W_{80}Ni_{10}Nb_{10}$ powder was reduced from 100 µm at 0 h to 45.6 nm at 10 h and 34.1 nm at 20 h of milling whereas lattice strain increases to 0.35% at 20 h of milling. The dislocation density shows sharp increase up to 5 h of milling and the rate of increase drops beyond 5 to 20 h of milling. The lattice parameter of tungsten in $W_{80}Ni_{10}Nb_{10}$ expanded upto 0.04% at 10 h of milling and contracted upto 0.02% at 20 h of milling. The SEM micrograph revealed the presence of spherical and elongated particles in $W_{80}Ni_{10}Nb_{10}$ powders at 20 h of milling. The particle size decreases from 100 µm to 2 µm with an increase in the milling time from 0 to 20 hours. The crystallite size of W in milled $W_{80}Ni_{10}Nb_{10}$ alloy as evident from bright field TEM image was in well agreement with the measured crystallite size from XRD. Structure of W in 20 h milled $W_{80}Ni_{10}Nb_{10}$ alloy was identified by indexing of selected area diffraction (SAD) pattern. Formation of NbNi intermetallic was evident from XRD pattern and SEM micrograph of sintered alloy. Maximum sinterability of 90.8% was achieved in 20 h milled sintered alloy. Hardness and wear study was also conducted to investigate the mechanical behaviour of the sintered product. Hardness of $W_{80}Ni_{10}Nb_{10}$ alloy reduces with increasing load whereas wear rate increases with increasing load. The evaluated hardness value in the present study for all loads is lower than the literature reported hardness of nanostructured W.

1. Introduction

Tungsten is a candidate material for electrical, electronic, nuclear and space vehicle application owing to highest melting point (3410°C) and density (19.3 g/ml) of all engineering materials, excellent mechanical strength at elevated temperature, and tensile elastic modulus of 411 GPa. [1]. However poor fabricability and high ductile-brittle transition temperature of tungsten limited the wide range of application. In order to improve the properties and to increase the spectrum of applicability, much effort has been directed towards the development of tungsten alloys in recent years. The work is

[1] patraa@nitrkl.ac.in/anspat.met@gmail.com (Prof. A. Patra)

essentially centered on the goals such as to improve fabricability, particularly at elevated temperatures and to lower the ductile brittle transition temperature of W based alloys. As a result of increasing demand for higher and better mechanical properties, amorphous and nanocrystalline tungsten alloys have attracted enormous attention in recent years [2, 3]. Mechanical properties of nanostructured materials are known to be improved by the refinement of microstructures by mechanical alloying which is a convenient solid state synthesis method alternative to melt spinning and similar rapid quenching techniques to develop crystalline and amorphous alloys [4]. Among the various solid state methods, high-energy ball milling has gained quite popularity in recent years because of its simplicity, ease scale up production, high productivity, relatively inexpensive equipment, and applicability to a wide variety of materials [5]. Addition of nickel facilitates liquid phase sintering if sintering is carried out at a temperature higher than the melting point of Nickel (1455°C) whereas Nb addition improves high temperature response of W [1].

Present investigation aims at lowering the sintering temperature of $W_{80}Ni_{10}Nb_{10}$ alloy by formation of nanostructure by mechanical alloying prior to sintering. Characterization of milled and consolidated $W_{80}Ni_{10}Nb_{10}$ alloy was also carried out.

2. Materials and methods

A planetary ball mill (Fritsch Pulverisette P5) was used to mill elemental W, Ni, Nb powder at a mill speed of 300 r.p.m using chrome steel as grinding media and ball to powder weight ratio of 10:1. Milled samples were taken out after 1, 5, 10, 15, 20 h for characterization purpose. The details of the selected alloy and milling parameters are presented in the table 1.

Table 1. Details of alloy composition and milling parameters

Composition (wt.%)	Grinding Medium	Ball to powder weight ratio	Mill Speed (rpm)	Milling Duration (h)	Milling Medium
$W_{80}Ni_{10}Nb_{10}$	Chrome Steel	10:1	300	20	Toluene

A high resolution X- ray diffractometer (Make: Rigaku Japan, Model: Ultima IV) was employed to record the X-ray diffraction pattern (XRD) of the mechanically alloyed powders at different stages of milling as well the sintered alloy using Cu-kα radiation (λ=1.541874 Å). Evolution of phases during mechanical alloying was detected by matching the XRD results with JCPDS data bank [6]. Measurement of the crystallite size and lattice strain was carried out by evaluation of the peak position and broadening of peak from the X- ray diffraction pattern [7].

Dislocation density of the milled powder was calculated from the equation below:

$$\rho_{\rm d} = 2\sqrt{3}\frac{\left(\varepsilon^2\right)^{1/2}}{D \, X \, b} \qquad (1)$$

Where, b is the burgers vector of dislocations, b = (a√3)/2 for the bcc structure, a= cell parameter = lattice parameter, D = crystallite size, ε = lattice strain, [8].

The lattice parameter was measured from the X-ray diffraction pattern by stripping of $K\alpha_2$ of XRD pattern using precise lattice parameter calculation method [7]. The precise measurement of d at cos (θ) tending to 0 was carried out by extrapolating the d values of the high angle diffraction peaks against the function $\cos^2(\theta)/\sin(\theta)$, $\cos^2(\theta)$ [6]. It was evident that $\cos^2(\theta)$ function displayed the best possible fit. Scanning electron microscope (SEM) (Make: JEOL, Model: JSM-6084LV) was used to investigate the morphology of the mechanically alloyed (MA) powders at different periods of milling and the microstructure of the sintered product. A test tube containing the 20 h milled powder was dissolved in acetone and place in an ultrasonic vibrator for vigorous stirring. The powder was positioned on a carbon grid before TEM study. Detection of crystallite size of nanostructured W in mechanically alloyed $W_{80}Ni_{10}Nb_{10}$ powder at 20 h of milling was carried out by transmission electron microscope (TEM) (Make : JEOL, Japan, Model- JEM 2100). Crystal structure was identified by selected area diffraction (SAD) pattern using appropriate aperture and tilt. The 20 h milled powders were subjected to compaction in a uniaxial press at 500 MPa pressure for 5 mins followed by sintering at 1500°C for 2 h with Argon purging at the rate of 100 ml/min. Microhardness of the sintered product was evaluated by a Microhardness tester (Make: Leco, Model: LM248AT) with varying load of 50 gf, 100 gf, 500 gf, 1000 gf for 10 sec. A ball on plate wear tester (Make: Ducom, Model: TR-208-M1) was employed to determine the wear depth with sliding distance of the indenter on the sintered sample with 20 N and 30 N load. The sliding speed and time was kept at 25 r.p.m and 10 mins respectively.

3. Result and Discussion

3.1. X-ray Diffraction (XRD) analysis of milled powder

Figure 1 shows the XRD patterns of as milled $W_{80}Ni_{10}Nb_{10}$ powder at different milling times (1, 5, 10, 15, 20 h). Refinement of crystallite size of W with increase in milling time is due to continuous broadening and reduction in the intensity of the XRD peaks. Formation of solid solution (Ni and Nb undergoes into W matrix) with increase in milling time is evident from figure 1. Decrease in crystallite size and increase in plastic strain is due to the increase in full-width at half-maximum with increasing milling time [9]. Scherrer equation was used to analyse the peak broadening after removing the strain and instrumental error components [6].

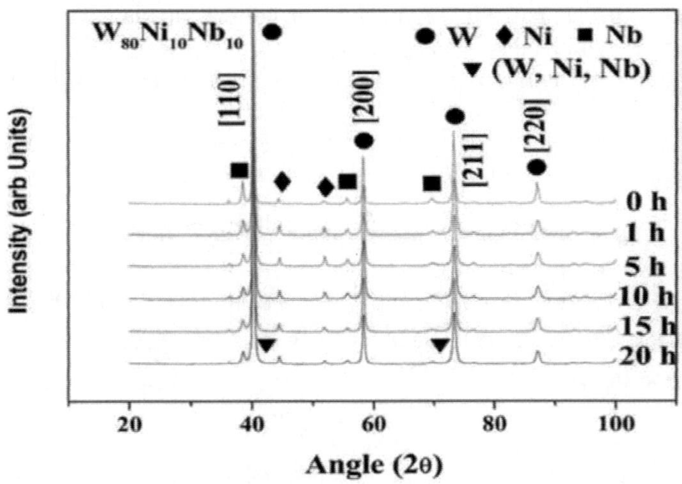

Figure 1. XRD pattern of $W_{80}Ni_{10}Nb_{10}$ alloy milled for 20 h.

The crystallite size and lattice strain was calculated by using Williamson–Hall equation [10] as:

$$\beta \cos\theta = \frac{0.94\lambda}{D} + 4\eta \sin\theta \qquad (2)$$

where, β is the full width half maxima (FWHM), D is the crystallite size and η is the lattice strain.

Figure 2 displays the variation of crystallite size and lattice strain with milling time. The crystallite size is 34 nm at 20 h of milling for $W_{80}Ni_{10}Nb_{10}$ composition. Recent literature suggest that the crystallite size of W based alloys can be reduced more (30 nm) than the present study at 20 h of milling using tungsten carbide (WC) grinding media during milling [1]. Tungsten carbide possess higher hardness as compared to chrome steel resulted in more reduction in crystallite size [1].

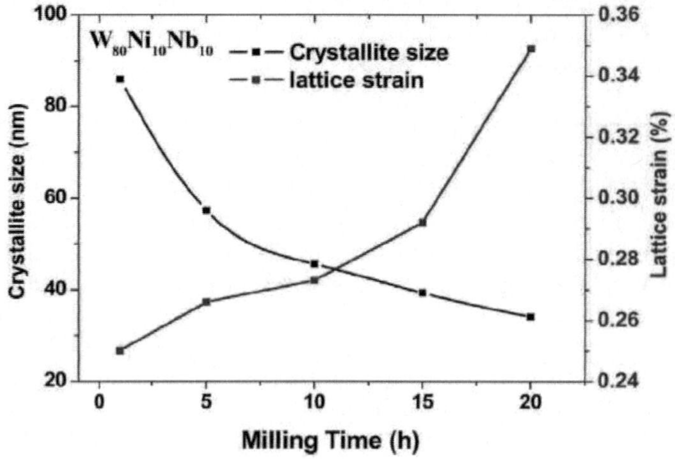

Figure 2. Variation of crystallite size and lattice strain of W with milling time in $W_{80}Ni_{10}Nb_{10}$.

Figure 3 (a) illustrates that dislocation density induced in $W_{80}Ni_{10}Nb_{10}$ alloy increases sharply at 5 h of milling, however moderate increase is evident beyond 10 h of milling. Expansion of lattice parameter of W upto 0.04% at 10 h and contraction upto 0.02 % beyond 10 to 20 h of milling is evident from figure 3 (b). Recent report suggests that exertion of negative hydrostatic pressure due to the formation of nano-crystallites during mechanical alloying of pure Nb resulted in significant expansion of Nb lattice [11]. The present investigation also displays a similar trend. Nb and Ni form substitutional solid solution with W matrix. The atomic radius of W (0.193 nm) is larger than Ni (0.149 nm) and smaller than Nb (0.198 nm). Therefore the resultant atomic radius shows a decrease beyond 10 h of milling [12]. Predominant reduction in crystallite size at 10 h of milling is responsible for sharp increase in dislocation density whereas formation of solid solution (alloying of Nb and Ni with W) dominates over crystallite size reduction beyond 10 to 20 h of milling which results in marginal increase in dislocation density and decrease in lattice parameter. However the decrease in lattice parameter in the present study is lower than recent investigation on mechanically alloyed $W_{80}Ni_{10}Mo_{10}$ as atomic radius of both Ni (0.149 nm) and Mo (0.190 nm) is lower as compared to the atomic radius of W [13].

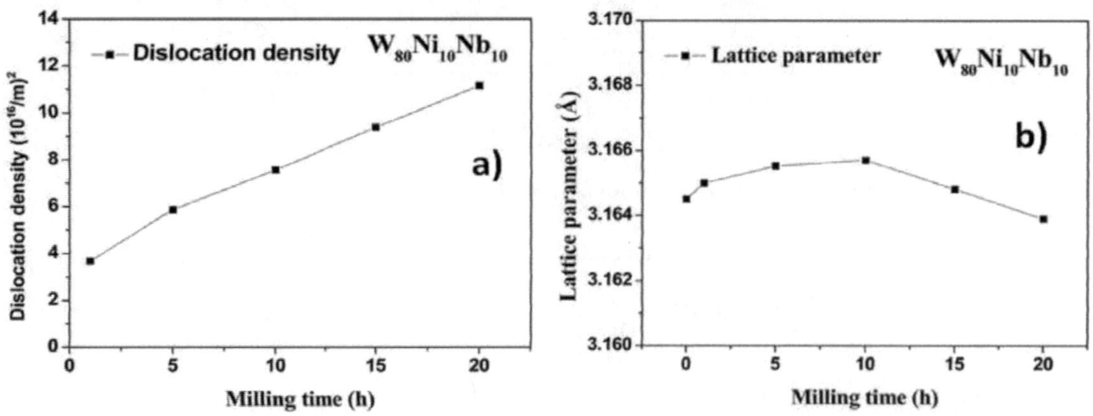

Figure 3. Variation of a) Dislocation density and b) Lattice parameter of W in $W_{80}Ni_{10}Nb_{10}$ with milling time.

IOP Conf. Series: Materials Science and Engineering **115** (2016) 012026 doi:10.1088/1757-899X/115/1/012026

3.2. Scanning Electron Microscope (SEM) analysis of milled powder

Figure 4 (a-d) displays the variation of particle morphology and size of the milled powder at different milling time. The particle size changes from ~100 μm to ~2 μm with increase in milling time. The particles are continuously strain hardened and fractured with increase in milling time. The spherical shape of W, Ni and Nb particles at 0 h change to elongated shape at 20 h of milling. Welding of some particles is also evident at 10 h of milling (figure (4c)). The elongated nanoparticles at 20 h of milling results in poor flowability due to large interparticle friction therefore responsible for poor green densification property [13].

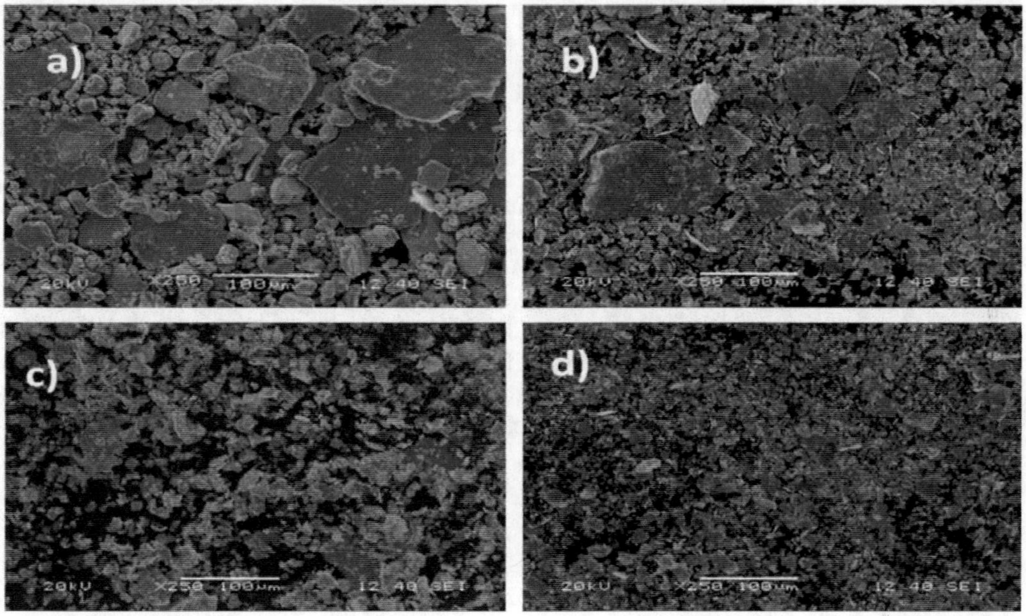

Figure 4. SEM images of powder morphology of $W_{80}Ni_{10}Nb_{10}$ alloy after different milling times: (a) 0 h, (b) 5 h, (c) 10 h, and (d) 20 h.

3.3. Transmission Electron Microscope (TEM) analysis of milled powder

Figure 5 (a) and (b) displays bright field TEM image along with corresponding SAD pattern of $W_{80}Ni_{10}Nb_{10}$ powder milled for 20 h. Formation of nano-crystallites at 20 h of milling with size varying from 20 to 30 nm is evident from figure 5(a). The observed crystallite size nearly matches with the crystallite size derived from XRD analysis (figure 2). Continuous rings evident from SAD pattern confirm the presence of fine polycrystallites. Indexing of SAD pattern shows the presence of (110), (200), (222) planes of W. The measured interplaner spacing (d) values of BCC-W are 0.22 nm for (110), 0.15 nm for (200), and 0.09 nm for (222) respectively. The interplaner spacing (d) values matches well with the standard values of BCC-W from JCPDS data files.

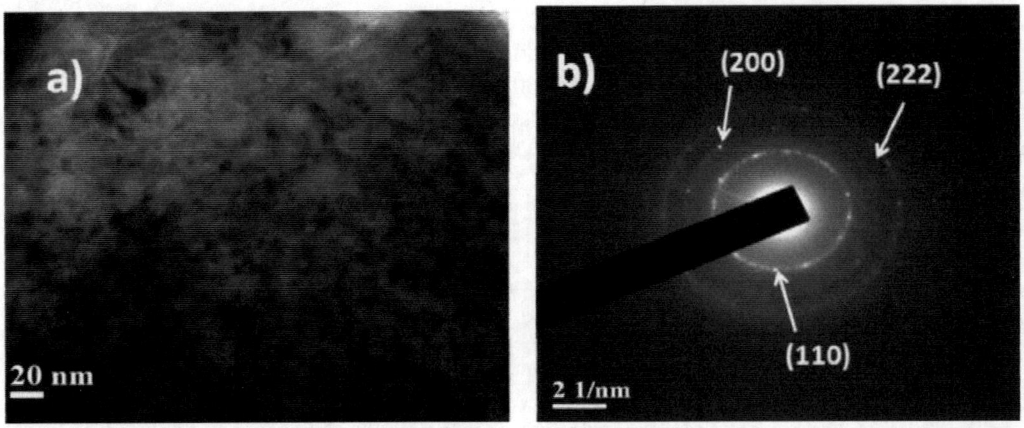

Figure 5. TEM image of $W_{80}Ni_{10}Nb_{10}$ powder milled at 20 h: (a) Bright field TEM image and (b) corresponding SAD pattern.

3.4. X-ray Diffraction (XRD) analysis of sintered alloy

Figure 6 displays the XRD pattern of sintered $W_{80}Ni_{10}Nb_{10}$ alloy. Formation of brittle hard intermetallic NbNi is evident from the XRD pattern. The strength of the W matrix could be enhanced by the presence of NbNi intermetallic. Formation of NbNi intermetallic is also reported by Li et al. [14]. Any unwanted phase transition apart from formation of NbNi intermetallic is not evident from the XRD pattern.

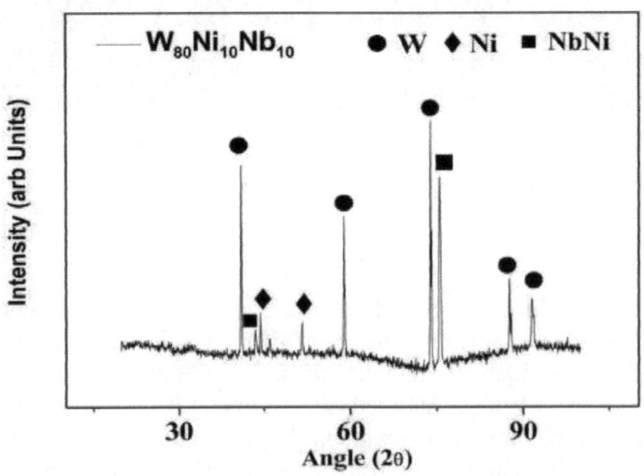

Figure 6. XRD pattern of $W_{80}Ni_{10}Nb_{10}$ alloy milled for 20 h and sintered at 1500°C for 2 h.

3.5. Scanning Electron Microscope (SEM) analysis of sintered alloy

Figure 7 displays the SEM micrograph of $W_{80}Ni_{10}Nb_{10}$ alloy sintered at 1500°C for 2 h. The micrograph reveals the presence of dark phase (Ni) in W matrix (bright phase). The formation of NbNi intermetallic phase as evident from figure 7 is due to the lower melting point of Ni (1455°C) as compared to the sintering temperature (1500°C) selected for the present investigation. The phases evident from SEM micrograph match well with the identified phases from the XRD analysis.

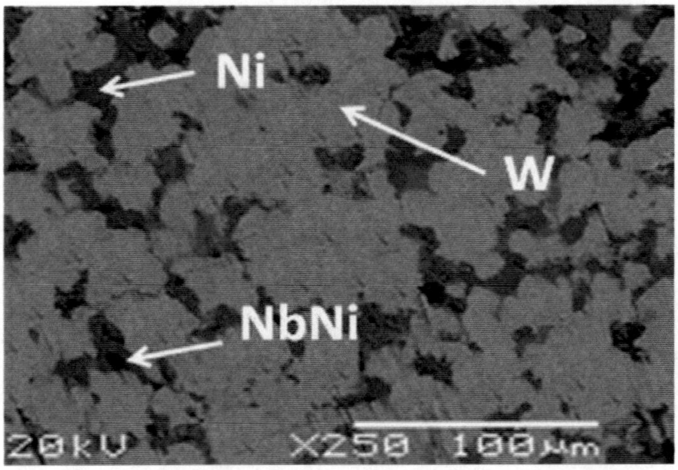

Figure 7. SEM micrograph of 20 h milled $W_{80}Ni_{10}Nb_{10}$ alloy sintered at 1500°C for 2 h.

3.6. Sinterability study

Sinterability is defined as the percentage ratio of sintered sample density to theoretical density of the sintered sample. Sintered alloy density can be calculated by Archimedes' principle [15] as below:

$$\rho_s = \frac{W_a}{(W_{sat} - W_{susp})} \; X \, \rho_w \frac{\text{gm}}{\text{cm}^3} \qquad (3)$$

Where W_a is weight of the sintered sample in air. W_{sat} is the weight of the sample with all the open porosity saturated with water, W_{susp} is the weight suspended in water. ρ_w is the density of water. The sinterability of the 20 h milled sample sintered at 1500°C for 2 h is recorded 90.8% by pressing at 500 MPa pressure for 5 mins.

3.7. Hardness study

The microhardness can be measured by using the equation as [16]:

$$\text{HV} = 1.8544 \frac{P}{d^2} \qquad (4)$$

where P is the applied load and d is the diagonal length of the indentation.

Figure 8 shows the variation in hardness along the cross section of the $W_{80}Ni_{10}Nb_{10}$ alloy with varying load. It is evident from the figure that hardness increases with decrease in applied load. This is attributed to appreciable elastic recovery at lower load and error in appropriate detection of diagonal length of the indentation due to smaller indentation at lower loads [17]. The variation in hardness for a specific load at different point of the sample may be due to variation in composition through the cross section. The average hardness for all loads is lower than recently investigated nanostructured W due to softening of W lattice by addition of Ni and Nb [1].

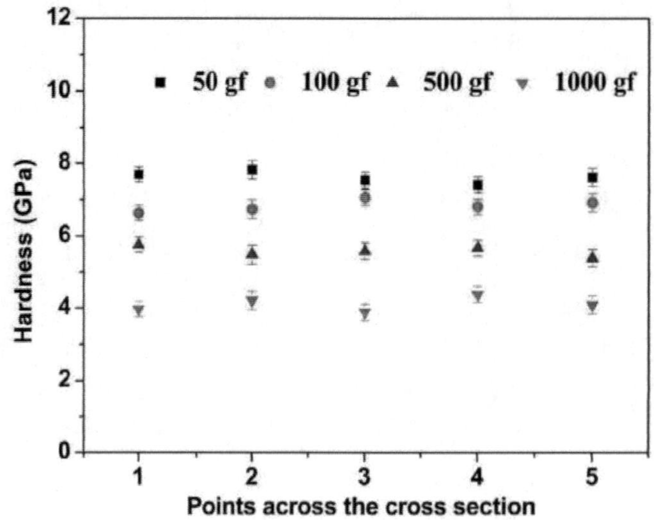

Figure 8. Variation of Hardness of $W_{80}Ni_{10}Nb_{10}$ alloy milled for 20 h and sintered at 1500°C for 2 h.

3.8. Wear study

Figure 9 illustrates the variation of wear depth with sliding distance traversed by the indenter on the sample surface. The sliding distance can be calculated as [18]:

$$\text{Sliding Distance (S. D)} = \left(\frac{R}{60}\right) \times t \times 2\pi r \quad (5)$$

where, R is the number of rotation per minute traversed by the indenter on sample surface, t is time in sec (600 s), r is track radius (4 mm) measured from the center of the sample to the track. It is evident from figure 9 that wear depth increases with increasing load. The results can be attributed to pushing out the intermetallic particles from matrix to the wear track during abrasion which can eventually enhance the wear process. The wear track width also increases with increase in the load as revealed from figure 10. Nuthalapati et al. have reported that three body motion mechanism is responsible for increase in the wear due to increase in abrasion effect at higher load [19].

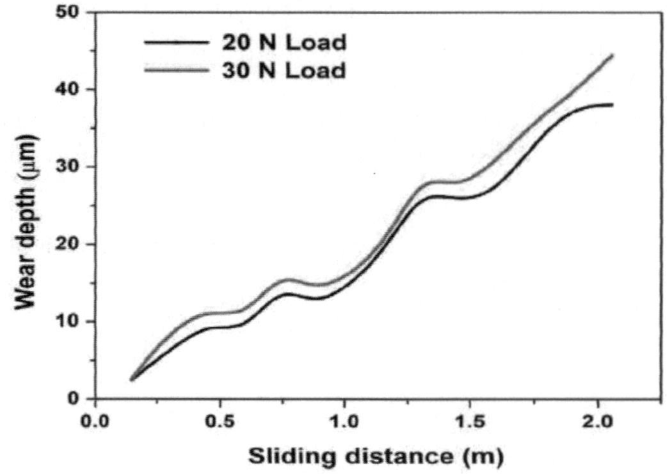

Figure 9. Variation of wear depth of $W_{80}Ni_{10}Nb_{10}$ alloy milled for 20 h and sintered at 1500°C for 2h.

Wear rate at maximum wear depth for both 20 N and 30 N load was evaluated by using archard equation [20]. Wear rate is the wear volume per unit load per sliding distance. The measured wear rate at maximum wear depth was recorded 4.01 ($\times 10^{-15}$) m^3/Nm for 20 N load and 4.43 ($\times 10^{-15}$) m^3/Nm for 30 N load respectively.

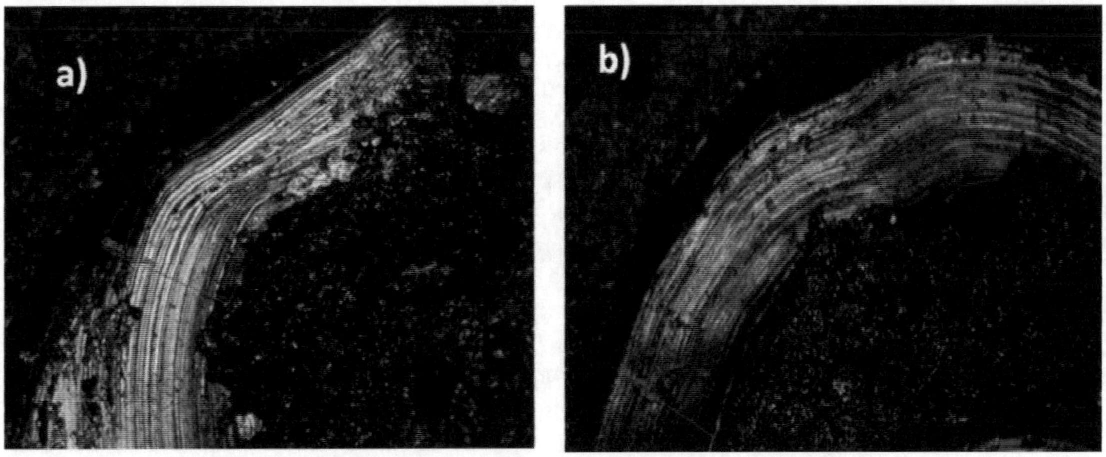

Figure 10. Micrograph of wear track at a) 20 N b) 30 N load.

4. Conclusions

From the fabrication, microstructural characterizations and evaluation of mechanical properties of the investigated alloy, the following conclusions can be drawn:

- $W_{80}Ni_{10}Nb_{10}$ alloy powder can be effectively produced by mechanical alloying.
- Crystallite size of tungsten in $W_{80}Ni_{10}Nb_{10}$ reduces with increasing milling time and minimum crystallite size of 34 nm is achieved at 20 h of milling.
- Expansion of the lattice parameter of tungsten in $W_{80}Ni_{10}Nb_{10}$ records 0.04% at 10 h and contraction upto 0.02% at 20 h of milling.
- Bright field TEM image and corresponding SAD pattern reveals the presence of nanocrystalline BCC-W phase with 20-30 nm in size at 20 h of milling.
- Presence of NbNi intermetallic is evident in the sintered alloy due to Ni rich liquid phase formation at the selected sintering temperature.
- Microhardness value increases with decrease in load due to smaller indentation and detection error of the diagonals of the indentation at lower loads.
- Wear rate at maximum wear depth increases with increase in load due to enhanced abrasion effect at higher load.

References

[1] Patra A 2012 *Microscopy and Analysis* **26 (5)** 14-15.
[2] Glieter H 2001 *Acta. Mater.* **48** 1.
[3] Koch C 1997 *Nanostruct. Mater.* **9** 13.
[4] Suryanarayana C 1998 *Mechanical Alloying* ASM Handbook **7** 80.
[5] Suryanarayana C 2001 *Prog. Mater. Sci.* **46** 1-184.
[6] Cullity B 1978 *Elements of X-ray diffraction* **2** (Addison Wesley: Philippines) 86-88.

[7] Williamson G and Hall W H 1953 *Acta. Mater.* **1** 22.
[8] Zhao Y H, Sheng H W and Lu K 2001 *Acta. Mater.* **49** 365–375.
[9] Nuthalapati M, Karak S K, Majumdar J D and Basu A 2014 *Metall. Mater. Trans. A* **45A** 3748-3754.
[10] Jenkins R and Snyder R L 1996 *Introduction to X-ray Powder Diffractometry* (J Wiley & Sons Inc: New York).
[11] Chattopadhayay P, Nambissan P M G, Pabi S K and I Manna 2001 *Appl. Surf. Sci.* **182** 308-312.
[12] Singh V 2002 *Physical Metallurgy* **1** (Standard Publisher Distributor: New Delhi) 63-64.
[13] Patra A, Karak S K and Pal S 2015 *Proc. Nat. Conf. on Processing and Characterization of Materials (Rourkela)* vol 75 (Bristol: UK/IOP Science) 012032.
[14] Li X, Cheng Z, Hu K, Chen H and Yang C 2015 *Vacuum* **114** 93-100.
[15] Ozkan N and Briscoe B J 1994 *J. Eur. Ceram. Soc* **14** 143-151.
[16] Ayers L J 2012 *The Hardening of Type 316 L Stainless Steel Welds with Thermal Aging* Massachusetts Institute of Technology.
[17] Dieter G 1928 *Mechanical Metallurgy* SI Metric Edition (McGraw-Hill Co: Singapore) 189-93.
[18] Patra A, Karak S K and Pal S 2015 *Microscopy and Analysis* **29(4)** 18-22.
[19] Nuthalapati M, Karak S K and Basu A 2015 *Materials Today: Proceedings* **2** 1109-1117.
[20] Archard J F 1953 *J. Appl. Phys* **24 (8)** 981-988.

Scheming of microwave shielding effectiveness for X band considering functionalized MWNTs/epoxy composites

SMRUTISIKHA BAL[1], SUNIRMAL SAHA[2]

[1]Department of Metallurgical and Materials Engineering, National Institute of Technology Rourkela, India.
[2]Department of Physics and Astronomy, National Institute of Technology Rourkela, India.

Corresponding Author:
Smrutisikha Bal, Department of Metallurgical and Materials Engineering, NIT Rourkela-769008, Odisha, India.

Email:* balss@nitrkl.ac.in, Telephone number: 08763115908.

Abstract. Present typescript encompasses anextraordinary electrical and mechanical behaviors of carboxylic (–COOH) functionalized multiwall carbon nanotube (MWNTs)/epoxy composites at low wt.% (0.5, 0.75, 1wt.%). Functionalization on the surface of the nanotube assists MWNTs in dispersing it into epoxy polymer in a respectable manner. Fabricated composites are exposed to different characterization techniques in order to examine the overall physical properties. Microwave shielding effectiveness (SE) for X band (8-12 GHz) and the flexural properties have been premeditated to predict the electrical and mechanical performances.It was found that the total SE of the nanocomposites was increased with the positive gradient of MWNT contents. The best result was recorded for 1 wt.% MWNT loading (SE of about 51.72 dB).In addition, incorporation of nanofillers enhanced the flexural modulus, flexural strength and micro-hardness of the resulting composites while comparing with neat epoxy. Nanocomposites with 0.75 wt.% MWNT loading demonstrated an incrementof 101% in modulus than that of neat epoxy. Theincrement in mechanical properties was due to achievement of good dispersion quality, effective bonding between MWNTs and epoxy polymer analyzed by micrographs of fracture surfaces.

Keywords:MWNT/epoxy composite; functionalization; shielding effectiveness; flexural modulus; hardness

1. Introduction

Polymer-based composites are gaining more importance compared to metals and other commercial materials due to their light weight, low cost and easy processing [1]. Since the discovery of carbon nanotubes (CNTs), polymer-based composites find a new dimension for the application of this material in various structural parts, automobiles, civil engineering, electrical circuits etc. CNTs being such nanofiller combined with exceptional material properties like high specific strength and stiffness, high aspect ratio, high flexibility and high electrical conductivity. Further CNTs possesses diameter dependent specific surface area that offers available surface for interaction while incorporated into polymermatrix [2]. This interaction in fact is responsible in order to modulate the matrix properties in a significant manner.

However, using CNTs(MWNTs) as reinforcing agents in polymer matrix can be apretty challenging task. Because, the presence of strong inter-tube interaction offered by the van der Waals' forces creates a genuine difficulty to achieve qualitative dispersion of MWNTs into polymer. Till date, a number of techniques are available to overcome the problem of dispersionof CNTs into polymer. These include mechanical stirring, magnetic stirring, Ball milling, Calendering process, stir and extrusion, ultra sonication and high shear mixing etc. [3, 4]. The method of sonication among them has been most commonly used to disperse MWNTs in polymer like epoxy resin. Now the fact is that like dispersion interfacial adhesion between the CNTs and polymer like epoxy is also a critical issue tailoring which one can have sufficient stress transfer from the matrix to the MWNTs. That is, a strong interfacial adhesion between MWNTs and polymer is highly desirable for the effective performance of the resulting composites. A number of researcheshave been worked since a decade to improve this interfacial adhesion between MWNTs and polymer. According to them, attachments of some functional groups on the surface of MWNTs not only improve the adhesion rather helps in achieving uniform dispersion of MWNTs into polymer [4].

Now, in the present era, careful investigation reveals a critical need to develop versatile and effective novel materials that would suit to applications like EMI (Electromagnetic Interference) shielding and ESD (Electrostatic Discharge). Current material options that provide active SE are metals, metal-fibre filled plastics, polyacrylonitrile (PAN) nickel coated reinforced polymers, aluminium structures, nickel and copper metalized fabrics, and more recently, nano reinforced polymer composites (NRPCs) [5, 6]. Typical limitations found with materials used for shielding to prevent EMI are associated with corrosion susceptibility, lengthy processing times, high equipment cost for production and difficulty of material utilization mainly as conductive layers. NRPCs, like carbon nanotube (CNTs) reinforced polymer matrices seem to overcome some of these limitations because they are lightweight materials with design flexibility, corrosion resistant and suitable for mass production through conventional plastic manufacturing technologies such as extrusion and injection moulding [1].
Though a quantitative number of literatures are available on EM shielding i.e. SE of the non-functionalized MWNTs/polymer composites, a little work has been reported on SE of functionalised MWNTs/polymer composites. For example, polyaniline /functionalized coated MWNTs have been added to polymer matrices for EMI shielding materials and tested in the frequency range of 8.2-12.4 GHz (X band). According to their report, SE was about 23.3 dB for 7% MWNTs loading in polystyrene (PS) [7]. S.H. Park et al. [8] worked on the –COOH attached single walled CNT/RET polymer and they got 9-25 dB of SE taking 0.5-4.5 wt.% of fillers content. Grimes et al. [9] reported that SWNT-polymer composites possess high real permittivity as well as imaginary permittivity in the 0.5-2GHz range. They found that the permittivity decreases rapidly with increase in frequency.

In this present schoolwork the novelty lies with a different curing mechanism at the time of fabrication compared to the conventional techniques. Fabricating very low wt.% –COOH functionalized MWNT/epoxy composites, a details study of the EMI shielding has been performed in the X band frequency range (8-12 MHz). In addition, mechanical studies have been performed to observe the overall strength of the fabricated composites. So the prime goal here is to produce an effective shielding material along with the significant mechanical properties that can be used in marine, military, scientific electronic devices and communication instruments.

2. Materials and methods

In this present study, –COOH functionalised MWNT as reinforcing component and epoxy resin (Ciba-Geigy, araldite LY-556 based on bisphenol A, India) as matrix material have been used for the synthesis of the nanocomposites. Functionalised MWNTs were commercially purchased from NanoAmour, USA; having diameter of around 20–30 nm and lengths of about 10–30 μm with a carbon purity of > 95%. During fabrication epoxy resin (5.3-5.4 equiv/kg) has been cured with an aliphatic primary amine hardener (HY-951) in the wt. ratio 100/12.

Functionalised MWNTs with 0.5, 0.75, and 1 wt.% were taken for the fabrication of MWNTs/epoxy nanocomposites. The technique adopted for preparing the composite was solution reduction following ultra-sonication. Prior to the mixing of nanotubes to the epoxy, MWNTs were sonicated with ethanol solution to remove entanglement of the MWNTs bundle. Ethanol was added to epoxy resin also and mixed manually for reducing the viscosity of the epoxy. The obtained untangled solution of MWNTswas then dispersed with epoxy mixture using an ultrasonic processor. This procedure has been adopted following sufficient optimisation. After 6 hours of continuous sonication, the mixture was allowed to keep under vacuum chamber for removing air bubbles, generated due to sonication process for 12 hours. The suspensions were then mixed with the curing agent in the specified wt. ratio and mixed manually for 10 minutes. Thereafter, the final mixture were poured into open moulds and kept at low temperature (4^0C) for 2-3 days for the effective settling purpose. Neat epoxy samples without adding nanotubes were also fabricated in the same method as above for comparison purposes. The obtained cured composite in this process was then post cured at a temperature around 90 ^{0}C for a few hours. A detailed description on the processing of the nanocomposites is given in our earlier work [10]. Prepared composites were designated as follows:

Neat epoxy composite: E_p
Nanocomposite with 0.5 wt.% of functionalized MWNTs loading: $FC_{0.5}$
Nanocomposite with 0.75 wt.% of functionalized MWNTs loading: $FC_{0.75}$
Nanocomposite with 1 wt.% of functionalized MWNTs loading: FC_1

3.Characterisation details

Shielding effectiveness (SE) of all the nanocomposites was experimentally investigated over a frequency range 8-12 GHz (X band), using rectangular waveguide with the help of Agilent PNA-L Network Analyzer (N5230C) and Agilent software 85071E. The operational range of the analyser was 10 MHz-20 GHz. Using S- parameters generated from the instrument, SE of the nanocomposites has been calculated. Method used for real and imaginary parts of the dielectric permittivity was 'Transmission Epsilon Fast' method with the accuracy of 1-2%. Flexural tests were accompanied according to ASTM D790–02 ''Standard Test Methods for Flexural Properties of Un-reinforced and Reinforced Plastics and Electrical Insulating Materials" on a universal testing machine (Instron 5967). The mode of test was three point bending and the rate at which the load was imposed was 2.0 mm/min. The span between two supports, width and thickness of the specimen were60 mm, 5mmand 12.5 mm respectively. In order to reproduce the results five specimens were tested in each group. The hardness of all composite specimens was measured using a micro-hardness tester. The measurement was done by the Vickers hardness test method, which is all about indenting the test material with a diamond indenter. In this method, the hardness is calculated by measuring the load applied and the average diagonal length of the indentation. A total of 10 points on each of the specimens were measured in order to get the average readings. Finally, field emission scanning electron microscope (Nova NANOSEM 450) was used to observe the fracture surface topography of all the composites specimens.

4.Results and discussions

4.1.Electromagnetic shielding effectiveness

The shielding effectiveness SE (dB) is defined as the ratio of power of the incident to the transmitted electromagnetic (EM) wave. It is expressed as [11],

$SE_T\,(dB) = 10\log_{10} P_T/P_I = 20\log_{10} E_T/E_I = \log_{10} H_T/20\,H_I$;--------------(1)

where, P_I (E_I or H_I) and P_T (E_T or H_T) are the power (electric field intensity) of the incident and transmitted EM wave, respectively.
When a microwave radiation is incident on a material surface, its energy is partly reflected, absorbed, and transmitted. The MWNTs/epoxy composites can attenuate the microwave radiation and absorb its energy. According to the classic transmission line theory, the reflectance (R), absorbance (A) and transmittance (T) can be expressed with the scattering parameters S_{11} (S_{22}) and S_{12} (S_{21}) of vector network analyser (VNA) i.e., $T = |E_T/E_I|^2 = |S_{12}|^2 = |S_{21}|^2$, $R = |E_R/E_I|^2 = |S_{11}|^2 = |S_{22}|^2$. The total electromagnetic interference shielding effectiveness (EMI SE) i.e., SE_T is the sum of contributions from absorption (SE_A), reflectance (SE_R), and multiple reflection (SE_M); for example, $SE_T = SE_A + SE_R + SE_M$. When $SE_T < 15$ dB, it is usually assumed that $SE_T = SE_A + SE_R + SE_M$ and when $SE_T > 15$ dB then $SE_T = SE_A + SE_R$ (i.e., SE_M is negligible). These two attenuation factor i.e., SE_R and SE_A can be conveniently expressed as:
$SE_R = 10\log_{10}(1-R)$; --------------------(2) and
$SE_A = 10\log_{10}(1-A) = 10\log_{10}T/(1-R)$;----------------(3)

Here in the fig. 1, EMI SE of various –COOH functionalised MWNTs/epoxy with different loadings of MWNTs (0.5, 0.75, 1 wt.%) in the measured frequency region (8 to 12 GHz, X band) have been portrayed. The neat epoxy showed very less attenuation with $SE_T \approx 0.4$ dB that is why not shown in this fig. It is clearly visible that addition of functionalised MWNTs to the epoxy increased the total shielding effectiveness. In addition, the SE was increased upon increasing the wt.% of MWNTs. The highest EMI SE ≈51.72 dB was provided by the nanocomposite FC_1, whereas, $FC_{0.5}$ and $FC_{0.75}$ showed maximum SE of 51.03 dB and 51.55 dBrespectively. Sung-Hoon Park et al. [8] also reported the increment of SE for the gradual increment of contents of CNTs. They have obtained SE of low wt.% functionalised CNT epoxy composites maximum up to 30 dB in the same frequency range. Saib et al. [12] reported the EMI SE of CNT/polyethylene composites with low wt.% of CNT loading at 11 to 28 dB in a frequency range from 20–40 GHz. Singh et al. [13] measured the EMI SE in the range ≈11-20 dBwith greater wt. % of CNTs (4.2, 15.1, 20.4 wt.%). So it can be concluded that, in our case,

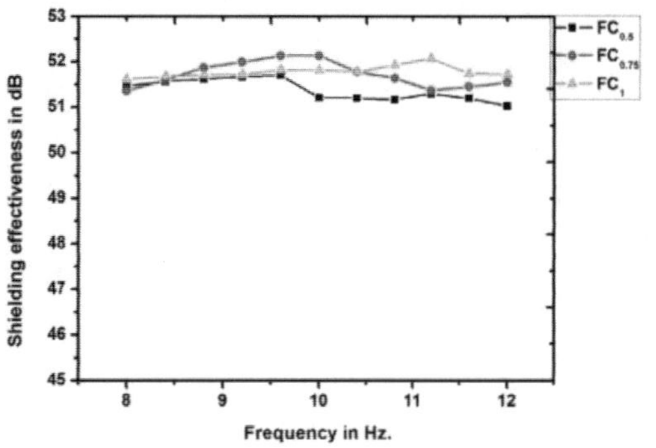

Fig.1:- Total shielding effectiveness of functionalised nanocomposites

inclusion of lower wt.% of MWNTs to the epoxy polymer greatly improved the EM wave attenuation and hence contributed profoundly to the total SE. This kind of enhancement of SE may occur mainly for the achievement of qualitative dispersion of the MWNTs in the epoxy. In addition, formation of strong secondary bonds between MWNTs and the induced functional group may be the cause of such enormous increment.

On the line of investigation, fig. 2 revealed that the contribution of absorption attenuation in total SE was intense than the reflection attenuation for all the cases. This was due to the fact that MWNTs has free electrons in its skeleton, so it can absorb energy and helps in attenuating the microwaves [14]. Further the high surface area of MWNTs also may be a reason for its good microwave absorption behaviour. In addition, the individual MWNT and MWNT bundles in the epoxy composite samples can absorb the microwave energy and attenuate the radiation via the interaction between interior electrons and exterior microwave radiation. It is also well established that compared with any other CNTs like single and double walled CNTs, MWNTs have more defects due to their complicated structures which facilities in greater absorption of EM waves [15]. Apart from these, even if covalent attachment of functional groups to the surface of MWNTs can improve the efficiency of load transfer, it must be noted that these functional groups may introduce defects on the walls of the perfect structure of nanotubes [16]. These defects in turn may help in absorption of EM waves acting as polarization centres.

When EMI shielding materials are exposed in an external EM field, the radiation would induce two kinds of electric currents. One is the displacement current due to the presence of polarization effect including unpaired point defects and polarized interfaces. The displacement current contributes to the real part of permittivity (ε'). The other one is the conduction current due to the presence of free electrons in the material. The conduction current contributes to the imaginary part of permittivity (ε'') and causes dielectric loss [17]. The frequency dependence of permittivity in the current investigation is represented in fig. 3. This frequency dependence reflects the fact that a material's polarization does not respond instantaneously to an applied field. The response must always be causal (arising after the applied field) which can be represented by a phase difference. For this reason, permittivity is treated as a complex function of the frequency of the applied field. The real and imaginary part represents the stored energy in the material and energy dissipation respectively by the material itself. Fig. 3(a) and (b) showed the real and imaginary permittivity spectra of the epoxy as well as nanocomposites

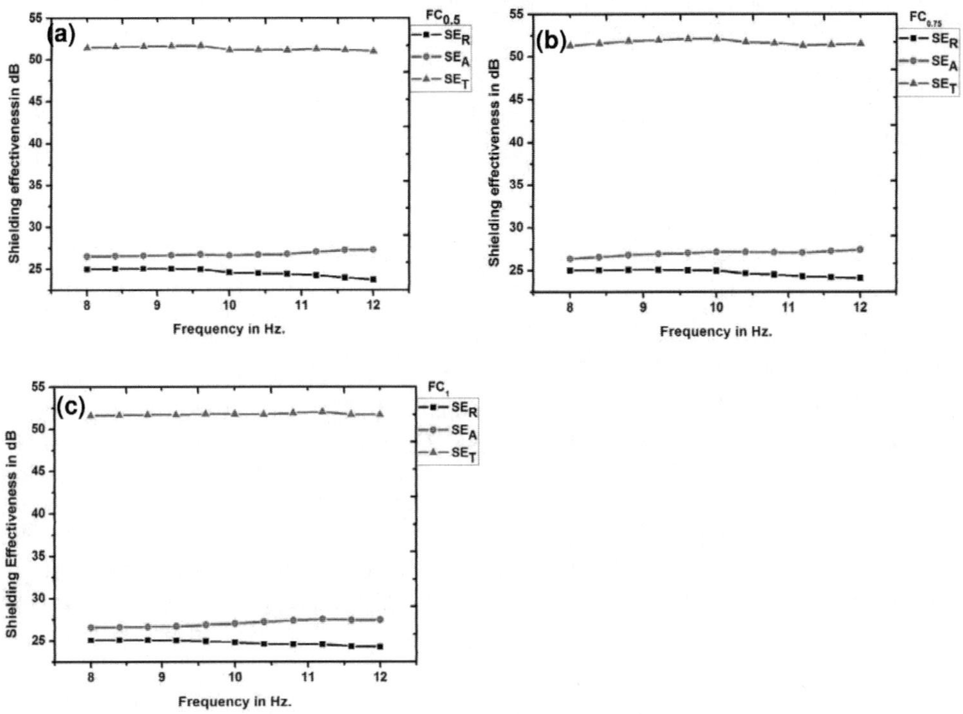

Fig.2:- Shielding effectiveness of (a) $FC_{0.5}$ sample (b) $FC_{0.75}$ sample, (c) FC_1 sample

samples respectively. The spectra showed that ε' (Fig. 3a) value decreased with increase in frequency. This can be ascribed to the inability of the polarization vector to maintain in phase movement with incident high frequency electromagnetic radiation. Further ε'' increases with loading of higher amount of CNT which may be due to the Maxwell-Wegner interfacial polarization [18].

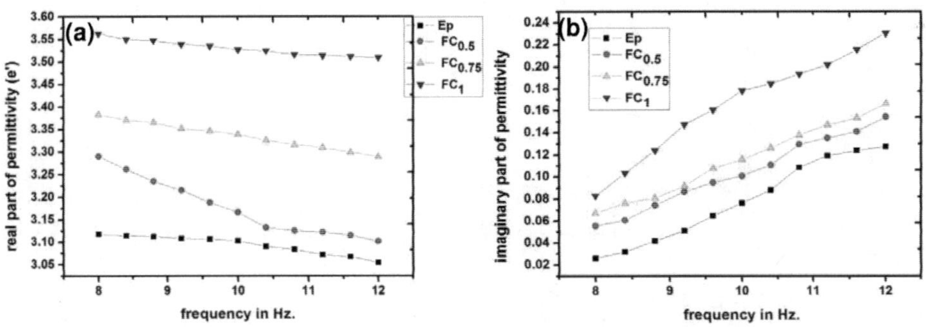

Fig.3:- Real part of permittivity vs. frequency (a) imaginary part of permittivity vs. frequency (b) of epoxy and nanocomposites

4.2.Flexural properties

Fig. 4 illustrates the flexural properties of epoxy and functionalized MWNTs/epoxy composites. The addition of any kind of nanofiller increases the stiffness and strength of the matrix material, as long as a proper impregnation with the matrix and a sufficient adhesion is ensured. Concordantly, in this case, epoxy containing 0.5 and 0.75 wt.% of MWNTs i.e., $FC_{0.5}$ and $FC_{0.75}$ were showing 89% & 101% modulus enhancement in comparison to pure epoxy with modulus 4501 MPa while relatively lower (40%) improvement in modulus was observed in FC_1. The strength values were in the order like $FC_1 < FC_{0.5} < FC_{0.75}$ with 84, 93, 148 MPa respectively in comparison to epoxy having strength of 66 MPa (Fig. 4).The functionalized nanotubes turned out to be more efficient concerning animprovement of the flexural modulus, thus again proving the relevance of surface functionalization. Good dispersion of MWNTs and cross-linking of functional groups present on the surface of MWNTs with polymer that restricts the mobility of polymer chains in turn enhanced strength and modulus [2].

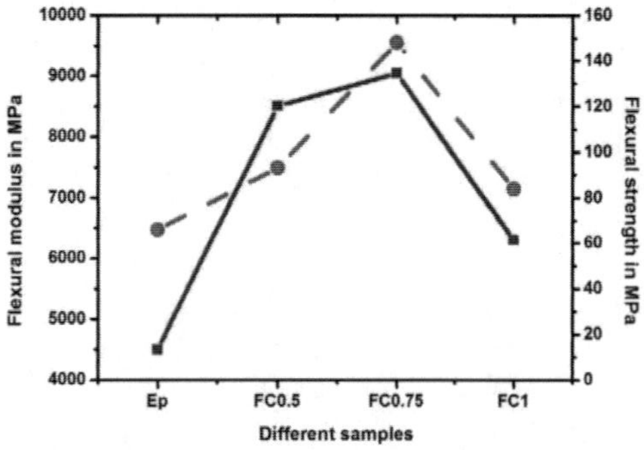

Fig.4:- Flexural properties of all the composite samples

In addition, good interfacial adhesion between the functionalized MWNTs and matrix, high aspect ratio and high modulus and strength of MWNTs, also contributed to the incrementin overall properties. However, decrease in strength at higher MWNTs content can be attributed to the following effects: firstly, creation of voids and agglomeration of MWNTs in the sample that may have decreased the strength;secondly, improper dispersion of MWNTs at higher loading can reduce the strength. In addition, Agglomerations of MWNTs act as areas of weakness and can cause stress concentration leading to lower flexure properties of epoxy samples modified with MWNTs at higher loading.

4.3.Hardness

The strength of all the composites is studied using a micro hardness tester. The harness value is calculated using the Vickers hardness (HV) method is given by HV= 2F *sin* $(136^0/2)/D^2$= 1.854 F/D^2 (approximately). Where, F= load in kgf and d = arithmetic mean of the two diagonals, d_1 and $d_2$in mm.

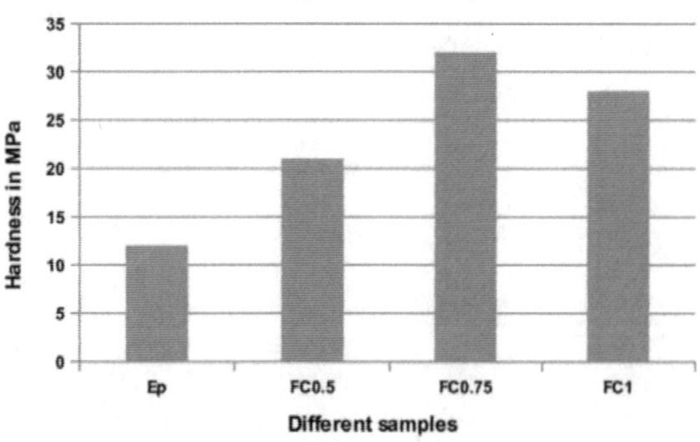

Fig.5:- Hardness of all the composite samples

The obtained hardness value of epoxy matrix was 12 MPa while addition of functionalized MWNTs increased it to a great extent. An increment of 75%, 166% and 133% hardness values were observed respectively for $FC_{0.5}$, $FC_{0.75}$, FC_1 nanocomposites with respect to neat epoxy. There may be a number of reasons for which hardness of the nanocomposites can be increased. Firstly, materials having higher crystallinity have greater hardness. Secondly, the reinforcement effect which is caused due to the addition of high aspect ratio, high modulus, and high strength nanofillers (MWNTs). These reinforced fillers contributed to the increment in hardness value. Third is the homogeneous dispersion of the fillers (MWNTs). Since CNTs are resilient materials, homogeneously dispersed CNTs can add rigidity and hardness via interfacial interaction with the epoxy matrix. Thus, the soft polymer matrix holds together when challenged by an applied force and the mechanical properties like modulus, hardness are improved [19]. Fourth is the formation of long range ordered structure which can be accomplished by the functionalization of MWNTs. So, all of these said reasons are equally accountable for the dramatically enhancement in hardness values in this case. Hardness results agreed well with flexural test results with the finding that among the specimens $FC_{0.75}$ showed the maximum improvement. However, for higher concentration in case of FC_1, the hardness value got deteriorated than $FC_{0.75}$ though it was still higher than other specimens. Reduction in hardness value at FC_1 may be due to the formation of voids due to addition of higher nanoparticle contents [20]. In addition, higher concentrated MWNTs may act as stress concentrator in the composite which can be the cause behind the decrement in mechanical properties.

4.4.Fracture surface analysis

The fracture surfaces of composites give first information about fracture mechanisms and the influence of particle modification on the fracture behaviour. To investigate the failure mechanism of epoxy as well as functionalized MWNTs/epoxy composites, the fracture surfaces of failed specimens were examined. The micrograph of neat epoxy in fig.6(a) demonstrates a typical feature of brittle fracture behaviour, thus accounting for the low fracture toughness of the unfilled epoxy. The surface is smooth, with very directionally deformation lines, which means that the crack propagated is not interrupted. However, the fracture surfaces of composites in fig.6(b) seem to be comparatively rougher. The ridge patterns and rivers marks indicate that the composites presented a more ductile fracture. This means that the energy required for the propagation of the crack increased. This is in accordance to the obtained results on mechanical characterisation (Fig.4), where it was concluded that the sample with best mechanical properties (with highest strength and modulus) was with 0.75 wt. % functionalized MWNTs.

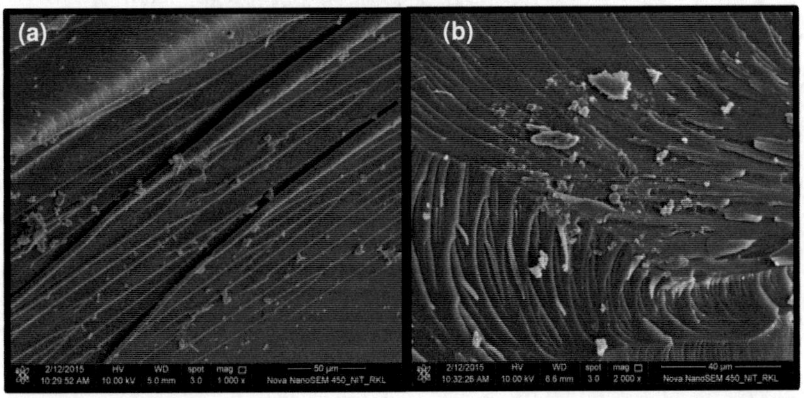

Fig.6:- Comparison of fracture surfaces between (a) neat epoxy, (b) nanocomposite ($FC_{0.75}$)

The degree of functionalization has a certain influence on the mechanical properties of this resulting composite. The presence of carboxylic acid groups on the surface enables a better dispersion of MWNTs and reduces the amount of impurities. These introduced functional groups lead to covalent bonds with the epoxy resin. Higher magnification in fig.7 (a),(b) shows that in $FC_{0.5}$ and$FC_{0.5}$ system, nanotubes were well separated and uniformly embedded in the epoxy resin. Stronger adhesion at the MWNTs/epoxy interface was obtained due to the reaction between carboxyl group and DGEBF (hardener used in present work). Although, several deboned interface was observed in fig.7(b) superior adhesion at the functionalized MWNTs/epoxy interface resulted in the failure of nanotubes without pulling out from epoxy matrix. A spatial bonding of the reactive groups at the interface, enables a partial debonding of the interface, but allows a crack bridging.

When the MWNTs loading increased to 1 wt%, the failure mode of nanocomposite changed (Fig. 8). In most locations, the crack damage started from the interface due to which MWNTs were pulled out. In these areas, MWNTs suffered from external force leaving the smooth resin matrix exposed weak interfacial bonding. This type of stress concentration could be due to agglomeration of several MWCNTs and formation of bubbles leading to voids due to viscous mixture of epoxy and MWCNTs. However, strong bonding was noticed in some areas that may be attributed to functionalization process. Agglomeration facilitates network formation or CNT-to-CNT interactions through surface contact (or tunnelling) and hence facilitates the electrons flow through the formed CNT network [21]. Hence, best shielding effectiveness obtained in this specimen.

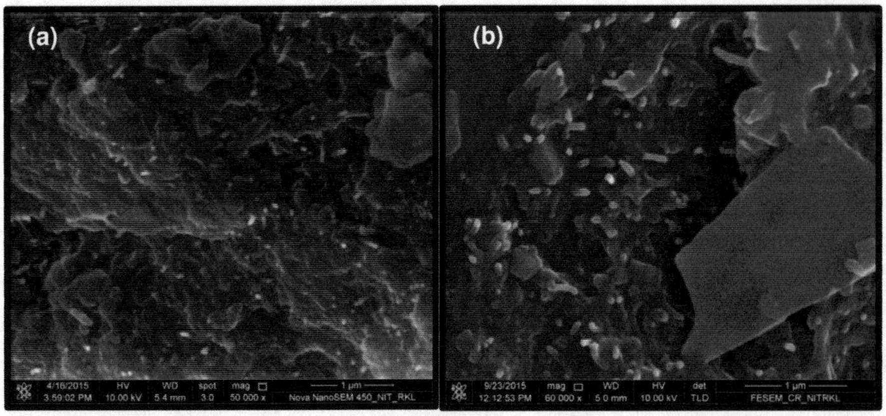

Fig.7:- Fracture surfaces of (a) $FC_{0.5,}$ (b) $FC_{0.75}$

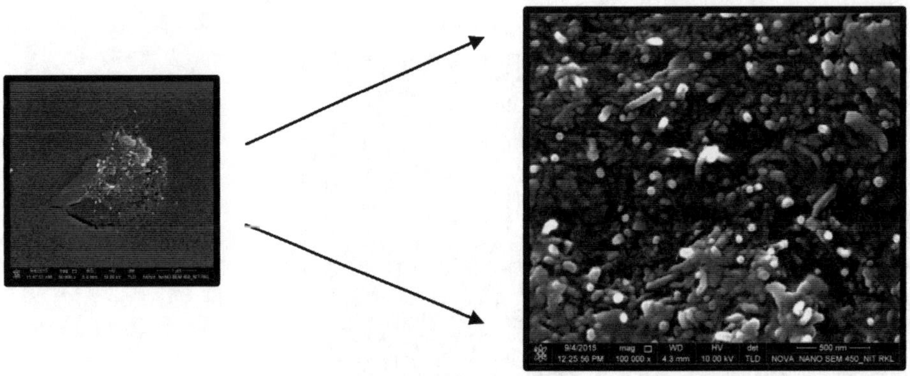

Fig.8:- Low and high resolution of the fracture surface of FC_1

5. Conclusion

Due to the low temperature curing and incorporation of functionalized MWNTs to the epoxy matrixan improved dispersion as well as strong interfacial bonding was realized in the final composites. This facilitates significantly the total shielding effectiveness and flexural properties of epoxy. The highest EMI SE of51.72 dB was provided by the nanocomposite FC_1. According to the flexural data $FC_{0.5}$ and $FC_{0.75}$ were showing 89% & 101% modulus enhancement in comparison to pure epoxy while relatively lower (40%) improvement in modulus was observed in FC_1. The strength values were in the order $FC_1 < FC_{0.5} < FC_{0.75}$ with 84, 93, 148 MPa respectively in comparison to epoxy having 66 MPa. Nanocomposites containing 0.5, 0.75 and 1wt.% of CNTs were demonstrated an increment of 75%, 166% and 133% hardness values respectively with respect to neat epoxy. Micrographical analysis confirms better dispersion in $FC_{0.75}$ resulting maximum enhancement in mechanical properties and found agglomerates in FC_1 resulting maximum improvement in electrical properties.

Acknowledgement

The authors thank the Naval Research Board, New Delhi for providing the financial support for this research work.

References

[1] C.S. Xiang, Y.B. Pan, X.J. Liu, X.W. Sun, X.M. Shi, J.K. Guo, Microwave attenuation of multiwalled carbon nanotube–fused silica composites. *Applied Physics Letters* 87(12) (2005) 1231031–3.

[2] M.B.A. Salam, M.V. Hosur, S. Zainuddin, S. Jeelani, Improvement in Mechanical and Thermo-Mechanical Properties of Epoxy Composite Using Two Different Functionalized Multi-Walled Carbon Nanotubes. *Open Journal of Composite Materials* 3 (2013) 1-9.

[3] H. Mahfuz, S. Zainuddin, M.R. Parker, T. Al-Saadi, V.K. Rangari, S. Jeelani, Improvement in Mechanical and Thermo-Mechanical Properties of Epoxy Composite Using Two Different Functionalized Multi-Walled Carbon Nanotubes. *Journal of Materials Science*44(4) (2009) 1113-1120. doi:10.1007/s10853-008-3161-5.

[4] P.C. Maa, N.A. Siddiqui, G. Marom, J.K. Kim, Dispersion and functionalization of carbon nanotubes for polymer-based nanocomposites: A review. *Composites Part A* 41(2010) 1345–1367.

[5] P. Glatkowski, P. Mack, J.L. Conroy, J.W. Piche, P. Winsor, Electromagnetic shielding composite comprising nanotubes. *Patent No*. 6,265,466, U.S. (2001).

[6] P.R. Nahass, S.O. Friend, R.W. Hausslein, High strength conductive polymers *US Patent* 5 (1997) 651,922.

[7] P. Saini, V. Choudhary, Enhanced Electromagnetic Interference Shielding Effectiveness of Polyaniline Functionalized Carbon Nanotubes Filled Polystyrene Composites. *Journal of Nanoparticle Research* 15 (2013) 1415. DOI 10.1007/s11051-012-1415-2.

[8] S.H. Park, P. Theilmann, P. Asbeck, P.R. Bandaru, Enhanced Electromagnetic Interference Shielding Through the Use of Functionalized CarbonNanotube-Reactive Polymer Composites. *Transactions on nanotechnology* 9(4) (2010).

[9] C.A. Grimes, C. Mungle, D Kouzoudis, S Fang, PC Eklund, The 500 MHz to 5.50 GHz complex permittivity spectra of single-wall carbon nanotube-loaded polymer composites. *Chemical Physics Letters* 319 (2000) 460.

[10] S. Bal, S. Saha, Comparison and analysis of physical properties of carbon nanomaterial-doped polymer composites. *High Performance Polymer* 26(8) (2014) 953–960.

[11] P.Saini, V. Choudhary, B.P. Singh, R.B. Mathur, S.K. Dhawan, Polyaniline-MWCNT nanocomposites for microwave absorption and EMI shielding. *Materials Chemistry and Physics* 113 (2009) 919–926.

[12] A. Saib, L. Bednarz, R. Daussin, C. Bailly, L. Xudong, J. Thomassin, C. Pagnoulle, C. Detrembleur, R. Jerome, I. Huynen, Microwave Theory and Techniques. *IEEE Transactions on microwave theory and techniques* 54(6) (2006) 2745-54.

[13] B.P. Singh, V. Prasanta, V. Choudhary, P. Saini, S. Pande, V.N. Singh, R.B. Mathur, Enhanced microwave shielding and mechanical properties of high loading MWCNT–epoxy composites. *Journal of Nanoparticle Research* 15 (2013) 1554. DOI 10.1007/s11051-013-1554-0.

[14] J. Wu, L. Kong, High microwave permittivity of multiwalled carbon nanotube composites. *Applied Physics Letter* 84 (2004) 4956–4958. DOI: 10.1063/1.1762693.

[15] Z. Wang, G.L. Zhao, Microwave absorption properties of carbon nanotubes-epoxy composites in a frequency range of 2 - 20 GHz. *Open Journal Of Composite Materials* 3 (2013) 17-23.

[16] P Liu, Modifications of carbon nanotubes with polymers. *European Polymer Journal* 41 (2005) 2693–2703.

[17] T.N. Narayanan, V. Sunny, M.M. Shaijumon, P.M. Ajayan, M.R. Anantharaman, Enhanced microwave absorption in nickel-filled multiwall carbon nanotubes in the S band.*Electrochemical and Solid-State Letters* 12 (4)(2009) K21-K24.

[18] X. Zhao, A.A. Koos, B.T.T. Chu, C. Johnston, N. Grobert, P.S. Grant, Spray deposited fluoropolymer/multi-walled carbon nanotube composite films with high dielectric permittivity at low percolation threshold. *Carbon* 47 (2009) 561-5969.

[19] S.H. Hsu, M.C. Wu, S. Chen, C.M. Chuang, S.H. Lin,W.F. Su, Synthesis, morphology and physical properties of multi-walled carbon nanotube/biphenyl liquid crystalline epoxy composites. *Carbon*50(3) (2012) 896–905.

[20] A. Zafar, F. Bertocco, J. Schjødt-Thomsen, J.C. Rauhe, Investigation of the long term effects of moisture on carbon fibre and epoxy matrix composites.*Composites Science and Technology*72 (2012) 656-666.

[21] J.O. Aguilar, J.R. Bautista-Quijano, F. Avilés, Influence of carbon nanotube clustering on the electrical conductivity of polymer composite films. *eXPRESS Polymer Letters* 4 (5) (2010) 292–299. DOI: 10.3144/expresspolymlett.2010.37.

Dry sliding wear behavior of epoxy composite reinforced with short palmyra fibers

Somen Biswal and **Alok Satapathy**

Department of Mechanical Engineering, National Institute of Technology, Rourkela, 769008, India

somenbiswal7@gmail.com, alok@nitrkl.ac.in

Abstract. The present work explores the possibility of using palmyra fiber as a replacement for synthetic fiber in conventional polymer composites for application against wear. An attempt has been made in this work to improve the sliding wear resistance of neat epoxy by reinforcing it with short palmyra fibers (SPF). Epoxy composites with different proportions (0, 4, 8 and 12 wt. %) of SPF are fabricated by conventional hand lay-up technique. Dry sliding wear tests are performed on the composite samples using a pin-on-disc test rig as per ASTM G 99-05 standards under various operating parameters. Design of experiment approach based on Taguchi's L_{16} Orthogonal Arrays is used for the analysis of the wear. This parametric analysis reveals that the SPF content is the most significant factor affecting the wear process followed by the sliding velocity. The sliding wear behavior of these composites under an extensive range of test conditions is predicted by a model based on the artificial neural network (ANN). A well trained ANN has been used to predict the sliding wear response of epoxy based composites over a wide range.

1. Introduction

The use of various natural fibers as substitutes for synthetic functional fibers is increasing day-by-day owing to their several advantages like bio-degradability, renewability, low-cost and low-toxicity. These composites find several applications starting from household appliances to wear resistant components. A number of studies on friction and wear characteristics of natural fiber composites have been reported in recent past. Chand et. al. [1] studied the influence of fiber orientation on the wear characteristics of sisal fiber reinforced composites with epoxy as the matrix material. It was found that the wear rate decreased considerably by addition of sisal fiber to the epoxy composites. Yousif et. al. [2] studied and compared the mechanical and tribological properties of oil palm fiber reinforced polyester composites (OPRP) and chopped strand mat glass fiber reinforced composites (CGRP). Similarly, Tong et. al. [3] studied the sliding wear behavior of bamboo fiber against grey iron (HT200). Chin et. al. [4] studied the potential of kenaf fiber as reinforcement for application against wear. Tayeb [5] studied sugarcane fiber reinforced polyester composites for tribological characterizations. Nirmal et. al. [6] studied the effect of betelnut fiber on the adhesive wear of polyester composites. A superior wear performance was observed in anti-parallel oriented betelnut fiber followed by parallel orientation and normal orientation of the fiber. Siva et. al. [7] studied the mechanical and tribological characteristics of woven coconut fibers and found the wear resistance to be maximum when a combination of silane treated glass fiber and coconut fiber was used. Chand et. al. [8] studied the effect of silane coupling agent and applied load on the frictional and sliding wear behavior of sisal fiber reinforced polyester composites. Most of the published works on palmyra fibers have reported on the strength properties of the composites and not on the effects of fibers on friction and wear behavior. Moreover, although a lot of work has been done on the wear characteristics of composites with other natural fibers, no work has been reported so far on epoxy based composites reinforced with short palmyra fiber. In view of this, the present work includes the investigation of dry sliding wear behavior using a pin-on-disc set up for epoxy filled with short palmyra fibers with an objective to find ways to use these locally available inexpensive fibers as a substitute for highly expensive synthetic fibers for wear resistance applications. Taguchi technique and Artificial Neural Network (ANN) have been used in an integrated manner in this work for parametric analysis of the wear process and to predict the sliding wear response of epoxy-palmyra fiber composites within and beyond the experimental limits.

2. Experimental details

2.1. Sliding Wear Test

In the present work Epoxy LY556 which belongs to the epoxide family has been used as the matrix material. Epoxy LY556 is a low-temperature curing resin and it is mixed with hardener HY951 in the ratio of 10:1. The short palmyra fiber (SPF) used in the present work is extracted from the palmyra tree leaf stalk and this fiber is thoroughly mixed with the epoxy resin (LY556) in four different weight proportions (0, 4, 8 and 12 wt%) to prepare composite samples. The samples are subjected to sliding wear test using a pin-on-disc type friction and wear monitoring test rig as per ASTM G 99-05. The counter body is a disc made of hardened ground steel (EN-32, hardness 72 HRC, surface roughness 0.6 μm *Ra*). The specimen is held stationary while the disc rotates. Normal load is applied through a lever mechanism. A series of tests are conducted with four sliding velocities of 63, 125,190 and 250 cm.s−1 each under four different normal loads of 5, 10, 15 and 20 N. The material loss from the composite surface is measured using a precision electronic balance with accuracy ±0.1 mg and the specific wear rate (mm^3/N-m) is expressed on a volume loss basis as:

$$W_S = \frac{\Delta m}{\rho \times t \times V_s \times F_n} \quad [1]$$

where Δm is the mass loss of the composite in the test duration (g), ρ is the density of the composite (g/mm^3), t is the test duration (s), Vs is the sliding velocity (cm.s−1), and Fn is the average normal load (N).

2.2. Taguchi Experimental Design

Taguchi experimental design is a very powerful tool to make parametric appraisal of any multi variable engineering process. In the present study for sliding wear test of epoxy-SPF composites, four major factors have been considered such as sliding velocity, applied load, sliding distance and fiber content each at four levels according to Taguchi's L_{16} orthogonal array as shown in Table 1. Use of Taguchi L_{16} reduced the number of experiments from 4^4= 256 conventional runs to just 16 runs thereby saving a lot of experimental time and cost. These experimental observations are again converted into signal-to-noise (S/N) ratios. The S/N ratio for minimum wear rate can be expressed as "smaller is better" which is calculated as logarithmic transformation of loss function as shown below:

$$\frac{S}{N} = -10\log\frac{1}{n}\left(\sum y^2\right), \quad [2]$$

where n is the number of observations and y is the observed data.

Table 1. Control Factors and their Selected Levels for Dry Sliding Wear

Control factors	Levels				
	1	2	3	4	Units
Sliding velocity (A)	63	125	190	250	Cm/sec
Normal load (B)	5	10	15	20	N
Sliding distance (C)	250	500	750	1000	m
Fiber content (D)	0	4	8	12	Wt %

Table 2. Specific Wear Rates Obtained for Different Test Conditions with S/N Ratios

Test Run	Sliding Velocity A (cm/sec)	Normal Load B (N)	Sliding Distance C (m)	SPF Content D (wt %)	Sp. Wear Rate ($10^{-4}mm^3$/N-m)	S/N Ratio (dB)
1	63	5	250	0	6.3600	-16.0691
2	63	10	500	4	5.4690	-14.7582
3	63	15	750	8	3.6250	-11.1862
4	63	20	1000	12	0.9165	0.7574
5	125	5	500	8	3.8560	-11.7227
6	125	10	250	12	2.3320	-7.3546
7	125	15	1000	0	7.8180	-17.8619
8	125	20	750	4	5.8230	-15.3029
9	190	5	750	12	1.4440	-3.1913
10	190	10	1000	8	2.8280	-9.0296
11	190	15	250	4	6.4300	-16.1642
12	190	20	500	0	8.8180	-18.9074
13	250	5	1000	4	7.4690	-17.4652
14	250	10	750	0	9.4840	-19.5398
15	250	15	500	12	2.6660	-8.5172
16	250	20	250	8	3.6560	-11.2601

2.3. Prediction Using Artificial Neural Network

ANN is widely used in the field of classification and control of dynamic system, medical science, image and speech recognition, etc [9]. In the present study ANN serves as a very helpful tool for predicting the specific wear rate as it is very difficult to get a mathematical formulation for the wear behavior. This proposed approach not only yields a sufficient understanding of the effects of process parameters but also produces an optimal parameter setting to ensure that the composites exhibit the best wear performance characteristics. The details of this methodology are described by Kadi [10].

1. Results and discussion

1.1. Sliding Wear Test Results

The specific wear rates and their corresponding S/N ratios obtained by Taguchi L_{16} orthogonal array for all the 16 experiments are presented in Table 2. From the Table 2 it is found that the overall mean of the S/N ratio is -12.4429 dB for epoxy based composites reinforced with short palmyra fibers. A commercial software Minitab 14 was used for the application of design of experiment. From the S/N ratio output, it is observed that among all the four factors considered for the analysis, the specific wear rate is mostly affected by the SPF content in the epoxy composite followed by the sliding velocity, sliding distance and normal load. The analysis showed that at sliding velocity (A) of 63 cm/s, normal load (B) of 20 N, sliding distance (C) of 1000 m and SPF content (D) of 12 wt%, the specific wear rate is minimum as evident from the response table for S/N ratios (Table 3).

1.2. Artificial Neural Network for Prediction

ANN is a very useful tool to predict the input and output pattern of nonlinear problems such as sliding wear. In the present analysis sliding velocity, normal load, sliding distance and SPF content are taken as the four input parameters. The input variables are normalized so that it lies in the range 0-1. The output layer has one neuron which represents the specific wear rate. Different structures of ANN were tested by varying the number of neurons in the hidden layer, gradient value and friction coefficient value (*mu* value). Finally one structure with minimum error was selected for training. The selected structure was trained rigorously by selecting maximum number of cycles. The predicted values are compared with the experimental values and the results are shown in Table 4 with the associated error percentages.

The Figure 1-2 below shows the effect of two most dominant factors i.e. SPF content and sliding velocity on the specific wear rate of the specimens and a simulated specific wear rate has been predicted using ANN.

Table 3. Response Table for Minimum Sliding Wear for Composite with Short Palmyra Fiber

Level	A	B	C	D
1	-10.314	-12.112	-12.712	-18.095
2	-13.061	-12.671	-13.476	-15.924
3	-11.823	-13.432	-12.305	-10.800
4	-14.196	-11.178	-10.900	-4.576
Delta	3.882	2.254	2.577	13.518
Rank	2	4	3	1

IOP Conf. Series: Materials Science and Engineering **115** (2016) 012028 doi:10.1088/1757-899X/115/1/012028

Table 4. Comparison of Experimental Results with ANN Predicted Values

	Specific wear rate (10^{-4} mm^3/N-m)		
Test Run	Experimental	ANN Predicted	Error %
1	6.360	6.359136	0.013
2	5.469	5.468302	0.012
3	3.625	3.601238	0.655
4	0.917	0.979166	6.837
5	3.856	4.343180	12.634
6	2.332	2.328292	0.159
7	7.818	7.773308	0.571
8	5.823	5.818631	0.075
9	1.444	1.443930	0.004
10	2.828	2.805736	0.787
11	6.430	6.425846	0.064
12	8.818	8.119834	7.917
13	7.469	7.382000	1.520
14	9.484	9.433750	0.529
15	2.666	2.654097	0.446
16	3.656	3.660072	0.111

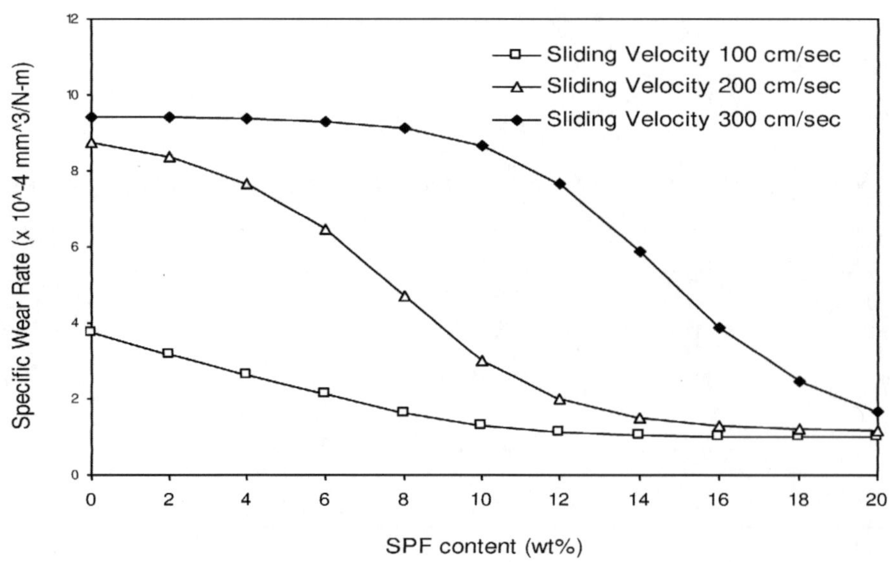

Figure 1. ANN prediction of variation in specific wear rate with SPF content for SPF-epoxy composite

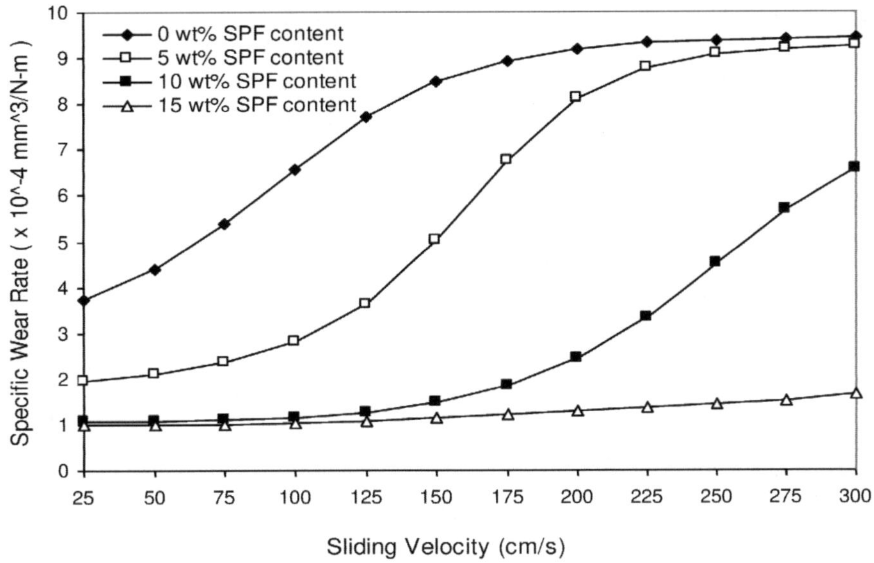

Figure 2. ANN prediction of variation in specific wear rate with sliding velocity for SPF-Epoxy composites.

2. Conclusions

The present experimental effort shows that successful fabrication of short palmyra fiber reinforced epoxy composites is possible and such composites can be employed in dry sliding wear situations. It is seen that by increasing the fiber loading, the wear resistance of neat epoxy can be significantly improved. A parametric appraisal of the wear process revealed that the fiber content and the sliding velocity are the most significant factors that affect the wear performance of the composites in an interacting environment. This work also illustrates the gainful use of artificial neural networks for predicting the specific wear rate of the composites with increase in fiber content and sliding velocity beyond the experimental domain.

3. References

[1] Chand N and Dwivedi UK 2007 *Polymer Composites* **28** 437-441.

[2] Yousif B F and Tayeb N S M 2009 *International Journal of Precision Technology* **1** 213-222.

[3] Tong J, Arnell R D and Ren L Q 1998 *Wear* **221** 37–46.

[4] Chin C W and Yousif B F 2009 *Wear* **267** 1550–1557.

[5] Tayeb N S M 2008 *Wear* **265** 223–235.

[6] Nirmal U, Yousif B F, Rilling D and Brevern P V 2010 *Wear* **268** 1354–1370.

[7] Siva I, Jappes J T W and Suresha B 2012 *Polymer Composites* **33** 723-732.

IOP Conf. Series: Materials Science and Engineering **115** (2016) 012028 doi:10.1088/1757-899X/115/1/012028

[8] Chand N and Dwivedi U K 2008 *Polymer Composites* **29** 280-284.

[9] Padhi P K and Satapathy A 2013 *Tribology Transactions* **56** 789-796.

[10] Kadi H E 2006 *Composite Structures* **73** 1–23.

Water absorption behavior and residual strength assessment of glass/epoxy and glass-carbon/epoxy hybrid composite

S C Mohanty, B P Singh, K K Mahato, D K Rathore*, R K Prusty, B C Ray

Composite Materials Group, Department of Metallurgical and Materials Engineering, National Institute of Technology, Rourkela-769008

*Email: dineshrathore2603@gmail.com

Abstract. Present investigation is aimed to study the water absorption behaviour and evaluation of residual strength of glass fibre/epoxy (GE) and alternate plies of glass-carbon/epoxy (GCE) hybrid composite. Both the composite systems were exposed to water at 70°C. Specimens were weighed after certain time periods to study the water uptake kinetic. Flexural tests were conducted after 4, 100 and 450 hours of ageing to evaluate the effect of hot water ageing on the mechanical properties of these potential materials. The water uptake kinetic was found to follow Fickian diffusion kinetic for GE as well as GCE hybrid composite but the rate of diffusion was higher for GE composite over GCE composite. The water content was also higher in GE composite over GCE composite after 450 hours of ageing. Significant decrement in flexural strength was observed with the increase in ageing time. Presence of water in the composite also imparted significant embrittlement to the matrix as reflected in the decrease in strain at peak for both the composite systems.

1. Introduction

Fibre reinforced polymer (FRP) composites are promising materials in the marine and offshore applications due to their relatively lighter weight and higher corrosion resistance over their metallic counterparts. Further their low maintenance cost and larger life span also makes them potential candidate for this application area. During their service life these materials are exposed to different environments which may include moisture, different low and high temperatures and may be more complex under changing humid conditions. Environmental degradation and damages are quite advanced issues in FRP composites[1]. Polymeric materials are susceptible to degradation when interact with water during their service life. The water absorption kinetic and the mechanism of degradation for fibrous polymeric materials under the exposure to humid ageing or hot water are still far away from conclusion. Presence of moisture in the composite leads to swelling, hydrolysis and plasticization of the polymer matrix[2]. The durability and integrity of fibrous polymeric composites is primarily governed by the interface/interphase which is most stressed and non-uniform region[3]. This interface also becomes susceptible upon the interaction with the environmental moisture. The concept of fiber hybridization is s recent advancement in the field of composite materials, where two or more types of reinforcements are used to take advantages of both reinforcements[4]. Out of the synthetic fibres, glass and carbon are the most commonly used materials. Glass fibers are fairly cheap and exhibit moderate strength and strain to failure, but lack in stiffness when compared with carbon fibers. Carbon fibers exhibit very good strength and stiffness, but at the same time they are quite brittle and very costly. Hence the motivation behind using both carbon and glass fiber in a single polymer matrix composite is to derive the advantages of both material, hence optimizing the properties of the resulted composite[5], [6]. In addition to these mechanical properties, carbon/polymer and glass/polymer composites show different moisture absorption tendency.

Present investigation is focused on the water uptake kinetic study of glass/epoxy (GE) composite and glass/carbon/epoxy (GCE) hybrid composite and its subsequent effect on the flexural performance of these materials.

2. Materials and Experimental methods

The materials used in the present investigation were woven fabric E-glass fibre supplied by Owens Corning and the epoxy was Diglycidyl Ether of Bisphenol A (DGEBA) from Atul Industries Ltd under the trade name Lapox l-12 and K-6 hardener which is triethylenetetramine (TETA). The carbon fibres were also 3K plain weave supplied by Soller composites. Some specifications of the used materials are given in table-1. The resin to hardener ratio were taken as 10:1 as per the manufacture's specification. Both the composite laminates (GE and GCE hybrid) were prepared by hand lay-up method with total 7 layers of fibers. Alternate layers of glass fibre and carbon fibre (4 glass and 3 carbon layers) were used in the fabrication of GCE hybrid laminate. The laminates were consolidated in the hot press at 5MPa pressure and 60°C temperature for 20 minutes. The final thickness of the laminates was 1.8±0.1 mm.The specimens for mechanical testing were cut from the laminate by diamond tipped cutter. The dimension of the specimens for flexural testing was taken as per ASTM standard D790-10. All the specimens after cutting were post cured in an oven at 140°C temperature for 6 hours[7]. After post curing 5 specimens of each GE and GCE composite were tested for having the reference flexural properties without any hot water ageing at 1 mm/min loading rate. Remaining samples were then marked and their dry weight has been recorded. All these samples were then immersed in hot water at 70°C temperature. After a regular interval of time the weight of these immersed samples has been taken to study the moisture uptake kinetics. Flexural testing of samples was then carried out after 4, 100 and 450 hours of hot water ageing.
Mechanical properties of E-glass, carbon fibers and epoxy are shown in table 1.

Material	Tensile strength (MPa)	Young's Modulus (GPa)	Density (g/cm^3)
E-glass	2000	80	2.58
Carbon	2900	525	1.85
Epoxy	85	3.5	1.2

Table 1. Mechanical Properties of E-glass, carbon fibers and epoxy.

3. Results and Discussions

3.1 Water uptake kinetic study

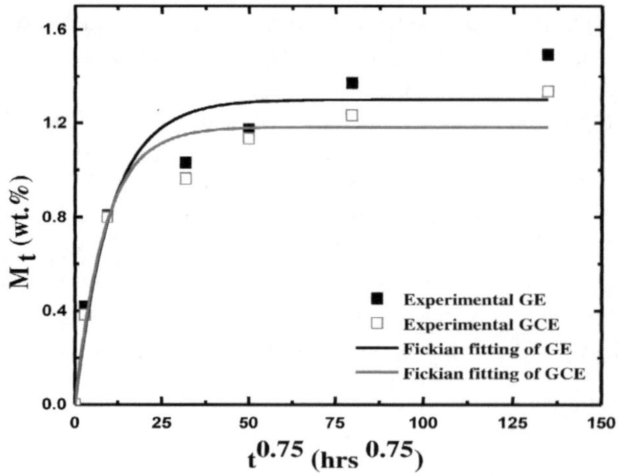

Figure 1: Water uptake kinetic analysis of GE and GCE composites

To study the water absorption rate in the composite specimens the water content in the samples were measured and then plotted against the ageing time as shown in figure 1. It has been seen that for both the composites the rate of water uptake was quite high at the initial time and there is no significant difference in the water content for GE and GCE composite. As the outer layers of both the composites are of glass plies, the water intake behaviour remains similar in the initial time.

The water molecules then start diffusing into the bulk material. In case of GCE composite the presence of carbon plies next to the outer glass plies retards its further water diffusivity inside the material and this has been replicated in the graphs where the rate of water uptake in both the composite starts deviating from each other. In both the composite systems a linear trend has been noticed till an absorption time of 100 hours after which a remarkable deviation from linearity was observed. Fick's diffusion kinetic has been fit (eq. 1) for both the composite and it was observed that the water uptake kinetic in this linear portion resembles with Fick's law of diffusion.

$$\frac{M_t}{M_\infty} = 1 - \exp\left[-7.3\left(\frac{Dt}{h^2}\right)^{0.75}\right] \quad ...(1)$$

Where, M_t is the water content in the composite at time t, M_∞ is the saturation water content. D represents the diffusivity of the material at the testing conditions and h is the thickness of the sample.

3.2 Flexural Behaviour

Flexural properties of all GE and GCE samples are presented in fig.2 with hot water ageing time. Fig.2 (a) represents the stress-strain behaviour of the samples with 450hours of ageing and without ageing. It is evident from figure that there is significant change in the stress-strain response of both the composites after 450 hours of ageing. Although the stress-strain pattern of aged and unaged samples resembles.

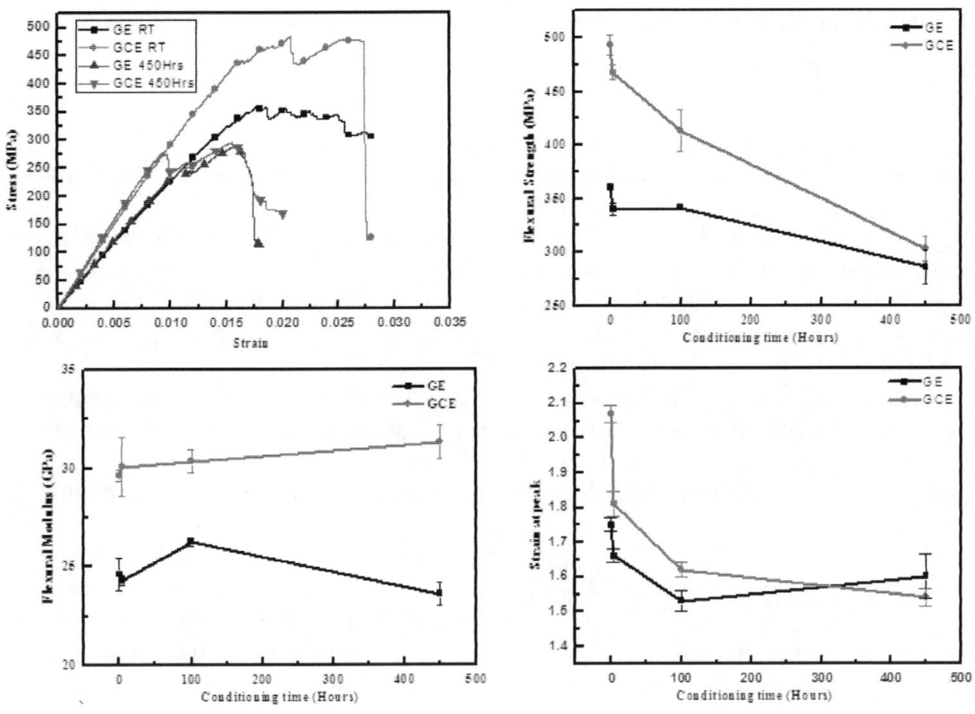

Figure 2: Variation of flexural properties of GE and GCE hybrid composites with hot water ageing time.

From these stress-strain plots other flexural properties have been derived and plotted in fig.2 (b), (c) and (d).A decrease of about 5% in the flexural strength was observed for both the composites only after four hours of hot water ageing. This decrease was remains same upto 100 hours of ageing for GE composite whereas there was further decrement in strength of about 16% for GCE composite. After 450 hours of exposure to hot water the strength of GE composite reduced by 20% whereas GCE hybrid experienced 38% reduction in its flexural strength. The variation of flexural modulus with hot water ageing upto 450 hours was not significant. For GE composite 4% decrease was observed whereas for GCE hybrid composite there was an increment of about 5% in its modulus. The variation of strain at peak can provide an indication about the relative ductile or brittle nature of the composites under consideration. Under the exposure to hot water there was a noticeable change in the strain at peak for both the composite systems indicating that the presence of moisture has modified the deformation behaviour of the composites. On the initial exposure to hot water for a short period of time (100 hours) there was nearly 5% and 12% decrease was observed for GE and GCE composite respectively. Further ageing of nearly 450 hours to GE composite did not significantly altered the strain at peak but the GCE hybrid composite experienced dramatic decrease of nearly 25%. The rate of decrease in strain at peak was higher of GCE composite over GE composite which depicts that the rate of embrittlement due to presence of moisture is higher in this composite.

4. Conclusion

It can be concluded that at the initial period, the water uptake behaviour is similar for both the composite system as the outer layers of both composites comprise of glass plies. Again in both materials, the initial water uptake follows Fick's law. Remarkable difference in the water content appears after certain time period and the hybrid composite seems to be more water resistive due to presence of carbon plies. With increase in water content, the flexural modulus for both the composite materials remains mostly unaffected; whereas a significant reduction in the flexural strength was observed as the water uptake time increases. The rate of strength degradation was found to be more in case of hybrid composite than GE composite.

References

[1] R. K. Prusty, D. K. Rathore, M. J. Shukla, and B. C. Ray, "Flexural behaviour of CNT-filled glass/epoxy composites in an in-situ environment emphasizing temperature variation," *Compos. Part B Eng.*, vol. 83, pp. 166–174, Dec. 2015.
[2] C.-H. Shen and G. S. Springer, "Moisture Absorption and Desorption of Composite Materials," *J. Compos. Mater.*, vol. 10, no. 1, pp. 2–20, Jan. 1976.
[3] B. C. Ray and D. Rathore, "Durability and integrity studies of environmentally conditioned interfaces in fibrous polymeric composites: Critical concepts and comments," *Adv. Colloid Interface Sci.*, vol. 209, pp. 68–83, Jul. 2014.
[4] M. J. Shukla, D. S. Kumar, D. K. Rathore, R. K. Prusty, and B. C. Ray, "An assessment of flexural performance of liquid nitrogen conditioned glass/epoxy composites with multiwalled carbon naotube," *J. Compos. Mater.*, p. 0021998315615648, Nov. 2015.
[5] C. Dong and I. J. Davies, "Optimal design for the flexural behaviour of glass and carbon fibre reinforced polymer hybrid composites," *Mater. Des.*, vol. 37, pp. 450–457, May 2012.
[6] C. Dong, J. Duong, and I. J. Davies, "Flexural properties of S-2 glass and TR30S carbon fiber-reinforced epoxy hybrid composites," *Polym. Compos.*, vol. 33, no. 5, pp. 773–781, May 2012.
[7] D. S. Kumar, M. J. Shukla, K. K. Mahato, D. K. Rathore, R. K. Prusty, and B. C. Ray, "Effect of post-curing on thermal and mechanical behavior of GFRP composites," in *IOP Conference Series: Materials Science and Engineering*, 2015, vol. 75, p. 012012.

Production and Compressive Characterization of Aluminium MMC Foam Manufactured Using Dual Foaming Agent

S Haidar[1]*, S Ansary[2] AND A Rahman[2]
[1]Asst. Professor, Department of Mechanical Engineering, Aliah University, New Town, Kolkata-156, West Bengal. INDIA
[2]PG Scholar, Department of Mechanical Engineering, Aliah University, New Town, Kolkata-156, West Bengal. INDIA

*E-Mail: shamimhaidar@yahoo.com

Abstract. Aluminium foams, produced by melting Aluminium alloy (LM6) containing blowing agent(s) and vigorous stirring. TiH_2 is a known agent for this. As TiH_2 begins to decompose into Ti and gaseous H_2 when heated above about 465°C, large volumes of hydrogen gas are rapidly produced, creating bubbles that leads to a closed cell foam. A novel Strategy to enhance the mechanical properties of Al-MMC foams is discussed here, and it is demonstrated that titanium hydride (TiH_2) in the form of 10-15 μm diameter particles can be pre-treated by selective oxidation to produce more uniform foams having better compressive properties (yield strength and energy absorption). It is found that the mechanical properties of the foams and the uniformity of cell size distribution is improved when the foam is blown with an optimized mixture of $CaCO_3$ and pre-treated TiH_2. In order to define the relationship of mechanical properties with relative density of this material, correlations which uniquely defines the compressive behaviour of this modified Al-MMC foam has been developed.

Keywords: Al-Si MMC foam, mechanical behaviour of cellular materials, Dual foaming agent.

1. Introduction

Metal foam is a type of cellular solids, having a combination of properties such as high stiffness with very low density and a capability to absorb impact energy. These unique combinations of properties indicate various potential applications such as packaging materials for protection sensitive devices, machinery enclosures, automobiles, and as sound absorbing material under difficult situations. Mechanical testing of aluminium foams is a prerequisite for any application. The study of compressive properties of metallic foams is necessary as its major applications are primarily load-bearing and energy absorption. Even aluminium foams whose main properties are functional require minimal mechanical properties to prevent damage or failure. The compressive stress-strain diagram of metal foam as defined by Gibson and Ashby [1] consists of three distinct regions namely linear elastic region, collapse region and densification region. Fig. 1 shows a representative stress – strain curve of metal foam under compressive loading. The first zone (linear elastic zone) is recorded up-to small strain (about 2-3%). The second zone i.e. plateau region, continues up to about 70% of strain, characterized by a small slope of the stress-strain curve. In some cases the curve is even horizontal. In second zone collapsing of cell continues till the foam behaves like a solid material. The third zone (densification zone) shows a rapidly increasing stress, here the cell walls become pressed together and the material attains bulk-like properties.

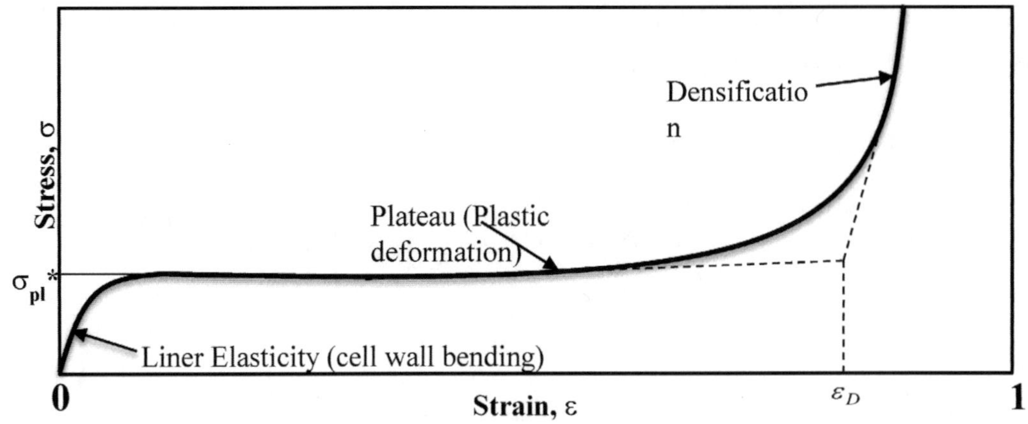

Figure 1: Stress strain curve for metal foams [1].

The main aim of the present investigation is to determine the compressive characteristics of the closed cell Aluminium Metal Matrix Composite (Al-MMC) foams developed in the laboratory. The outcome of the experimental investigation are compared with the established theoretical models developed by different researchers [1-9] so that it can be further referred for different industrial applications.

2 Experimental Method

2.1 Synthesis of Aluminium Foam

The material under investigation is closed cell aluminium foam, manufactured through liquid metallurgy route using aluminium alloy (LM6: consisting of 0.1% Cu, 0.1% Mg, 0.13% Si, 0.6% Fe, 0.5% Mn, and trace amount of Zn, Pb, Sn and rest Al). The aluminium alloy used is of density ($\boldsymbol{\rho_s}$), 2.7gm/cm^3, having compressive elastic modulus (E_s) of 69 GPa and compressive yield strength ($\boldsymbol{\sigma_s}$) of 120 MPa. The ingot is melt in a tilting resistance furnace. The formation of foam requires a high melt viscosity which is achieved by the addition of Silicon Carbide (SiC) particulate to the melt. The amount of Aluminium is 1000 gm. 5% SiC (pre-heated) are added to the melt, which also increases the mechanical strength of the foamed component. For homogeneous mixture of SiC in Al matrix, continuous stirring is required. The achieved high viscosity allows liquid Aluminium to be stable at a temperature of TiH_2-decomposition (465°C) which is much lower than the freezing temperature of liquid Aluminium.
The homogeneous Al-SiC mixture is then poured into a pre heated mold (which is fitted with a stirring arrangement) after removal of slag as much possible. 2.5% blowing agent (Titanium Hydride) is added to the mold. TiH_2 begins to decompose into Ti and gaseous H_2 when heated above about 465°C. By adding titanium hydride particles to the aluminum melt, large volumes of hydrogen gas are rapidly produced, creating bubbles that leads to a closed cell foam. It is needed to stir the mold with constant speed for good foaming.

Figure 2: Pouring of Al-SiC melt into mold.

As TiH_2 powder is very costly, so, manufacturing of Al-SiC foam by this method is not so cost effective. The solution to this problem is Calcium Carbonate (which is very cheap in cost). So, instead of adding 2.5% TiH_2, a dual foaming agent (2% $CaCO_3$ and 0.5% – 1.0% TiH_2) is added separately and this produces same result with minimum cost. Addition of Ca in Al matrix slightly changes the mechanical properties but it is nearly identical.

The properties of metal foams depend on many morphological features, such as pore size distribution, cell wall curvature, defects, etc.[3] Although the exact interrelationship between properties and structure is not yet sufficiently known, one usually assumes that a uniform distribution of convex pores free of defects is highly desirable. The task for the experimentalist is to produce such structures. A short look at existing foams shows that there is still much potential for development since these often tend to be irregular [4].

Thus, the foam fabricated by this method are usually non-uniform which leads to inferior mechanical properties. The reason for this can be non-adoption of TiH_2 to the melting range of the alloy to be foamed. This is avoided by pre treatment of titanium hydride (TiH_2) in the form of 10-15 μm diameter particles by selective oxidation.

2.2 Pre Treatment of Foaming Agent

The pre treatment of TiH_2 was first introduced by B. Matijasevic-Lux and J. Banhart for manufacturing of Aluminium foam through powder metallurgy [10]. The same method is followed here for melt route also.

TiH_2 powder supplied by LOBA chemical, India (purity 98.9%) was used in this study. The powder was supplied in the ''untreated'' state. Pre-treatments of the TiH_2 powder were carried out isothermally at various temperatures (450, 480, 510 °C) and times (30, 60, 120 and 180 min) under air in a chamber furnace. For heating, the ceramic crucible (with required amount of TiH_2) is placed into a volume chamber muffle furnace and is left there for the time specified. After pre-treatment all powders were gently homogenized by tumbling in a container.

Hydrogen starts to be released from TiH_2 at about 405 – 470 °C with some variations between powders of different origin. However, most of TiH_2 powder starts decomposing at 465°C.
As heating is carried out under air, an oxide layer grows which is roughly 100 nm thick [10] after 180 min at 480°C and contains an outer shell of TiO_2 and an inner shell of Ti_3O.[10]. Pre-treatment under air also reduces the amount of hydrogen and shifts the temperature of decomposition by 160°C. Using pre-treated TiH_2 for foaming Al alloys delays foaming and leads to a more uniform distribution of rounder pores. The best parameters found are close to 60 min at 480°C. It is noted that at higher pre-treating temperature (510°C), the amount of available hydrogen is not sufficient to produce uniform foam (Fig.4).

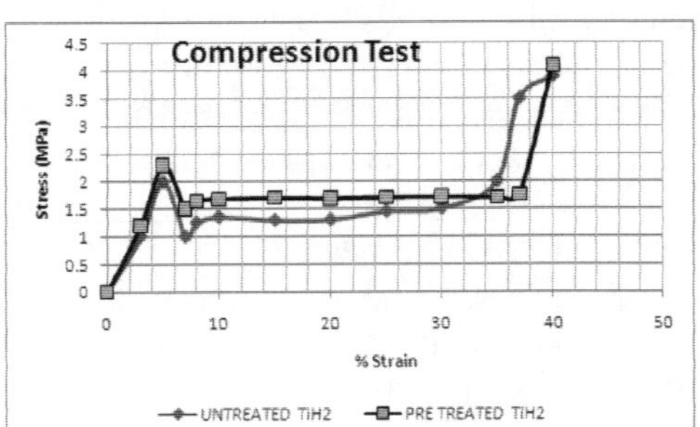

Figure 3: Comparison of stress-strain curves (compression) of Al foam using untreated and pre-treated TiH_2.

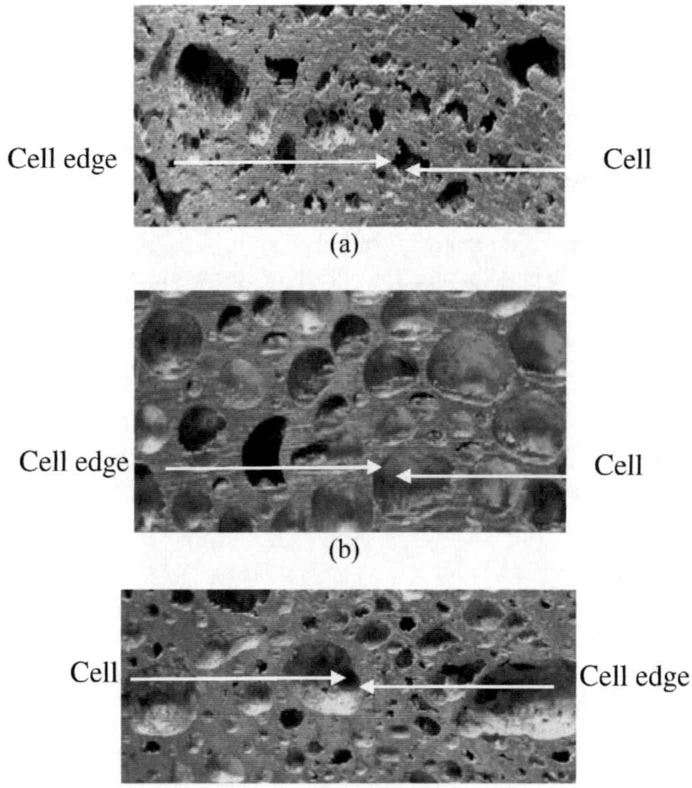

(c)

Figure 4: Optical microscopy image of the Al-MMC foam samples at 10x magnification. (a) using untreated TiH_2, (b) TiH_2 treated at 480°C for 60 Minutes, (c) TiH_2 treated at 510°C for 60 minutes.

3. Results and Discussion

3.1 Yield Strength Characterization

Aluminium foam is specially designed for load bearing applications; therefore efforts are made to evaluate the compressive properties of the developed Al-MMC foam. Compressive tests are performed on Al-MMC foams of various relative densities. All the compression tests are performed in a computer controlled 100KN compressive testing machine. The specimens for compressive tests are of 10x10x20 mm sizes. The specimen surfaces are polished before the test and during test the surfaces are greased to reduce friction. Most of the specimen tested exhibit a significant drop of the stress after the end of the elastic deformation at ++different strains. This is because of failures in the structure of the foam (fractures of cell walls) at several locations, manifested in the form of sudden stress drop. The yield stress ($\boldsymbol{\sigma_{pl}^*}$) are determined from the curves and the results is presented in Table 1. The scatter in the experimental results can be explained from the fact that Al-MMC foams are produced by random dispersion of gases in the melt, as such structure i.e. cell size and cell shape will vary. Such non-uniformities in the structure of the material yield a scattered result.

It is observed that the use of Dual foaming agent not only reduces the cost, but also it enhances the mechanical properties. The mechanical properties are even better (definite plateau region) if pre-treated TiH_2 is used (Table- 1). And the optimum pre treatment temperature is 480°C, holding time is 60 minutes.

Untreated TiH2		TiH2 at 450C 60		TiH2 at 480C 60		TiH2 at 510C 60	
RD	Yield Strength (MPa)	RD	Yield Strength (MPa)	RD	Yield Strength (MPa)	RD	Yield Strength (MPa)
0.1522	2.4	0.1162	1.7	0.1015	2.1	0.1272	2
0.1718	1.8	0.119	1.75	0.1176	2.5	0.1225	1.9
0.1937	2.3	0.143	2.1	0.1265	2.7	0.147	2.3
0.1988	2.4	0.1624	2.8	0.132	2.8	0.153	2.4
0.2	2.5	0.1832	2.7	0.15	3.2	0.179	2.8
0.2015	2.6	0.19	2.9	0.158	3.4	0.188	2.9
0.2111	2.67	0.2047	3.1	0.187	3.9	0.2005	3.1
0.2151	2.9	0.21	3.2	0.198	3.9	0.205	3.2
0.2177	2.1	0.2125	3	0.2015	3.95	0.211	3.3
0.2267	3	0.22	3.3	0.215	4.05	0.22	3.4

Table 1: Compressive properties of Al-MMC foams of different relative density.

The plot of Yield stress (σ_{pl}^*) versus relative density is shown in Figure 5. The results showed a large amount of scatter. This can be explained from the fact that aluminium foams are produced by random dispersion of gases in the melt. As such structure i.e. cell size, cell shape, imperfections will vary from specimen to specimen. There is also a probability of size effects and strain rates effects. It can be also explainable from Figure 5 that, the strength is higher in case of pre treated samples because of its evenness in the porous structure.

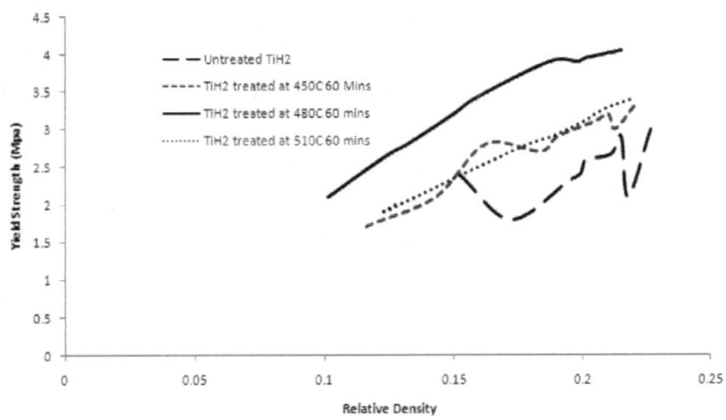

Figure 5: Compressive yield stress of aluminium foams for different relative density

It is evident from figure 6 that the pre treatment temperature of TiH_2 plays an important role in the Yield strength of Aluminium foam. This can be explained from the fact that, the degree of oxidation is higher at high temperature and there is a deficiency of hydrogen gas to produce better foam. At low temperature the oxidation level is so less, that it can not initiate the 'delayed gas formation'.

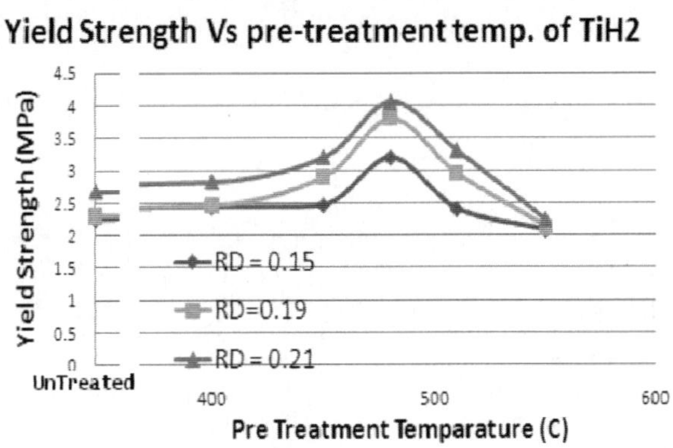

Fig. 6: Variation of Yield Strength with respect to heat treatment temperature of TiH_2

3.2 Comparison with existing theoretical models of compressive behavior

Gibson and Ashby [1] proposed a model for closed cell metal foam of density ρ_0 and made from solid aluminium alloy of density ρ_s with a structure consisting of cubes of solid struts and plates, in which the struts meet at their midpoints,. When closed cell foam is deformed the cell edges bend, and the cell wall carry membrane stresses. The contribution from cell wall stretching to the overall stiffness and strength of the foam is represented by a term which is linear in the relative density, (**RD=** $\boldsymbol{\rho_0/\rho_s}$) , while the contribution from cell edge bending is non-linear in the relative density. Thus according Gibson and Ashby [1], the yield stress of metallic foam $\boldsymbol{\sigma_{pl}^*}$ in compression is related to the yield stress of the cell wall material σ_s as:

$$\frac{\sigma_{pl}^*}{\sigma_s} = 0.3\ \varphi^{3/2}.\mathrm{RD}^{3/2} + 0.4\,(1-\varphi).\mathrm{RD} \quad (1)$$

where, φ is the "distribution constant", is the fraction of solid in the foam which is contained in the cell edges; the remaining fraction (1- φ) is in the faces. And RD is the Relative Density.

$$\mathrm{RD} \leq \varphi \leq 1 \quad (2)$$

Gibson and Ashby [1] developed this model without considering the distribution of the cell-size, cell edge and cell-wall thickness, or any other the irregularities present in the structure of the closed cell metal foam.

Similar efforts have been made by various researchers. Andrew et al. [5] and Simone and Gibson [6] generalized Eq. (1) as:

$$\frac{\sigma_{pl}^*}{\sigma_s} = 0.33\,.\mathrm{RD}^2 + 0.44\,.\mathrm{RD} \quad (3)$$

Lu and Ong [7] described compressive behaviour of closed-cell aluminium foams as,

$$\frac{\sigma_{pl}^*}{\sigma_s} = 0.98\,.\mathrm{RD}^{1.5} \quad (4)$$

Tzeng and Ma [8] described compressive behaviour of closed-cell aluminium foam developed by them as,

$$\frac{\sigma_{pl}^*}{\sigma_s} = 0.911\,.\mathrm{RD}^{1.45} \quad (5)$$

Grenestedt [9] described the compressive behaviour of closed-cell aluminium foam considering the cell wall imperfections as,

$$\frac{\sigma_{pl}^*}{\sigma_s} = 0.3\,.\mathrm{RD} \quad (6)$$

The normalized compressive yield stresses are measured by finding the ratio of the measured compressive yield stress of the Al-MMC foams (Table 1) to the compressive yield stress of the solid aluminium alloy. The normalized compressive yield stress of the Al-MMC foams is then compared with the models. Fig. 7 shows the normalized yield strength (σ_{pl} */σ_s) of closed-cell aluminium for different relative density (RD = ρ_0/ρ_s) based on various theoretical models. All the models excepting that defined by Eq. 6 is developed assuming that cell edges and faces are uniform, they have also neglected morphological defects like curves and wiggles. The normalized compressive yield strength of the Al-MMC foam is also plotted in Fig. 7. It is evident from Fig. 7 that the properties of closed cell

Al-MMC foams are well below that suggested by models of closed cell Al-MMC foams. The theoretical results seem to be higher compared with the current work. Such difference can be attributed to the fact that the model assumes identical cube cells, without considering the distribution of the actual cell shape, cell-size, cell edge and cell-wall thickness or the different imperfections (broken cell-walls, cell-edge and wall curvature etc.).

However Gibson and Ashby model defined by Eq. 1 with high value of φ=0.7 is much closer to the experimental values [12]. High value of φ means fraction of solid in the foam which is contained in the cell edges is higher than the remaining fraction (1- φ) in the cell walls. However none of the model could exactly define the compressive deformation behaviour of the material developed. Hence an empirical equation is used to fit the experimental data to correlate the normalized yield stress of the developed Al-MMC foam with its relative density (Fig. 5). The developed correlations for foam using untreated TiH_2 and pre-treated TiH_2 are:

a. Using Dual foaming agent, $CaCO_3$ & TiH_2 (untreated):

$$\frac{\sigma^*_{pl}}{\sigma_s} = 0.11\,.\,RD \qquad (7)$$

b. Using Dual foaming agent, $CaCO_3$ & TiH_2 (heat treated 450°C, 60 mins):

$$\frac{\sigma^*_{pl}}{\sigma_s} = 0.125\,.\,RD \qquad (8)$$

c. Using Dual foaming agent, $CaCO_3$ & TiH_2 (heat treated 480°C, 60 mins):

$$\frac{\sigma^*_{pl}}{\sigma_s} = 0.175\,.\,RD \qquad (9)$$

d. Using Dual foaming agent, $CaCO_3$ & TiH_2 (heat treated 510°C, 60 mins):

$$\frac{\sigma^*_{pl}}{\sigma_s} = 0.13\,.\,RD \qquad (10)$$

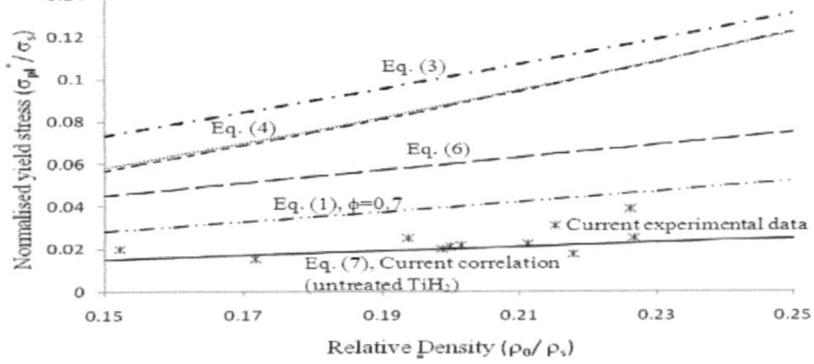

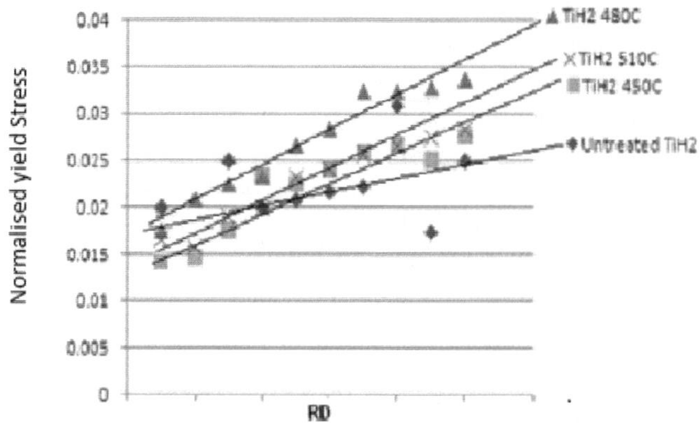

Figure 7: Normalised yield strength (σ_{pl}^*/σ_s) of aluminium foam of different relative density RD (ρ_0/ρ_s)

An important aspect of deformation mechanism can be inferred from Eq. 7-10, is that the linear density term dominates. This implies that cell wall stretching is the more significant mechanism during plastic deformation of this type of closed cell foams.

It is also observed that heat treatment of TiH_2 at 480°C for 60 minutes results in higher slope, which implies higher compressive strength for same RD value, thus, more usable foam (Fig. 5).

TiH_2 is also pretreated for different time durations other than 60 minutes, but considerable enhancement in the mechanical properties are not found.

3.3 Energy Absorption Characterization

The stress-strain response revealed that the material can undergo large amount of deformation, as such the energy absorption capability of this material is expected to be high. The energy absorbed by the material during the three above mentioned uniaxial compressive tests are calculated from the curves (Figure 1). Absorption energy per unit initial volume (E) of the aluminium foam specimen can be calculated by integrating the area under the stress-strain (σ -ε) curve,

$$E = \int_0^{\varepsilon} \sigma(\varepsilon)\, d\epsilon \qquad (11)$$

The areas under the curves (Figure 3) are measured using commercial plotting software (Origin). The energy absorbed during each compressive deformations are calculated. The results reveal that significant amount of energy is absorbed by the material. Therefore the developed material can be applied for those application which requires large energy absorption like designated crash elements of automobiles/trains, impact prone hollow automobile parts, support structures of automobiles, packaging material for of sensitive equipment etc.

Figure 8 shows the plot of the amount energy absorbed by aluminium foam versus different relative densities obtained from different foam samples by different pre-treatment temperature of TiH_2 under compressive deformation up to densification strain. It is clear from the figure that the energy absorbed increases with increase in relative density, however it is expected that this increase would continue up to certain relative density and after that it is expected to absorb lesser energy. This can be explained from the fact that with increase in relative density the plateau region decreases (i.e. densification strain decreases) hence after certain relative density the area under the curve will also decrease.

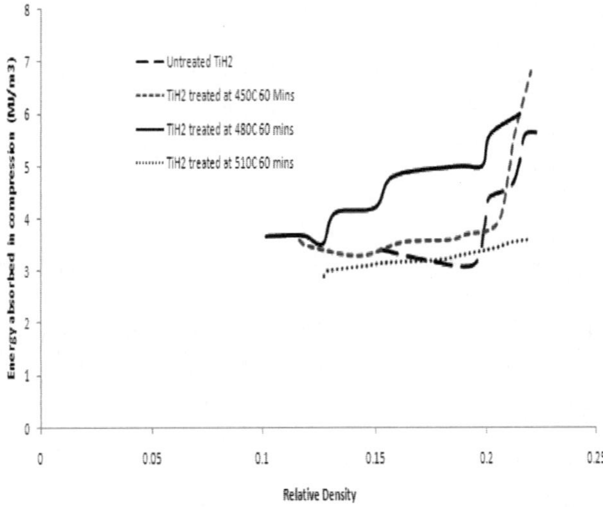

Figure 8: Energy absorbed by aluminium foam of different relative densities under compressive deformation

4. Conclusions

Uniaxial compressive tests are carried out find the compressive properties of Al- MMC foam. The material properties calculated based on existing theoretical models yield result which is higher compared with the experimentally measured properties of the developed material. The theoretical models therefore, inadequately describe the compressive properties of the developed Al-MMC foam. The reasons for this can be attributed to the fact that the model assumes identical cube cells, which is far from the actual case. An important observation that can be made from the comparisons of the experimental data is: pre-treating of TiH_2 at 480°C for 60 minutes produces best result. In order to define the relationship of different compressive properties with relative density of this material, correlations which uniquely defines the compressive behaviour of this material has been developed. Such correlations are very much essential in order to properly design and apply this material.

5. References

[1] Gibson, L.J., and Ashby, M.F., 1997, Cellular solids, Cambridge University Press.
[2] Ashby, M.F., Evans, A.G., Fleck, N.A., Gibson, L.G., Hutchinson, J.W., and Wadley, H.J.N.G., 2000, Metal foams: A Design Guide, Butterworth– Heinemann, Boston.
[3] Daxner T, Bo¨hm HJ, Seitzberger M, Rammerstorfer FG. Modelling of cellular metals. In: Degischer H-P, Kriszt B, editors. Handbook of cellular metals. Weinheim: Wiley-VCH; 2002. p. 245–80.
[4] Degischer HP, Kriszt B. Handbook of cellular metals. Weinheim: Wiley-VCH; 2002.
[5] Andrews, E., Sanders, W., and Gibson, L.J., 1999, "Compressive and tensile behaviour of aluminum foams," Materials Science and Engineering, A270, pp 113–124.
[6] Simone, A.E., and Gibson, L.J., 1998, "Aluminum foams produced by liquid state processes," Acta Mater, 46, pp 3109–3123.

[7] Lu, T.J., and Ong, J.M., 2001, "Characterization of close-celled cellular aluminum alloys," J Mater Sci, 36, pp 2773–2786.

[8] Tzeng, S.C., and Ma, W.P., 2006, "A novel approach to manufacturing and experimental investigation of closed-cell Al foams," Int J Adv Manuf Technol, 28, pp 1122–1128.

[9] Grenestedt, J.L., 1998, "Influence of wavy imperfections in cell walls on elastic stiffness of cellular solids," J Mech Phys Solids, 46, pp 29–50.

[10] B. Matijasevic-Lux, et al., Modification of titanium hydride for improved aluminium foam manufacture, Acta Materialia, 54 (2006) 1887–1900

[11] Sutradhar G. & Kumar S., "Metallic Foam – A Potential Material for the Future", Indian Foundry Journal, Vol. 54, No.2, Feb. 2008.

[12] Ghose J, Sharma V, Kumar S., "Compressive behavioral analysis of Al-MMC foam", International Journal of Industrial and production engineering & Technology, Vol 1, No. 1 (2011), pp 35-43.

Parametric Optimization of Simulated Extrusion of Square to Square Section Through Linear Converging Die

S. K. Mohapatra and K. P. Maity*

Department of Mechanical Engineering, National Institute of Technology, Rourkela-769008, Orissa, India

*E-mail : kpmaity@gmail.com,

Abstract. The effect of various process parameters for determining extrusion load has been studied for square to square extrusion of Al-6061 alloy, a most used aluminium alloy series in forming industries. Parameters like operating temperature, friction condition, ram velocity, extrusion ratio and die length have been chosen as an input variable for the above study. Twenty five combinations of parameters were set for the investigation by considering aforementioned five parameters in five levels. The simulations have been carried out by Deform-3D software for predicting maximum load requirement for the complete extrusion process. Effective stress and strain distribution across the billet has been checked. Operating temperature, extrusion ratio, friction factor, ram velocity and die length have the significant effect in decreasing order on the maximum load requirement.

Keywords: Extrusion, DEFORM-3D, Aluminium alloy, Optimization

1. Introduction

Extrusion, a demanded metal forming process is versatilely admitted for production of long straight metal products having complex cross sections because of its energy saving, near net shape production with better mechanical properties along with high production rate characteristics [1]. Control over the process is influenced by number of internal variables (chemical composition of billet material, grain size, segregation, metallurgical structure and prior strain history) and number of state variables (operating temperature, friction condition, shape factor, deformation degree and rate) [2]. Effect of few state variables like extrusion ratio and ram velocity, affecting equivalent strain and dynamic recrystallization of Mg-Zn-Y alloy at the time of extrusion has been studied by Hirano et al. [3]. At the same time aforementioned parameters affect the temperature condition at various positions in the deformed zone [4]. Increase of temperature at die exit directly affects load requirement and microstructural changes.

For an efficient production process the combination of variable parameters is needed to be optimized [5]. For a good mechanical property of an extruded product, maintaining a temperature range at the time of extrusion is most vital. The range of temperatures for Al-6061 + Al2O3p alloy was found 500-560°C as an exemplary condition [6]. Operating temperature, friction factor and ram speed have the significance in decreasing order for maximum extrusion load [7].

Estimation of exact load requirement for deformation is very difficult but some analytical and numerical techniques are there for finding out the approximate load [8]. Due to the complexity of the numerical analysis, few empirical methods are staying good for the prediction of maximum load in most of the industries. Computerized simulations are now a better alternative for predicting the process outcomes than expensive empirical process. Abundant amount of work is going on for predicting various extrusion parameters like maximum load [9] , minimum surface cracks [10] etc. by FEM simulations.

kpmaity@gmail.com (Prof. Kalipada Maity)

In this investigation effect of various variable process parameters on maximum extrusion load has been studied with the help of commercial simulation package DEFORM-3D. Linear converging die for square to square reduction of the well deformable material Al-6061, has been considered. The process was analysed by Taguchi optimization technique.

2. Finite Element Analysis

Finite element modeling and simulation techniques have been necessitated these days for predicting the effect of variable parameters on metal forming process before real production process to reduce the experimental trials because of its cost and time efficiency. For the simulation of metal forming process, numbers of commercial finite element codes such as Forge, Hyperextrude, LS-DYNA, SUPERFORM, DEFORM and ABACUS etc. are available. Various metal forming investigations have been carried out successfully with the application of codes in various fields [11-13]. DEFORM^{TM}-3D finite element code was used to analyse the effect of different variable parameters for extrusion of Al-6061 alloy through linear converging rigid die. The simulation process follows pre - processor, Run-engine and Post-processor sequentially for a complete analysis.

Pre-processor: A new problem can be created in this step by positioning the objects, defining mesh, defining thermal and mechanical boundary conditions. With all specific inputs the pre-processer generates a database file for simulation engine for further processing.

Run-engine: All the numerical calculations on the generated database are processed here. Depending on the number of elements and numerical method, the engine takes time to solve the problem.

Post processor: The outcome results after simulation run can be analysed in this step with a very suitable user friendly graphic interface. The step wise change of the various factors in the material model can be visualized along with the graphical presentations.

A square billet, punch and dies of varying dimensions have been designed with solid works software and imported to DEFORM interface as .stl file. Al-6061 billet of dimension (40×40×100) mm^3 was divided into 130000 numbers of tetrahedral mesh elements. The billet was allowed only to flow through the linear converging die for different reductions with different die length. The properties of work material selected from the software database are given in Table 1. In the present investigation, the die, billet and container were modeled as rigid bodies, whereas the billet was modeled as rigid-plastic object.

Table 1. Properties of Al-6061

Material	Young's Modulus (MPa)	Poison's ratio	Thermal expansion (1/°C)	Thermal Conductivity (W/m/°C)	Heat capacity (N/mm^2/°C)	Emissivity
Al-6061	68947.6	0.30	22×10^{-6}	180.181	2.43369	0.7

The flow stress adopted for simulation is effective strain, effective strain-rate and temperature dependent:

$$\bar{\sigma} = \bar{\sigma}(\bar{\varepsilon}, \dot{\bar{\varepsilon}}.T) \quad (3)$$

Von-Mises yield criterion model is used for finding effective stress of the isotropic billet metal i.e.

$$\bar{\sigma} = 1/(\sqrt{2}) \sqrt{\{((\sigma1-\sigma2)^2+(\sigma2-\sigma3)^2+(\sigma3-\sigma1)^2\}} \quad (4)$$

where σ1, σ2 and σ3 are the principal stresses. Interface friction model is considered as shear type (constant friction) and the friction factor m ($o \le m \le 1$) is expressed as

$$\tau = m\,(\bar{\sigma})/(\sqrt{3})\ [11] \tag{5}$$

Where τ is frictional shear stress, $\bar{\sigma}$ is the effective flow stress of the deforming material. The process parameters and friction factors considered for the simulation are listed in Table 2. The value of friction condition for the die - billet interface and the other interface is assumed as a variable parameter.

3. Taguchi optimization technique

Five variable parameters have been considered for optimizing the process. Those are operating temperature (T), extrusion ratio (R), ram velocity (V), die length (L) and friction factor (m). The values of the factors have been considered in five levels and an orthogonal array of 25 combinations of parameters was set for the simulation process.

Table 2. Parameters considered for the simulation

Parameters	Values
Billet length (mm)	100
Billet cross section area (mm^2)	40*40
Operating temperature (T) (°C)	250, 300, 350, 400, 450
Extrusion ratio (R)	2, 5, 10, 15, 20
Ram velocity (V) (mm/sec)	0.1, 1, 2, 3, 4
Die length (L) (mm)	15, 20, 25, 30, 35
Friction factor at die-billet interface (m)	0.3, 0.4, 0.5, 0.6, 0.7

4. Results and Discussion

Figure 1 shows post processor result analysis with a load-stroke plot for run-3 (R=2, L=25mm, m=0.5, V=2 mm/sec, T=350°C) combinationl set. The figure also shows the effective stress distribution of the work material. Figure 2 shows (a) the effective strain and (b) velocity distribution of the billet material across the die profile. The value of the effective stress as well as the couresponding strain at the die exit is maximum. The value is gradually increasing from inlet to die exit. From the velocity profile it is clear that there is no dead metal zone, so there is no extra energy consumption to overcome redundant work which only minimizing the maximum load requirement.

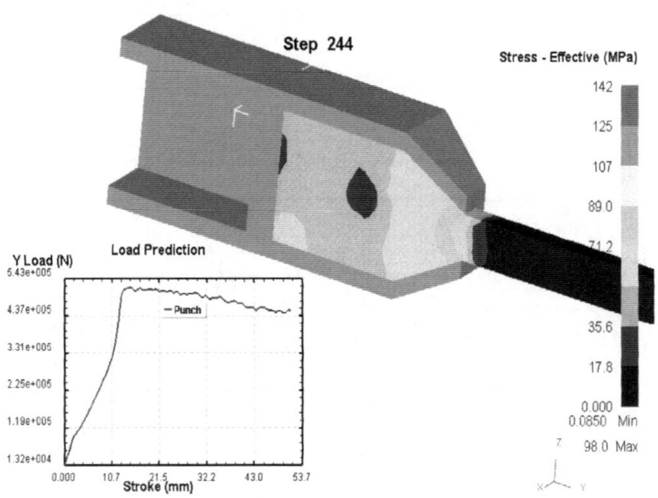

Figure 1. Prediction of Load-stroke plot and Effective stress distribution from the post-processor analysis

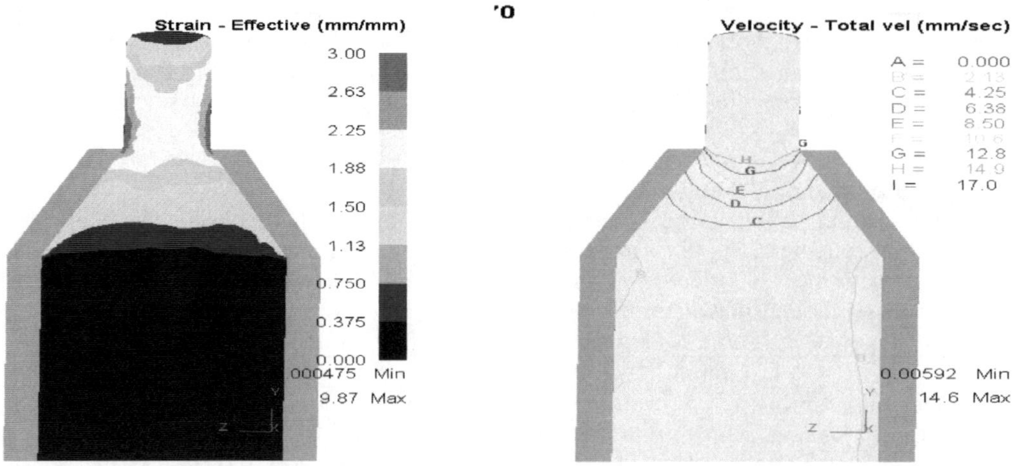

(a) (b)

Figure 2. (a) Effective strain distribution and (b) velocity profile distribution across thedie profile

The load versus stroke curves have been determined for twenty five sets of simulations. Maximum extrusion load for each run in L_{25} orthogonal array is mentioned in Table 3.

Table 3. Set of 25 combinations of parameters along with output

Run	Extrusion Ratio	Die Length (mm)	Friction Factor (m)	Ram Velocity (mm/sec)	Opcraing Temperature (°C)	Manimum Extrusion Load (N)
1	2	15	0.3	0.1	250	589000
2	2	20	0.4	1	300	555000
3	2	25	0.5	2	350	517000
4	2	30	0.6	3	400	401000
5	2	35	0.7	4	450	402000
6	5	15	0.4	2	400	427000
7	5	20	0.5	3	450	404000
8	5	25	0.6	4	250	1030000
9	5	30	0.7	0.1	300	906000
10	5	35	0.3	1	350	546000
11	10	15	0.5	4	300	873000
12	10	20	0.6	0.1	350	740000
13	10	25	0.7	1	400	639000
14	10	30	0.3	2	450	410000
15	10	35	0.4	3	250	1030000
16	15	15	0.6	1	450	512000
17	15	20	0.7	2	250	1340000
18	15	25	0.3	3	300	836000
19	15	30	0.4	4	350	702000
20	15	35	0.5	0.1	400	569000
21	20	15	0.7	3	350	867000
22	20	20	0.3	4	400	557000

23	20	25	0.4	0.1	450	400000
24	20	30	0.5	1	250	1280000
25	20	35	0.6	2	300	1150000

The value of loss function used for measuring performance characteristics which deviates form required one, recommended by Taguchi is further converted into a signal-to-noise (S/N) ratio. As it is a minimization problem, the performance category for smaller the better is chosen.

For minimization case: smaller –is-the better: $S/Ns = -10 \log \left(\frac{1}{n}\sum_{i=1}^{n}\frac{1}{yi^2}\right)$ (6)

Where n is the number of observations and y is observed result.

For all the results SN ratio has been calculated and main effect plot for load required for extrusion by linear converging die profile is shown in Figure 3. The optimum combinations for minimum load is the highest level of SN ratio i.e 'T'=450°C, 'R'=2, 'V'=0.1mm/sec, 'L'=15 mm and 'm'=0.1.

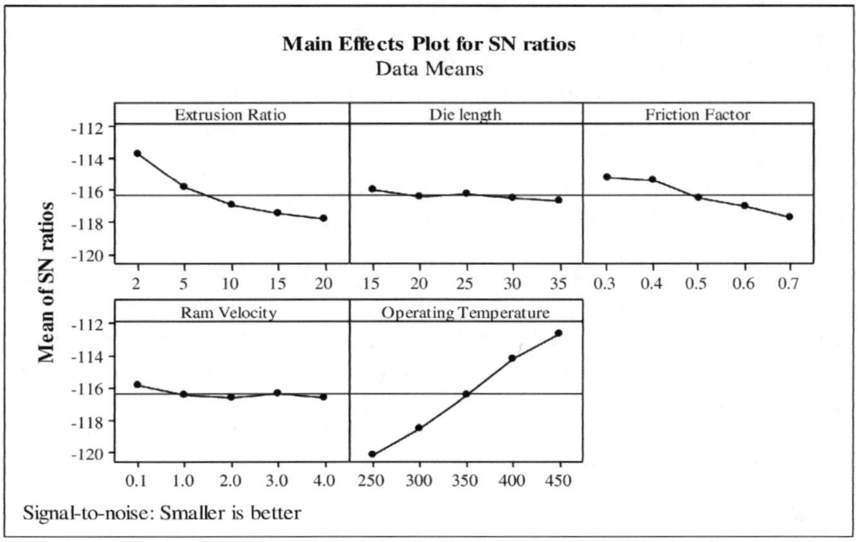

Figure 3. Main effects plot for maximum load in extrusion

The significant effects of individual parameters on the process was studied. It was observed that operating temperature; extrusion ratio, friction factor, die length and ram velocity have the effect on extrusion load in decreasing order.

4.1 Confirmatory test

By considering the optimal setting of parameters simulation has been carried out for checking the maximum extrusion load. The maximum extrusion load was found as 228000N. Also by changing only extrusion ratio and keeping other parameters same in the optimal combination the simulations have been conducted to confirm the process. For extrusion ratios R=5, 10, 15, and 20 with all other optimum combinations the extrusion load requirement was 299000, 352000, 380000, 401000N respectively. The result given by the confirmatory test is quite satisfactory. It was found a great amount of energy saving with considering the optimum parameter setting.

5. Conclusion

Modeling and simulation of extrusion of Al-6061alloy has been carried out successfully by considering a wide range of variable operating parameters. An optimum combination of parameters set successfully for the extrusion of square section from square billet in concern to minimize the maximum extrusion load. Operating temperature, extrusion ratio and friction factor have the maximum significance in decreasing order, whereas ram velocity and die length have minimum effect, but not negligible for load requirement. At high operating temperature condition, effect of friction and strain rate is more pronounced.

References

[1] Zhang C, Zhao G, Chen H, Guan Y and Kou F 2011 Numerical simulation and metal flow analysis of hot extrusion process for a complex hollow aluminum profile *Int. J. Adv. Manuf. Technol.* **60** 101–10

[2] Svyetlichnyy D S, Majta J and Nowak J 2013 A flow stress for the deformation under varying condition—internal and state variable models *Mater. Sci. Eng. A* **576** 140–8

[3] Hirano M, Yamasaki M, Hagihara K, Higashida K and Kawamura Y 2010 Effect of Extrusion Parameters on Mechanical Properties of Mg97Zn1Y2 Alloys at Room and Elevated Temperatures *Mater. Trans.* **51** 1640–7

[4] Shahzad M and Wagner L 2009 Influence of extrusion parameters on microstructure and texture developments, and their effects on mechanical properties of the magnesium alloy AZ80 *Mater. Sci. Eng. A* **506** 141–7

[5] Jurkovic Z, Jurkovic M and Buljan S 2006 Optimization of extrusion force prediction model using different techniques **17** 353–6

[6] Luan B F, Hansen N, Godfrey a., Wu G H and Liu Q 2011 High strength Al–Al2O3p composites: Optimization of extrusion parameters *Mater. Des.* **32** 3810–7

[7] Venkatesh C and Venkatesan R 2014 Optimization of Process Parameters of Hot Extrusion of SiC/Al 6061 Composite Using Taguchi's Technique and Upper Bound Technique *Mater. Manuf. Process.* **30** 85–92

[8] Valberg H S 2010 *Applied metal forming: including FEM analysis* (Cambridge University Press)

[9] Altan T, Oh S-I and Gegel G 1983 Metal forming fundamentals and applications *Am. Soc. Met. 1983,* 353

[10] Mihelič A and Štok B 1998 Tool design optimization in extrusion processes *Comput. Struct.* **68** 283–93

[11] Li L, Zhou J and Duszczyk J 2004 Prediction of temperature evolution during the extrusion of 7075 aluminium alloy at various ram speeds by means of 3D FEM simulation *J. Mater. Process. Technol.* **145** 360–70

[12] Sheppard T 2013 *Extrusion of aluminium alloys* (Springer Science & Business Media)

[13] Duan X, Velay X and Sheppard T 2004 Application of finite element method in the hot extrusion of aluminium alloys *Mater. Sci. Eng. A* **369** 66–75

Grain boundary character modification employing thermo-mechanical processing in type 304L stainless steel

S K Pradhan* and S Mandal

Department of Metallurgical and Materials Engineering,
Indian Institute of Technology, Kharagpur-721302, W.B, India

*E-mail: sumanta14julymet@gmail.com

Abstract. Grain boundary engineering (GBE) approach has been employed to modify the boundaries character of a type 304L stainless steel through thermo-mechanical processing (TMP) route, which combined a low level of cold deformation (5, 10 and 15 %) followed by annealing at 1173K and 1273K for 1hour. Employing Electron Back Scatter Diffraction based Orientation Imaging Microscopy, the fraction and distribution of low Σ CSL boundaries (Σ≤ 29) and its effect on random high-angle grain boundaries connectivity and triple junction distribution of as-received (AR) and GBE specimens were evaluated. It was possible to increase the fraction of low Σ CSL boundaries up to 75% following GBE treatment (as compared to 50% in AR specimen). The GBE specimens also contained maximum number of percolation resistant triple junctions which could render better resistance against percolation related phenomena.

Keywords: Grain boundary engineering; Thermo-Mechanical processing; Triple junction distribution; CSL boundaries Introduction

1. Introduction

Grain boundary engineering (GBE) is relatively a new concept on the basis of controlling the grain boundary character distribution (GBCD) in many alloy systems [1-4]. Generally low Σ coincidence site lattice (CSL) boundaries (Σ ≤ Σ29) exhibit some special behaviors in comparison to random high angle grain boundaries (HAGBs). Proportion of Σ3 and its variant like Σ9 and Σ27 boundaries have been enhanced in the past to modify boundary related properties in low stacking fault energy FCC materials [5,6] like resistance against solute segregation [7], inter-granular corrosion (IGC) [8-10], stress corrosion cracking (SCC) [11], creep and fatigue [12,13]. The main objective of GBE is to enhance the materials properties by increasing the fraction of low CSL grain boundaries and to disrupt the connectivity of random HAGBs.

Thermo-mechanical processing (TMP) is the most effective approach to achieve GBE microstructure in complex alloys. The combination of deformation amount as well as annealing temperature and time needs to be optimized for a material under investigation [14]. Type 304L stainless steel is widely used because of its better mechanical properties along with excellent corrosion resistance over a large temperature range. However, at a particular temperature range (sensitization temperature), the material is susceptible to inter-granular corrosion (IGC) and inter-granular stress corrosion cracking (IGSCC). Population, distribution and connectivity along with the nature of the boundaries are mainly responsible for inter-granular corrosion. The objective of this study is to understand the evolution of special boundaries and its effect on the connectivity of random HAGBs as well as triple junction distribution (TJD) in as-received (AR) and GBE-treated specimens of type 304L stainless steel.

2. Experimental Details

The material used in this study is type 304L stainless steel and its chemical compositions (in wt %) is given in Table 1. AR specimens were subjected to three different thickness reductions (viz. 5, 10 and 15pct) at room temperature by laboratory cold rolling mill. Deformed specimens were subsequently annealed at two different temperatures (1173 K and 1273 K) for 1 hour followed by water quenching (WQ). The TMP treatment employed in this investigation has been summarized in Table 2.

Table 1: Chemical composition (in wt %) of 304L stainless steel

C	Cr	Ni	P	S
0.0147	17.7	13.7	0.0032	0.0028

Table 2: Thermo-mechanical processing conditions adopted in this work

Specimen name	Processing conditions
AR	As-Received
R5-1173K	5% cold rolled + 1173 K for 1 h + WQ
R5-1273K	5% cold rolled + 1273 K for 1 h + WQ
R10-1173K	10% cold rolled + 1173 K for 1 h + WQ
R10-1273K	10% cold rolled + 1273 K for 1 h + WQ
R15-1173K	15% cold rolled + 1173 K for 1 h + WQ
R15-1273K	15% cold rolled + 1273 K for 1 h + WQ

The AR and GBE-treated specimens were polished up to 0.25 μm grit diamond paste using the standard metallographic polishing procedure. Specimens were further polished with colloidal silica (0.04 μm) suspension to ensure removal of any residual surface deformation. Electron Back Scatter Diffraction

(EBSD) based Orientation Imaging Microscopy (OIM) scans were performed on all the processed specimens to determine the GBCD in the materials using a TSL-OIM system attached to Zeiss Merlin scanning electron microscope operating at 20 kV. A step size of 0.5 μm was used for the orientation imaging. Misorientations above 2° were considered as grain boundaries. To identify CSL boundaries, Brandon's criterion [15] is used. Random HAGBs are defined as those with misorientation $\theta > 15°$ and which are not low Σ ($\Sigma \leq 29$) CSL boundaries. The grain size was measured using linear intercept method (average of horizontal and vertical intercept lengths) and a misorientation of 5° was used for determining the grain size.

3. Results & Discussion

3.1. *Characterization of GBs*

The grain boundary plus image quality map of the AR and GBE-treated specimen (R5-1273K) is shown in Fig. 1(a) and Fig. 1(b), respectively. In addition to increase in the fraction of Σ3, the frequency of Σ9 and Σ27 has been increased significantly in GBE-treated specimen. In AR specimen, most of the Σ3 boundaries lie within the grains and are not part of the grain boundary network (see Fig. 1a). However, the random HAGBs have been substantially replaced by a large number of twin and its variants (i.e. Σ9 and Σ27) in GBE-treated specimen with moderate increase in grain size. It is to be noted that average grain size (considering twin as grain boundary) of AR and R5-1273K specimen is 14 μm and 21 μm, respectively.

Formation of annealing twins is mostly favored in low stacking fault energy FCC metals or alloys like 304L stainless steel [5]. Twin boundaries may form in two ways such as by "growth accident" mechanism during recrystallization [16] and/or by the movement of grain boundaries during GBE type TMP [5,6,9,17]. It is observed (Fig. 1a) that the Σ3 boundaries in AR specimen are straight as well as parallel and terminate within a single grain. Further, the fractions of Σ9 and Σ27 boundaries are very small which indicates that the interaction between Σ3 is insignificant. Therefore, it is logical to assume that the annealing twins in AR specimen have been formed by "growth accident" mechanism [9,16,17]. The morphology of Σ3 boundaries in GBE-treated specimen is different – these boundaries are part of grain boundary network (see Fig. 1b). In addition to Σ3 boundaries, the fraction of other annealing twin variants (i.e. Σ9 and Σ27) is also significantly high in GBE specimen. Maximum numbers of triple junctions are formed via interaction between twin and its variants and this is only possible by "Σ3 regeneration mechanism" initiated through strain induced grain boundary migration (SIBM) [12,14,17]. Hence, it may be inferred that the GBE microstructures have been evolved via grain boundary migration initiated through TMP [6,9,17].

Figure 2 shows the 2D grain boundary networks excluding low Σ ($\Sigma \leq 29$) CSL boundaries in order to qualitatively asses the degree of connectivity of random HAGBs. In AR specimen, random high angle grain boundary length density (BLD_{RHAGB} length of RHAGBs per unit area) value is very high (0.062 μm^{-1}) and random HAGBs networks are interconnected (see Fig. 2a). In contrast to this, the BLD_{RHGB} (0.021μm^{-1}) has been decreased as well as connectivity of random HAGBs has been substantially

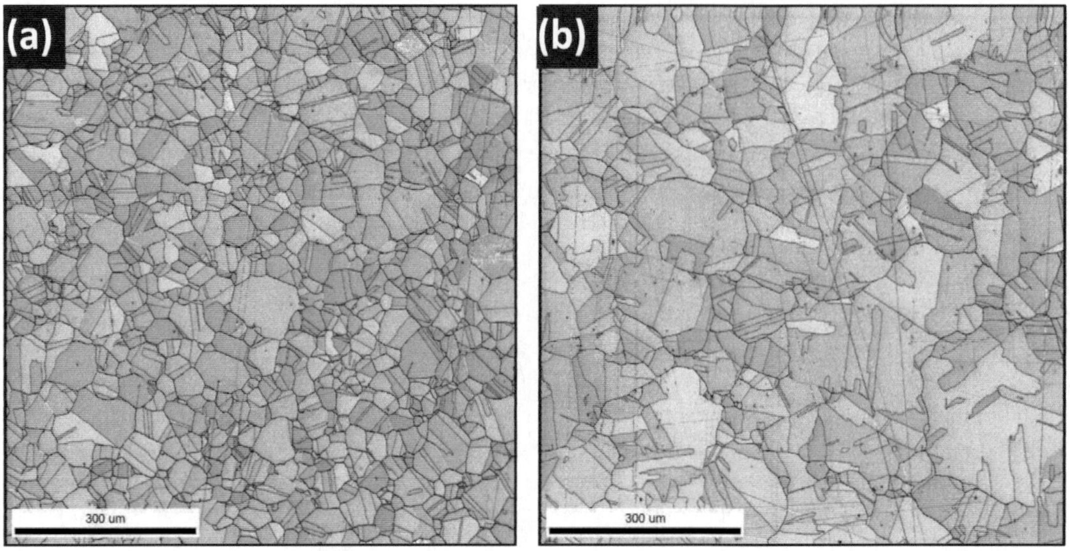

Figure 1: Grain boundary plus image quality (IQ) map of the (a) as-received and (b) GBE treated (R5-1273K) specimen. In orientation maps GBs are colour-coded as follows: Σ3—red; Σ9—blue; Σ27—green; random HAGBs—black)

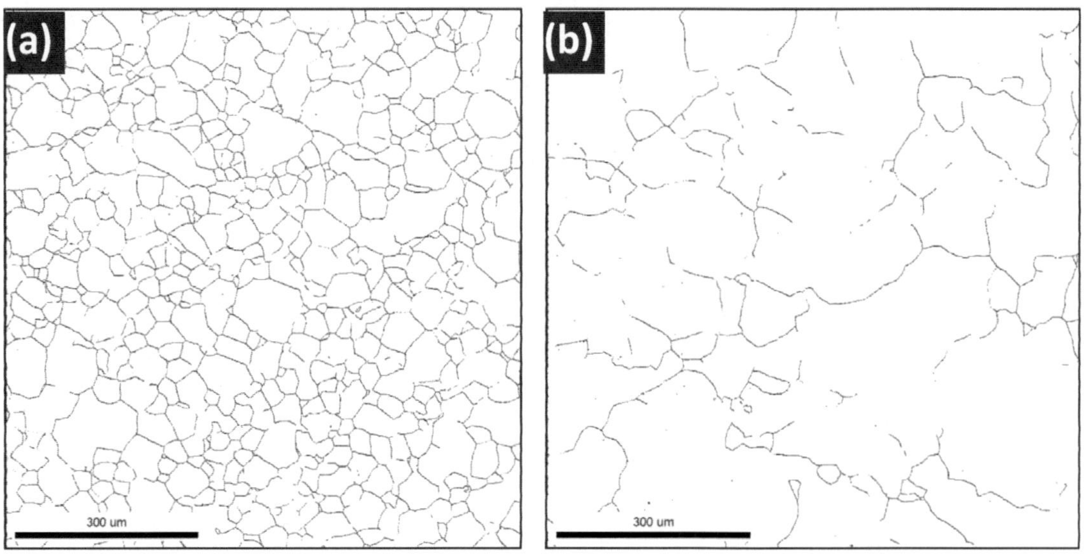

Figure 2: Grain boundary map showing the connectivity of random HAGBs in (a) as-received and (b) GBE treated (R5-1273K) specimen.

disrupted in GBE treated specimen (see Fig. 2b). It is important to note here that one has to increase the fraction of special boundaries and also need to break the random HAGBs networks in order to achieve

GBE microstructure [18]. This is due to the fact that boundary related properties of materials strongly depend on the type, density and connectivity of boundaries [7,8,10-13,19].

3.2 *GBCD and Triple junction*

Figure 3 shows the GBCD in AR and GBE-treated specimens. It could be observed the fraction of low Σ CSL boundaries have been significantly increased in all the specimens following GBE treatment. Especially in the R5-1273K specimen, low Σ CSL boundaries have increased to75% from 50% in AR condition whereas Σ3 boundaries have increased to 65% from the 44%. The

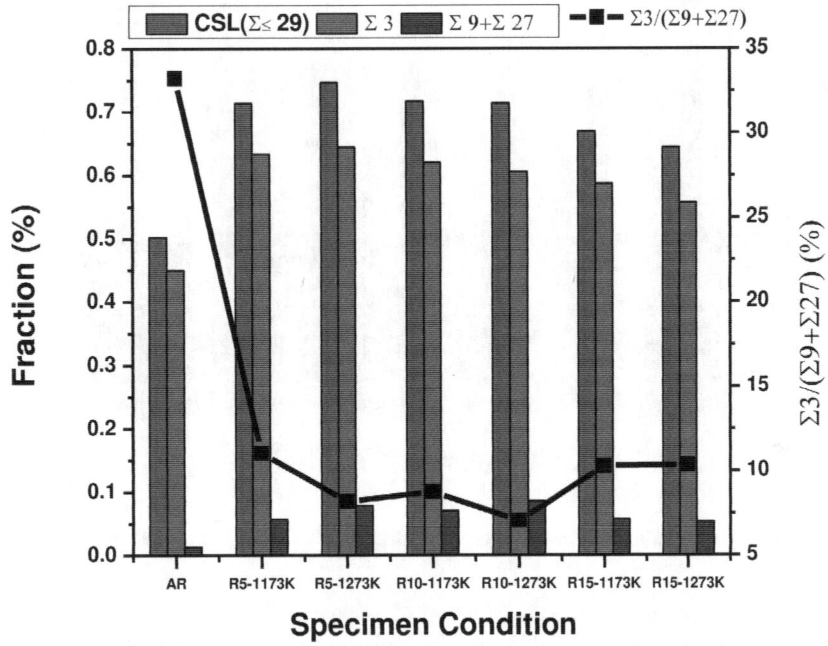

Figure 3: GBCD shown as fraction of CSL (≤ Σ 29), Σ3 and (Σ9+Σ27) in AR and GBE-treated specimens. The ratio of Σ3/(Σ9+Σ27) is also shown to asses random HAGBs connectivity

Increase in Σ3 following deformation and annealing emphasize the role of strain induced grain boundary migration necessary for the creation of special boundaries [14,17,19]. However, percentage of low Σ CSL boundaries marginally decreases towards higher level of cold deformation (see Fig. 3). It is also seen from Fig. 3 that the proportions of second and third order twins (i.e. Σ9+Σ27) are more than 5% in the GBE-treated specimens (with 5pct and 10pct pre-strain and annealed at 1273K). These higher order twin boundaries are preferable because they have the ability to generate large number of Σ3 boundaries on the basis of ''Σ3 regeneration mechanism'' which states that $\Sigma3^{n+1} + \Sigma3^{n} = \Sigma3$ occurs more frequently at triple junctions than $\Sigma3^{n+1} + \Sigma3^{n} = \Sigma3^{n+2}$ [3,5,17]. Further, these higher order Σ3 boundaries take part in reconfiguration of the existing grain boundary networks. Hence, the ratio of Σ3 to (Σ9+Σ27) parameter provides better indication of the fragmentation of random HAGBs networks [14,17]. This ratio found to decrease from 34 in AR specimen to 7 and 4 in R5-1273K and R10-1273K specimen, respectively (Fig. 3). This correlate well with OIM generated random HAGBs connectivity map of R5-1273K specimen (see Fig. 2b) clearly indicating substantial disruption in random HAGBs connectivity in GBE treated specimen.

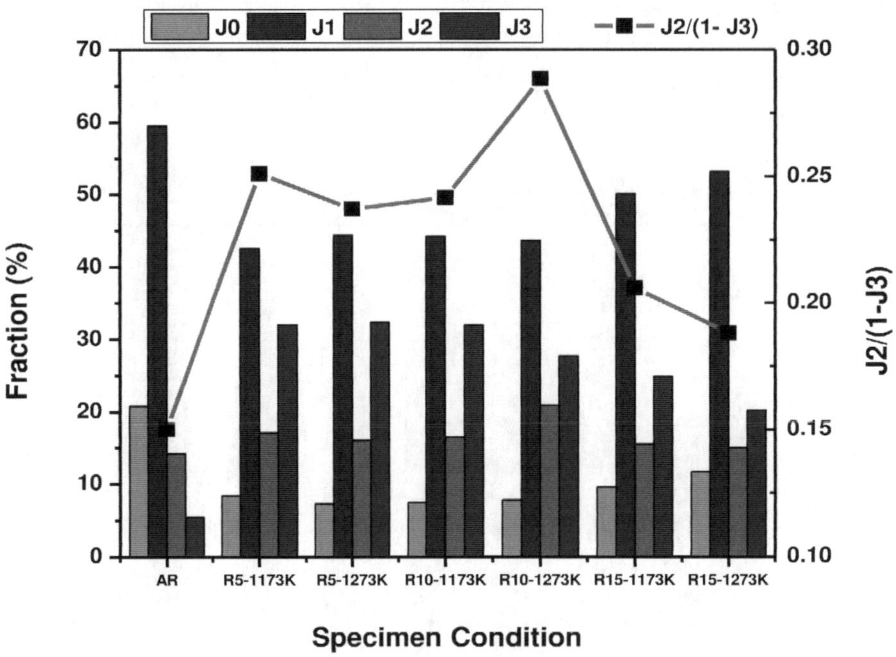

Figure 4: Triple junction distribution in AR and GBE-treated specimens

Triple junction (J_n) (where J_n indicates the fraction of triple junctions which have n low Σ CSL boundaries, $n = 0, 1, 2$, or 3) distribution for AR and GBE specimens is shown in Fig. 4. It is to be noted here that J_3-triple junctions are fully resistant against percolation related phenomena [6]. In this investigation, approximately 34% J_3-triple junctions are achieved in R5-1173K, R5-1273K and R10-1173K specimens in contrast to 5% in AR condition. Therefore, it could be easily inferred that the substitution of random HAGBs through formation and interaction of special boundaries increase the percolation-resistant triple junction. According to Schuh et al, $J_2/(1-J_3)$ could be used as an important parameter to measure the probability of crack arrest for inter-granular crack propagation as well as resistance to percolation related phenomena [20]. It could be observed that $J_2/(1-J_3)$ is maximum for the specimen R10-1273K owing to the large population of J_2-triple junction with intermediate number of J_3-triple junction (see Fig. 4).This value is also high in R5-1273K specimen. Hence, it could be envisaged here that both the R5-1273K and R10-1273K specimens may exhibit better resistance against percolation-related phenomena.

4. Conclusions

Thermo-mechanical processing involving small strain (5% and 10% cold rolling) and high-temperature annealing has led to GBE microstructure in a type 304L stainless steel. The maximum fraction of low Σ CSL boundaries (75% as compared 50% in AR specimen) has been obtained with 5% pre-strain following annealing at 1273K for 1 hour. The GBE specimens have also found to contain maximum number of percolation resistant J_2 and J_3-triple junctions. Hence the GBE specimens could exhibit better resistance against percolation related phenomena.

References

[1] Watanabe T 1984 *Res. Mech.* **11** 47
[2] Watanabe T 1988 *Mat. Forum* **11** 284
[3] Randle V and Owen G 2006 *Acta Mater.* **54** 1777
[4] Randle V 2006 *Scripta Mater.* **54** 1011
[5] Randle V 1999 *Acta mater.* **47** 4187
[6] Randle V 2004 *Acta Mater.* **52** 4067
[7] Kobayashi S, Tsurekawa S, Watanabe T and Palumbo G 2010 *Scripta Mater.* **62** 294
[8] Xia S, Li H, Liu T G and Zhou B X 2011 *J. Nucl. Mater.* **416** 303
[9] Mandal S, Bhaduri A K and Sarma VS 2012 *Mat. S. Forum* **702** 714
[10] Hu C, Xia S, Li H, Liu T G, Zhou B X, Chen W and Wang N 2011 *Corr Sc.* **53** 1880
[11] Telang A, Gill A S, Tammana D, Wen X, Kumar M, Teysseyre S, Mannava S R, Qian D and Vasudevan V K 2015 *Mat. Sci. and Eng. A* **648** 280
[12] Randle V 1996 *The Role of the Coincidence Site Lattice in Grain Boundary Engineering*, The Institute of Materials, London.
[13] Kobayashi S, Kamata A and Watanabe T 2015 *Acta Mater.* **91** 70
[14] Mandal S, Sivaprasad P V, Raj B and Sarma V S 2008 *Metal. Mater.Trans. A* **39** 3298
[15] Brandon D G 1966 *Acta Metall.* **14** 1479
[16] Gleiter H 1969 *Acta Mater.* **17** 1421
[17] Mandal S, Bhaduri A K and Sarma V S 2011 *J. Mat. Sci.* **46** 275
[18] Lee S L and Richards N L 2005 *Mat. Sci.and Eng. A* **405** 74
[19] Shimada M, Kokawa H, Wang Z J and Karibe I 2002 *Acta Mater.* **50** 2331
[20] Schuh C A, Kumar M and King W 2003 *Acta Mater.* **51** 687

Improvement of thermal properties of water and ethylene glycol by metallic and ceramic dispersion

Sripooja Mishra, Manaswini Chinara, Mohan Nuthalapati, A. Basu*

Department of Metallurgical and Materials Engineering, National Institute of Technology, Rourkela 769008, Odisha, India

*E-mail: basua@nitrkl.ac.in, anindya.basu@gmail.com

Abstract. Owing to rising need for more energy efficient coolant systems, the current work focuses on synthesis of Alumina-water, Alumina-Ethylene Glycol (EG), Zirconium-water, Zirconium-Ethylene Glycol particle dispersion systems for use as quenchants, with varying volume of particles obtained using a top–down approach of 10h ball milling. Characterisation of the particles was performed using XRD analysis, also measuring the crystallite size. Particle size measurement was also undertaken by using particle size analyzer, further corroborated by SEM analysis. A zeta-potential study was carried out to obtain the iso electric point for maintaining stability. Synthesis of the dispersions at different volume % of powders were followed by thermal conductivity measurements with water and EG base fluids, with and without addition of oleic acid as a surfactant. The stability studies were performed by visual observation. From the results it was observed that Al_2O_3 and Zr can increase conductivity of EG more prominently than water and stability of Zr based suspensions are better than Al_2O_3. Oleic acid did not play any positive role in these systems.

Keywords: Dispersion; Thermal conductivity; Water; Ethylene glycol; Surfactant

1. Introduction

The manifold growth in electronics, communication and computing technologies through miniaturization and enhanced rate of operation and storage of data result in enormous amounts of heat, thus demanding cooling capabilities with better performance. Fluids are often used as heat carriers in heat transfer equipment. Instances of significant usage of heat transfer fluids (conventionally water, ethylene glycol and engine oil) are in automotive, aeronautical cooling systems and manufacturing process heating and cooling systems. For use as coolant, thermal conductivity of the fluid plays a vital role in the heat transfer apparatus. The addition of metallic and metallic-oxide particles to coolants used in thermal control systems can intensely increase the thermal conductivity of the solvent.

In dispersion, movement of particles is random, thus carrying comparatively large volumes of enveloping liquid with them. This micro-scale interfacing [1] may occur between areas with a temperature gradient, subsequently resulting in a lowering of local temperature gradient for a given heat flux compared with the pure liquid case. Thus, as a consequence of Brownian motion, the thermal conductivity increases [2]. As heat transfer is prominent at the surface of the particle, greater surface area is desirable. Given that the stability criterion is fulfilled, microparticles offer high surface areas and reduced viscosity, as opposed to macroparticles. Further, microparticles do not pose a threat of erosion of the channels, unlike macroparticles. The surface to volume proportion is high implying that a very small volume fraction of dispersed particles (very low particle loading) can have a very high surface area (believed to significantly improve the heat transfer characteristics and stability of dispersions) [3]. In the same line, nano sized particle dispersed "nano fluid" has gathered attention in recent years. Many researchers have shown manifold improvement of thermal properties of water and Ethylene Glycol based nano fluids [3-6].

The current work focuses on fabrication of Alumina-water, Alumina-Ethylene Glycol (EG), Zirconium-water, Zirconium-Ethylene Glycol particle dispersion systems for use as quenchants, with varying % by volume of particles obtained using top–down approach. Alumina is a cheap and readily available ceramic oxide and exhibits inherent inertness. A dispersion of alumina in water does not lead to any unwanted chemical interaction owing to its inertness. Pure Zr develops a layer of ZrO on it resulting in its inertness in aqueous media. Addition of common surfactants such as oleic acid was also experimented to influence stability [7].

2. Materials and Methods

2.1. Powder synthesis and characterization

To reduce the size of Al_2O_3 and Zr powders, they were taken in two separate vials of Fritsch Pulverisette planetary ball mill (dual vial system) along with the grinding media i.e. the chrome steel balls. The vials and balls were cleaned with acetone and dried before using them. The ball to powder weight ratio was maintained 10:1. The powders were then milled in a planetary at 300 rpm for 10 hours. The milling was carried out in a wet medium using toluene.

The milled and unmilled powders were characterised by X-ray diffraction (XRD) to identify the phase and to determine the crystallite size. XRD was carried out by Rigaku Ultima-IV X-ray diffractometer using Cu K_α (α = 0.15406 nm) radiation. Crystallite size of both the milled powder was calculated from the Scherrer equation as given below:

$$D = \frac{0.9\lambda}{\beta \cos\theta}$$

Where, D is crystallite size, λ is X-ray wavelength, β is line broadening at half the maximum intensity and θ is Bragg's angle.

The particle size and zeta potential of the Al_2O_3 and Zr particles were measured using Malvern Nano ZS model particle size analyzer. For that, very small amount of the milled powders were dispersed in distilled water and later *pH* was changed systematically to find the zeta potential at different *pH* values and thus to calculate the Iso Electric Point (IEP). The size distribution and microstructure of the milled and unmilled powders was studied by scanning electron microscopy (SEM-JEOL-JSM-6480LV).

2.2. Synthesis of dispersion and characterization

Following the synthesis and characterization of the powders dispersion of the powders were made with distilled water and ethylene glycol with varying degree of dispersion (0.05 vol%, 0.1 vol%, 0.4 vol%, 0.8 vol%). Powders were added in 20 ml test tubes filled with water/ethylene glycol and test tubes were sonicated for 30 minutes. One drop of oleic acid was added as surfactant to prepare stable dispersion by increasing the magnitude of the zeta potential [7], and then they were sonicated again.

The thermal conductivity of the prepared dispersions of Al_2O_3 and Zr was measured using thermal conductivity probe (KD2 pro). The experimental value of thermal conductivity of the dispersions was compared with the theoretical thermal conductivity as calculated using the rule of mixture and Maxwell's effective medium theory [8].

The stability of the prepared dispersions of Al_2O_3 and Zr in water and ethyl glycol was studied over 24 hours of time. The dispersions were kept stationary for 24 hours after which the settling condition of the particles was checked. After the addition of oleic acid the stability of the dispersions were again observed and recorded.

3. Result and discussion

3.1. Particle size analysis and zeta potential measurements by particle size analyser

Fig. 1 (a) shows the particle size distribution of Al_2O_3 particles after milling. The particle size of Al_2O_3 was found to be in range of 700nm to 1700nm.The peak is observed at 1281nm. After milling the Zr, it shows bimodal particle size distribution as evident from Fig. 1(b). Two peaks were observed at 712nm and 5560nm. The particle size lies in the range of 400-1500nm and 4800-6500nm.

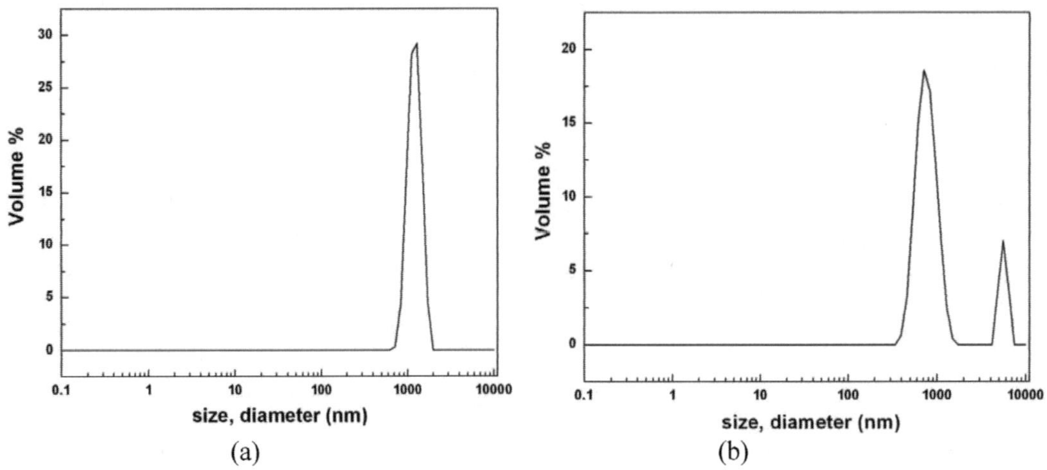

Fig. 1:- Particle size distribution (in terms of diameter) of (a) Al_2O_3 and (b) Zr by volume percentage

Table 1 shows the zeta potential of both the powders at different *pH* values and from extrapolation of the data the IEP of Al_2O_3 was found to be at *pH* ~4 and for Zr *pH*~8. During synthesis of the dispersed fluids the *pH* values were checked so that the values were away from the IEP, which increases the stability of the suspension by repulsion between each other and reduces the chance of agglomeration.

Table 1: pH vs. Zeta potential of Al_2O_3 and Zr

Al_2O_3		Zr	
pH	Zeta potential (mV)	pH	Zeta potential (mV)
3.1	10.00	5.1	-9.00
3.5	6.73	6.6	-4.39
4.4	-6.53	8.1	-1.44
6.5	-12.00		
9.2	-21.70		

3.2. Characterization by XRD

Fig. 2 shows the XRD plot of unmilled and milled Al_2O_3 and Zr powder. The XRD plot of the milled Al_2O_3 shows significant reduction in intensity of the peaks and a subsequent broadening of the peaks.

The decrease in the intensity of the peaks and the broadening of the peaks after milling can be attributed to the reduction in grain size of $Al_2O_3$and, incorporation of lattice strain due to ball milling. The average crystallite size of milled Al_2O_3 as calculated from Scherrer formulae was in the range of 9nm - 38nm. In case of XRD plot of Zr also, the intensity and number of peaks were found to be reduced. Considerable broadening of the peaks was also observed. The crystallite size of milled Zr as calculated from Scherrer formula was in the range of 8nm - 35nm.

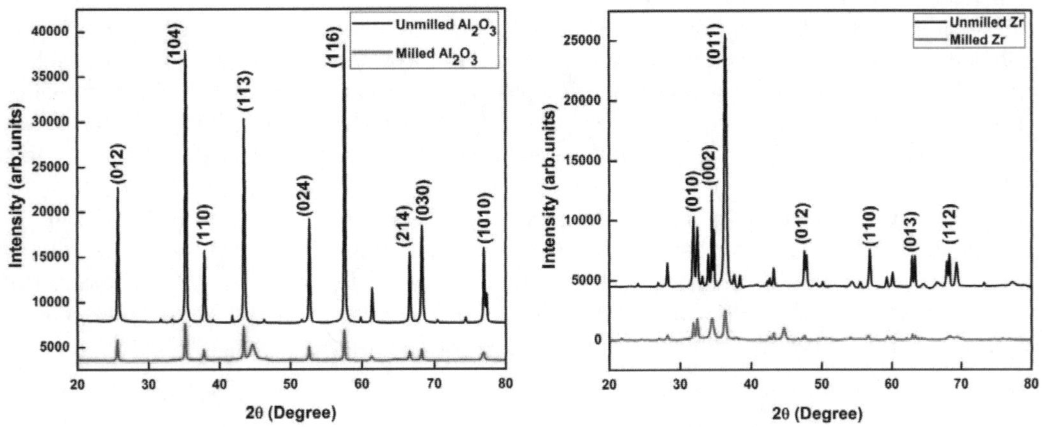

(a) (b)

Fig. 2:- XRD profiles of milled and unmilled (a) Al_2O_3 and (b) Zr powders

3.3. Scanning electron microscopy

Fig. 3 shows SEM images of milled powders and from that it is evident that milled powders have homogeneous size distribution. From Fig. 3(a) it seems that the average particle size of milled alumina powder was approximately 2μm and the same for Zr powder was found to be finer (~1 μm) (Fig. 3(b)). The difference in particle size obtained from SEM analysis and particle size analyser can be attributed towards agglomeration of the powder during test.

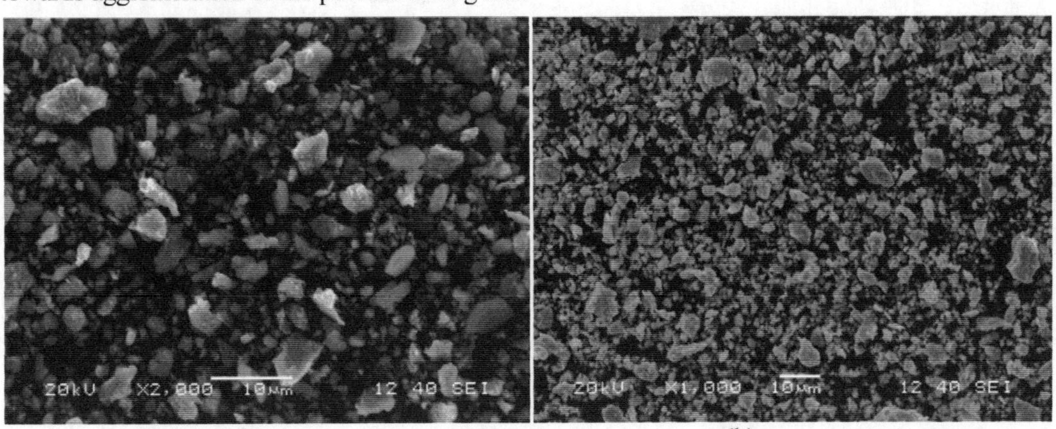

(a) (b)

Fig. 3:- SEM images of milled (a) alumina and (b) Zirconium powders

3.4. Thermal conductivity

Thermal conductivity of the dispersions is plotted against volume fraction of the particles. Such plots are displayed in Fig. 4 and 5. The horizontal dotted lines in the plots at k=1.4W/mK and k=0.3W/mK represent the measured conductivities of distilled water and pure EG respectively, taken as references. In the figures theoretical plot represents data based on rule of mixture. Plots also include theoretical values calculated based on Maxwell's equation. In both the figures the theoretical conductivities were found to increase with volume % of the particles.

From Fig. 4(a) it can be observed that with increasing alumina amount the conductivity decreases, which may attributed to the agglomeration factor. In case of surfactant addition also similar problem was observed. But, the observed conductivity of 0.05% and 0.1% dispersions are higher than that of water. Thus, minor addition of alumina particle can be helpful to increase conductivity of water. When alumina was dispersed in EG, it was observed that conductivity of all dispersions were higher than that of EG (Fig. 4(b)). But effect of surfactant was not positive as that was observed in water. But conductivity trend was increasing in nature with increase in alumina vol. %. Due to higher density of EG over water it can effectively suspend more particles and thus the suspension shows higher conductivity. Moreover, in EG, chance of agglomeration may be less than water. The theoretical values are completely based on the idea that particles should be there in suspension only and due to this fact the theoretical values are higher than the observed values reported here.

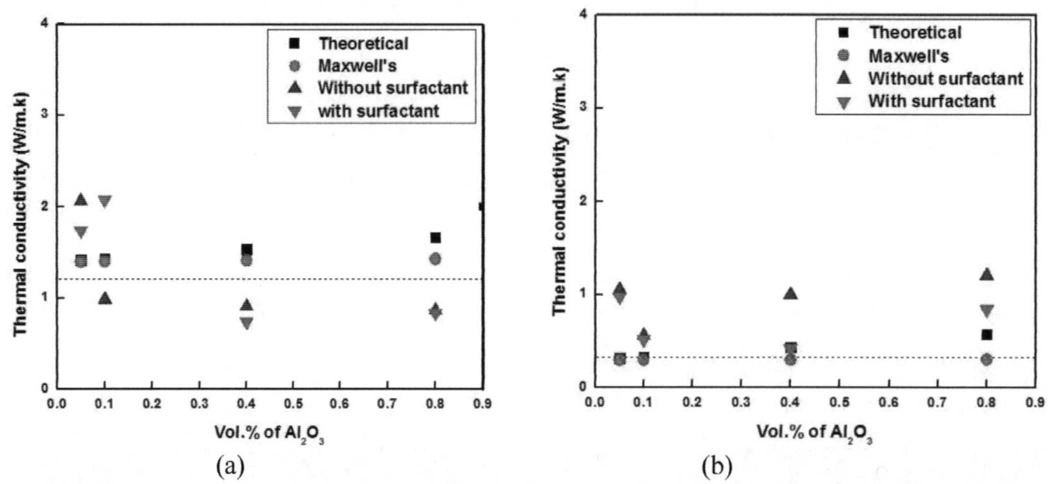

Fig. 4:- Thermal conductivity vs. vol. %of Alumina plot after dispersion in (a) water and (b) EG.

Fig. 5 shows the thermal conductivity nature of Zr based fluids. From Fig 5(a), it was observed that conductivity in water media without surfactant decrease with increasing vol. % of Zr particles possibly due to agglomeration. The trend is similar as was observed in alumina dispersion. However, the observed conductivity with added surfactant shows no such fixed trend. The observed conductivity of 0.05% dispersion alone is higher than that of water.

As shown in Fig 5(b), the observed conductivity of EG based fluid without surfactant rapidly increases from 0.05% to 0.1%. After that, it decreases with further increase in vol. % of Zr particles. The observed conductivity with added surfactant shows no such fixed trend. The observed conductivity of all dispersions is higher than that of EG.

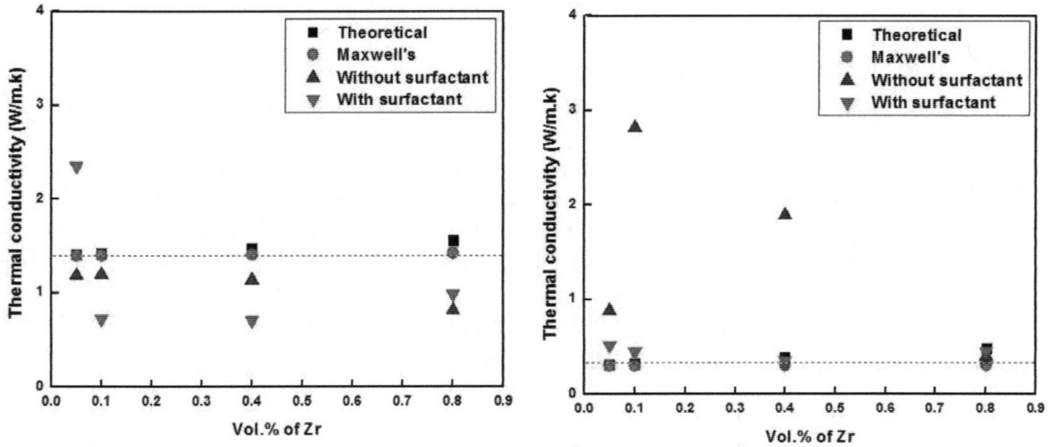

Fig. 5:- Thermal conductivity vs. vol. % of Zr plot after dispersion in (a) water and (b) EG

3.5. Stability study

The stability of the dispersed fluids was checked after 24 hours and the results are summarized in Table 2. From the table it can be observed that only zirconium based fluids having dispersion content up to 0.4% were stable even after 24 hours. This may be attributed towards smaller particle size of Zr compared to Al_2O_3. Moreover, higher addition of Zr makes it unstable.

From stability study and earlier section it can be observed that oleic acid has negligible effect, mainly in water based dispersion. The solubility of oleic acid is more in EG than in water. For better stability, water soluble surfactants can be tried out [9]. Moreover, though oleic acid can be helpful for better dispersion distribution, its conductivity may also play a negative role in heat transfer as it forms a coating around the dispersed particles [10].

Table 2: Observed stability of the dispersions after 24

Sample	Without oleic acid	With oleic acid
0.05%Al_2O_3 + Water	Unstable	Unstable
0.1% Al_2O_3 + Water	Unstable	Unstable
0.4% Al_2O_3 + Water	Unstable	Unstable
0.8% Al_2O_3 + Water	Unstable	Unstable
0.05% Al_2O_3 + EG	Unstable	Stable
0.1% Al_2O_3 + EG	Unstable	Unstable
0.4% Al_2O_3 + EG	Unstable	Unstable
0.8% Al_2O_3 + EG	Unstable	Unstable
0.05% Zr + Water	Stable	Stable
0.1% Zr + Water	Stable	Stable
0.4% Zr + Water	Stable	Stable
0.8% Zr + Water	Unstable	Unstable
0.05% Zr + EG	Stable	Stable
0.1% Zr + EG	Stable	Stable
0.4%Zr + EG	Stable	Stable
0.8%Zr + EG	Unstable	Unstable

4. Conclusion

From the present study following conclusions may be drawn:

(i) Al_2O_3 and Zr powders were ball milled and milled powder shows reduction in particle and crystallite size.
(ii) Thermal conductivity of water was enhanced only by small addition of Al_2O_3 and Zr (0.05-0.1%).
(iii) Thermal conductivity of EG was enhanced with addition of any amount of Al_2O_3 and Zr.
(iv) Thermal conductivity improvement was better in EG than water in both Al_2O_3 and Zr.
(v) Stability of the suspension after 24 hour was found to be better in case of Zr due to its finer particle size.
(vi) Oleic acid did not show any remarkable positive effects on these systems.

References

[1] Keblinski P, Phillpot SR, Choi SUS, Eastman JA 2002 *Int. J. Heat Mass Tra.* **45(4)** 855.
[2] Jang SP, Choi SU 2004 *Appl. Phy. Let.* **84(21)** 4316.
[3] Eastman JA, Choi SUS, Li S, Yu W, Thompson LJ 2001 *Appl. Phy. Let.* **78(6)** 718.
[4] Murshed SMS, Leong KC, Yang C 2005 *Int. J. Therm. Sci.* **44(4)** 367.
[5] Garg J, Poudel B, Chiesa M, Gordon JB, Ma JJ, Wang JB, Chen G 2008 *J. Appl. Phy.* **103(7)** 074301.
[6] Hwang KS, Jang SP, Choi SU 2009 *Int. J. Heat Mass Tra.* **52(1)** 193.
[7] Hwang Y, Lee JK, Lee JK, Jeong YM, Cheong SI, Ahn YC, Kim SH 2008 *Powder Technol.* **186(2)** 145.
[8] Maxwell JC 1904 *A Treaties on Electricity and Magnetism* Oxford University Press Cambridge.
[9] Yu W, Xie H 2012 *Journal of Nanomaterials* **1**.
[10] Manna I 2012 *Journal of the Indian Institute of Science* **89** pp 21-33.

Synthesis CNTs Particle Based Abrasive Media for Abrasive Flow Machining Process

Sonu Kumar[a], Q.Murtaza[b*], R.S Walia[a], S. Dhull[a]. P. K. Tyagi[c]

[a]Precision Engineering Laboratory, Mechanical & Production Engg., Delhi Technological University (formerly Delhi College of Engineering), Delhi, India
[b]Mechanical Engineering Department, Aligarh Muslim University, Aligarh, India
[c]Applied Physic Department, Delhi Technological University (formerly Delhi College of Engineering), Delhi, India

[*]E-mail: qasimmurtaza@gmail.com, Phone: +91-9654364948

Abstract. Abrasive flow machining (AFM) is a modern fine finishing process used for intricate and internal finishing of components or parts. It is based on flowing of viscoelastic abrasive media over the surface to be fine finished. The abrasive media is the important parameter in the AFM process because of its ability to accurately abrade the predefined area along it flow path. In this study, an attempt is made to develop a new abrasive, alumina with Carbon non tubes (CNTs) in viscoelastic medium. CNTs in house produced through chemical vapour deposition technique and characterize through TEM. Performance evaluation of the new abrasive media is carried out by increasing content of CNTs with fixed extrusion pressure, viscosity of media and media flow rate as process parameters and surface finish improvement and material removal as process responses in AFM setup. Significantly improvement has been observed in material removal and maximum improvement of 100 % has been observed in the surface finish on the inner cylindrical surface of the cast iron work piece.

Keywords:AFM, CNTs, TEM

1. Introduction

Abrasive flow machining (AFM) is used to fine machining by flowing of viscoelastic abrasive media over predetermined surface areas. AFM process has capability of fine finishing of intricate shapes and holes which are not accessible easily [1]. In addition AFM process has many attractive advantages, such as self--sharpening, self-adoptability, controllability and finishing tool requires neither compensation nor dressing. AFM can be classified based on motion or flow of abrasive media as: One way AFM, Two-way AFM and Orbital AFM process [2-7]. In AFM process parameters, extrusion pressure has a major role in material removal and reduction of surface roughness [5-15]. On the other hand, abrasive media acts as deformable grinding tool. It is the key of Abrasive Flow machining. In general, abrasive media flow volume is increased than material removal increases. Theoretically it can be understand that as media flow volume is increased, more number of abrasive particles comes in contact with work piece and more abrasion takes place. Rhoades [10] has reported that media flow rate is less influential parameter in respect to material removal. Slower slug flow rates are best for uniform material removal and high slug flow rates produce large edge radii. The media flow rate is insignificant with regard to material removal. On the other hand, it has been claimed by Jain and Jain [7] Singh [11] the media flow rate influences both of the material removal and surface roughness. Williams and Raju [5] and Williams et. al. [12] have reported that viscosity of the media is one of the significant parameters of the AFM process. Keeping all other parameters constant, an increase in viscosity improves both material removal and reduction of surface roughness. According to one thumb rule [14] finer abrasives should be used when the initial roughness of the work surface is less. The reason for a decrease in material removal is that with an increase in mesh size (or decrease in grain size in mm) the depth of penetration as well as width of penetration, decreases. There are various abrasive particles like Al2O3,SiC, CBN, diamond powder and so on are available which are being used for abrasive finishing process [6-14]. AL2O3 and SiC are most suitable abrasives for many applications but Cubic boron nitride (CBN) and diamond are specifically used for special applications. Abrasive particles to base material ratio can be varying from 2 to 12. The abrasives have limited life. The additives are used to enhance the base carrier to get the desired flowability and rheological characteristic of the media. In the present study, a new type of abrasive media is developed to insure faster finishing and reduction of surface roughness. The new type of abrasive flow media is based on synthesis of iron filled CNTs with AL2O3 and its performance evaluation of cast iron workpiece in AFM as CNTs is known very strong mechanical properties. CNTs has been synthesised through chemical vapour deposition technique as it is fast and economical viable process [15-18].

2. Materials and Methods

The newly developed abrasive media, as shown in fig. 1, which having AL2O3 (180 mesh size), polymer and hydrocarbon gel (Polymer to gel ratio:1:1) is mixed properly in predetermined proportion to get the viscous elastic media which can machine effectively. Hydrocarbon gels are commonly used lubricants in the media. All additives including CNT particles (0, 5grams and 10 grams) in different percentages are carefully blended in predetermined quantities to obtain consistent formulation with 810 Pas media viscosity. CNTs with filled Fe was synthesized by Chemical Vapour Deposition. In this process, heating the solution of Ferrocene(5gm) and Toluene(25ml) in the oven, maintaining the oven temperature around 800°C and argon gas(100 sccm) is passed to get inert atmosphere (setup fig. 2). The performance evaluation of the newly developed abrasive media was conducted in Precision Manufacturing Laboratory developed vertical double acting AFM machine setup (table 1), of Delhi Technological University. Cast Iron work piece (fig. 3) were prepared with initial surface roughness and weight were taken. The surface roughness was measured in five different locations using Taylor Hobson. The internal cylindrical surface was finished by AFM process. Each workpiece was machined for a predetermined number of cycles (8 cycles) and extrusion pressure.

Fig. 1:- Abrasive laden Media (carrier+Al_2O_3+CNT) Fig.2. Thermal CVD Setup Fe-filled CNTs.

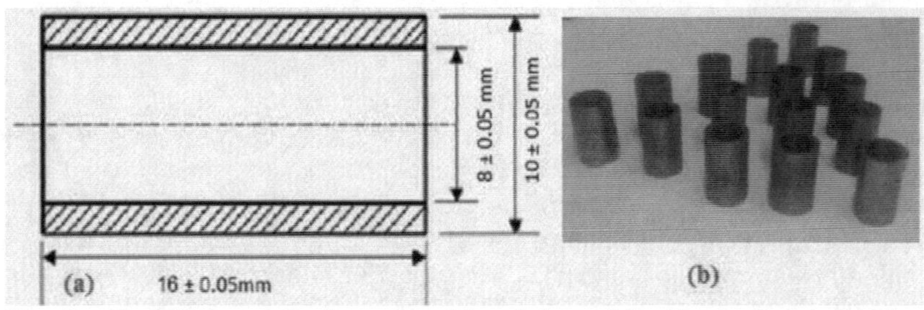

Fig. 3:- Cast Iron work pieces used in trials and its dimensional details.

Table1:- Process parameter of AFM setup.

S. No.	Process Parameter	Range	Unit
1	Extrusion Pressure	5-35	MPa
2	No. of Cycle	1-9	Number
3	Temperature	32± 2	°C
4	Media Flow volume	290	cm^3
5	Capacity	25 + 25	Ton
6	Stroke length	96	mm
7	Hydraulic cylinder Bore diameter-2 No.	130	mm
8	Hydraulic cylinder Stroke	90	mm
9	Working Pressure	210	kg/cm^2
10	Maximum Pressure in the Cylinder	35	MPa
11	Stroke Length of Piston	300	mm

3. Results and discussion

3.1. Characterization of CNTs filled with Iron

Fig. 4 shows the TEM micrographs of Fe filled CNTs at different locations and it shows that the growth of CNTs is uniform and there is very less amount of other carbon impurities such as amorphous carbon, soot or carbon particles.

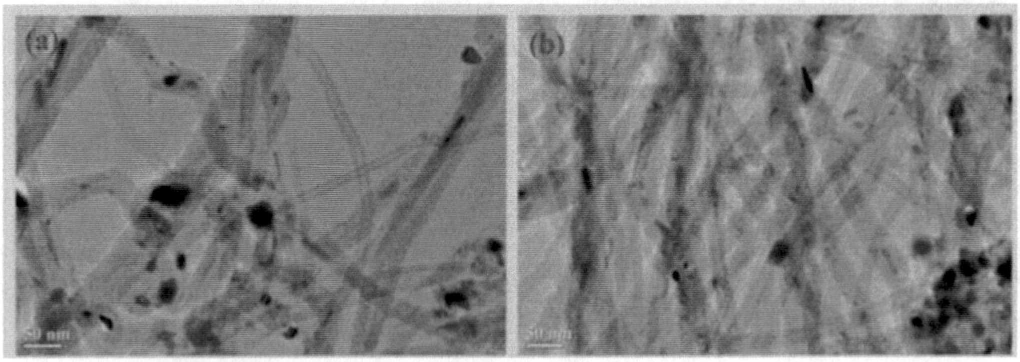

Fig. 4:- TEM images of CNTs at different locations.

3.2. Characterization of developed abrasive Media

The three abrasive media has developed. The first abrasive media has aluminium oxide, polymer and hydrocarbon gel is mixed properly in predetermined proportion to get the viscous elastic media with no content of CNTs particles. The second and third abrasive media added with 5 and 10 grams CNTs with iron filled respectively. Fig. 5 is the SEM micrograph of the media (Al_2O_3 and Carrier only) before addition of CNTs where, (a) is the SEM micrograph at scale bar of 100 μm in which abrasives edges are not clearly visible, (b) is the SEM micrograph at a scale bar of 5μm in which abrasive edges are visible. This SEM micrograph (Fig. 5 (a)) of the media (Al_2O_3 +carrier+ Fe-filled CNTs) shows the sharp edges of alumina particle are clearly visible which is responsible for machining the work surface.

Fig. 5:- SEM image of media (Al_2O_3 +carrier+ Fe-filled CNT.

3.3. Effect of CNTs on material removal

Fig. 6 shows that as the amount of CNT particles increased, the material removal of cast iron workpiece increased when other AFM machining parameters are constant. As we know that CNT particles are harder as compare to the aluminium oxide and CNT has very sharp cutting edges. So when the extrusion pressure is applied, it easily cuts the peak of the surface material and causes more material removal.

3.4. Effect of CNT on surface finish

It can be seen from the fig. 7, that as the amount of CNT increases the % increase in surface finish (Ra) slightly increases but after level 2 (10 grams CNTs), % improvement in Ra sharply increased (100%).

This is due to more number of abrasive particles taking part in machining process and also continues to remove fresh materials from the work surface which lead to increase in MR and surface finish.

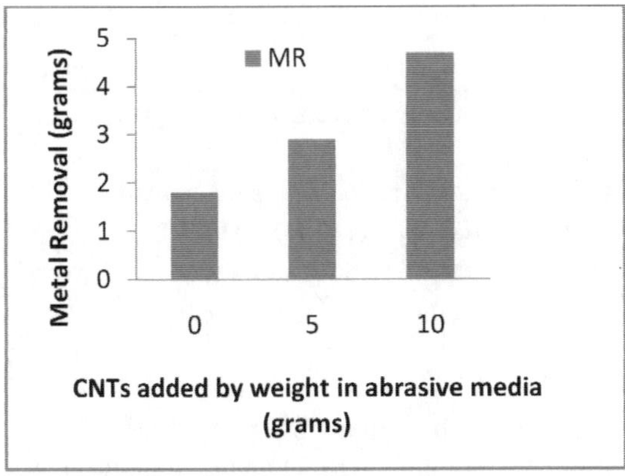

Fig. 6:-Effect of CNT particles on MR.

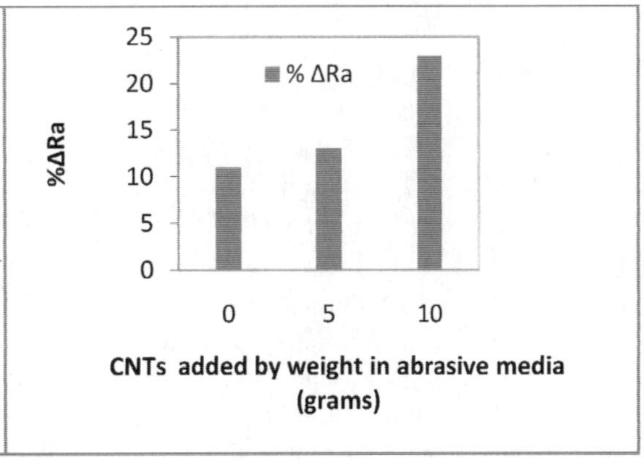

Fig. 7:- Effect of CNT on Ra.

4. Conclusion

The present study was carried out to develop CNTs based abrasive media for AFM successfully. It has been found and showed that the use of CNTs with abrasive media led to an improvement in the response parameter of percentage improvement in surface finish and material removal. At selected parameters a maximum improvement of 100% has been observed in the Surface finish on the inner cylindrical surface of the cast iron work piece.

References

[1] L.J. Rhoades, Abrasive Flow Machining, *Technical Paper of the Society of Manufacturing Engineers (SME),MR*(1989) 89-145.

[1] R.E.William and K.P. Rajurkar L.J, Rhoades, Performance Characteristics of abrasive flow machining, *SME Technical paper No. FC*(1989) 89-806.

[3] Tzeng, Yan, Hsu and Lin, Self-modulating abrasive medium and its application to abrasive flow machining for finishing micro channel surfaces, *Int. J. of Advanced Manufacturing Technology* 32(2007) 1163-1169.

[4] S. Rajesha, G. Venkatesh, A.K. Sharma and K. Pradeep, Performance study of a natural polymer based media for abrasive flow machining, *Indian Journal of Engineering & materials Sciences* (2010) 407-413.

[5] R.E. William R.E and K.P. Rajurkar, Stochastic modeling and analysis of abrasive flow machining, *Trans. ASME, J Engg. For Ind,*114 (1992), 74-81.

[6] R.K. Jain and V.K. Jain, Abrasive fine finishing process-a review, *Int. J. of Manufacturing science and Production* 2(1) (1999) 55-68.

[7] R.K. Jain, V.J.C. Jain and P.M. Dixit, Modelling of material removal and surface roughness in abrasive flow machining process, *Int. J. of Machine Tool and Manufacture*39(1999)1903-1923.

IOP Conf. Series: Materials Science and Engineering **115** (2016) 012034 doi:10.1088/1757-899X/115/1/012034

[8] U.S. Shan and A.K. Dubey, Micro machining by flow of abrasives, *Proceedings 17th MMTDR Conference, Warangal, India* (1997) 269-275.

[9] G.F. Benedict, Nontraditional Manufacturing Processes, *Marcel Dekker, New York* (1987).

[10] L.J. Rhoades and T.A. Kohut, Reversible Unidirectional AFM, (1991) *US patent number 5,070,652.*

[11] S. Singh, Studies in metal finishing with magnetically assisted abrasive flow machining, *Ph.D. Thesis* (2002) I IT Roorkee, India.

[12] Williams R.E., and Rajurkar K.P, Monitoring of abrasive flow machining using acoustic emission, *Proc. First S.M. Wu Symposium on Manufacturing Sciences*1 (1994) 35-41.

[13] K. Przyklenk, Abrasive flow machining- A process for surface finishing and deburring of work pieces with a complicated shape by means of abrasive laden media, *Advances in Non-Traditional machining, PED, ASME* 22 (1986) 101-110.

[14] A. Agrawal, V.K. Jain and K. Muralidhar, Experimental determination of viscosity of abrasive flow machining media, *Int. J. of Manufacturing Technology and Management* 7(2005) 142-156.

[15] S. Devanshu, K. Amulya, Q. Murtaza(CNT reinforced aluminium matrix composite-A Review, *Materials Today: Proceedings*2 (2015) 2886–2895.

[16] R. Kumar, Q.Murtaza, R.S.Waila (2014), Three start helical abrasive flow machining for ductile material, *Procedia Materials Science*6 (2014)1884–1890.

[17] S. R, Ruoff ,D. Qian , W.K. Liu, Mechanical properties of carbon nanotubes: theoretical predictions and experimental measurements, C. R. Physique 4 (2003) 993–1008.

[18] V. N. Popov, (2014), Carbon nanotubes: properties and application, *Materials Science and Engineering* 43 (2014) 61–102.

Ultraviolet photodetection characteristics of Zinc oxide thin films and nanostructures

S P Ghosh[1], K C Das[1], N Tripathy[1], G Bose[2], D H Kim[3], T. I. Lee[4], J.M. Myoung[3], J.P. Kar[1*]

[1]Department of Physics and Astronomy, National Institute of Technology, Rourkela, India 769008
[2]FST, IFHE University, Hyderabad, India 501203
[3]Department of Materials Science and Engineering, Yonsei University, Seoul, Republic of Korea' 120749
[4]Department of Bio-Nanotechnology, Gachon University, Gyeonggi-do, Republic of Korea, 461701

[*]E-mail:karjp@nitrkl.ac.in, Phone: +919438532157

Abstract. ZnO thin films were deposited by using RF sputtering technique at 150W and 4×10^{-3} mbar pressure. Post-deposition rapid thermal annealing of ZnO thin films were carried out at 1000°C for 600sec. ZnO nanostructures (nanowires/nanorods) were hydrothermally grown by using equimolar solution of Zinc nitrate and hexamethyltetramine. Morphology of ZnO thin films and nanostructures were investigated using scanning electron microscope. A comparative study on the ultraviolet photodetection behaviour of ZnO thin films and nanostructures were carried out by adopting current-voltage measurement technique at room temperature. ZnO annealed films have shown relatively lower transient photocurrent decay as compared to as-deposited film, which may be due to the faster evacuation of the charge carries. However, in case of ZnO nanorods transient photocurrent decay is relatively slow than that of nanowires, which is attributed to delayed readsorption of oxygen molecules onto the nanorods surface due to its larger width.

Keywords: Thin films; Nanostructures; Photodetection; Ultraviolet; Zinc Oxide

1. Introduction

In recent years, ultraviolet (UV) light is used in many commercial fields such as telecommunications, ozone and pollution monitoring, high temperature flame detection, and missile warning systems [1–3]. The well-established CMOS compatible semiconductor such as silicon (Si) has a limitation for UV detection due to its low bandgap. Since the bandgap energy of Si is 1.1eV, costly high pass optical filters and phosphors are needed to stop low energy photons [4-9]. Therefore, the Si based UV detector performance are significantly reduced due to its efficiency and increase of dark currents [10-11]. However, the bandgap of ZnO (3.37eV) corresponds to the intermediate region between UV and visible light. In addition, outstanding physical and chemical properties of ZnO nanostructures such as large surface area, radiation hardness [12], superior spatial resolution [13] and high electron mobility provide a platform for versatile applications such as in food processing, sterilization of medical equipment, research laboratories, semiconductor processing industry etc., where an installation of UV detector is highly required. In addition, ZnO is also highly desired for thin film transistors, light emitting diodes, solar cells, surface acoustic wave and energy harvesting devices [14-15]. Herein, synthesis of ZnO thin films/nanostructures and study of their UV detection behavior were systematically carried out.

2. UV detection mechanism of ZnO thin films and nanostructures

Intrinsic defects of ZnO that is oxygen vacancies and/or zinc interstitials plays a vital role in UV detection mechanism. The conductivity of ZnO is extremely sensitive to UV light exposure. Basically, the photosensitivity in ZnO is regulated by chemisorption of oxygen [15-19]. In the absence of UV light, oxygen molecules get adsorbed on the surface as negatively charged ions by confining the free electrons, thereby create a depletion layer with low conductivity near the surface:

$$O_2\ (g) + e^- \rightarrow O_2^-\ (ads.) \quad \textbf{(1)}$$

When the sample is illuminated with UV light, whose photon energy is more than the bandgap of ZnO, it causes generation of electron-hole pairs. Adsorbed oxygen ions combine with the holes to produce oxygen molecules, which desorbs from the surface. Under bias, the unpaired electrons are collected at the anode; thereby an increase in conductivity takes place with the decrease in the width of depletion layer [20-22].

$$h^+(h\vartheta) + O_2^-\ (ads.) \rightarrow O_2 \quad \textbf{(2)}$$

After the UV exposure, oxygen gets readsorbed on the surface until equilibrium is restored. This readsorption is a slow process and significantly lengthens the relaxation time constant for the devices [23-27]. This slow transient can be expressed by an exponential function as [22]

$$I(t) = I_d + (I_p - I_d)e^{-\frac{t}{\tau}} \quad \textbf{(3)}$$

Where I (t) is the transient current, I_d is the current measured in dark, Ip is the current measured under UV illumination, t is the time after turning off UV light. Using exponential fit to the experimental data, the decay time constant (τ) can be calculated for all the samples. The details of the samples used for UV detection are given in Table 1.

3. Experimental

For UV detection measurement ZnO thin films were deposited by using RF sputtering technique at 150W and 4×10^{-3}mbar pressure. Post-deposition rapid thermal annealing (RTA) of ZnO thin films were carried out at 1000°C for 600sec. ZnO nanostructures (nanowires/nanorods) were hydrothermally grown by using equimolar solution of Zinc nitrate and hexamethytetramine. Morphology of ZnO thin films and nanostructures were investigated using scanning electron microscope (SEM).

Top electrodes were deposited by thermal evaporation of aluminum. Ultraviolet photodetection properties were investigated by adopting current-voltage (I-V) and current-time (I-t) measurement. The nature of electrical contact between metal (aluminum) and semiconductor (ZnO) samples has been observed by taking the current-voltage measurement before studying the photodetection behavior. The nature of electrical contact between metal (aluminium) and semiconductor (ZnO) samples has been observed by taking the current-voltage measurement before studying the UV detection behaviour. Linear enhancement of current with sweep in voltage of ±5V for all samples depicted good Ohmic nature of the electrical contact. Fig. 1 represents the current voltage characteristic of ZnO thin film based UV detector. The details of the samples used for UV detections are given in Table 1.

Table 1. Comparative study of UV detection behaviour of various samples.

Sample No.	Sample Type	Growth Condition	UV on/off ratio	Decay Constant (τ in sec.)
S1	ZnO thin film	150W, 4×10^{-3}mbar, Ar/O_2-2:3	1.5	47
S2	Annealed film	1000°C, 600sec	1.7	21
S3	Nanowires	95°C, 0.01M	6.1	39
S4	Nanorods	75°C, 0.05M	2.9	57

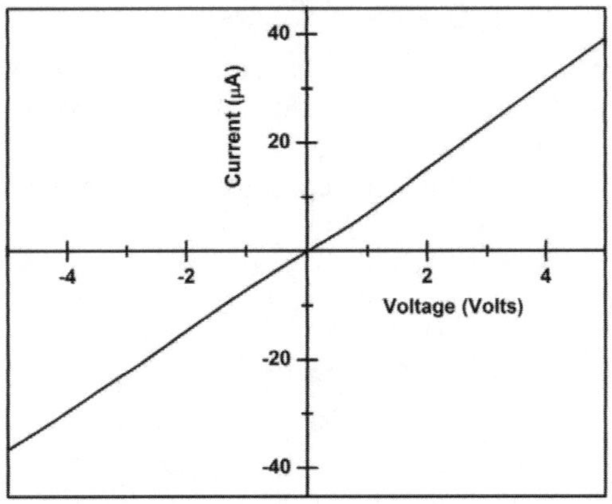

Fig. 1:- Current -voltage plot of ZnO thin film based UV detector in air ambient.

4. Results and discussions

4.1. Morphological properties of ZnO thin films and nanostructures

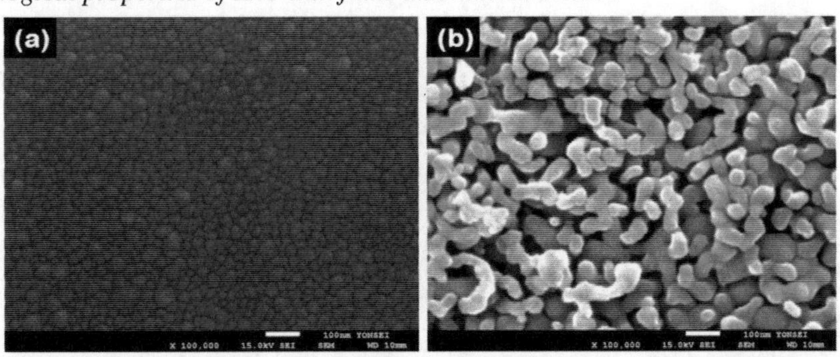

Fig. 2:- SEM images of ZnO thin films (a) As-deposited (S1), and (b) annealed at 1000°C in air ambient for 600sec (S2). The scale bar represents 100nm.

Fig. 2 (a) shows the SEM images of rf sputtered ZnO thin films deposited at (150W, 4×10^{-3}mbar, Ar/O_2-2:3). Smaller granular structure with uniform and continuous morphology of ZnO film has been observed. Fig. 2 (b) shows the SEM images of ZnO thin film annealed at 1000°C in air ambient for 600sec. Longer annealing period provides sufficient time and thermal energy to the ZnO microstructures for subsequent growth and as a result rough and porous surface is obtained [28].

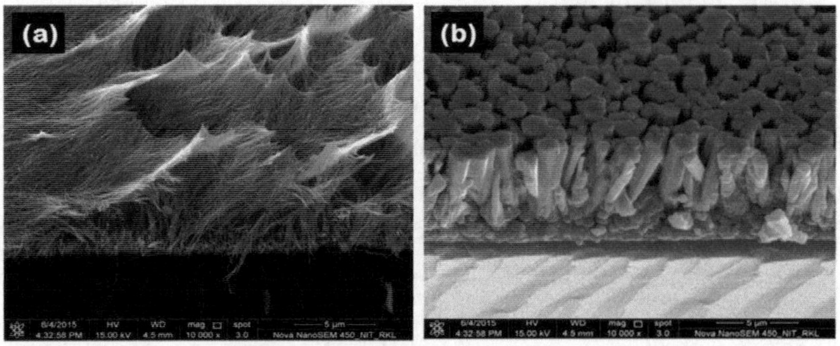

Fig. 3:- FESEM images of ZnO nanostructures (a) ZnO Nanowire (S3), and (b) ZnO nanorods (S4). The scale bar represents 5μm.

Fig. 3 (a) depicts the tilted cross-sectional SEM images of ZnO nanowires. The nanowires were grown at lower concentration (0.01M at 95°C) and results thin and long nanowires with a diameter < 30nm, and the length around 5-6μm. Fig. 3 (b) depicts a nanorod-like morphology with a concentration of 0.05M at 75°C. Although low temperature favours the formation of nucleation sites, it does not have enough energy for the subsequent growth of nanostructures. It indicates that the growth temperature is not sufficient for faster growth along c-axis. Therefore, the length of nanostructures was restricted and as a result short nanorods are obtained.

4.2. UV detection of ZnO thin films

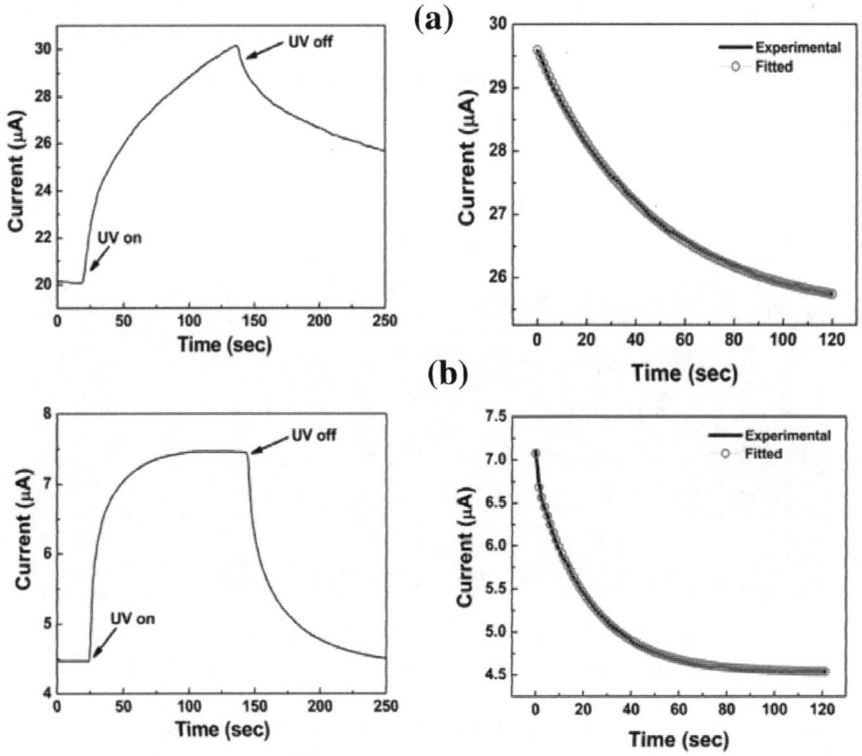

Fig. 4:- Current-time and decay curve of (a) as-deposited (S1), and (b) RTA processed ZnO thin film (S2) under UV exposure.

Fig. 4 (a) and (b) shows current-time and decay curve of (a)as-deposited (S1), (b)RTA processed (S2) ZnO thin film. The on/off ratio for the sample S1 is 1.5, whereas it is found to be 1.7. for S2. Fig. 2 (a) and (b) also shows the transient photocurrent decay of samples S1 and S2 with decay time constant (τ) value of 47sec and 21sec, respectively. Annealed films have relatively lower decay constant (τ), which may be due to the faster evacuation of the charge carries. The relatively better response of sample S2 may be due to the annihilation of surface trap states during RTA process. In addition, the sample S2 has shown the quasi saturation of photo current.

4.3. UV detection of ZnO nanostructures

Fig. 5 (a) and (b) shows current time and decay curve of ZnO nanowires (S3) and ZnO nanorods (S4) under UV excitation. The on/off ratio of ZnO nanowires (S3) photodetector is about 6.1, whereas it is found to be 2.9 for sample S4. From experimental results, the decay time constants $\tau 1$ and $\tau 2$ are estimated to be around 39sec and 57sec, respectively by fitting the decay curve using the exponential function. However, the photocurrent was not well saturated for both sample S3 and S4, which may be due to the presence of surface states in the aqueous grown nanowires/nanorods. Secondly, in case of nanorods (S4) the decay time constant is relatively slow, which is attributed to delayed evacuation of charge carriers.

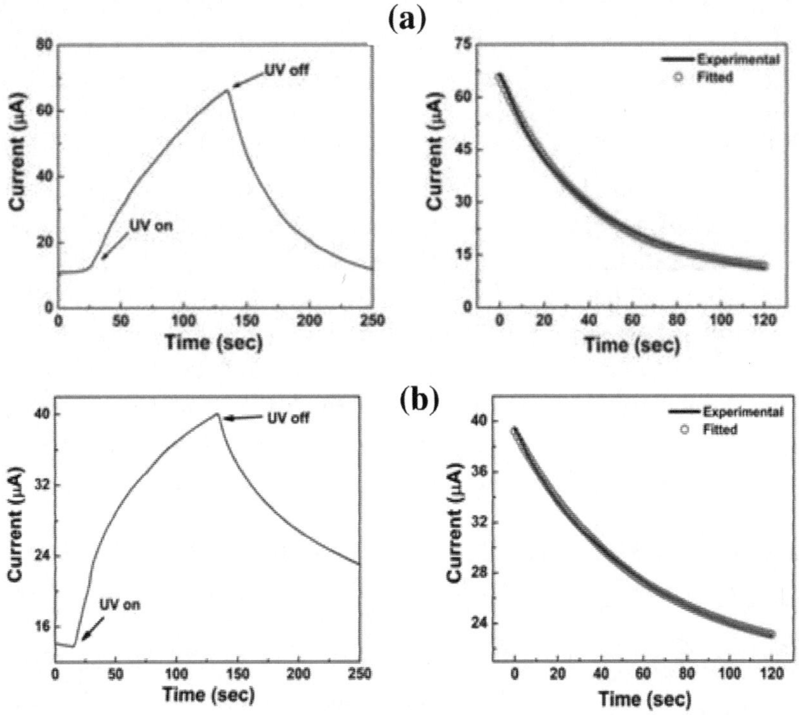

Fig. 5:- Current- time and decay curve of (a) ZnO nanowires (S3), and (b) ZnO nanorods (S4) under UV exposure.

5. Conclusion

Ultraviolet photodetection properties of ZnO thin films and nanostructures were obtained by adopting current-voltage (I-V) and current-time (I-t) measurement. Annealed films have shown relatively lower decay constant (τ) than that of as-deposited ZnO films, which may be due to the faster evacuation of the charge carries. Among different nanostructures, ZnO nanowires have shown higher on/off ratio as compared to ZnO nanorods which is attributed to their higher aspect ratio. From experimental results, the decay time constants τ_1 and τ_2 are estimated to be around 39sec and 57sec, for nanowires and nanorods, respectively by fitting the decay curve using the exponential function. The above experimental results can provide a platform in order to achieve room temperature operated next generation of metal oxide UV detector.

Acknowledgments
This work was supported by the Department of Science and Technology (DST), India sponsored Indo-Korea project (INT/Korea/P-16/2013). This research was also supported by the International Research & Development Program of the National Research Foundation of Korea (NRF) funded by the Ministry of Science, ICT & Future Planning (Grant number: 2012K1A3A1A19038371).

References

[1] C. S. Rout, M. Hegde, A. Govindaraj and C.N.R. Rao, Ammonia sensors based on metal oxide nanostructures, *Nanotechnology* 18 (2007) 9.

[2] B. Timmer, W. Olthuis and A. Berg, Ammonia sensors and their applications—a review *Sensors and Actuators B*, 107 (2005) 666.

[3] H. Nanto, T. Minami and S. Takata, Zinc-oxide thin film ammonia gas sensor with high sensitivity and excellent selectivity, *Journal of Applied Physics,* 60 (1986) 482.

[4] G. Sberveglieri, S. Groppelli, P. Nelli, A. Tintinelli and G. Giunta, Stable and highly sensitive gas sensors based on semiconducting oxide nanobelts, *Sensors and Actuators* B 24 (1995) 588.

[5] M. Aslam, V. A. Chaudhary, I. S. Mulla, S. R. Sainkar, A. B. Mandale, A. A. Belhekar and K. Vijayamohanan, A highly selective ammonia gas sensor using surface-ruthenated zinc oxide, *Sensors and Actuators A* 75 (1999)162-167.

[6] M. S. Wagh, G. H. Jain, D. R. Patil, S. A. Patil and L. A. Patil, Modified zinc oxide thick film resistors as NH_3 gas sensor, *Sensors and Actuators B* 115 (2006) 128-130.

[7] T. Tut, T. Yelboga, E. Ulker and E. Ozbay, Solar-blind AlGaN-based p-i-n photodetectors with high breakdown voltage and detectivity, *Applied Physics Letters* 92 (2008) 103502.

[8] Z. G. Ju, C. X. Shan, D. Y. Jiang, J. Y. Zhang, B. Yao, D. X. Zhao, D. Z. Shen and X. W. Fan, $Mg_x Zn_{1-x} O$ based photodetectors covering the whole solar-blind spectrum range, *Applied Physics Letters* 93 (2008) 173505.

[9] C. Soci, A. Zhang,B. Xiang, S. A. Dayeh, D. P. R. Aplin, J. Park, X. Y. Bao,Y. H. Lo, and D. Wang, ZnO nanowire UV photodetectors with high internal gain, *Nanoletters* 7 (2007) 1003-1009.

[10] M. E. Swanwick, S. M. Pfaendler, A. I. Akinwande and A. J. Flewitt, Near-ultraviolet zinc oxide nanowire sensor using low temperature hydrothermal growth, *Nanotechnology* 23 (2012) 344009.

[11] J. G. Lu, P. Chang, Z. Fan, Quasi-one-dimensional metal oxide materials synthesis, properties and applications, *Material Science and Engineering*, R 52 (2006) 49–91.

[12] A. Liu, Towards development of chemosensors and biosensors with metal-oxide-based nanowires or nanotubes, *Biosensors and Bioelectronics* 24 (2008) 167–177.

[13] Y. J. Chen, L. Nie, X. Y. Xue, Y. G. Wang and T. H. Wang, Linear ethanol sensing of SnO_2 nanorods with extremely high sensitivity, *Applied Physics Letters* 88 (2006) 083105.

[14] S. Park, S. An, H. Ko, S. Lee, and C. Lee, Synthesis, structure, and UV-enhanced gas sensing properties of Au-functionalized ZnS nanowires, *Sensors and Actuators B: Chemical*, 188 (2013) 1270–1276.

[15] Y. Takahashi, M. Kanamori, A. Kondoh, H. Mimoura and Y. Ohya, Photoconductivity of ultrathin zinc oxide films, *Japanese Journal of Applied Physics*, 33 (1994) 6611.

[16] L. Liu, C. R. Goria, S. Liang, N. Emanetoglu, Y. Lu, H. Shen and M. Wraback, Ultraviolet detectors based on epitaxial ZnO films grown by MOCVD, *Journal of Electronics Material* 29 (2000) 69.

[17] C. Y. Lu, S. P. Chang, S. J. Chang, T. J. Hsueh, C. L. Hsu, Y. Z. Chiou and I. C. Chen, A lateral ZnO nanowire UV photodetector prepared on a ZnO:Ga/glass template, *Semiconductor Science Technology*, 24 (2009) 075005.

[18] M. H. Mamat, Z. Khusaimi, M. M. Zahidi, and M. R. Mahmood, Performance of an Ultraviolet Photoconductive Sensor Using Well-Aligned Aluminium-Doped Zinc-Oxide Nanorod Arrays Annealed in an Air and Oxygen Environment, *Japanese Journal of Applied Physics* 50 (2011) 06GF05

[19] J. D. Prades, F. Hernandez-Ramirez, R. Jimenez-Diaz, M. Manzanares, T. Andreu, A. Cirera, A. Romano-Rodriguez and J. R. Morante, The effects of electron–hole separation on the photoconductivity of individual metal oxide nanowires, *Nanotechnology* 19 (2008) 46.

[20] D. C. Kim, B. Jung, J. H. Lee, H. K. Cho, J. Y. Lee and J. H. Lee, Dramatically enhanced ultraviolet photosensing mechanism in a n-ZnO nanowires/i-MgO/n-Si structure with highly dense nanowires and ultrathin MgO layers, *Nanotechnology* 22 (2011) 26.

[21] S. Hullavarad, N. Hullavarad, D. Look and B. Claflin, Persistent Photoconductivity Studies in Nanostructured ZnO UV Sensors, *Nanoscale Research Letters*, 4 (2009) 1421–1427.

[22] B. Claflin, D. C. Look, S. J. Park and G. Cantwell, Persistent n-type photoconductivity in p-type ZnO, Journal of Crystal Growth, 287 (2006) 16–22.

[23] R. Laiho, Y. P. Stepanov, M. P. Vlasenko and L. S. Vlasenko, Persistent photoconductivity of ZnO, *Physica B*, 404 4787–90 (2009).

[24] T. E. Murphy, K. Moazzami and J. D. J. Phillips, Trap related photoconductivity in ZnO epilayers, *Journal of Electronic Materials*, 35 (2006) 543–549.

[25] J. Nayak, J. Kasuya, A. Watanabe and S. Nozaki, Persistent photoconductivity in ZnO nanorods deposited on electro-deposited seed layers of ZnO, *Journal of Physics: Condensed Matter*, 20 (2008) 19.

[26] C. James Moore and V. Cody Thompson, A Phenomenological Model for the Photocurrent Transient Relaxation Observed in ZnO-Based Photodetector Devices, *Sensors* 13 (2013) 9921-9940.

[27] A. Lalchand Patil, S. Lalita Sonawane, G. Dhanashri Patil, Room Temperature Ammonia Gas Sensing Using MnO_2-Modified ZnO Thick Film Resistors, *Journal of Modern Physics* 2 (2011)1215-1221.

[28] S. P. Ghosh, K. C. Das, N. Tripathy, G. Bose, D. H. Kim, T. I. Lee, J. M. Myoung and J. P. Kar, Microstructural evolution of sputtered ZnO thin films with rapid thermal annealing, *ournal of Material Science.: Material Electron,* 26 (2015) 7860–7866.

IOP Conf. Series: Materials Science and Engineering **115** (2016) 012036 doi:10.1088/1757-899X/115/1/012036

A Study On the effect of Surface treatment on the Physical and Mechanical properties of date-palm stem fiber embedded epoxy composites

SatchidanandaTripathy[1], Janaki Dehury[2], DebasmitaMishra[3]

[1]Department of Mechanical Engineering, VSSUT, Burla, Odisha, 768018
[1]Department of Mechanical Engineering, VSSUT, Burla, Odisha, 768018
[1]Department of Mechanical Engineering, VSSUT, Burla, Odisha, 768018

E-mail - debasmita.mishra01@gmail.com

Abstract Natural fiber reinforced polymer composites are being used frequently for variety of engineering applications due to many of their advantages like ease of availability, low density, low production cost and good mechanical properties but natural fibers are more or less hydrophilic in nature. Therefore, an investigation has been carried out to make better utilization of a class of natural fiber that is date palm stem fiber, for making a wide range of products. Attempts have been made in this research work to study the effect of fiber loading on the physical, mechanical and water absorption behaviour of treated and untreated short fiber based epoxy composites. Composites of various compositions of different amounts of fiber loading are fabricated by simple hand lay-up technique. It has been observed that there is a significant effect of surface treatment of fibers on the overall properties of composites. Further enhancement of properties with lower water absorption rate was attained with glass fiber-epoxy based hybrid composites.

1. Introduction

Lately, the concept of 'eco-materials' has gained key importance due to the need of preservation of our environment. The meaning of eco-material includes 'safe' material systems for human and other life forms at all times. Recently, the world is facing a serious problem in developing proper methods of decomposing the solid wastes through different chemical processes with minimal energy which are not cost-effective and may subsequently produce harmful gases during the process. This environmental issue has forced the scientists and material designers to focus on the development of recyclable and biodegradable materials. Past experiences have shown that it is necessary to characterize materials and determine those which are safe for both short and long-term utilization. Selection of a material system that satisfies not only industrial requirements but also this wider definition of eco-materials, as described above, is an urgent necessity. Therefore, due to the growing global energy crisis and ecological risks, natural fiber reinforced polymer composites have attracted more researchers. The advantages of natural fibers are their availability, biodegradable and renewable nature, low density, high specific properties, good thermal properties, low energy consumption, enhanced energy recovery, non-abrasive nature and low cost. A great deal of work has been carried out to measure the potential of natural fibers like jute, coir, bamboo, sisal, banana and date palm leaf as reinforcement in polymers due to their renewable nature and good mechanical properties. Similarly, date palm fiber has also a great potential to be used as reinforcement in polymer composites. The date palm tree, a member of the palm tree family (Phoenix dactylifera), is normally found in the Middle East, Northern Africa, the Canary Islands, Pakistan, India, and in the United States (California). Traditionally, the leaves of date palm tree are used in making ropes, baskets and mats in many parts of the world. Date palm stem fiber is a lingo-cellulosic fiber having high toughness and aspect ratio and its composites have high impact strength with moderate tensile and flexural properties compared to other fibers like coir, sisal, pineapple, banana etc. The in -born properties of date palm fiber such as low density, low elongation at break and its specific stiffness and strength is comparable to those of glass fiber. The efficient utilization of this natural resource in making natural fiber composite will have a positive impact on environment and may improve the economic standard of rural people.

Few works have been undertaken by researchers on polymer composites using fibers obtained from different parts of date palm tree. Sbiai et al. [1] have investigated various properties of date palm leaf fiber reinforced in a poly-epoxy thermoset resin. Mahdavi et al. [2] studied the morphological and mechanical properties of composite fabricated by reinforcing the fibers obtained from trunk, rachis and petiole of date palm tree in high density polyethylene matrix. It has been observed that there is a significant difference between trunk and petiole on fiber length but rachis has no significant difference relative to other parts and that every part of date palm like trunk, rachis and petiole gives higher strength with increasing fiber loading. The tensile strength of fibers obtained from leaves and stems of date palm tree have also been investigated and compared with other natural fibers by Rao et al. [3]. Mohanty et al. [4] studied the Effective Mechanical Properties of Polyvinyl alcohol Bio-Composites with date palm Leaf fiber reinforcement. In their study, they found that acrylic acid treated date palm fiber has shown desirable tensile properties in single fiber pull out test. They have also observed that the tensile strength of date palm fiber is comparable to that of other fibers such as banana, bamboo, coconut and sisal. Alawar et al. [5] have studied the effect of different chemical treatment processes on date palm fiber surrounding the stems of date palm tree. From their work, it has been concluded that 5% NaOH treated date palm fiber shows optimum mechanical properties.

Mechanical properties of various date palm fiber composites produce significant strength according to various researchers. Al-Sulaiman & A. Faleh, [6] studied the mechanical properties of date palm fiber reinforced composite at high temperature curing with phenolic and bisphenol resin and observed that long fibers laminates (300 mm) have more strength than wood. Salah Amroune et al.[7] investigated the tensile properties and surface chemical sensitivity of technical fibers obtained from date palm fruit branches (Phoenix dactylifera L.) and found that the tensile tests clearly show that use of post-processing NaOH-based chemical treatments allow significant increase in stress at failure and the Young's modulus with a low influence on strain at failure. T. Alsaeed, B.F. Yousif, and H. Ku [8] investigated the potential use of date palm fiber as reinforcement for polymeric composites and investigation on interfacial adhesion of date palm fiber with epoxy matrix was experimentally done by using single fiber pull out technique. The influence of NaOH concentrations (0–9%), fiber length and fiber diameter on the interfacial adhesion was also considered in this study.

Luo and Netravali [9] studied the effect of fiber content on tensile and flexural properties of pineapple fiber reinforced poly (hydroxybutyrate-co-valerate) resin composites. Chawla and Bastos [10] investigated the effect of fiber volume fraction on Young's modulus, tensile strength and impact strength of untreated jute fiber - unsaturated polyester resin composites.

Date palm wood flour as filler in linear low-density polyethylene matrix was studied by Seyed Mohammad Mirmehdi et al. [11]. Nabila Saadaoui et al. [12] studied the characterization of date palm lingo-cellulosic by-products and self-bonded composite materials. They found the possibility of valorising four date palm tree by-products: leaflets, rachis, leaf sheath and fibrillum as self-bonded materials. The mechanical behaviour of unidirectional hemp fiber reinforced epoxy composites was studied by Hepworth et al. [13]. The comparison showed high contents in lignin and cellulose for fibrillum, high contents in protein and hot water extracting agent for leaflets and high amount of hemi-cellulose and cellulose for rachis. Mechanical and thermal properties of date palm leaf fiber reinforced recycled poly (ethylene terephthalate) composites was studied by Alireza Dehghani et al. [14]. They [14] carried out Dynamic mechanical analysis (DMA) which indicated that the addition of DPLF to PETR matrix increased the toughness of the composites. The crystallization behaviour of the samples, analyzed by Differential Scanning Calorimetry (DSC) indicated an increase in the onset crystallization temperature and showed a higher degree of crystallinity of the composites as compared to PETR, demonstrating that DPLF particles could act as nucleating agents.

Though there is an extensive research on mechanical properties of date palm leaf fibers, there is very little work carried out on date palm stem composites. Again, the effect of surface treatment on the fibers on the physical and mechanical characteristics is hardly analysed.

Therefore, this paper describes about a detailed analysis of physical and mechanical characteristics of date palm-stem epoxy composites and date palm-stem hybrid composites. It also throws light on the improvement of hydrophobic properties of the composites by surface treatment methods of the reinforcing fibers.

2. Materials and Methods

2.1. Materials

1. Date palm steam fiber (DPS)

2. 5% NaOH solution

3. Glass fiber

4. Epoxy (Araldite LY-556)

5. Hardener (Aliphatic amine HY-951

2.2. Processing of the composites

The fabrication of composite is carried out by conventional hand lay-up technique. The natural fibers of the stem of date palm (short fibers) collected from local sources are used as reinforcement and epoxy is chosen as the matrix material. The epoxy resin and the corresponding hardener are supplied by Ciba Geigy India Limited. The low temperature curing epoxy resin and hardener are mixed in a ratio of 10:1 by weight percentage. The natural date palm stem of 15 mm length is used as the reinforcement. Then the short fibers are chemically treated by 5% NaOH solution for 1hour .The treated fiber is then cleaned with water and then oven dried at 60^0C for about 24 hours.

Three composite samples are fabricated at different fiber loading conditions. Composites of compositions 5%, 10% and 15% by weight are fabricated. NaOH treated date palm stem fibers were then fabricated for the same weight fractions. A hybrid composite (Date palm stem + glass fiber) is

IOP Conf. Series: Materials Science and Engineering **115** (2016) 012036 doi:10.1088/1757-899X/115/1/012036

also fabricated and the properties are recorded and compared with earlier observations. The detailed composition and designation of the composites are presented in Table 1, 2,3 and 4.The cast of each composite is cured under a load of about 50 kg for 24 hours. Finally, the specimens of suitable dimensions are prepared for characterization and testing.

Table 1 Designation of Composites (Un-treated DPS-epoxy composites)

Un-Treated Composites	Compositions
5%wt of fiber	Epoxy (95%) + 5% NaOH treated date palm stem fiber (DPSF) (5%)
10%wt of fiber	Epoxy(90%) + 5% NaOH treated date palm stem fiber (DPSF) (10%)
15%wt of fiber	Epoxy (85%) + 5% NaOH treated date palm stem fiber (DPSF) (15%)

Table 2 Designation of Composites (Treated DPS-epoxy composites)

Treated Composites	Compositions
5%wt of fiber	Epoxy (95%) + 5% NaOH treated date palm stem fiber (tDPSF) (5%)
10%wt of fiber	Epoxy(90%) + 5% NaOH treated date palm stem fiber (tDPSF) (10%)
15%wt of fiber	Epoxy (85%) + 5% NaOH treated date palm stem fiber (tDPSF) (15%)

Table 3 Designation of Composites (Treated Hybrid composites)

Treated Hybrid Composites	Compositions
5%wt of DPS fiber	Epoxy (90%) + 5% NaOH treated date palm stem fiber (tDPSF) + 5% NaOH treated glass fiber (tGF)
10%wt of DPS fiber	Epoxy(85%) + 10 % NaOH treated date palm stem fiber (tGF) (5% NaOH treated glass fiber (tGF)
15%wt of DPS fiber	Epoxy (80%) + 15% NaOH treated date palm stem fiber (tGF) (5% NaOH treated glass fiber (tGF)

Table 4 Designation of Composites (Treated Hybrid composites)

Treated Hybrid Composites	Compositions
5%wt of glass fiber	Epoxy (90%) + 5% NaOH treated date palm stem fiber (tDPSF) + 5% NaOH treated glass fiber (tGF)
10%wt of glass fiber	Epoxy(85%) + 5 % NaOH treated date palm stem fiber (tGF) (10% NaOH treated glass fiber (tGF)
15%wt of glass fiber	Epoxy (80%) + 5% NaOH treated date palm stem fiber (tGF) (15% NaOH treated glass fiber (tGF)

3. Results and Discussion

3.1 Physical and mechanical characterisation of composites

3.1.1 Effect of fiber content on density of composites

Density is a material property which is of prime importance in several weight sensitive applications. There is always a difference between the theoretical and the measured density values of a composite due to the presence of voids and pores. Presence of voids is one of the major factors influencing the mechanical performance of composites and the knowledge of void content is desirable for estimation of the quality of the composites. The presence of void content in the composites significantly reduces the mechanical and physical properties of the composites. Therefore, voids are undesirable in composites as they reduce the overall properties of the composites. Table 2 presents the theoretical density, experimental density and their corresponding void content of all the composite specimens. It can be observed from the table that the void content of composites decreases with increase in fiber loading. The similar trend of decrease in void content with increase in fiber loading and length has already been reported by previous researchers.

Table 2 Void fraction of composites

Composite	Density (Theoretical)	Density (Experimental)	Void content (%)
5% wt treated(DPS)	1.141	1.117	2.100
10%wt treated (DPS)	1.133	1.108	2.206
15%wt treated(DPS)	1.124	1.078	4.600
Untreated (DPS)	1.133	1.087	4.060
Untreated hybrid	1.173	1.152	1.790
Treated hybrid	1.173	1.164	0.760

***Note:** Here DPS stands for Date palm stem*

It was also found out that compared to the untreated specimen; the chemically treated specimen had less number of voids in the composite. In the hybrid specimen sample shows least amount of void fraction in comparison to untreated and treated date palm stem/glass composite.

3.1.2 Effect of fiber content on tensile strength of the composites

The influence of fiber content on the tensile strength of the composites is shown in Figure 1. There is an improvement in mechanical properties of composites with increase in fiber content due to the fact that the interfacial adhesion between the fiber and matrix is high through an efficient load transfer between the fiber and matrix. It is evident from the Figure1 that the tensile strength of the composites increases with increase in fiber content upto 10 % by weight and goes on decreasing. This is because of the fact that up to 10 % fiber content, there is a good adhesion between fiber and matrix. As there is further increase in fiber content, there is difficulty in compatibility between fiber and matrix adhesion.

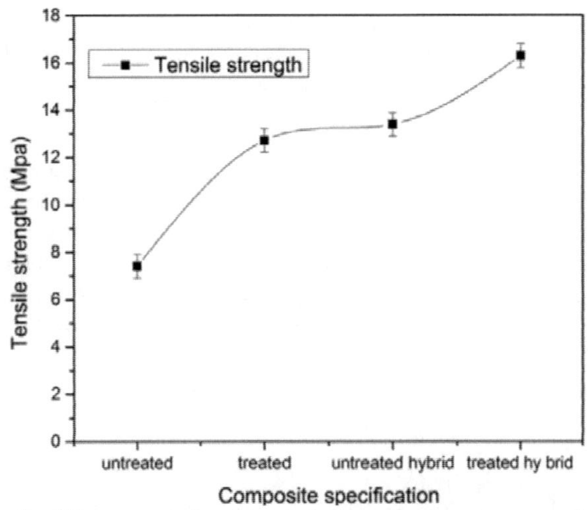

Fig.1:- Tensile strength of the untreated and treated date palm epoxy composite & date palm - glass fiber epoxy composite with 10% fiber content

3.1.3 Effect of fiber content on flexural strength of the composites

The determination of flexural strength is an important characterization of any structural material. It is the ability of a material to withstand bending forces before reaching the breaking point. The flexural strength of composites is increased from 23.20MPa to 29.03MPa for 10 wt. % and then decreased from 29.03MPa to 14.16MPa as shown in Figure 2 i.e. up to 15wt. %. It shows a linearly increasing trend up to a certain value of fiber content (10 wt. %) and suddenly drops due to breakage and pullout of individual fibers from the resin matrix. An improvement in the bonding between the reinforcement and the matrix is done by the alkali treatment which subsequently increases the flexural properties of the composites. This is because alkaline treatment improves the adhesive property of date palm fiber surface by extruding hemi- cellulose, thereby producing rough surface topography. It is also observed that the flexural strength is maximum in hybrid composite which has undergone surface treatment compared to all other fabrications.

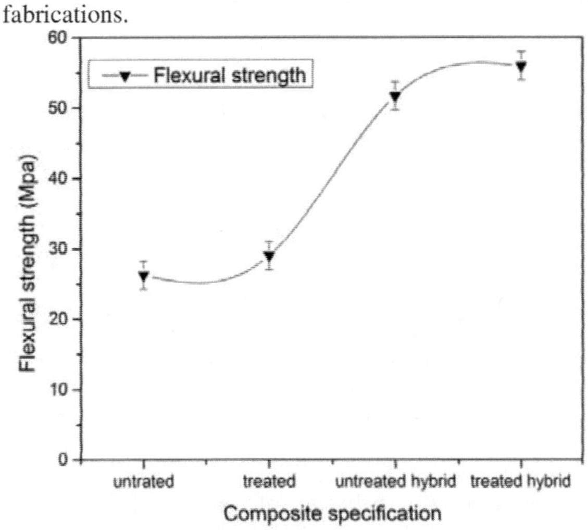

Fig. 2:- Flexural strength of the untreated, treated, date palm - glass fiber epoxy hybrid composite with 10% fiber content.

IOP Conf. Series: Materials Science and Engineering **115** (2016) 012036 doi:10.1088/1757-899X/115/1/012036

3.1.4. Effect of fiber content on water absorption of the composites

The water absorption behaviour of treated and untreated date palm epoxy composites as well as glass hybrid composites was studied. It was observed that the amount of water absorption decreases with increase in filler content. The water absorption of alkali treated composites is also considerably less than the untreated composites. In case of untreated composites the poor wettability and adhesion between untreated date palm towards epoxy resin are attributed to the hydrophilic nature of date palm fiber. This hydrophilicity is responsible for the higher percentage water uptake in untreated date palm composites. The water absorption of glass reinforced date palm epoxy composites is less than that of date palm epoxy composites. In case of glass reinforced date palm epoxy composites, the fibers are arranged randomly and in closely packed manner in which water impermeable glass fibers act as barriers and prevent the contact between water and hydrophilic date palm fibers, and hence prevent the water absorption of date palm fibers.

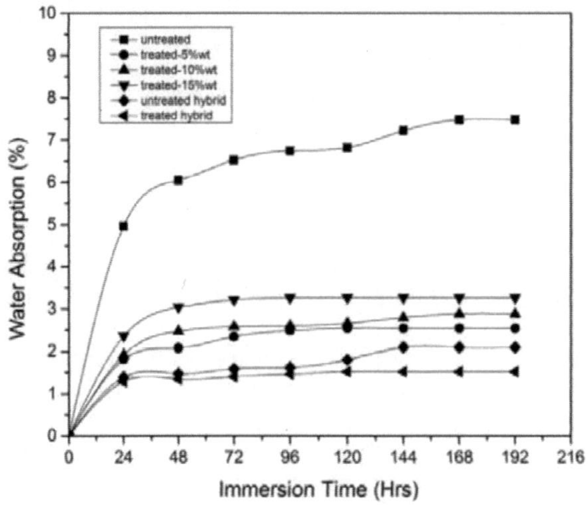

Fig. 3:- Effect of immersion time on water absorption properties of composite

4. Conclusions

The experimental investigation on the physical, mechanical and water absorption behaviour of treated and untreated date-palm composites as well as its hybrid composites led to the following conclusions:

1. The successful fabrication of date-palm stem composites and its hybrid (date palm stem-glass epoxy composites) by simple hand lay- up technique were possible.

2. The present investigation revealed that increase in fiber content; surface modification of fibers by chemical treatment and use of secondary inorganic filler with desirable properties significantly influences the properties of the composites. The maximum tensile strength and flexural strength is obtained for composites reinforced with fibers treated with 10% NaOH. Hybrid composites using only 5% of chemically treated fibers exhibit significant enhancement in mechanical properties.

3. The water absorption rate gradually increases with increase in fiber content irrespective of the orientation of the fibers. The maximum amount of water absorption is distinguished for composites with 10 wt.% date palm stem composites. The water absorption rate gradually decreases as from untreated specimen to chemically treated specimen as there is reduction in

number of voids. Again, due to NaOH treatment the hydrophilic nature of DPS fiber also decreases. The water absorption of glass reinforced DPS epoxy hybrid composites is less than that of DPS –epoxy composites. In case of glass reinforced date palm epoxy composites, the fibers are arranged randomly and in closely packed manner in which water impermeable glass fibers act as barriers and prevent the contact between water and hydrophilic DPS fibers and hence prevent the water absorption of date palm fibers.

References

[1] A. Sbiai, A. Maazouz, E. Fleury, A. Sautereau & H. Kaddamu, Effective mechanical properties of polyvinylalcohol biocomposites with reinforcement of date palm leaf fibers , *Bio-Resources*, 5(2), (2010) 672.

[2] S. Mahdavi, H. Kermanian & A. Varshoei , Comparison of mechanical properties of date palm fiber – polyethylene composite *Bio-Resources* 5 (2010) 2391.

[3] H.R.Rao , M.K.Khamar & G.R.Reddy , Hybrid composites: Effect of fibres on Mechanical properties, *International Journal of Macromolecular Science* 1(2011) 9-14.

[4] Jyoti R. Mohanty, Sankar N. Das, Harish C. Das & Sarat K. Swain , Effective Mechanical Properties of Polyvinyl alcohol Bio-composites with Reinforcement of Date Palm Leaf Fibers *Polymer composites* 34(6) (2013) 959-966.

[5] A. Alawar, M.A. Hamid & A. Kaabi, The effect of different chemical treatment processes on date palm fiber surrounding the stems of date palm tree, *Composite -Part B Engineering*, 40 (2009) 601.

[6] Al-Sulaiman & A. Faleh, Mechanical Properties of Date Palm Leaves, *Journal of Reinforced Plastics and Composites*,19(2000) (17).

[7] Salah Amroune et al., Tensile mechanical properties and surface chemical sensitivity of technical fibres from date palm fruit branches (Phoenix dactylifera L.), *Composites: Part A* 71(2014) 95–106 .

[8] T. Alsaeed , B.F. Yousif & H. Ku, The potential of using date palm fibres as Reinforcement for polymeric composites, *Materials and Design* 43(2012) 77–184.

[9] S Luo & A.N.Netravali, Mechanical and thermal properties of environmentally friendly green composites made from pineapple leaf fibres and poly (hydroxybutyrate-co-valerate) resin, *Polymer Composites* 20(3) (1999) 367-378.

[10] K. K. Chawla & A.C. Bastos, The mechanical properties of jute fibres and polyester/jute composites , *In Proceedings of the third international conference on mechanical behaviour of materials , Cambridge, UK: Pergamon Press* (1979)191-196.

[11] Seyed Mohammad Mirmehdi , Farhad Zeinaly & Fatemeh Dabbagh, Date palm Wood flour as filler of linear low-density polyethylene , *Composites: Part B* 56 (2012) 137–141 .

[12] Nabila Saadaoui , Antoine Rouilly , Khalid Fares & Luc Rigal ,Characterization of Date palm lingo cellulosic by-products and self-bonded composite materials obtained there of , *Materials and Design* 50(2013) 302–308 .

[13] D. G. Hepworth, R. N .Hobson, D. M. Bruce & J. W. Farrent, ,The use of unretted hemp fibrein composite manufacture, *Composites Part A: Applied Science and Manufacturing* 31(11) (2000)1279-1283.

[14] Alireza Dehghani et al., Mechanical and thermal properties of date palm leaf fiber reinforced recycled poly (ethylene terephthalate) composites , *Materials and Design* 52 (2013) 841–848.

IOP Conf. Series: Materials Science and Engineering**115** (2016) 012037 doi:10.1088/1757-899X/115/1/012037

Effect of pre-strain on ratcheting behavior of A668 Class D steel

S. Tripathy*, A. K. Mondal and K. Dutta

Department of Metallurgical and Materials Engineering,
National Institute of Technology, Rourkela- 769008, India

*E-mail: subhra359@gmail.com

Abstract. The aim of this investigation is to study the effect pre-strain (0%, 2%, 4% and 8%) on ratcheting behavior of ASTM A668 Class D steel in different heat treatment conditions (normalized and hardened-tempered). Ratcheting tests were carried out at room temperature on cylindrical specimens having 12.5 mm gauge length and 6 mm gauge diameter. The results include reduced strain accumulation with increasing pre-strain level due to work hardening of the pre-strained samples. Further, cyclic hardening takes place during ratcheting deformation.

1. Introduction

ASTM A668 class D steel is a kind of plain carbon steel which is potentially being used in various engineering sectors that include aerospace, automotive and defense applications. In all these kinds of applications, the much needed property requirement is good fatigue and fracture behavior of the material. It is known that almost 90% of all service failures are due to fatigue [1] and hence it is indeed important to know the fatigue behavior of a material before being considered for any application. Traditionally, strain-controlled low cycle or stress-controlled high cycle fatigue studies are being performed to assess the fatigue behavior of a material, but in recent days, considerable attention is being paid by investigators towards generating information regarding the stress-controlled low cycle fatigue behavior of a material; the phenomenon is known as ratcheting. Ratcheting deformation is pronounced when the studies are done in presence of some positive or negative mean stress. Accumulation of ratcheting strain during cyclic loading reduces low cycle fatigue life of a material. Therefore it is one of the important factors that should be considered in the design of any structural component [2]. Further, it is known that application of pre-strain causes variations in the mechanical properties of a material. In last few years, effects of mean stress, stress amplitude etc. on ratcheting behavior of a material were being reported [3-6], however, not much reports are existing considering the effect of pre-strain on ratcheting behavior of metallic materials. This investigation aims to fulfill this gap.

2. Experimental details

The material selected for this investigation is class D steel, of ASTM standard A668 which was subjected to two types of heat treatments viz. normalizing and quenching followed by tempering at 600°C. For initial characterization of the material, microstructural analyses, hardness and tensile tests were done.

Tensile and fatigue test specimens were fabricated as per ASTM standards E8M [7] and E-606 [8] respectively. Ratcheting tests were done at a constant stress rate of 50 MPa/s on specimens subjected to varying pre-strain levels of 2%, 4% and 8% (strained at a nominal strain rate of 0.001 s^{-1}) for both, normalized and hardened-tempered conditions. In both the cases, the maximum stress values for cyclic loading were chosen as 50% of the ultimate tensile strength value. The nature of strain accumulation was examined for specimens without pre-strain also, which is designated as 0% pre-strain level in this paper.

3. Results and discussion

3.1 Microstructure, hardness and tensile properties

The optical microstructures of the investigated material under normalized and hardened-tempered conditions are shown in figure 1. Figure 1(a) shows the microstructure in normalized condition which clearly shows ferrite and pearlite phases whereas figure 1(b) illustrates the microstructure of the hardened-tempered steel showing tempered martensite. This also contains some amount of precipitates.

Figure 1. Optical microstructures of (a) normalized (b) hardened-tempered steel

The results of hardness tests indicate that the average hardness of normalized steel was 230 VHN while that for the hardened- tempered steel was 274 VHN. Figures 2(a) and (b) show the engineering stress-strain curves of the steel in normalized and hardened-tempered condition respectively. Results of tensile tests are summarized in Table 1. In both the cases it was observed that there are no sharp yield points; thus yield strength values were determined by the stress corresponding to the intersection of the stress-strain curve and an offset line parallel to the elastic portion of the curves at a strain of 0.2%.

3.2 Effect of pre-strain on ratcheting behavior in normalized steel

The results of ratcheting tests conducted on normalized samples up to 100 cycles under various pre-strain conditions (0%, 2%, 4% and 8%) are shown in figure 3(a) and typical nature of obtained hysteresis loops (for 0% pre-strain condition) are shown in figure 3(b). It can be noted from figure 3(a) that accumulation of ratcheting strain decreases with increasing the level of pre-strain, and it is maximum when there is no

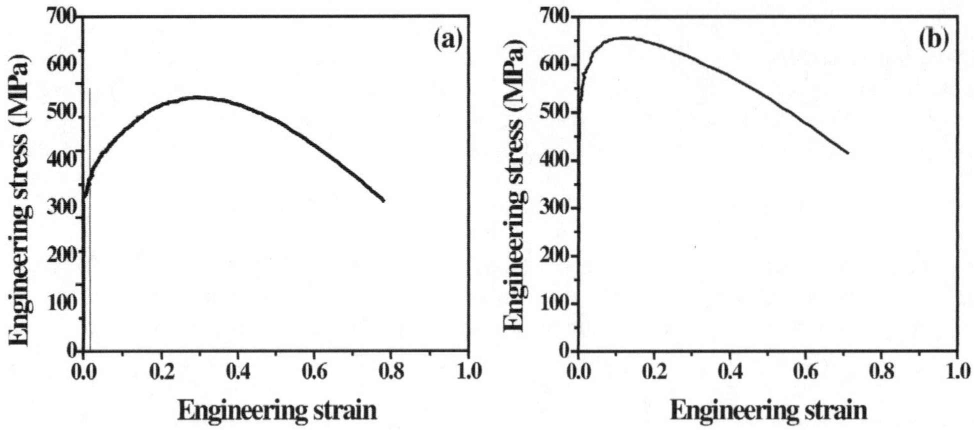

Figure 2. Engineering stress-strain curves for (a) normalized and (b) hardened- tempered steel

Table 1. Tensile properties of ASTM A668 Class D steel in normalised and hardened-tempered conditions

Property	Normalized	Hardened-Tempered
Yield strength (MPa)	355	592
Ultimate tensile strength (MPa)	530	655
Uniform elongation	0.29	0.10
Total elongation	0.77	0.71
Strain hardening exponent (n)	0.32	0.11

pre-strain (0%, as indicated in the figure). It is noted that the strain accumulation can vary from 0.21 % to 0.44% with decreasing pre-strain. Wang et al. [9] reported that if the level of pre-strain is increased, dislocation gets piled up and as a result back stress is generated. The back stress generated during pre-straining causes hindrance to accumulate strain during subsequent ratcheting deformation. The extent of back stress increases with the level of pre-strain and accordingly accumulation of ratcheting strain gets reduced during a test done with positive mean stress. The decrease of ratcheting strain with increase in pre-strain is in accordance with the results reported by Wang et al. [9] and De et al. [10].

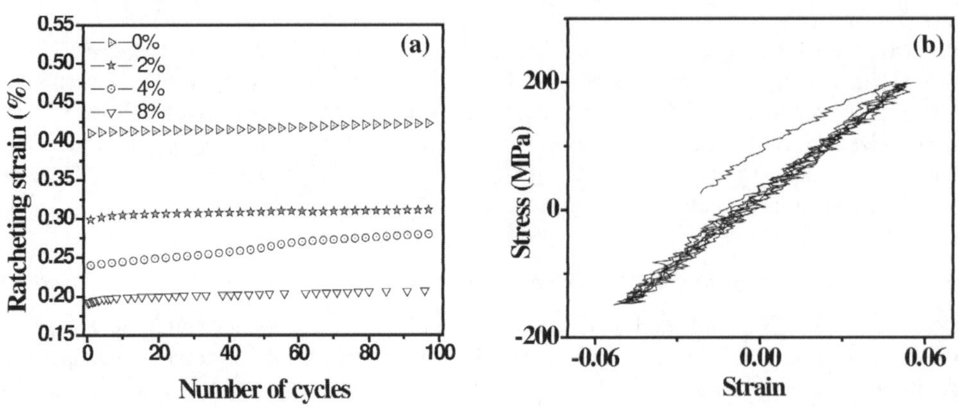

Figure 3. (a) Effect of pre-strain on ratcheting behavior and (b) hysteresis loops produced during ratcheting (for 0% pre-strain condition) in normalized steel.

IOP Conf. Series: Materials Science and Engineering **115** (2016) 012037 doi:10.1088/1757-899X/115/1/012037

A set of hysteresis loops are illustrated in figure 3.3 (b) which were produced from a ratcheting test conducted without any pre-strain. One can note that the loops shift from its initial position towards right. This fact indicates that a considerable amount of plastic strain gets accumulated during each cycle. When the loop area was calculated individually, it was seen that loop area decreases with increase in number of cycles. This indicates that the selected material shows cyclic hardening behavior.

3.3 Effect of pre-strain on ratcheting behavior in hardened-tempered steel

It is understood that the microstructure of a material can alter its deformation behavior and thus the nature of strain accumulation due to ratcheting must be dependent on the microstructural constituents. One of the major applications of the steel is in hardened-tempered condition for making hydro turbines, where ratcheting deformation should be a major concern. Therefore, a set of steel samples were hardened-tempered and tested for the effect of pre-strain on strain accumulation due to ratcheting up to 100 cycles. The nature of strain accumulation and a set of hysteresis loops (0% pre-strain) are shown in figure 4 (a) and (b), respectively. Here also, the strain accumulation reduces with increasing the level of pre-strain. It may be inferred that with pre-strain, the strength of a material increases due to work hardening and thus the material accumulates lower amount of ratcheting strain.

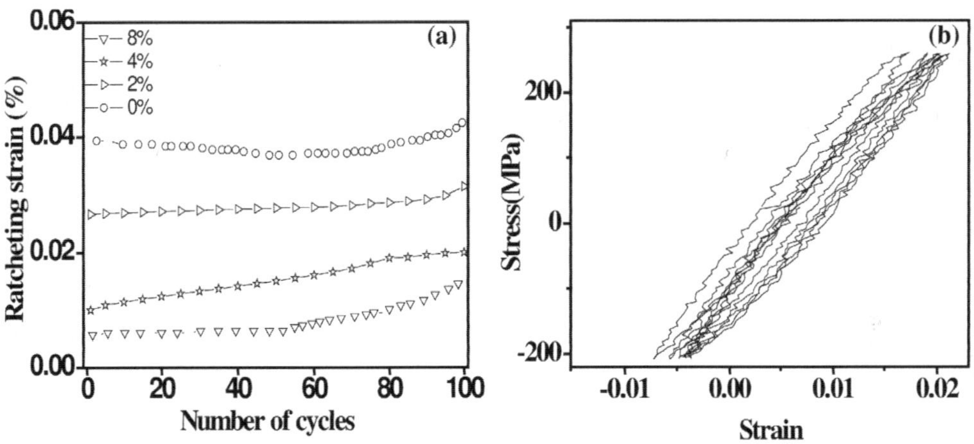

Figure 4. (a) Effect of pre-strain on ratcheting behavior and (b) hysteresis loop produced during ratcheting in hardened- tempered steel.

Although considerable amount of strain accumulation takes place both in normalized and hardened-tempered conditions, severity of ratcheting was on normalized samples; in association to this, pre-strain can cause seizing of strain accumulation.

4. Conclusions

The ratcheting behavior of ASTM A668 class D steel has been investigated in this study in two different heat treatment conditions along with varying levels of pre-strain. The conducted experiments and their pertinent analyses lead to the following conclusions:

1. Accumulation of ratcheting strain decreases with increase in the level of pre-strain and it is maximum when there is no pre-strain (i.e. at 0% pre-strain), both in normalized and hardened-tempered samples of ASTM class D A668 steel. The decrease in strain accumulation with increasing pre-strain can be explained with formation of dislocations during cyclic loading. The

nature of strain accumulation is similar in normalized and hardened-tempered samples; however, the extent is more in normalized samples.

2. The area under the stress-strain hysteresis loops decrease continuously which indicates that the steel is cyclically hardenable in nature.

References

[1] Dieter, G. E. 1987, Mechanical Metallurgy, SI metric edition, McGraw-Hill Book.
[2] Rider R J, Harvey S J, Chandler H D, *Int. J. Fatigue* 1995 **17** 507–511.
[3] Xia Z, Kujawski D, Ellyin F, *Int. J. Fatigue* 1996 **18** 335–341.
[4] Ozgen U C, *Mater. Des.* 2007 **29** 1575–1581.
[5] Ruggles M D, Krempl E, *J. Mech.Phys. Solids* 1990 **38** 575–585.
[6] Kobayashi M, Ohno N, Igari T, *Int. J. Plast.* 1998 **14** 355–372.
[7] E8M-12, 'Annual book of ASTM standards', Standard test method for tension testing of metallic materials (Metric), 2012, West Conshohocken, PA.
[8] E606M-12, Standard test method for strain-controlled fatigue testing, ASTM International, 2012, West Conshohocken, PA.
[9] Wang Y, Dunji Y, Chen G, Chen X, *Int. J. Fatigue* 2013 **52** 106–113.
[10] De P S, Kundu A, Chakraborti P C, *Mater. Des.* 2014 **57** 87–97.

IOP Conf. Series: Materials Science and Engineering **115** (2016) 012038 doi:10.1088/1757-899X/115/1/012038

Equilibrium state of anatase to rutile transformation for nano-structured Titanium Dioxide powder using polymer template method.

Tapabrata Dam, Sidhartha S. Jena and Dillip K. Pradhan*

Department of Physics & Astronomy, National Institute of Technology Rourkela, Rourkela Odisha – 769008, India

E-mail: `dillip.pradhan79@gmail.com`

Abstract. In this work we report the novel synthesis procedure of phase-pure nano-structured titania in anatase phase using polyacrylamide gel based polymer template method. The evolution of rutile phased titania with increasing temperature has also been investigated. The synthesized nano-materials are characterized using X-ray diffraction, Brunauer - Emmett - Teller surface analysis technique and Scanning electron microscopy. We have used dual phase Rietveld refinement method to analyse the X-Ray diffraction data to get clear picture of crystallographic information of the prepared samples.

1. Introduction

Titanium dioxide also known as Titania (TiO_2) in different polymorphs is an important electronic ceramic material for its potential use in various device applications such as photocatalysis and photovoltics, electrolysis of water to generate hydrogen, dye-sensitised solar cells, gas sensors, purifier of environmental pollutants, ceramic membranes, self cleaning materials etc [1, 2]. Titania show three different polymorphs namely, (*i*) anatase, (*ii*) rutile and (*iii*) brookite. Among these three phases rutile is stable whereas anatase and brookite are metastable phase. Therefore anatase and brookite titania are difficult to synthesize. From crystallographic point of view anatase ($I4_1/amd$) and rutile ($P4_2/mnm$) are of tetragonal crystal structure, whereas brookite ($Pbca$) is of orthorhombic crystal structure [3, 4].

The physical and chemical properties of different polymorphs of TiO_2 are affected by size and related surface effects. The reduction of size down to nano scale induces a marked differences in band gaps, refractive index and the charge transfer characteristics across the granular interface. As a result remarkable improvement of properties, for example photo-catalytic properties in anatase phase can be observed [5]. Literature suggests that enhancement of photo-catalytic activity is even larger for nano-meter sized systems containing mixed polymorphs of TiO_2 [3]. The reason behind this may be the effective separation of charge carriers in the different phases, which suppresses the electron-hole recombination. In brief pure phased polymorphs as well as mixed phased polymorphs of TiO_2 are equally important form the application point of view.

The present work aims to clarify the transformation in between the two main polymorphs of TiO_2 i.e. anatase and rutile. It also clarifies the principle of controlling the compositional phase percentage in a particular TiO_2 sample with different synthesis temperature in gel template synthesis technique. Keeping these in mind in the present study we have successfully prepared phase pure anatase polymorph of TiO_2 using poly-acrylamide gel based polymer template synthesis technique and thereafter with increasing temperature we have studied the transformation of anatase to rutile phase of titania.

2. Experimenal Details

2.1. Synthesis Technique

Titanium isopropoxide ($Ti(OCH(CH_3))_4$) is taken as the source of titanium. Two and half ml of titanium isopropoxide is added drop wise in a beaker filled with 50 ml of deionized water kept at 4°C under vigorous stirring. Instantaneously yellowish white precipitate of titanic acid ($Ti(OH)_4$) is observed. This precipitates are collected and dissolved in dilute nitric acid keeping temperature of the solution below 4°C. With this clear solution of titanyl nitrate ($TiO(NO_3)_2$) is obtained. Optimized amount of acrylamide, N,N-methylenebisacrylamide are dissolved to this solution and mixed thoroughly. To initiate the gelation 0.5 ml ammonium persulfate (10 wt.%) solution is added and mixed again. The mixture is kept at 40°C in a water bath for 5 *h* to get acrylamide hydrogel. Acrylamide hydrogel is then heat treated at 300°C for 2 *h* to get the black network like structure. This network like structure is then ground to get fine powder using mortal pestle and calcined at different temperatures ranging from 450°C to 650°C to obtain TiO_2 samples. Figure.1 is representing the synthesis procedure of nano-crystalline TiO_2 in form of flow chart.

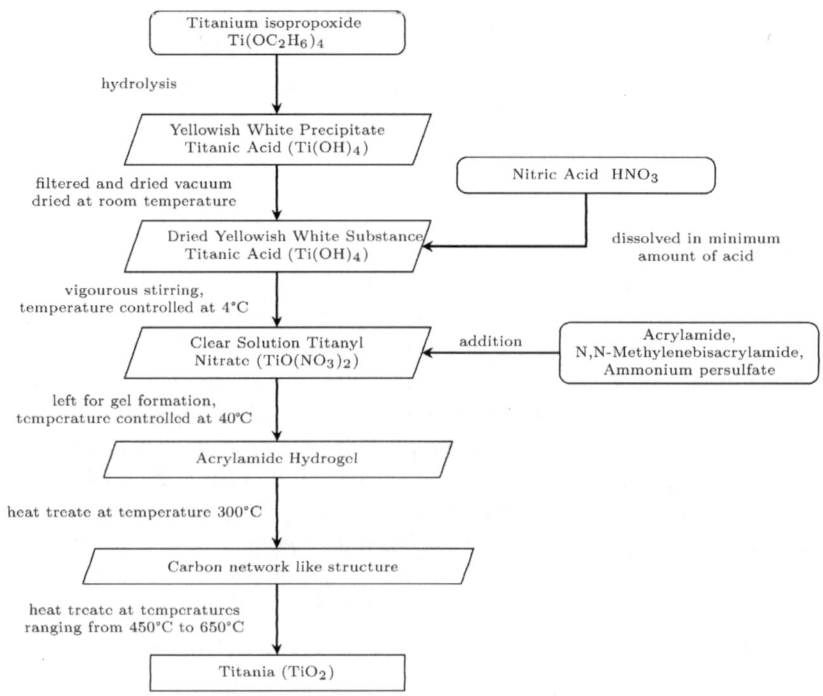

Figure 1. Flow chart for the synthesis of nano-crystalline TiO_2 using polyacrylamide gel template method.

2.2. Characterization Techniques

Crystallographic properties of the synthesized titania samples are investigated using X-ray diffraction (XRD) technique. XRD patterns of TiO_2 samples are recorded using Rigaku Ultima-IV X-Ray diffractometer operated at 40 kV and 40 mA current with Cu $K_{\alpha 1}$ radiation (1.5405Å). XRD patterns are recorded with a scanning step of 0.002° and rate 3° per minute with Bragg angle ranging from 15° to 75°. Surface morphology of the prepared samples are studied with Field Emission - Scanning Electron Microscopy (FE-SEM) micrographs recorded using Nova-Nano FE-SEM. Surface analysis are carried out with Autosorb iQ Quantachrome system. For surface analysis samples are degassed at 200°C for 1 h before commencing the measurement.

3. Results & Discussion

3.1. Crystallographic study

In Figure.2 we have shown the XRD patterns of synthesized TiO_2 samples for different calcination temperatures. The XRD patterns suggests that upto 500°C, only anatase phase of TiO_2 is synthesized, whereas with further increase in temperature the presence of secondary rutile phase of TiO_2 is observed. Again it can be noticed that with increasing calcination temperature the XRD peaks are becoming sharper and stronger, indicating increase in crystallinity for both anatase and rutile polymorphs. The relative intensity of strongest peak of rutile phase (110) is found to increase as compared to the strongest peak of anatase (101) phase of TiO_2. Anatase being a metastable phase compared to stable phase of rutile, any transition from anatase to rutile will be irreversible in nature[3, 6]. From XRD patterns we can conclude that the irreversible transition between anatase to rutile polymorphs of of TiO_2 starts after 500°C.

Rietveld refinement analysis of XRD pattern of titania synthesized at 650°C using FULLPROF package is shown in Figure.3. Similar refinements are carried out for other samples of TiO_2 synthesized at different temperatures. The refined parameters, obtained from Rietveld analysis are presented in Table.1. From Rietveld analysis percentage of rutile phase in the prepared samples are evaluated and found to increase with increase in temperature. At 550°C

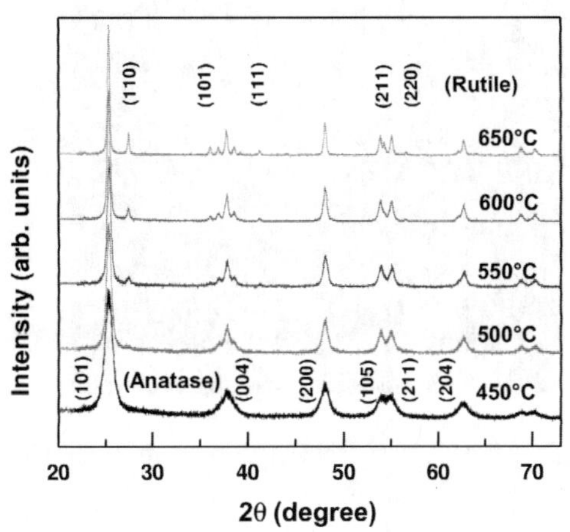

Figure 2. X-Ray diffracttion pattern for titania synthesized at different temperatures.

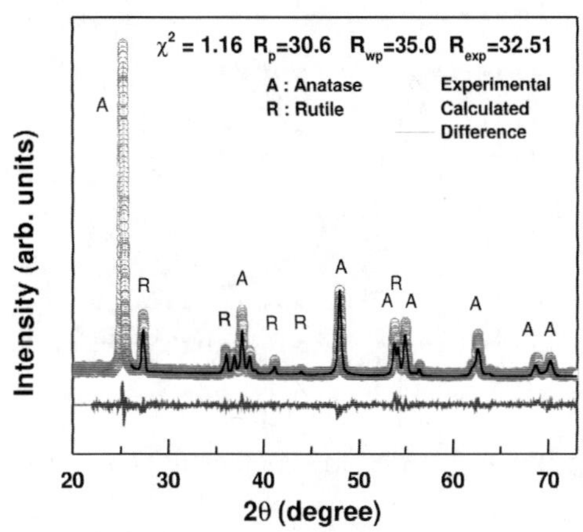

Figure 3. The Rietveld refined XRD pattern of titania synthesized at 650°C.

the percentage of rutile phase is 3.29 and it increases to 16.37 percent for sample synthesized at 650°C.

The diffraction profile from polycrystalline materials of all size and shape in different orientation is represented by the Fourier series coefficients which are considered to be the natural characteristic of the profile[7]. According to convolution theorem, the Fourier transform of the nth order pure diffraction profile will be the product of nth order Fourier transform of particle size $A_L^P(n)$, strain $A_L^S(n)$ and faulting $A_L^D(n)$ respectively i.e. $A_L(n) = A_L^P(n) \times A_L^S(n) \times A_L^D(n)$. Consequently,

$$\left[\frac{dA_L}{dL}\right]_{L=0} = \left[\frac{dA_L^P}{dL}\right]_{L=0} = -\frac{1}{p}$$

Here, $L = na$ with a representing the real distance in the crystallite and can be obtained using the following equation

$$\frac{2a(sin\theta_1 - sin\theta)}{\lambda} = \frac{2a(sin\theta - sin\theta_2)}{\lambda} = \frac{1}{2}$$

Here, θ refers to Bragg angle corresponding to peak position of intensity distribution for wavelength λ. θ_1 and θ_2 are Bragg angles corresponding to peak position in the intensity distribution where tail merges in to the background and p is the crystallite size in the crystallographic direction considered [8]. The line broadening parameters of (101) and (200) diffraction lines of anatase are obtained from fitting experimental data with Pseudo-Voigt function. The parameters of Pseudo-Voigt function are then used to calculate crystallite size and crystallite size distribution using BREADTH Programme. Figure.4 show Fourier size coefficients (A_s) of the X-ray line profile versus coherent length for (101) and (200) reflections of anatase TiO_2. In the inset of Figure.4 we have shown crystallite size distributions (P_v) for different TiO_2 samples. It can be observed that the average crystallite size of the prepared samples increases and the sample become poly-disperse in nature with increasing synthesis temperature.

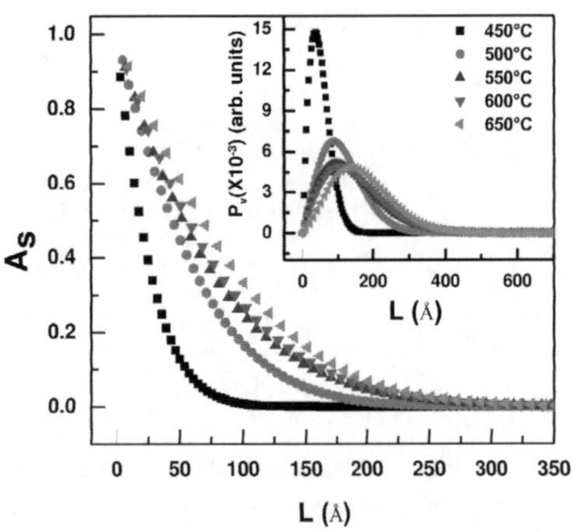

Figure 4. Fourier size coefficients (A_s) of the X-ray line profile versus coherent length for (101) & (200) reflection of anatase TiO_2. Inset show volume-weighted domain-size-distribution (P_v) as a function of coherent length.

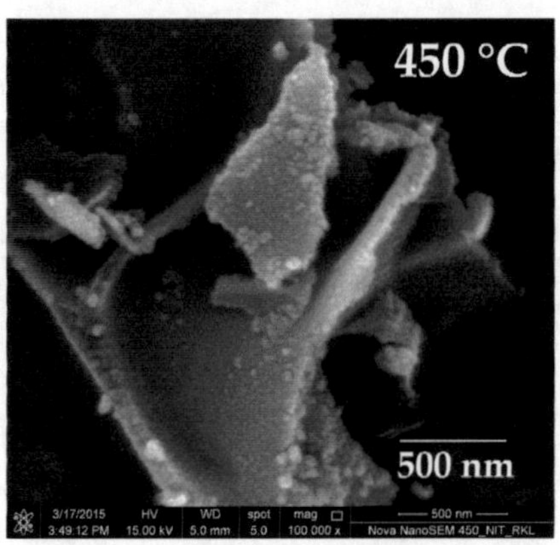

Figure 5. FE-SEM micrographs of titania sample synthesized at 450°C.

Table 1. Rietveld refined structural parameters of TiO_2 samples synthesized at different temperatures obtained using FULLPROF software.

Parameters		450°C	500°C	550°C	600°C	650°C
			Anatase(I4$_1$/amd)			
Lattice Constant (Å)	a	3.7834(6)	3.8026(4)	3.7945(4)	3.7933(3)	3.7916(3)
	c	9.5043(7)	9.5527(3)	9.5341(2)	9.5371(2)	9.5395(2)
Atomic Position	Ti(4a)	0, 0.75, 0.125	0, 0.75, 0.125	0, 0.75, 0.125	0, 0.75, 0.125	0, 0.75, 0.125
	O(8e)	0, 0.75, 0.331	0, 0.75, 0.336	0, 0.75, 0.332	0, 0.75, 0.331	0, 0.75, 0.338
Phase Percent		100	100	96.71	92.16	83.63
			Rutile(P4$_2$/mnm)			
Lattice Constant (Å)	a	–	–	3.7945(4)	3.7933(3)	3.7916(3)
	c	–	–	9.5341(2)	9.5371(2)	9.5395(2)
Atomic Position	Ti(2a)	–	–	0, 0, 0	0, 0, 0	0, 0, 0
	O(4f)	–	–	0.311, 0.311, 0	0.311, 0.311, 0	0.303, 0.303, 0
Phase Percent		0	0	3.29	7.84	16.37
χ^2		1.85	1.77	1.30	1.13	1.16

3.2. Surface morphology

Surface morphology and surface properties are studied using microscopy and BET analysis respectively. Figure.5 show the TiO_2 sample prepared at 450°C. FE-SEM micrographs of the synthesized samples reveal the flake like structure of prepared TiO_2 samples. Adsorption and desorption isotherm of TiO_2 samples prepared at 500°C are shown in Figure.6. On analysing with multipoint Brunauer - Emmett - Teller (BET) technique the effective surface area of the TiO_2 sample prepared at 500°C is found 39.876 m^2/g. With density functional theory [9] method applied to the adsorption and desorption isotherm confirms the highly porous nature of the samples with half pore width 21.714 Å. This is in close agreement with different literature producing nano-crystalline TiO_2 without any stabilizer in various other synthesis techniques[10, 11].

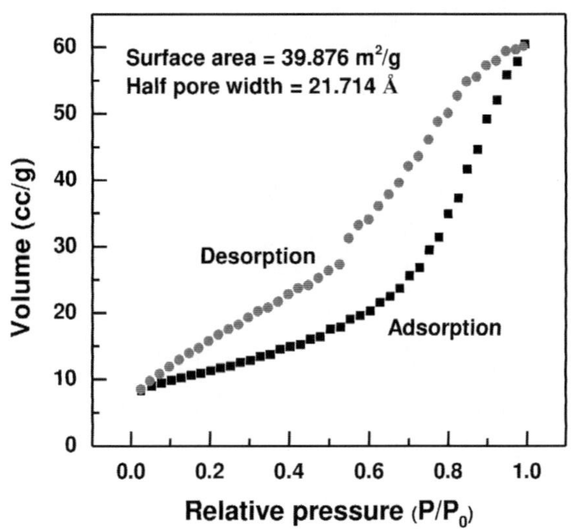

Figure 6. Adsorption and desorption isotherm of TiO_2 samples prepared at 500°C.

4. Conclusion

A series of titania samples are prepared using poly-acrylamide gel template method with different calcination temperatures ranging from 450°C to 650°C. X-Ray diffraction reveal upto 500°C temperature pure anatase polymorph of titania is synthesized whereas at temperatures beyond 500°C synthesis temperature a mixed phased titania of anatase and rutile is obtained. It is found that metastable anatase phase starts transforming to stable rutile phase at elevated temperatures. Size distribution of the synthesized nano-crystalline samples are analysed with Fourier technique and a good agreement is found between XRD and Microscopic analysis. Poly-acrylamide gel template synthesis technique provides ease in processing technique. In addition to this it also helps in controlling the percentage of different polymorphs of titania in a particular sample which is essential for large scale industrial and commercial applications of TiO_2 based systems.

Acknowledgments

T.D. thankfully acknowledge Ministry of Human Resource Development, India for research fellowship.

References

[1] Reyes-Coronado D, Rodríguez-Gattorno G, Espinosa-Pesqueira M E, Cab C, de Coss R and Oskam G 2008 *Nanotechnology* **19** 145605
[2] Gouma P I and Mills M J 2001 *J Am. Cera. Soc.* **84** 619 -622
[3] Hanao D A H and Sorrell C C 2011 *J Mater. Sci.* **46** 855 - 874
[4] Vásquez G C, Peche-Herrero M A, Maestre D, Gianoncelli A, Ramírez-Castellanos J, Cremades A, González-Calbet J M and Piqueras J 2015 *The Journal of Physical Chemistry C* **119** 11965–11974
[5] Ricci P C, Carbonaro C M, Stagi L, Salis M, Casu A, Enzo S and Delogu F 2013 *The Journal of Physical Chemistry C* **117** 7850 - 7857
[6] Satoh N and Yamamoto T N K 2013 *Scientific Reports* **3** 1959
[7] Tripathy S N, Mishra B, Shirolkar M M, Sen S, Das S R, Janes D B and Pradhan D K 2013 *Materials Chemistry and Physics* **141** 423 - 431
[8] Warren B E and Averbach B L 1950 *Journal of Applied Physics* **21** 595–599
[9] Ravikovitch P I, Haller G L and Neimark A V 1998 *Advances in Colloid and Interface Science* **76–77** 203 - 226
[10] Raj K J A and Viswanathan B 2009 *Indian Journal of Chemistry* **48A** 1378–1382
[11] Hu S, Wang A, Li X and Löwe H 2010 *Journal of Physics and Chemistry of Solids* **71** 156 - 162

Study on blast furnace cooling stave for various refractory linings based on numerical modeling

T R Mohanty, S K Sahoo* and M K Moharana
Department of Mechanical Engineering,
National Institute of Technology, Rourkela – 769 008, Odisha, India

* E-mail: sks@nitrkl.ac.in, Phone: +91 661 2462520

Abstract. Cooling technology for refractory lining of blast furnace is very important for the metallurgical industry, because it can substantially increase output and operation life of furnaces. A three dimensional mathematical model for the temperature field of the blast furnace stave cooler with refractory lining has been developed and analyzed. The temperature and heat dissipated by stave cooler is examined by using the finite element method. The cast steel stave is studied and computational analysis is made to know the effect of the cooling water velocity, temperature, and the lining material on the maximum temperature of the stave hot surface. The refractory lining materials, which are used in this experiment, are high alumina bricks with different stave materials (copper, aluminum and cast iron). The obtained numerical calculations are compared with that obtained from experiments performed at Rourkela Steel Plant, Odisha taking a stave in belly zone having maximum heat load shows very good agreement.

Keywords: Blast furnace; Stave; cooling; Refractory; Stave

1. Introduction

The blast furnace, mostly used in steel industries, is a multiphase reactor, which produces molten iron (Figure 1). It is basically a vertical shaft with height from 20 to 45 meter and diameter at bottom is about 12.0 meter. It is one of the most normally utilized iron producing procedure, for about 1800 cubic meters per year. The charging arrangements of blast furnace are at the top for ore, flux and coke and provisions at base to take out molten iron and slag. Hot air is blown through the tuyere, near the bottom of the furnace, increases the rate of combustion and sustains the required higher temperature for melting of iron.

The campaign life of the blast furnace very much affected by the properties of refractory Lining materials and decrease of its thickness due to erosion in the process. The cooling of refractory lining is the most contributing factor in deciding the Furnace campaign life. Lining cooling by stave is one of the results of such endeavours. A stave is a cooling gadget having one or more inside channel, and is introduced in numbers on the inward surface of a blast furnace. The staves are made conventionally of cast iron. But now days copper staves are used in place of cast iron staves, which is excellent in heat conductivity. Water is used as a medium for transfer of excess heat from inside the furnace to keep the lining cooled & prevent it from faster wearing out (Figure 1), indicates arrangement of stave cooler in a blast furnace for lining cooling. The intense load of heat within the blast furnace is the lower stave area of a blast furnace with increased smelting. Normal service life of the lining brick in this zone is shorter than in other zone. Damaged cooling staves are among the fundamental significant causes that prompt a remarkable redesign.

The major reason of overhaul or medium maintenance of a blast furnace is due to its damaged linings due to poor cooling staves. As a result, campaign period of a blast furnace principally depends on cooling stave life. Higher cost of copper staves in comparison to cast iron staves prohibits its use, although it has better thermal conductivity. Hence, steel staves are gaining wider acceptance because of its better properties, like, specific elongation, tensile strength, melting temperature and thermal conductivity. It is seen that thumb rules and/or wear of stave coolers are the major parameters on which it is designed and fabricated. With advance of finite element method, application of heat transfer calculations to the cooling stave of blast furnace is also become significant. There are different kind of mathematical models are applied by many researchers to know the effect of cooling stave, lining materials and cooling channels.

Pückoff and Knoche [1] have prescribed different guidelines and designed different plate coolers for heat extraction at blast furnace bosh, belly and stack regions. Changko et al. [2] have analyzed the thermal behaviour in upper tap hole area of the blast furnace and indicated that thermal properties of mud-core, convection heat transfer coefficient of cast and bricks of spool have a great effect on the top hole area temperature distribution. Shrivastava and Himte [3] have studied stave cooler of blast furnace using heat transfer analysis taking different skull and brick lining thickness. Steiger et al. [4] proposed a heat transfer model to calculate the temperature field of copper cooling plate and refractory lining. Gdula et al. [5] took different cooling system and lining material to analyze the heat transfer of the blast furnace hearth (bottom portion). A three-dimensional heat transfer model is reported by Wang et al. [6] to illustrate the temperature variation in the wall of the lower stack region of a blast furnace. Torrkulla and Saxen [7] have proposed a mathematical model for the prediction of wearing down and skull thickness of hearth in the blast furnace. They have used the thermocouples to measure the temperature at hearth and lining and show a model to predict the erosion of the lining and thickness of the skull. Verscheure et al.[8] have used a pyro-metallurgical processes to monitor and analyse the blast furnace temperature information to increase its campaign life. Peng et al. [9] used different lining thickness made of copper staves and sensor bars for their modelling. Their model showed the effect of skull and lining thickness on heat transfer from the staves. In their work, Chang et al. [10] gave emphasis on erosion in the hearth section of the blast furnace when molten metal is collected. They have analysed the process taking three different height for their observation; ie.,10 centimetres, 30 centimetres and 50 centimetres from the bottom of the blast furnace. Kumar et al. [11] have taken two different types of lining material, i.e., high alumina and silicon carbide and proposed a three dimensional model for the blast furnace stave. Lijun et al. [12] have used the finite element method to analyze the blast furnace and provide a three dimensional model. They observed that it is uneconomical to reduce the temperature of cooling water and increasing the velocity of water. David Roldan et al. [13] have performed a 3D analysis of furnace shell, refractory material, ram and hot metal. Cheng at al. [14] have designed copper stave for blast furnace and showed its effectiveness for blast furnace cooling. Fu et al. [15] have modelled the blast furnace considering all type of gas and mass in/out flowing and analysed the system using computational fluid dynamics.

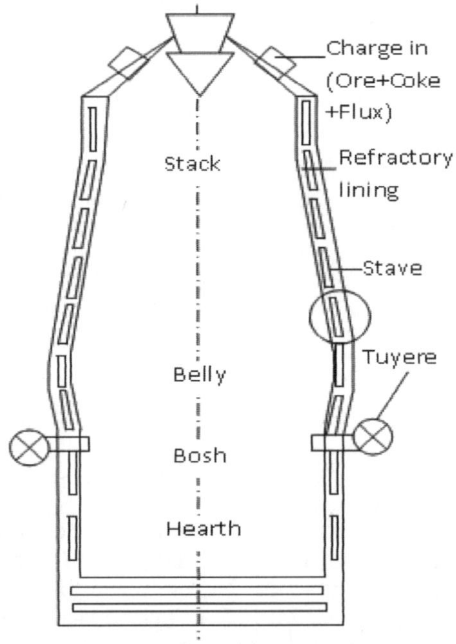

Fig. 1:- Schematic diagram of a blast furnace

Literature survey of previous works indicate that most of works gave attention to the design parameter of stave cooler and cooling coil using simulation and analysed the stave cooler at different thermal load to calculate the critical temperature of hot surface of blast furnace stave. Very few works are reported on suitability of any other alternative medium to cool the refractory lining or a different material except copper and cast iron for design stave cooler. The present work describes a three-dimensional analysis for blast furnace cooling stave with linings to explore the possibility of an alternative cooling medium.

2. Modelling and analysis

Cast steel cooling stave and lining of a blast furnace situated at Rourkela Steel Plant, Odisha is taken for the present analysis. Figure 2 illustrates the three-dimensional schematic representation of the total setup. The staves are made of cast steel having a network of tubes (Figure 3) in which the cooling fluid circulates. Ansys® Fluent [16] on workbench environment is used for geometry and meshing of the model.

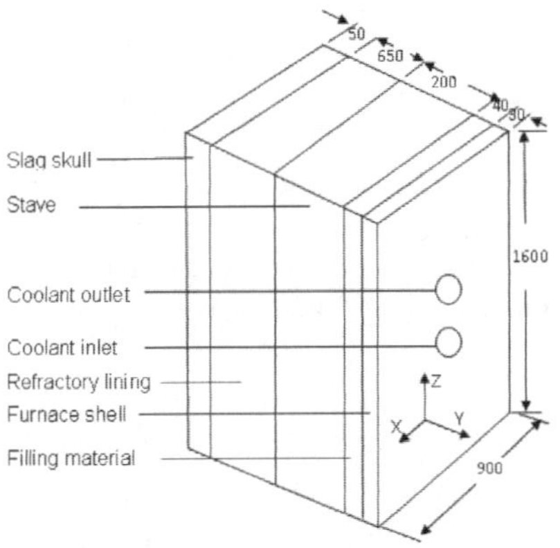

Fig. 2:- Isometric view of cooling stave with refractory linings

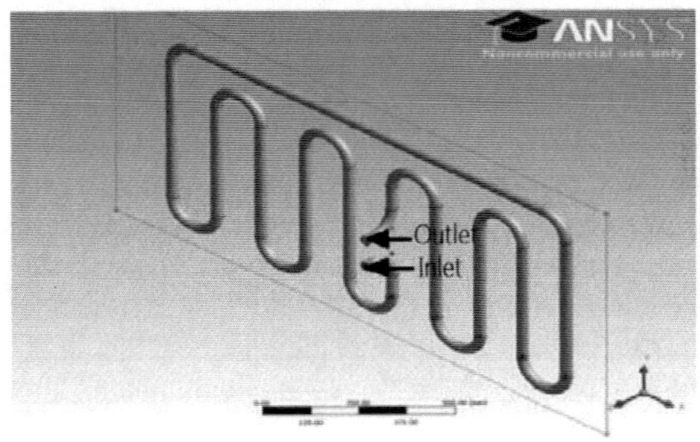

Fig. 3:- Cooling pipe of stave cooler

2.1. *Blast furnace cooling stave and refractory lining*

Dimensions of blast furnace cooling components and materials type are shown in Table 1 and Table 2 respectively. The cooling apparatus is usually comprised of slag skull, refractory lining, cooling stave, and filling materials at interfaces and furnace shell. There are passage holes for circulating cooling pipes inside the stave.

Table 1:- Dimension of parts.

Sl. No.	Part	Length	Breadth	Height	Diameter
1	Furnace shell	25	800	1600	-
2	Filling material	30	800	1600	-
3	Stave	180	800	1600	-
4	Refractory lining	600	800	1600	-
5	Slag skull	40	800	1600	-
6	Cooling pipe	8000	Inner distance between channel=150		ID-32, OD-40

2.2. *Present formulation*

Following assumptions are made to model the heat transfer of the lining and steel cooling stave for the present formulation:

(1) Steady-state conductive heat transfer process.

(2) It is assumed that skull is formed on the inner surface of refractory lining.
(3) Heat transfer resistances among furnace shell and filling material, stave and skull are neglected.
(4) Radiation heat transmitted from solid materials to inner surface of the stave is neglected.

Table 2:- Materials used for different parts.

Sl. No.	Part	Material
1	Furnace shell	Steel
2	Filling material	Alumina paste
3	Stave	Cast steel
4	Refractory lining	Alumina
5	Slag skull	Mixture of Iron & slag
6	Cooling pipe	Steel

The following equation is used to make a three-dimensional heat transfer analysis:

$$\frac{\partial}{\partial x_i}\left(k(T)\frac{\partial T}{\partial x_i}\right) = 0, \quad i = 1,2,3 \tag{1}$$

Where, k is the thermal conductivity coefficient in W/(m K), varies with temperature, T represents temperature in K, *i* represents x, y and z axis. Load and boundary conditions can be expressed as:

(a) Atmosphere temperature is 321 K and inlet water temperature is maintained at 305 K.

(b) Heat convection coefficients:

a) Between furnace shell and atmosphere-

$$k\frac{\partial T}{\partial x} = h_s(T_s - T_a) \tag{2}$$

Where, T_s and T_a are temperatures of furnace shell and atmosphere respectively; h_s is the overall coef. of heat convection between furnace shell and atmosphere, is taken as 12 W/(m^2 K).

b) Between water and inner sides of the cooling pipe-

$$k\frac{\partial T}{\partial n} = h_{pw}(T_p - T_w) \tag{3}$$

where, T_p the contact temperature between cooling stave and cooling water pipe, T_w cooling water temperature, h_{pw} the integration coefficient of heat convection between cooling stave and cooling water, is taken as 8000 W/(m^2 K). Where, $h_{pw} = \frac{1}{R}$ R is the total heat resistance between cooling water and stave body.

c) Between slag skull and hot air temperature –

$$k\frac{\partial T}{\partial x} = h_k(T_g - T_k) \tag{4}$$

where, T_k is the slag skull temperature, T_g the hot air temperature, h_k the coefficient of heat convection between slag skull and high temperature hot air is taken as 320 W/(m^2 K) as prescribed by Plyashkevich et al. [17].

d) The properties of materials used in the analysis are shown in Table 3.

To show the validity and reliability of the model, temperature field of the cast steel cooling stave and linings of a blast furnace situated at Rourkela Steel Plant, Odisha is identified and the stave temperatures estimated by heat transfer model are compared with the temperatures measured by thermocouples during an actual production stage.

Table 3:- Materials properties.

Sl. No.	Part	Density, ρ (kg/m^3)	Thermal conductivity, k, (w/m ^{0}C)	C_p (J/kg ^{0}C)	Young's modulus (Pa)	Poisson ratio	Coeff. Linear th. expansion, $^0C^{-1}$
1	Furnace shell	7840	52.2-0.25t	465	1.7×10^{11}	0.3	1.05×10^{-5}
2	Filling material	350	0.33	875	2.1×10^{10}	0.1	4.7×10^{-6}
3	Stave	7810	52.2-0.25t	500	1.7×10^{11}	0.3	1.05×10^{-5}
4	Refractory lining	2700	2.09+0.002t	1300	1.6	0.25	7.8×10^{-6}
5	Slag skull	2000	1.2	983	2.1×10^{10}	0.1	4.7×10^{-6}
6	Cooling pipe	7800	52.2-0.25t	480	1.7×10^{11}	0.3	1.05×10^{-5}

t is material temperature, ^{0}C.

3. Results and discussion

Figure 4 shows the temperature variation inside the cast iron stave body with water as a cooling agent and alumina refractory lining of 650 mm thickness. It displays that although the hot surface of refractory lining is 1440K, it reduces to 397K at stave when cooling water has inlet and outlet temperature are of 300K and 307.8K respectively. Figure 5 gives the temperature decrease from brick to stave, indicating cooling of stave is faster than bricks. From Figure 6, it is apparent that the temperature of stave decreases as refractory thickness increases when mass flow rate of coolant is 5 kg/s. It illustrates that both the factors, material of stave and as well as cooling channel position and coolant, affects the life of refractory lining and stave.

Figure 7 shows the temperature variation inside the stave body for different lining materials and thickness along middle of the stave. It illustrates that both factor affects the stave life as well as cooling channel position. Among the three materials copper was found to be the best material.

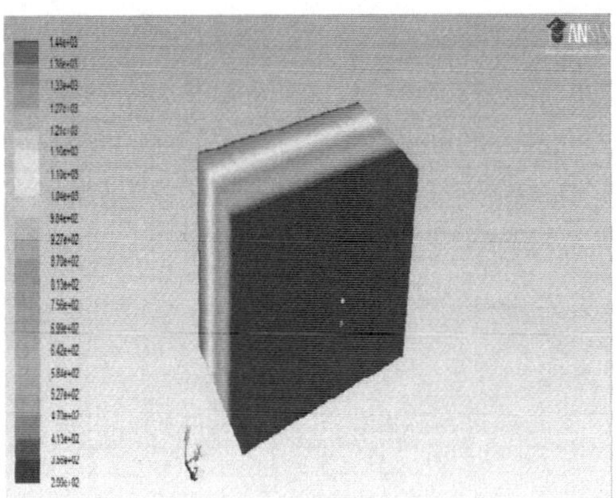

Fig. 4:- Temperature contour within stave cooler

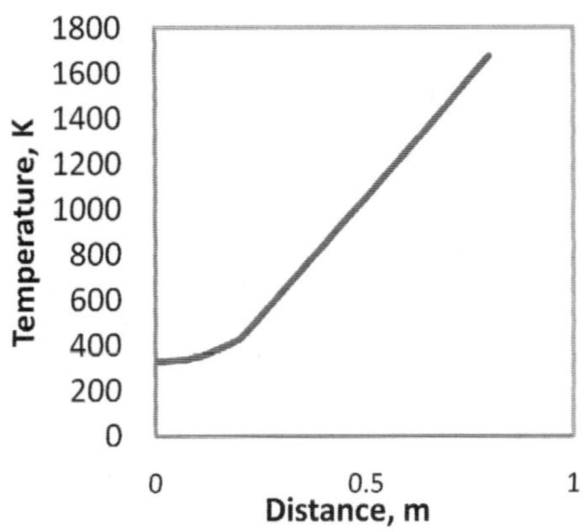

Fig. 5:- Temperature variation with distance

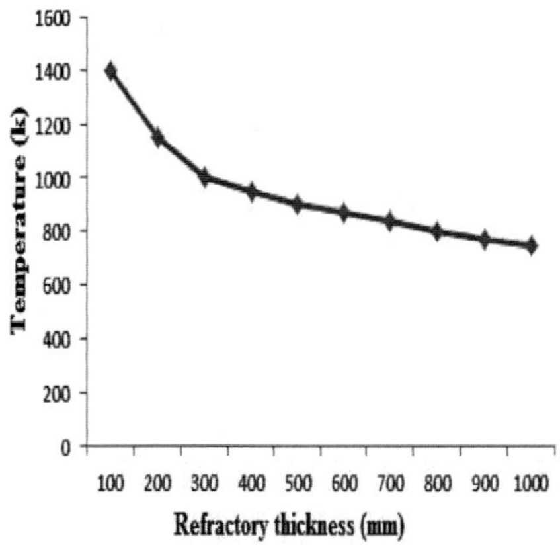

Fig. 6:- Temperature variation with refractory thickness where mass flow rate is 5 kg/s

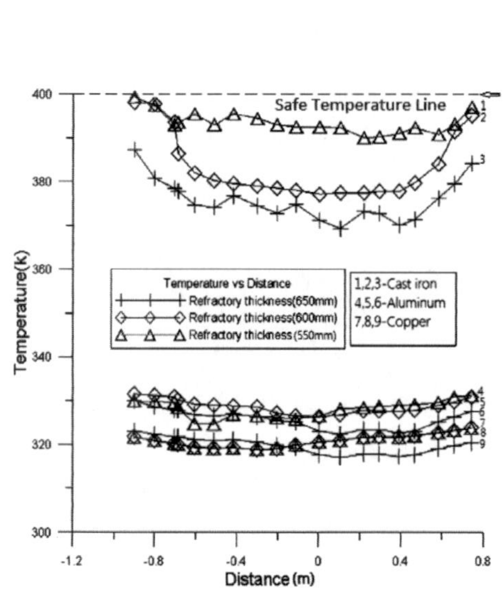

Fig. 7:- Temperature variation inside the stave body at the middle surface

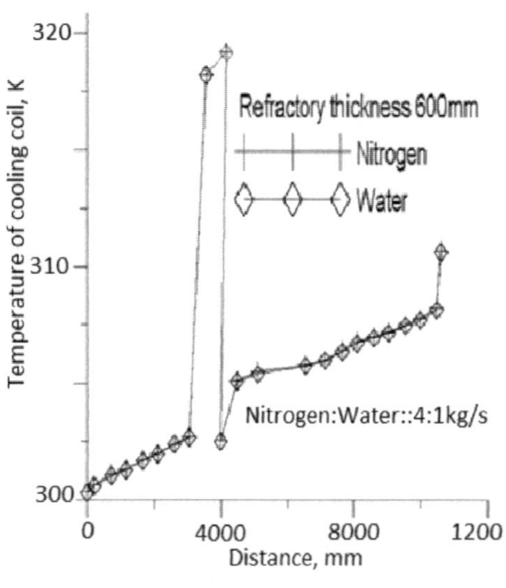

Fig. 8:- Coil temperature variation with distance for lining of 600 mm thickness

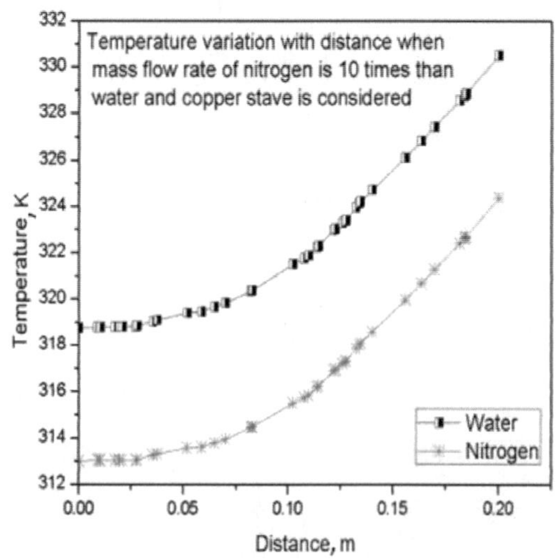

Fig. 9:- Temperature variation with mass flow for different cooling medium

Figure 8 demonstrate the variation of temperature with increase in length of the coil when nitrogen and water are used as cooling agent for refractory thickness of 600mm. It is observed that with increased mass flow rate of nitrogen by four times same cooling effect can be obtained as that of water. Comparison shown in Figure 9, when mass flow rate of nitrogen is 10 times than water one can get much advantage over water.

4. Conclusion

The three-dimensional mathematical model of temperature stress field of cooling stave and lining has been calculated by using finite element method based on the heat transfer analysis. The results indicate that lining materials and lining thickness have large influence on temperature field at stave, subsequently on its life. The findings obtained can be successfully used to optimize the lining thickness for a particular lining material. It is also found that nitrogen, which is a by-product of the steel plant can be used as an alternative cooling medium with increased flow rate.

References

[1] U. Pückoff, C. Knoche, 1986 Development of improved plate coolers (staves) for blast furnaces, *Directorate-General Science, Research and Development* (1986) 124-136.

[2] Y. Changko, K. Kenho, H. Tangkuo, The Thermal Behavior Analysis in Tap-Hole Area, *China Steel Technical Report* 21 (2008) 13－20.

[3] A. Shrivastava, R.L. Himte, Computational Study of Blast Furnace Cooling Stave using Heat Transfer Analysis, *International Journal of Innovative Technology and Exploring Engineering* 1 (2012) 72-78.

[4] R.W. Steiger, R.E. Braun, D.P. Grundtisch, Utilization of computer analysis in blast furnace refractory lining and shell design, *Proc. of Int. Conf. Iron making* (1985) 485–504.

[5] S.J. Gdula, R. Blaeecki, K. Kurpisz, A. Nowak, A. Sucheta, Mathematical model of steady state heat transfer in blast furnace hearth and bottom *Transactions ISIJ* 25 (1985) 381-387.

[6] G.X. Wang, A.B. Yu, P. Zulli, Three-dimensional modelling of the wall heat transfer in the lower stack region of a blast furnace, *ISIJ International* 37 (1997) 441–448.

[7] J. Torrkulla, H. Saxen, Model of the state of the blast furnace hearth, *ISIJ International* 40 (2000) 438–447.

[8] K. Verscheure, A.K. Kyllo, A. Filzwieser, B. Blanpain, P. Wollants, Furnace cooling technology in pyrometallurgical Processes, *In. Symposium Advanced processing of metals and materials* 4 (2006) 234-237.

[9] Y.C. Peng, H.C. Ken, Y.R. Jen, Conjugate heat transfer analysis of copper staves and sensor bars in a blast furnace for various refractory lining thickness, *Int. Communications in Heat and Mass Trans*fer 9 (2012) 58–65.

[10] C.M. Chang, W.T. Cheng, C.E. Huang, S.W. Du, Numerical prediction on the erosion in the hearth of a blast furnace during tapping process, *Int. Communications in Heat and Mass Transfer* 36 (2009) 480–490.

[11] A. Kumar, S. Bansal, R. Chandraker, Computational modeling of blast furnace cooling stave based on heat transfer analysis, *Materials Physics and Mechanics* 15 (2012) 46-65.

[12] W. Lijun, X. Xun, Z. Weiguo, S. Yunlong, L. Xiaojing, Heat transfer analysis of blast furnace stave, *Int. Journal of Heat and Mass Transfer* 51 (2008) 2824–2833.

[13] D. Roldan, Y. Zhang, R. Deshpande, D. Huang, Three-dimensional CFD Analysis for Blast Furnace Hearth, *Wear* 4 (2007) 43-50.

[14] C. Su-sen, Q. Liang, Z. Hong-bol, Monitoring method for blast furnace wall with Copper Staves, *Journal of iron and steel research international* 14 (2007) 01-05.

[15] D. Fu, Y. Chen, Y. Zhao, J. Alessio, K.J. Ferron, C.Q. Zhou, CFD modeling of multiphase reacting flow in blast furnace shaft with layered burden, *Applied Thermal Engineering* 70 (2014) 298-308.

[16] ANSYS® Fluent Release 15.0. *User manual* (2015).

[17] A.S. Plyashkevich, K.K. Streloi, A.S. Freudenderg, I.N. Sorokin, Thermo-technical analysis of optimum parameters of the brick linings and cooling system in blast furnace, *Steel, Russia* 3 (1975) 209–213.

Hygrothermal effects on the flexural strength of laminated composite cylindrical panels

Trupti R Mahapatra[1,*] and Subrata K Panda[2]

[1]KIIT University-Bhubaneswar, Odisha, India, 751024
[2]National Institute of Technology-Rourkela, Odisha, India, 769008

*E-mail: trmahapatrafme@kiit.ac.in

Abstract. In this article, the effect of hygrothermal conditions on the bending behaviour of laminated composite cylindrical panels has been investigated by using a nonlinear finite element micromechanical model. The panel model has been developed mathematically by introducing the geometrical nonlinearity in Green-Lagrange sense in the framework of the higher-order shear deformation theory. In order to achieve any general case, all the nonlinear higher order terms have been incorporated in the present formulation and the lamina material property variations due to hygrothermal effect are included via the micromechanical model. The governing equation is obtained using the variational principle and discretised using nonlinear finite element steps. The convergence behaviour of the present numerical model has been established and validated by comparing the responses with the existing results in the literature. The effects of thickness ratio, lamination scheme and support conditions under different sets of hygrothermal conditions are investigated in details.

Keywords: Laminated composite cylindrical panel; Hygrothermal effect; Green-Lagrange nonlinearity; micromechanical model

1. Introduction

Cylindrical panels made up of laminated composites are popularly used as structures/structural components in aerospace and marine applications because of their excellent compression and torsion resisting capabilities owing to high specific strength and stiffness and lightweight properties. However, throughout their service period these structures are subjected to combined hygro-thermo-mechanical loading. It is well known that the combined temperature and moisture environment induce residual stresses in the structural component and alter their thermo-elastic properties considerably. As a result of this not only the strength but also the mechanical responses like vibration, bending and stability behaviour of the structures/structural components are affected adversely and deviate from expected line. It is well known that for large deformation regime the basic geometry of panel is distorted and nonlinearity in geometry is induced. This in turn affects the structural behaviour of laminated structures. Hence, it is not only interesting but also challenging for design engineers to assess the nonlinear responses accurately for laminated composite shells of various geometries under hygrothermal environments.

Due to the expenses and difficulties associated with the experimental analysis of laminated structure under combined hygrothermal loading, many researchers have attempted to compute the linear and nonlinear flexural behaviour of laminated composites plates/shells under hygrothermal environment numerically/analytically using various existing and refined theories. The first order shear deformation theory (FSDT) and higher order shear deformation theory (HSDT) are more popular in comparison to refined/layer wise theories due to simplicity in formulation ([1], [2]). Attempts are also been made to include the degraded composite properties through macroscopic or microscopic approach. Sai Ram and Sinha [3] investigated the effects of temperature and moisture on the bending characteristics of laminated composite plates using finite element method (FEM). Naidu and Sinha [4] developed a FE model based on the FSDT and Green-Lagrange nonlinear kinematics to investigate the large deflection bending behavior of laminated composite shell panel under hygrothermal environments. Kundu et al. [5] studied the bending behaviour of laminated composite shells in hygrothermal environment by using the FSDT based geometrically nonlinear FE model. Lo et al. [6] investigated the effect of temperature dependent material properties and hygrothermal loading on multilayer composites using FEM based on a global–local HSDT mid-plane kinematics. Zenkour [7] studied the static responses of laminated composite plates using a refined (sinusoidal) shear deformation theory under hygrothermal load. Bending behaviour of laminated composite and sandwich shells in hygrothermal environment have been analyzed by Brischetto [8] by using Carrera's unified formulation. In all the referred cases the lamina material properties are considered through macroscopic model.

We note that, the analytical/numerical solutions of the linear/nonlinear free vibration ([9]), buckling ([10]) and flexural ([11], [12], [13], [14], [15] and [16]) behavior of the laminated composite plates and shells under elevated hygrothermal environment in conjunction with micromechanical material model to evaluate degraded composite properties are very rare in open literature. Moreover, in all the aforementioned cases the mathematical models have been developed using HSDT mid-plane kinematics in conjunction with von-Karman nonlinearity. To the best of the authors' knowledge, no work has been reported in literature on nonlinear bending analysis of laminated composite structure under hygro-thermo-mechanical loading by taking HSDT mid-plane kinematics with Green-Lagrange geometric nonlinearity. In this present work, the authors' aim to analyze the large deformation bending behavior of the laminated composite shell panel based on the proposed nonlinear kinematic model under combined hygro-thermo-mechanical loading. Here, the composite material properties are considered to be dependent on temperature and moisture and derived explicitly in terms of the properties of the fiber and the matrix and the fiber volume fraction. In addition to this, all the nonlinear higher order terms have been incorporated in the mathematical model to capture the original flexure of the structure. The nonlinear system governing equations are obtained using variational approach and discretised using suitable FEM. A direct iterative method is employed to solve the system governing equation to obtain the bending responses. The efficacy and accuracy of the model has been established by the convergence study and subsequent comparison of the responses with available published results. The effects of hygrothermal conditions, support conditions and curvature ratio on the bending response of laminated composite cylindrical panel are investigated and discussed in details. The outcome from this paper will be useful in the design of laminated structures for real life engineering (specifically in aircraft, aerospace, automobile, naval) applications.

2. Theory and Nonlinear FEM Scheme

In the present analysis, a typical laminated composite cylindrical panel is considered with length a, width b and uniform thickness 'h'. The shell panel consists of 'N' number of uniformly thick orthotropic layers as shown in Fig. 1. Here, R_1 and R_2 are the principal radii of curvatures of the shell panel along x and y directions at the mid-surface of the panel (z=0). The shell panel is subjected to the hygrothermal load in combination with uniform lateral load and the mid-plane displacement of the panel is modeled based on the HSDT kinematics [17]. Now, the displacement variable of any point within the shell panel with respect to the mid-plane along x, y, and z-directions are conceded as:

$$\begin{aligned}\bar{u}(x,y,z,t) &= u(x,y,t) + z\phi_1(x,y,t) + z^2\psi_1(x,y,t) + z^3\theta_1(x,y,t)\\ \bar{v}(x,y,z,t) &= v(x,y,t) + z\phi_2(x,y,t) + z^2\psi_2(x,y,t) + z^3\theta_2(x,y,t) \quad \left[-\frac{h}{2} \le z \le \frac{h}{2}\right]\\ \bar{w}(x,y,t) &= w(x,y,t)\end{aligned} \tag{1}$$

where, $(\bar{u},\bar{v},\bar{w})$ indicate the displacements at any arbitrary point on the plate along the (*x*, *y*, *z*) respectively. Similarly, (*u*, *v*, *w*) represent the corresponding displacements of the points on the mid-plane, ϕ_1 and ϕ_2 are the rotations of normal to the mid-plane relating to *y*-axis and *x*-axis, respectively. This displacement field represents the transverse shear strains as quadratic function of thickness coordinate at any point within the shell and also account for the parabolic distribution of shear stress across the thickness represented by ψ_1, ψ_2, θ_1 and θ_2, which are the higher order terms of Taylor series expansion defined at the mid-plane.

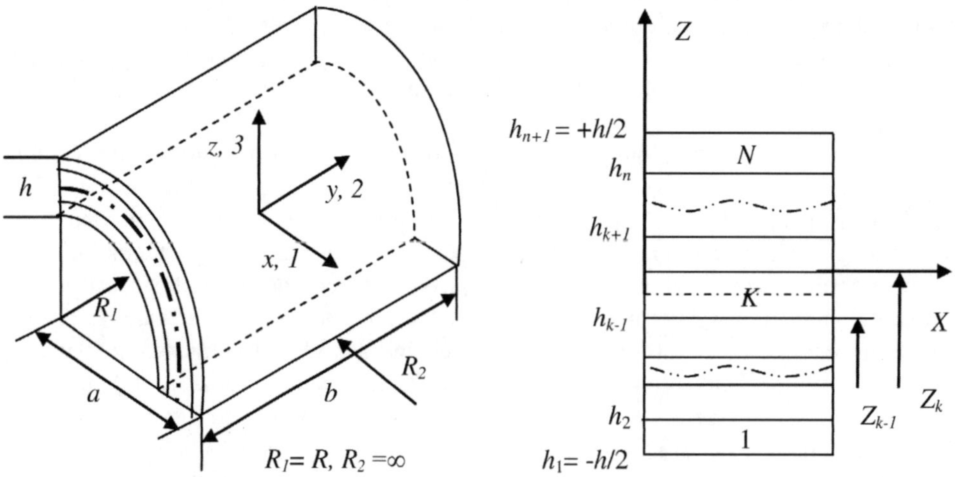

Fig. 1:- Geometry and Lay-up of laminated composite cylindrical shell panel.

The nonlinear strain displacement relations (Green-Lagrange type) considered for the present analysis are as in Mahapatra and Panda [18]:

$$\{\varepsilon\} = \{\varepsilon_L\} + \{\varepsilon_{NL}\} \tag{2}$$

Substituting Eq. (1) into Eq. (2), the linear and nonlinear strain terms are rearranged as:

$$\{\varepsilon\} = [\mathrm{H}]_L \{\bar{\varepsilon}_L\} + \frac{1}{2}[\mathrm{H}]_{NL}\{\bar{\varepsilon}_{NL}\} \tag{3}$$

where, $[\mathrm{H}]_L$, $[\mathrm{H}]_{NL}$ are the functions of thickness coordinate and $\{\bar{\varepsilon}_L\}$, $\{\bar{\varepsilon}_{NL}\}$ are the functions of *x* and *y*. The detail terms of $\{\bar{\varepsilon}_L\}$, $\{\bar{\varepsilon}_{NL}\}$, $[\mathrm{H}]_L$ and $[\mathrm{H}]_{NL}$ can be seen in [18].

In the present analysis, it is assumed that the deformation occurring due to unlike temperature and moisture change are not coupled. Hence, the constitutive matrix equation of generalized stress tensor for any general k^{th} orthotropic composite lamina with any fibre orientation angle θ is given by

$$\{\sigma_{ij}\}^k = \left[\overline{Q}_{ij}\right]^k \{\varepsilon_{ij} - \alpha_{ij}\Delta T - \beta_{ij}\Delta C\}^k \tag{4}$$

where, $\{\sigma_{ij}\}^k = \{\sigma_1 \ \ \sigma_2 \ \ \sigma_6 \ \ \sigma_5 \ \ \sigma_4\}^T$ and $\{\varepsilon_{ij}\}^k = \{\varepsilon_1 \ \ \varepsilon_2 \ \ \varepsilon_6 \ \ \varepsilon_5 \ \ \varepsilon_4\}^T$ are the stress and strain vectors respectively for the kth layer. $\left[\overline{Q_{ij}}\right]^k$ is the transferred reduced stiffness matrix for the kth layer, $\{\alpha_{ij}\}^k = \{\alpha_1 \ \ \alpha_2 \ \ 2\alpha_{12}\}^T$ is the thermal expansion/contraction coefficient vector and $\{\beta_{ij}\}^k = \{\beta_1 \ \ \beta_2 \ \ 2\beta_{12}\}^T$ is the moisture expansion/contraction coefficient vector. Here, $\Delta T = T - T_0$ is the temperature difference, where T is applied and T_0 is reference temperatures, respectively. Similarly, $\Delta C = C - C_0$ is the moisture difference between applied (C) and reference (C_0) values of weight percentage of moisture.

In order to obtain the realistic behaviour of laminated structure, the degraded hygro-thermo-elastic composite material properties (to be used in Eq. 4) are evaluated using a micromechanical model as presented in [14].

Here, a nine noded isoparametric quadrilateral Lagrangian element with nine degrees of freedom per node has been taken to discretize the present laminate model. The field displacement vector corresponding to any nodal point of the element can be expressed as

$$\{\delta^*\} = [u \ \ v \ \ w \ \ \phi_1 \ \ \phi_2 \ \ \psi_1 \ \ \psi_2 \ \ \theta_1 \ \ \theta_2]^T = \sum_{i=1}^{9}[N_i]\{\delta_i\} \tag{5}$$

where, $[N_i]$ and $\{\delta_i\}$ are the interpolation function and displacement vector for the ith node, respectively and the details of interpolation functions can be seen in [19].

Now, the mid-plane strain vector can be written as

$$\{\varepsilon\}_i = [B_i]\{\delta_i\} \tag{6}$$

where, $[B_i]$ is the strain displacement relation matrix.

The strain energy of the panel can be expressed as $U = \frac{1}{2}\int_v \{\varepsilon\}_i^T \{\sigma_i\}\, dV$ (7)

Using the expression of strain vectors and resultant stress from Eq. (3) and Eq. (4) and putting into Eq. (7) the strain energy can be expressed as

$$\begin{aligned} U &= \frac{1}{2}\int_V \{\varepsilon\}_i^T \left[\bar{Q}\right]\{\varepsilon\}_i dV = \frac{1}{2}\iiint \{\varepsilon_L + \varepsilon_{NL}\}^T \left[\bar{Q}\right]\{\varepsilon_L + \varepsilon_{NL}\}\, dx\, dy\, dz \\ &= \frac{1}{2}\int_A \left(\{\varepsilon_L\}_i^T [D_1]\{\varepsilon_L\}_i + \frac{1}{2}\{\varepsilon_L\}_i^T [D_2]\{\varepsilon_{NL}\}_i + \frac{1}{2}\{\varepsilon_{NL}\}_i^T [D_3]\{\varepsilon_L\}_i + \frac{1}{4}\{\varepsilon_{NL}\}_i^T [D_4]\{\varepsilon_{NL}\}_i \right) dA \end{aligned} \tag{8}$$

where, $[D_1] = \sum_{k=1}^{N}\int_{z_{k-1}}^{z_k} [H]_L^T\left[\bar{Q}\right][H]_L\, dz$, $[D_2] = \sum_{k=1}^{N}\int_{z_{k-1}}^{z_k} [H]_L^T\left[\bar{Q}\right][H]_{NL}\, dz$

$[D_3] = \sum_{k=1}^{N}\int_{z_{k-1}}^{z_k} [H]_{NL}^T\left[\bar{Q}\right][H]_L\, dz$ and $[D_4] = \sum_{k=1}^{N}\int_{z_{k-1}}^{z_k} [H]_{NL}^T\left[\bar{Q}\right][H]_{NL}\, dz$

Substituting Eq. (6) into Eq. (8) the expression for strain energy becomes

$$U = \frac{1}{2}\int_A \left(\begin{aligned} &\{\delta^*\}^T [B_L]_i^T [D_1][B_L]_i\{\delta^*\} + \frac{1}{2}\{\delta^*\}^T [B_L]_i^T [D_2][A]_i[G]_i\{\delta^*\} \\ &+ \frac{1}{2}\{\delta^*\}^T [G]_i^T [A]_i^T [D_3][B_L]_i\{\delta^*\} + \frac{1}{4}\{\delta^*\}^T [G]_i^T [A]_i^T [D_4][A]_i[G]_i\{\delta^*\} \end{aligned} \right) dA \tag{9}$$

where, $\{\varepsilon_L\}_i = [B_L]_i\{\delta^*\}$ and $\{\varepsilon_{NL}\}_i = \frac{1}{2}[B_{NL}(\delta)]_i\{\delta^*\} = \frac{1}{2}[A(\delta)]_i[G]_i\{\delta^*\}$.

[B_L] is the product form of the differential operator and nodal interpolation function in the linear strain terms. [A] is function of the displacements and [G] is the product form of differential operator and shape function in the nonlinear strain terms. The expressions of [A] and [G] arising due to the Green-Lagrange nonlinearity in the nonlinear stiffness matrices can be seen in [20].
The work done due to the external applied distributed transverse static load "q" can be expressed as

$$W = \int_A \{\delta\}^T q dA \tag{10}$$

where, the intensity of transverse static load is expressed in terms of the applied uniform lateral pressure as

$$q = \frac{Qh^4 E_{22}}{a^4} \tag{11}$$

3. System governing Equation

The governing equation of nonlinear bending for laminated composite flat panel is obtained by minimizing the total energy expression. This result in

$$\delta \Pi = 0 \tag{12}$$

where, $\Pi = (U\text{-}W)$
Using, Eq. (9) and Eq. (10) in Eq. (12) and applying finite element approximation, the system governing expression can be obtained as:

$$[K]\{\delta\} = \{q\} \text{ or } [K_L + K_{NL}]\{\delta\} = \{q\} \tag{13}$$

where, [K_L] and [K_{NL}] are the global linear and nonlinear stiffness matrices. The nonlinear stiffness matrix depends on the displacement vector linearly and quadratically, respectively.
Now, the numerical solutions are obtained using a direct iterative method and solutions scheme are depicted in [20].

4. Results and Discussion

A nonlinear finite element computer code have been developed in MATLAB 7.10.0 and based on the present formulation in order to obtain the hygro-thermo-mechanical linear/nonlinear bending responses of laminated composite shell panel. As a first step, the validation behaviour of the present model has been established. Subsequently, different parametric studies are also been carried out and their significance have been discussed in details. For computation, the material properties of the composite material are considered to be dependent on hygrothermal conditions. Their values corresponding to reference temperature 21°C and moisture concentration 0% are considered as in Upadhyay et al. [14] and remain unchanged for each case if not specified otherwise.
E_{f1}=220 GPa, E_{f2}=13.79 GPa, E_m=3.45 GPa, G_{f12}=8.97 GPa, υ_{f12}=0.2, υ_m=0.35, α_{f1}= -0.99×10^{-6}/°C, α_{f2}= 10.08×10^{-6}/°C, α_m=72×10^{-6}/°C, β_m=0.33, T_{g0}=216°C.
The boundary conditions used for analysis are given below:
(a) All edges simply support (SSSS):
$v = w = \Phi_2 = \Psi_2 = \theta_2 = 0$ at x = 0, a and $u = w = \Phi_1 = \Psi_1 = \theta_1 = 0$ at y = 0, b.
(b) All edges clamped (CCCC):
$u = v = w = \Phi_1 = \Phi_2 = \Psi_1 = \Psi_2 = \theta_1 = \theta_2 = 0$ for both x = 0, a and y = 0, b.
(c) All edges hinged (HHHH):
$u = v = w = \Phi_2 = \Psi_2 = \theta_2 = 0$ at x = 0, a and $u = v = w = \Phi_1 = \Psi_1 = \theta_1 = 0$ at y = 0, b.
The linear/nonlinear transverse central deflections are non-dimensionalized using the relations $Q = (q / E_2) * (a / h)^4$ and $w_{Central} = w_{\max}/h$, respectively. It is important to mention here that the value of E_2 is corresponding to actual (instantaneous) hygrothermal conditions.

4.1. Convergence and Comparison Study

Numerical results are provided to establish the convergence behavior of the present model under unlike environmental conditions. The linear and nonlinear flexural responses are computed using the present model by considering available flat panel examples in [13] and [14] and plotted with respect to mesh divisions in Fig. 2 (a) and (b), respectively. The geometrical, material properties, lamination scheme and support conditions are taken same as [13] and [14]. Analytical results of the individual reference are also shown for the comparison purpose. The results show very good convergence rate with respect to mesh refinement and also agree with existing results in the literature. Based on the convergence study, it is concluded that a (6×6) mesh is adequate to obtain desired responses throughout the analysis. For validation purpose the nondimensional linear and nonlinear central deflections of simply supported square anti-symmetric angle-ply ($[\pm45^0]_2$) laminated spherical panel (*R*/*a*=5, *a*/*h*=15) is computed and compared with the numerical results available reference [15]. The responses are computed for load parameter Q=150 by using the material (V_f =0.6) and geometrical parameters same as the reference and presented in Table 1. It is evident that the present results are showing excellent agreement with the available analytical and numerical results. It is worthy to mention that the difference in results exists because the present model is developed based on the HSDT and Green-Lagrange nonlinearity whereas all the references are modelled using HSDT and von-Karman nonlinearity. In addition to that, the present model have included all the nonlinear higher order terms in the formulation which makes the model more realistic in nature for laminated structures.

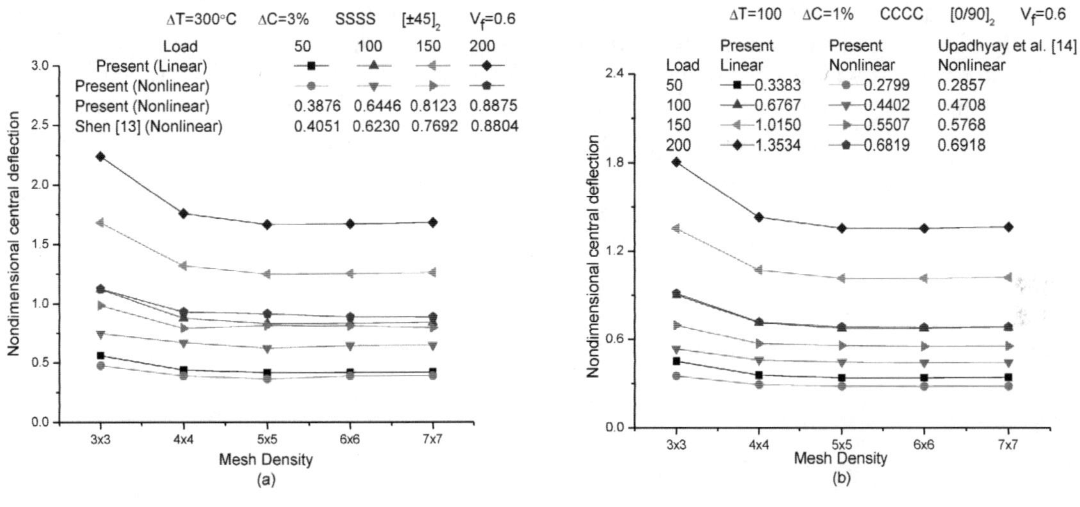

Fig. 2:- Convergence study of linear/nonlinear flexural responses of laminated composite flat panel.

Table 1. Comparison of linear and nonlinear flexural responses of simply supported laminated composite spherical panel ($[\pm45^0]_2$, *R*/*a*=5, *a*/*h*=15, V_f=0.6)

Hygrothermal Conditions	Present Linear	[15] Linear	Difference (%)	Present Nonlinear	[15] Nonlinear	Difference (%)
ΔT=50, ΔC=1%	0.9279	0.9576	-3.2008	0.6231	0.5144	17.4450
ΔT=50, ΔC=1.5%	0.9365	0.9583	-2.3278	0.6262	0.5299	15.3785
ΔT=100, ΔC=1%	0.9878	1.028	-4.0697	0.6434	0.5659	12.0454
ΔT=100, ΔC=1.5%	1.0052	1.0287	-2.3378	0.6489	0.5869	9.5546

4.2. Numerical illustrations

Now, in this section some more numerical experimentation have been presented to demonstrate the applicability of the proposed nonlinear finite element micromechanical shell panel model and the effect of parameters (hygrothermal conditions, curvature ratio and support conditions) on the hygro-thermo-mechanical bending behavior of laminated (cylindrical) structural components. If not stated otherwise, all the illustrations discussed here are computed by setting the fiber volume fraction (V_f) as 0.6 and for four sets of hygro-thermal loading conditions (ΔT=0-ΔC=0%, ΔT=100-ΔC=1%, ΔT=200-ΔC=2% and ΔT=300-ΔC=3%). The transversely applied load parameter are considered as (Q=100, 200, 300, 400 and 500).

4.2.1. Illustration 1. It is well known that the lamination schemes are one of the deciding factors for the strength of fiber reinforced composite structure. Hence, in this example the effect of lamination scheme on the nonlinear flexural response of laminated composite cylindrical panel under elevated hygrothermal environment is investigated and presented in Fig. 3 (a) and (b). The figure shows the load-deflection responses of a square clamped symmetric/anti-symmetric cross-ply/angle-ply laminated composite cylindrical shell panels (*a/h*=10, *R/a*=10) under four sets of the hygrothermal load as mentioned earlier. It is observed that the cylindrical shells with symmetric lamination are less affected under hygrothermal loads in comparison to anti-symmetric laminations. It is worthy to note the hard spring type of behaviour for all lamination schemes except the anti- symmetric angle-ply lamination, which shows soft spring type of behaviour with increasing hygro-thermo-mechanical loads. Nonlinear central deflections are decreasing with increasing in hygrothermal load for all sorts of lamination schemes. However, sudden increase in deflection values are noticed for ΔT=300°C and ΔC=3%. It is because the applied temperature is shifted beyond the glass transition temperature of the material and this may lead to such variation in the responses. It is also understood that, the original geometry of the laminated structure distorted considerably due to the high strain rate loading under the combined elevated environment and this in turn affect the final responses of the structural component due to the induction of the high residual stresses and degraded material properties.

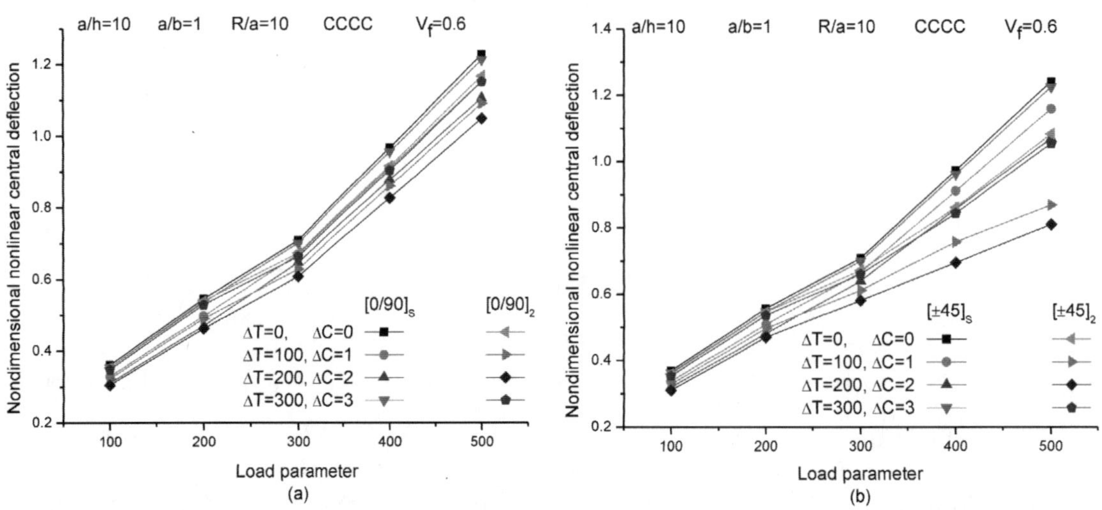

Fig. 3:- Effect of lamination scheme and hygrothermal conditions on the nonlinear central deflection of laminated composite cylindrical shell panel.

4.2.2. Illustration 2. It is true that the membrane strength of the shell panel structure greatly depends on its curvature. In order to understand the bending strength of laminated composite cylindrical panel under combined hygro-thermo-mechanical loading, this numerical example has been solved for different curvature ratios (*R/a*=10, 40, 70 and 100). The responses are computed for the simply supported square thin (*a/h*=100) anti-symmetric cross-ply laminated composite cylindrical panels and presented in Fig. 4 (a) and (b). It is observed that the nonlinear transverse central deflections increase as the curvature ratio increases, irrespective of the hygrothermal loading conditions. This is because of the fact that as the curvature ratio increases the panel becomes flat and the membrane stiffness decreases subsequently.

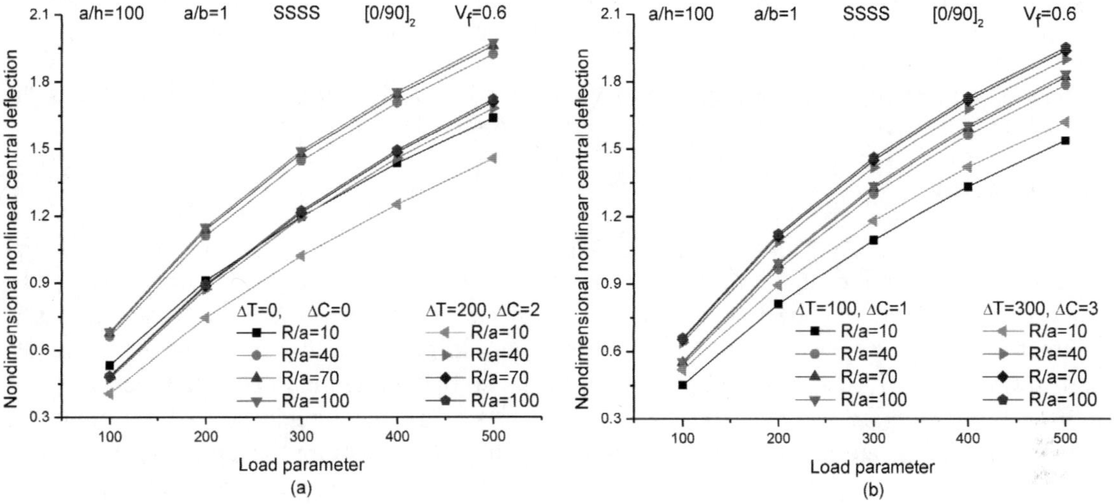

Fig. 4:- Effect of curvature ratio on the nonlinear central deflection of laminated composite cylindrical shell panel.

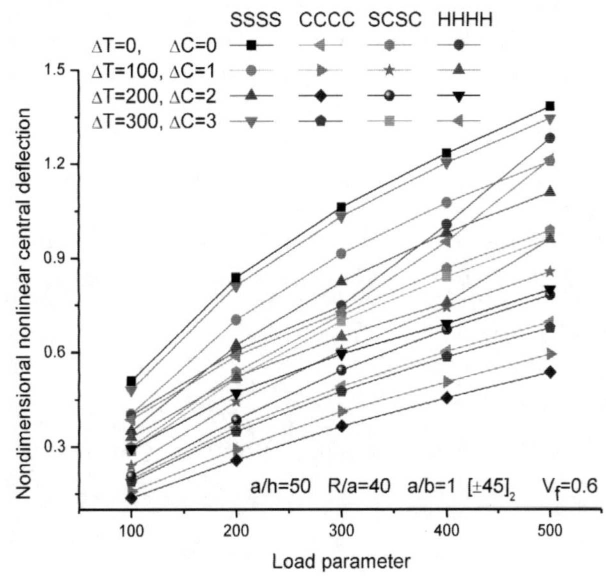

Fig. 5:- Effect of support conditions on the nonlinear central deflection of laminated composite cylindrical shell panel.

4.2.2. Illustration 3. The strength and stiffness behavior of the laminated structures are largely dependent on the support conditions which in turn affects the structural responses largely. In this example the effect of different support conditions (SSSS, CCCC, HHHH and SCSC) on the nonlinear central deflection of anti-symmetric angle-ply $[\pm 45]_2$ laminated square (a/b=1) cylindrical panel (R/a=40) under hygro-thermo-mechanical load have been investigated and the results are shown in Fig. 5. It is noted that the non-dimensional nonlinear central deflections are lowest for clamped and highest for simply-supported case. This is due to the fact that as the number of constraints increase or decrease, the stiffness of the structure decreases and/or increases and the deflection values decrease or increase monotonically. However, it is interesting to note that in all the cases the panels are showing soft spring type of behavior irrespective of hygrothermal load and the support conditions, except the HHHH support condition, for which the nonlinearity is observed to be more pronounced.

5. Conclusion

The effect of the hygrothermal conditions, curvature and constraint conditions on the flexural behaviour of laminated composite cylindrical shell panels have been investigated in the present study. A general mathematical model based on the HSDT mid-plane kinematics and Green-Lagrange geometric nonlinearity is developed to count the exact in-plane and out-of-plane shear stresses and strains. The effective material properties of the composite lamina due to temperature and moisture variation are incorporated via micromechanical model. In addition all the nonlinear higher order terms are included in the present formulation to capture the actual flexure of the shell structure. The system governing equations are obtained using the variational principle and discretised using isoparametric FE steps. The convergence and validation of the present model has been established. The significance of the present HSDT based model for the flexural analysis of laminated structures under combined hygrothermal loading condition is observed. The subsequent numerical illustrations indicate that considering the effect of hygrothermal conditions is inevitable for laminated composite structures, particularly when they are exposed to a severe environmental condition. Nonlinearity due to moisture absorption at high temperature is severe for anti-symmetric angle-ply in comparison to symmetric laminations. Irrespective of the hygrothermal load the nondimensional central deflections increase with curvature ratios. It worthy to note that as the number of restraints increase the nonlinear central deflections decrease irrespective of hygrothermal conditions. The outcome from this present work can be utilized in the optimum design of real structures.

Acknowledgement
Authors are thankful to AICTE (All India Council for Technical Education) for partial support through the sanction 8023/RID/RPS/56/11/12 Dated: 26/03/2012.

Reference

[1] A. A. Khdeir, J. N. Reddy, D. A. Frederick, Study of bending, vibration and buckling of cross-ply circular cylindrical shells with various shell theories, *International Journal of Engineering Science* 27 (11) (1989) 1337–1351.
[2] R. Khandan, S. Noroozi, P. Sewell, J. Vinney, The Development of Laminated Composite Plate Theories – A Review, *Journal of Material Science* 47(16) (2012) 5901- 5910
[3] K. S. Sai Ram, P. K. Sinha, Hygrothermal effects on the bending characteristics of laminated composite plates, *Computers and Structures* 40(4) (1991) 1009–1015
[4] N. V. S. Naidu, P. K. Sinha, Nonlinear finite element analysis of laminated composite shells in hygrothermal environments, *Composite Structures* 69(4) (2005) 387–395

[5] C. K. Kundu, D. K. Maiti, P. K. Sinha, Nonlinear finite element analysis of laminated composite doubly curved shells in hygrothermal environment, *Journal of Reinforced Plastics and Composites* 26(14) (2007) 1461–1478

[6] S. H. Lo, W. Zhen, Y. K. Cheung, Hygrothermal effects on multilayered composite plates using a refined higher order theory *Composite Structures* 92(3) (2010) 633-646

[7] A. M. Zenkour, Hygrothermal effects on the bending of angle-ply composite plates using a sinusoidal theory *Composite Structures* 94(12) (2012) 3685–3696

[8] S. Brischetto, Hygrothermoelastic analysis of multilayered composite and sandwich Shells, *Journal of Sandwich Structures and Materials* 15(2) (2013) 168–202

[9] X. L. Huang , H. S. Shen and J. J. Zheng, Nonlinear vibration and dynamic response of shear deformable laminated plates in hygrothermal environments, *Composites Science and Technology* 64(10–11) (2004) 1419–1435

[10] H. S. Shen, Hygrothermal effects on the postbuckling of shear deformable laminated plates, *International Journal of Mechanical Sciences* 43(5) (2001) 1259–1281

[11] V. V. S. Rao, P. K. Sinha, Bending Characteristics of Thick Multidirectional Composite Plates under Hygrothermal Environment, *Journal of Reinforced Plastics and Composites* 23(14) (2004) 1481–1495

[12] A. Sharma, A. K. Upadhyay, K. K. Shukla, Flexural response of doubly curved laminated composite shells, *Science China Physics Mechanics Astronomy* 56(4) (2013) 812–817

[13] H. S. Shen, Hygrothermal effects on the nonlinear bending of shear deformable laminated plates, *Journal of Engineering Mechanics* 128(4) (2002) 493–496

[14] A. K. Upadhyay, R. Pande, K. K. Shukla, Nonlinear Flexural response of laminated composite plates under hygro-thermo-mechanical loading, *Communications in Nonlinear Science and Numerical Simulation* 15(9) (2010) 2634–2650

[15] A. Lal, B. N. Singh, S. Anand Nonlinear bending response of laminated composite spherical shell panel with system randomness subjected to hygro-thermo-mechanical loading, *International Journal of Mechanical Science* 53(10) (2011) 855–866

[16] R. Kumar, H. S. Patil, A. Lal, Nonlinear flexural response of laminated composite plates on a nonlinear elastic foundation with uncertain system properties under lateral pressure and hygrothermal loading: micromechanical model, *Journal of Aerospace Engineering* 27(3) (2014) 529–547

[17] Reddy J N 2004 *Mechanics of laminated Composite Plates and Shells* Second Edition CRC Press

[18] T. R. Mahapatra, S. K. Panda, Thermoelastic vibration analysis of laminated doubly curved shallow panels using non-linear FEM, *Journal of Thermal Stresses* 38(1) (2015) 39-68

[19] Cook R D, Malkus D S and Plesha M E 2000 *Concepts and applications of finite element analysis* Third Edition John Willy and Sons (Asia) Pvt. Ltd., Singapore

[20] V. R. Kar, T. R. Mahapatra, S. K. Panda, Nonlinear flexural analysis of laminated composite flat panel under hygro-thermo-mechanical loading *Steel and Composite Structures An International Journal* 19(4) (2015) 1011-1033